# MODERN PRODUCTION TECHNOLOGIES IN
# HORTICULTURAL CROPS

# MODERN PRODUCTION TECHNOLOGIES IN HORTICULTURAL CROPS

## INNOVATIONS, PRACTICES AND OPTIMAL GROWTH

**Dr B. S. Tomar**
*Head, Division of Vegetable Science*
*Joint Director (Extension) Additional Charge*
*ICAR-Indian Institute of Agricultural Research, New Delhi-110 012, India*

**Dr A. K. Singh**
*Principal Scientist (Hort.-Fruit Science)*
*Central Horticultural Experiment Station (ICAR-Central Institute for Arid Horticulture),*
*Vejalpur-389 340, Panchmahals (Godhra), Gujarat, India*

**Dr. Praveen Kumar Singh**
*Head, Division of Plant Exploration & Germplasm Collection,*
*ICAR- National Bureau of Plant Genetic Resources, New Delhi*

**Dr. Prabhat Kumar**
*Horticulture Commissioner, GOI & Ex-National Coordinator*
*ICAR-National Agricultural Higher Education Project, Basti (U.P.)*

**Dr Gograj Singh Jat**
*Senior Scientist (Vegetable Science)*
*Division of Vegetable Science, ICAR- IARI, New Delhi.*

**Prawal Pratap Singh Verma**
*Senior Technical Officer (Agriculture)*
*Council of Scientific and Industrial Research-Central Institute of Medicinal and Aromatic Plants (CIMAP),*
*Regional Research Centre, Purara, Bageshwar, Uttarakhan*

**Dr Som Dutt**
*Formerly Editor*
*(Indian Journal of Agricultural Sciences and Indian Horticulture) ICAR- DKMA, New Delhi*
*Managing Editor, Current Horticulture, SHRDA, Ghaziabad, Uttar Pradesh, India*

**2026**

**Daya Publishing House®**
*A Division of*
**Astral International Pvt. Ltd.**
**New Delhi – 110 002**

ISBN: 978-93-5919-578-0

*Publisher's Note:*

*Every possible effort has been made to ensure that the information contained in this book is accurate at the time of going to press, and the publisher and author cannot accept responsibility for any errors or omissions, however caused. No responsibility for loss or damage occasioned to any person acting, or refraining from action, as a result of the material in this publication can be accepted by the editor, the publisher or the author. The Publisher is not associated with any product or vendor mentioned in the book. The contents of this work are intended to further general scientific research, understanding and discussion only. Readers should consult with a specialist where appropriate.*

*Every effort has been made to trace the owners of copyright material used in this book, if any. The author and the publisher will be grateful for any omission brought to their notice for acknowledgement in the future editions of the book.*



*Published by* : **Daya Publishing House®**
*A Division of*
**Astral International Pvt. Ltd.**
– ISO 9001: 2015 Certified Company –
4736/23, Ground Floor, Ansari Road,
Darya Ganj, New Delhi-110 002
Ph. 011-43549197, 8130496929
e-mail: info@astralint.com
Website: www.astralint.com

*Printed at* : **Replika Press Pvt Ltd**

त्रिलोचन महापात्र
अध्यक्ष
पौधा किस्म और कृषक अधिकार संरक्षण प्राधिकरण
(संसद के अधिनियम द्वारा निर्मित सांविधिक निकाय)
कृषि एवं किसान कल्याण मंत्रालय
भारत सरकार

Trilochan Mohapatra
Chairperson
Protection of Plant Varieties and Farmers' Rights Authority
(A Statutory body created by an Act of Parliament)
Ministry of Agriculture and Farmers Welfare
Government of India

# FOREWORD

The Horticultural crops comprising of a wide range of species belonging to fruits, vegetables, flowers and ornamentals species, plantation crops, medicinal and aromatic plants etc. constitute an important component of our food system. These crops provide the much-needed nutrition to the consumers and income security for the farmers. India is the second-largest producer of fruits and vegetables in the world, which is a testament to the Horticulture sector's growing prominence. This sector contributes approximately 33% to the country's total agricultural GDP that reveals its relevance and importance for the country's well-being.

Research efforts in horticulture have resulted in the development of high-yielding varieties, valuable genetic resources, and innovative technological interventions. Despite these advancements, there remains an urgent need to develop new hybrids, explore underutilized crops, and expand the horticultural frontier through sustainable technologies. The vast pool of knowledge generated across institutions remains scattered, and its effective compilation and dissemination are essential. In the light of these imperatives, the Society for Horticultural Research and Development (SHRD) successfully organized the **3rd Indian Horticulture Summit-cum-International Conference -2024** on *"Technological Intervention for Boosting Horticultural Production"*, held during February 1-3, 2024 at Rajasthan Agricultural Research Institute, S.K.N. Agriculture University, Jobner, Durgapura Campus, Jaipur, Rajasthan. The summit witnessed the participation of leading scientists, researchers, and specialists who presented insightful lectures and contributed expert chapters. The outcome of this international meet is the present book, ***Modern Production Technologies in Horticultural Crops: Innovations, Practices, and Optimal Growth.*** It is a comprehensive compilation of 32 rigorously reviewed chapters, contributed by subject-matter specialists across various horticultural domains. The book highlights production technologies for major horticultural crops, and invites attention to lesser-known crops with high nutritional and economic potential. Topics such as plant health management, pest control, and region-specific industrial opportunities have been thoughtfully integrated. Written in an accessible and readerfriendly style, this book is designed to serve a wide audience including students, scientists, educators, policy-makers, and practitioners in the field of horticulture. It reflects the dedicated efforts of its editors and contributors, especially Dr. B. S. Tomar, Dr. AK. Singh, Dr. P. K. Singh, Dr. Prabhat Kumar, Dr Gograj Jat, Dr P.P. Singh Verma and Dr Som Dutt. Their commitment to presenting cutting-edge information in a practical and usable format is truly commendable.

I congratulate the editors for their painstaking efforts. It is expected that this book will serve as a valuable resource and inspiration for continued innovation and development in Horticultural Science.

Date: 24th October, 2025

**(Trilochan Mahapatra)**

प्लांट अथॉरिटी भवन, एन.ए.एस.सी. काम्प्लैक्स के निकट, टोडापुर के सामने, नई दिल्ली-110012 (भारत)
Plant Authority Bhawan, Near NASC Complex, Opp. Todapur, New Delhi-110012 (India)
Tel.: 011-25748127, 25841696; Mob.: 91-7840000542; Fax: 011-25840478
Website: www.plantauthority.gov.in; Email: chairperson-ppvfra@nic.in

# PREFACE

The horticultural science has emerged as the leading discipline in total agricultural development In India. Now-a-days, horticultural science plays a very significant role as its total production has surpassed total food grain production in our country. Besides, horticultural science has also proven facts of protecting mankind from so many dreaded diseases and making hunger free in the crisis of food scarcity.

A number of ICAR institutes and State Agricultural Universities are devoted in the field of research and development of horticultural crops. There is a large network consisting of huge manpower working on research and developmental activities across the country. Since horticultural production has surpassed our food grain production, still we need further research. A number of new varieties, genetic resources, technological interventions and certain breakthroughs need to be updated, compiled and disseminated to the target audience and stakeholders. Still there are so many crops which have enormous potential to be tapped. A number of challenges and problems are also emerging in our day-to-day field of research. Biotic and abiotic stresses have also challenged our research system. To overcome all these problems, collection of adequate scientific information and its timely dissemination has become the need of the hour.

The Society for Horticultural Research and Development (SHRD), Ghaziabad, Uttar Pradesh, India, convened the Indian Horticulture Summit on Mitigating Climate Change and Doubling Farmers Income through Crop Diversification, and Webinar on Boosting Immunity through Horticulture during the year 2020. The researchers belonging to all streams of horticultural science participated with their pieces of research information. Besides, we have invited some relevant chapters from other innovative scientists. And all such information has been compiled in the book. A number of lead lectures belonging to very senior research workers have been compiled in this publication. The publication consists of recent information. All the contributors have tried their best to give their valuable information in their lead lectures and chapters. The Society compliments all researchers who have shared their information for publication in the form of a book.

The book comprises chapters on genetic resources, varietal wealth, crop improvement, plant health management, management of insect, pests and diseases, and value-addition of horticultural crops. Besides, some chapters dealing general horticulture have also been dealt at length.

We are also thankful to Dr A. K. Singh, DDG, Horticultural Science (ICAR) and Drs G. Kalloo, T. Janakiram, J. P .Sharma, U. S. Gautam, R. C. Agrawal, S. Rajan, P. L. Saroj, D. B. Singh, Vishal Nath, S. Uma, B. K. Pandey and V. Pandey and Shiv Shankar Singh for their inspiration and encouragement during compilation of the book. Our sincere thanks are due to Astral International Pvt Ltd, New Delhi, for timely publishing of the book.

The book will serve the purpose of all students, scientists, researchers, policy-makers and related stakeholders engaged in the field of horticultural science. The editors of the book have taken utmost care while editing and making the book in a very easy-to-understandable language.

**Dr B. S. Tomar**

**Dr A. K. Singh**

**Dr. Praveen Kumar Singh**

**Dr. Prabhat Kumar**

**Dr Gograj Singh Jat**

**Prawal Pratap Singh Verma**

**Dr Som Dutt**

# CONTENTS

# THE EDITORS

**Dr B. S. Tomar** has been actively involved in development of seed production technology of vegetable crops under open field condition as well as hybrid seed production under protected conditions. Besides, he worked on genetic improvement of brinjal, onion, okra and cucumber (under protected condition) from last two-and- a half decade. He is associated with the development of 30 vegetable varieties [brinjal (6), onion (3), okra(1), cucumber (3), musk melon (2), garden pea (1), bitter gourd (2), velayti palak(1), cow pea(1), dolichous bean (1) etc.].Most of them have been commercialized through MoU, National Seeds Corporation and NHRDF, IARI and its regional station Karnal. He has also organized live vegetable demonstrations on vegetable technologies at UVRD farm and *mela* site. He has been associated with the teaching in Seed Science and Technology and Division of Vegetable Science since last 24 years and guided 8 Ph.D. students and 6 M.Sc. students, and Dr R Y Vishwanath, one of his students, won the IARI gold medal for his outstanding contribution in his Ph.D. programme. Dr. Tomar has published 101 peer-reviewed research publications in vegetable crops and more than 100 popular articles in *Indian Horticulture/Phal-Phool/Khad Patrika* etc.

Dr Tomar has published 11 training manuals, 2 practical manuals and more than 30 book chapters. He has organized one each short training course and winter school sponsored by Agricultural Education Division of the ICAR, two Model Trainings sponsored by Directorate of Agricultural Extension, MoA&FW and other sponsored farmers training programme in vegetable and seed production. He has also organized *Sabzi Pathshala* at All India Radio sponsored by M/s. Dayal Seed, Meerut and delivered more than 500 talks on All India Radio, Delhi Doordarshan and DD Kisan programme for the benefit of farming community. Dr B. S. Tomar served as an expert member in UPSC, ASRB, NFL, NSC, CCSHAU-Hisar, PAU-Ludhiana, CSHPKV-Palampur, NDUAT-Faizabad, ICFR &E-Dehradun. He has served as Incharge Seed production Unit, 2011-13. He is Fellow of Indian Society of Seed Technology, Indian Society of Seed Spices, Indian Society of Vegetable Science and National Academy of Horticulture Sciences and got best teacher Award-2014 at IARI, New Delhi, and Dr G. Kallo Outstanding Horticulture Scientist Award-2018 by SHRD, Ghaziabad, Uttar Pradesh.

**Dr A. K. Singh**, Principal Scientist (Fruit Science) at Central Horticultural Experiment Station (ICAR-CIAH), Godhra, Gujarat, put in over 21 years for furtherance of dryland fruits of major and minor significance being grown in hot semi-arid and arid conditions, and proved mettle in *ex-situ* establishment and conservation of genetic diversity of dryland underutilized fruits in field gene bank, especially bael, 213 clonal and 121 seedling and have developed 8 varieties of bael (Goma Yashi, Thar Divya, Thar Neelkanth, Thar Srishti, Thar Prakriti, Thar Shivangi, Thar Gauri and Thar Bhavya) and 12 varieties of other fruit crops, viz. jamun (Goma Priyanka and Thar Kranti), tamarind (Goma Prateek and Thar Rashmi), chironji (Thar Priya), khirni (Thar Rituraj), phalsa (Thar Pragati), karonda (Thar Kamal), wood apple (Thar Gaurav and Thar Prabha) custard apple (Thar Amrit) and mahua (Thar Madhu) as co developer. Dr. Singh, specialized in dryland fruit culture, published 128 full length research papers (50 articles >6 NAAS rating), 13 review paper, 13 short notes, 133 popular articles, 57 book chapters, 29 leaflets, 16 technical bulletins, 3 training manuals, 57 full length papers and 151 Abstracts in Conference Proceedings and 14 books in his credit including *'The bael in India'*. Valuable scientific information was generated from 12 in-house and 9 externally funded projects. He has been rewarded several times by best oral and poster presentation and research paper awards, and recipient of awards from scientific professional societies, viz. *Young Scientist* (HTHS), *H. S. Mehta Memorial Best Young Scientist* (Mehta Foundation), *Distinguished Scientist* (AIASA), *Distinguished Horticultural Scientist* (SHRD), *Scientist of the Year* (AIASA), *Outstanding Research Scientist of Noni* (ISNS) *and Research Excellence Awards* (Global Research Foundation). He is recipient of national fellowships of CHAI, HSI and ISNS. He is member of various scientific societies and editor of *Indian Horticulture*, ICAR, New Delhi. He also organized many national and international conferences, symposiums, national horticulture fair and bael day as organizing secretary and convener. Dr Singh is a life member of many professional societies in the field of Horticulture and Agriculture.

Dr Singh has a distinction of being Nodal Officer for 'DUS testing Centre on bael and Co- Nodal Officer of DUS testing center of aonla, Nodal Officer chironji and tamarind as recognized by PPV&FRA, New Delhi. He is well recognized scientist of his area for furtherance of dryland fruits which is prized for health and economic security. Dr Singh is the person with a scientific attitude and trying to bring out the best in dryland horticultural research and development by dedicated and sincere efforts for economic, nutritional and health security to the stakeholders of country.

**Dr Praveen Kumar Singh** born in a farming family in District Hardoi, UP. Presently he is serving as Head, Division of Plant Exploration & Germplasm Collection, ICAR- National Bureau of Plant Genetic Resources, New Delhi. He has received his graduation from C.S. Azad University of Agriculture & Technology, Kanpur. He obtained his post graduation from Punjab Agricultural University, Ludhiana. Dr. Singh's journey has been marked by academic excellence and noteworthy contributions to the field of Horticulture (Vegetable Science).

Dr Singh started his early career in the seed industry before transitioning to academia. He served as an Assistant Professor (Horticulture) at Rajasthan Agricultural University, Bikaner and later joined as Senior Scientist (Hort: Vegetable Science) at Indian Institute of Vegetable Research, Varanasi, in 2008, where he contributed till 2015. He moved to the Center for Protected Cultivation Technology (CPCT), Indian Agricultural Research Institute, Pusa, New Delhi and served as the Principal Scientist (Hort: Vegetable Science) until May 2023. He is a recognized Vegetable Breeder and seed production expert, Dr. Singh has made significant contributions to the field of Vegetable Science. He had taught several courses of Vegetable Science, Seed Production, Plant Biotechnology, Tissue Culture, Protected Cultivation and Plant Genetic Resources to UG and PG students. He developed, standardized & registered petaloid-type of male sterile system in Asiatic carrot in 2008. He conducted several Plant Explorations in difficult land locked and border areas and acknowledged as a Certified Intellectual Property Manager by the Institute of Intellectual Property development & FICCI, New Delhi. Dr. Singh has developed several tomato varieties, including Tomat Pusa Protected-1, Pusa Golden Cherry no.2, Pusa Prasanskrit, Pusa Cocktail tomato, Pusa Cherry Hyb.-1 and Pusa Tomato hybrid-6. He has also contributed to the development of the Pak Choi variety (Pusa PakChoi-1), Capsicum (Pusa Preet) and Longmelon variety (Kashi Santushti).

Dr. Singh's involvement extents to the development of tomato varieties like Pusa Rakshit and H-81, Ridge gourd (Kashi Shivani), Satputia (Kashi Khushi), watermelon (Kashi Mohini) and Roundmelon (Kashi Hari). His extensive body of work includes three edited books, 16 book Chapters, 70 research papers and numerous other publications.

**Dr Prabhat Kumar**, Horticulture Commissioner, GOI &Ex-National Coordinator ICAR-National Agricultural Higher Education Project was born on 1st July 1973 in Basti(UP). He obtained his B. Sc. Degree from C. S. A. Univ.of Ag. & Tech. Kanpur (U.P.) and M. Sc. degree (Floriculture and Landscaping) from Punjab Agril. University, Ludhiana, Punjab. He completed his Ph. D degree (Floriculture and Landscaping) in the year 2002 from IARI New Delhi. He started his professional career as Assistant Professor at G.B.P.U.A.&T. Pantnagar in April 2003. He was selected to the post of Associate Professor at HNBGU, Srinagar, Garhwal (A Central University School of Agriculture)

in 2012. Further, he was selected to the post of Senior Scientist at IARI, New Delhi in 2013 and become Principal Scientist in 2015. Dr. Prabhat Kumar was actively engaged in teaching, research, extension and managerial assignments at all the places. He has guided 10 M.Sc. and 01 Ph.D. students as Major Advisor in the discipline of Floriculture and Landscaping & taught various courses at UG&PG level.

He has handled more than 04 externally funded and 6 institute research projects and published 112 research papers in various national and international journals, 35 popular articles, 15 book chapters, 11 technical bulletins and 10 books. He is associated in the development of 7 varieties namely; Pusa Deep, Pusa Bahar, Pusa Parv, Pusa Utsav(Marigold) &Pusa Sinduri, Pusa Shanti and Pusa Rajat. Established Model Floriculture Centre (MFC) under Horticulture Technology Mission at Pantnagar. In capacity of National Coordinator developed Monitoring and Evaluation cell to coordinate with all partners and facilitated to establish CAAST (Centre for Advanced Agriculture Science and Technology) centers (16) in cut edge areas like genomics, climate smart agriculture, digital agriculture, conservation agriculture, secondary agriculture, etc. He also organized many national and international conferences and symposiums as organizing secretary.

As, Horticulture Commissioner, Govt. of India he immensely promoted more innovative technology for sustainable development of horticulture and also promoting holistic development of bamboo sector as national mission director of National Bamboo Mission. He immensely contributed to give fillip to sweet revolution as mission director of National Honey and Bee keeping Mission. As CEO of Coconut development board, Kochi contributed a lot for development of sector particularly in value addition and climate resilient production system. He has been elected as Chairman, TWG (Technical Working Group) of coconut in ICC, Jakarta. Dr Kumar visited many countries like The Netherlands, Switzerland, Germany, Belgium, Spain, France, Indonesia, Japan, Singapore, Sri Lanka, Thailand and Maldives. He is a life member of many professional societies in the field of Horticulture and Agriculture. He has been honored with awards and fellow by many scientific societies. He is member of Board of Management.

**Dr Gograj Singh Jat** born on March 10, 1989 at Village Mukandpura First, Jaipur (Rajasthan). He obtained his B.Sc. degree in Horticulture (Hons.) from MPUA&T, Udaipur, Rajasthan, and M.Sc. and Ph.D. degree in Vegetable Science from ICAR-Indian Agricultural Research Institute, Pusa Campus, New Delhi. Currently, Dr Jat is working as a Senior Scientist in frontier areas of pre-breeding, genetic improvement and breeding for protected environment in bitter gourd, muskmelon and tomato crop at Division of Vegetable Science, ICAR-IARI, New Delhi. During his career spanning over nearly 10 years, Dr Jat is having his association in release of 6 varieties/ hybrids and 4 genetic stocks of bitter gourd and tomato. He has published 40 research papers in journals of international and national repute, 5 edited books, 16 book chapters, 7 technical bulletins, 6 training manuals, 7 extension folders, and 32 popular and technical articles in his credit. Dr Jat is the recipient of awards like

ICAR-Jawaharlal Nehru Award-2018, IAHS-Dr B. R. Barwale Young Researcher Award-2017, SPH-Dr G. S. Randhawa Award-2019, IAHS- Associateship-2021, SHRD-Young Scientist Award-2019, 13 best poster/paper/oral presentation awards. Recently, Dr Jat had visited Texas A&M University, College Station, Texas, USA for 3 months under NAHEP-CAAST project funded by World Bank-ICAR.

**Dr Prawal Pratap Singh Verma** is currently serving as Senior Technical Officer (Agriculture) at the Council of Scientific and Industrial Research-Central Institute of Medicinal and Aromatic Plants (CIMAP), Regional Research Centre, Purara, Bageshwar, Uttarakhand. He is a recognized expert in the field of high-altitude herbs and medicinal & aromatic crops in the Himalayan region of Uttarakhand. Mr. Verma has submitted his Ph.D. in Vegetable Science at Integral University, Lucknow, focusing on the fenugreek crop. Prior to this, he completed his M.Sc. (Ag.) in Horticulture with 1st Division from Babasaheb Bhimrao Ambedkar University (BBAU), Central University, Lucknow, and his B.Sc. (Ag.) with 1st Division from Dr. B.R. Ambedkar University, Agra (Uttar Pradesh). He has also qualified the ICAR-NET in Vegetable Science, conducted by ASRB, New Delhi.Since he joined CSIR-CIMAP in 2013, Mr. Verma has made remarkable contributions to R&D, significantly advancing the cultivation and management of high-altitude medicinal and aromatic crops. His work focuses on agro-technology development and post-harvest management of economically important herbs such as rosemary, geranium, oregano, marjoram, Himalayan thyme, and lemon balm.

He has successfully trained over 2,500 farmers across Uttarakhand in the cultivation of rosemary, damask rose, lemongrass, geranium, and lavender. As an active project member, he has contributed to more than five research and development projects. Mr. Verma has an extensive publication record, including 45 research papers in national and international peer-reviewed journals (NAAS-rated and impact factor journals), 10 review articles, 8 book chapters, 215 popular articles in Hindi and English magazines and newspapers, and 2 edited books. His academic engagement includes participation in 12 national and international conferences, delivering 8 seminar presentations, and giving 25 expert lectures at Kisan Melas and Gosthis. Through these platforms, he has directly addressed over 8,000 farmers, disseminating his expertise in medicinal and aromatic crop cultivation. Mr. Verma has been honored with several awards, including the Young Scientist Award and the Young Researcher Award from various professional societies. He also holds membership in 8 reputed agricultural and horticultural societies.

His key scientific contributions include the standardization of agro-technology and post-harvest management practices for high-altitude herbs, research on propagation methods and optimization of nutrient management of Rosemary, Oregeno, Thyme, Lemon Balm, Merjorum, stevia, pyrethrum, and geranium. His work has significantly advanced sustainable cultivation practices and promoted the economic viability of medicinal and aromatic herbs in the hilly regions of Uttarakhand.

**Dr Som Dutt** did his M.Sc. and Ph.D. in Botany from the CCSU, Meerut. He started his professional career as a Research Scholar at the IARI, New Delhi in 1982. Since Dr Som Dutt had keen interest in popular science writing, he joined Publications and Information Division (the then Directorate of Information and Publications of Agriculture (DIPA), and now Directorate of Knowledge Management in Agriculture (DKMA) at the ICAR Headquarters, New Delhi, in 1984. Dr Som Dutt edited the *Indian Journal of Agricultural Sciences* and *Indian Horticulture,* a national semi-technical magazine disseminating all research and developmental activities taking place in all horticultural crops across the country. Besides, he has edited the *Handbook of Horticulture* (the Encyclopaedia of Horticulture), the first ever published the bulkiest book of the ICAR and other important horticultural publications of the ICAR, New Delhi.

Dr Som Dutt has written about 500 articles on agricultural sciences, book chapters, and one book, *'Phoolon Ki Vyavsayik Kheti'*, in Hindi. In the field of electronic media, he has made a tremendous impact on farmers by giving radio talks in different programmes of All India Radio on agriculture. He has given several lectures to farmers, extension specialists, young scientists and Horticulture Development Officers of various states, that has contributed in increasing the agricultural production of farmers significantly. Dr Som Dutt has been awarded with the *Chaudhary Charan Singh Award for Excellence in Journalism in Agricultural Research and Development* 2008 (ICAR National Award), *Raizada Memorial Award for Young Information Scientist* 1990, and the *Best Editor Award* 2008, for the popularization of agricultural sciences in different media. At Present, Dr Dutt is Managing Editor of *Current Horticulture,* a peer reviewed research Journal published by SHRD, Ghaziabad, Uttar Pradesh, and also organized the *Indian Horticulture Summit*-2020 and Webinar on *Boosting Immunity through Horticulture* during 2020 as Secretary SHRD, Ghaziabad, Uttar Pradesh.

CHAPTER 1

# HIGH-DENSITY PLANTING IN TEMPERATE FRUIT CROPS FOR ENHANCING PRODUCTIVITY

**M.K. Verma, O.C. Sharma, J.I. Mir, W.H. Raja, S.U. Nabi and M.A. Shah**

*ICAR – Central Institute of Temperate Horticulture,*
*Old Air Field, Rangreth, Srinagar, Jammu and Kashmir*
*Corresponding author: mahenicar10@gmail.com*

---

Sustainable and maximum quality fruit production is the dream of every fruit grower. The yield and quality of the fruits are dependent factor of the number of trees /ha, ideal tree architecture (training system) and the annual required practices like pruning, fertilization and plant protection measures. Such results are expected from high density plantings (HDPs). High density planting is the accommodation of maximum number of plants per unit area for producing the maximum yield of quality fruits. This technique was first commercially exploited for apple in Europe during sixties and now majority of the apple orchards in Europe, America, Australia and New Zealand are grown under this system.

Increasing the number of trees/ha increases yields, up to a point, in young orchards. However, as trees matures, too many trees per hectare results in over crowding and shading with commensurate declines in yield and fruit quality and increases pest problems. Therefore, the ideal tree density varies greatly depending on soil type and topography, varieties, rootstocks, training systems, the presence of irrigation and the grower's management capabilities. Traditionally, it was common to set an orchard "on the square," in which the distance between trees in-a-row equalled the distance between-rows. Oftentimes, the orchard floor was kept free of grasses and weeds by cultivating down the rows and across rows. High fruit quality could be attained because mutual shading was not a problem. Yields were low to moderate because a relatively high percentage of the orchard space was devoted to drive areas and, hence, not covered with fruiting wood. Planting trees close enough in the row that they almost touch at maturity and leaving an extra eight feet between rows to accommodate equipment movement resulting in more trees per acre, which promises potentially higher early yields plus greater yield potential in the mature orchard.

Orchard establishment is a long term process. A good orchard gives profitable returns and unplanned orchards are unprofitable venture. Therefore, it is essential for orchardists to plan their new plantings according to the availability of resources judiciously. Among the resources, land is the most limiting factor and must be utilized to the fullest of its potential. This is only possible by adoption of the closer plantings. Accommodation of maximum plants per unit area to get the maximum possible profit per unit of the tree volume without impairing the soil fertility status is called the high density planting. It is defined as planting a density in excess of that which suffices to give the maximum crop yield at maturity if the individual trees grow to its full natural size. The underlying principle of high density planting is to make the best use of vertical and horizontal space and to harness the maximum possible return per unit of inputs and natural resources. In other words, it is nothing but accommodating more number of plants per unit area than optimum through manipulation of tree size. This gives better light distribution within tree canopy increases the number of well illuminated leaves and promotes the rate of photosynthesis that leads to high yield per unit area. This system of planting is going to revolutionize the fruit industry by enhancing the production of quality fruits with minimum damage coupled with reduction in production costs. Basically, the availability of a dwarf plant is the first and foremost prerequisite for establishing high density or meadow orchard. Fertilizer doses, spacing, growth regulation by the training and pruning, use of the mechanical devices etc., may also be tried either singly or coupled with other crop management practices for its successful adaptation.

## Principles of HDP

In the process of evolution in planting density several important principles have been discovered which have guided the shifting from wider planting to high density plantings. These mainly include-

- To achieve higher yield, orchard canopies must intercept a high proportion of available light (70-75%). Pedestrian orchards do not intercept more than 55% of available light.
- Dense canopies have too much heavily shaded area with poor fruit quality. Narrow canopies or planar canopies with a depth of not more than 3 feet have better light distribution. This has led the effort to narrow the canopy of modern orchards to no more than 3 feet deep.
- High early yield to pay back the initial investment to plant the orchard through the use of feathered trees, planting higher tree densities, maximizing tree growth after planting with irrigation and fertigation, minimizing pruning at planting and in the first 3 years and branch bending to induce early cropping.

- Simple and sparse tree canopies are more adaptable to partial mechanization of some orchard management practices than thick and complex tree canopies.
- Orchard planting density is limited by the economic law of diminishing returns. As planting density is increased, the additional benefit in yield is smaller and smaller with each additional tree. At some point the cost of additional trees is greater than the gain in yield.

## Impact of low planting density

When the planting density is too low, each individual plant may perform at its maximum capacity, but there are not enough plants as a whole to reach the optimum yield. Therefore, total yield of the crop becomes a limiting factor. Also, when there is too much space left between plants, causing more weed growth and increases the weeding cost.

## Impact of high planting density

If the planting density is too high, plants may compete against each other, known as intra-specific competition. Under such circumstances, the performance of individual plants becomes a limiting factor to achieve maximum crop yield.

## Using right density

As the planting density increases, the total crop yield increases and reaches to its maximum, at this point further increase in planting density results in reduced yield. It is, therefore, important to use appropriate planting density, keeping in mind the fact that the density that provides the highest yield and the density that provides the highest revenue may be different. Growers should consider both optimum yield and revenue when balancing their planting density.

## What is optimum planting density?

In general, it is established that the greater the planting density, the greater the investment cost to establish the orchard. However, due to higher early yield and higher cumulative yield, profitability is generally increases with increase in tree density up to a point. Nevertheless, the law of diminishing returns which results in less gain in cumulative yield as more trees are planted per hactare, meants that very high tree densities are not more profitable than more moderate densities. In addition, economists suggest that risk increases with increasing level of investment, thus making the very high-density systems riskier. In a study conducted by Robinson (2013) on net present value (NPV) of the accumulated profit over 20 years showed that the per unit land area the greatest profitability was at a tree density of 1,050 trees/acre when feathered trees were used.

## Tree density guide

- Semi-dwarf orchards have densities of 123-311 trees per acre at spacings of 16′ x 22′ to 10′ x 14′.
- Dwarf orchards are planted with spacing's as tight as 2′ x 10′, but more commonly 4′ x 14′ and up to 7′ x 16′ for densities of 388-777 trees per acre.
- Standard orchards are no longer planted because they take too long to produce.
- High density means more than 500 trees to the acre and will exceed 1000+ with multi row systems.

## Categories of HDP

The establishment of an orchard is a long term investment, with orchards having a life expectancy of more than 20 years. It is important that the establishment of an orchard be well planned including the site planning, planting and tree care. The high density plantings can require a grower investment in excess of $20,000 per acre for the establishment. The extensive planning for high density plantings will pay off in maximizing yields and returns over the productive life of the orchard. Mistakes made in the planning and planting can be very costly and are difficult to correct in later years. It is paramount that careful attention be paid to the site selection, soil preparation, tree quality, cultivar selection and early tree care. Within this range, the decision of how close to plant a new orchard should be made with consideration of the Vigour of the variety, vigour of the rootstock, and soil strength. With vigorous scion cultivars, growers should use a more dwarfing clone of M.9 or B.9 rootstock and/or greater planting distances. With weak scion cultivars, a more vigorous clone of M.9 rootstock should be used and/or closer planting distances. For weak and moderate growing cultivars such as Honeycrisp, Delicious, Braeburn, Empire, Jonamac, Macoun, Idared, Gala, NY674, Golden Delicious, etc it is suggested an in-row spacing of 3′. For vigorous varieties such as McIntosh, Spartan, Fuji, Jonagold, Mutsu, etc. and tip bearing varieties such as Cortland, Rome, Granny Smith and Gingergold an in-row spacing of 4′ is suggested. Between-row spacing should be 12′-13′ on slopes and 10′-11′ on level ground.

High density plantings were initially categorized in to four planting densities for apple viz., low HDP (< 250 trees/ha), moderate HDP (250-500 tree/ha), high HDP (500 to 1250 trees/ha) and ultra-high HDP (>1250 trees/ha). Recently, super high density planting system has been also established in apple orchards with a plant population of 20,000 trees per ha. In some orchards, still closer, planting of apple trees are followed (70,000 trees/ha) which is often referred as 'meadow orchards'. Corelli and Sansavini (1989) proposed 9 categories of planting densities in apple is given in Table 1.

**Table 1. Classification of high density orchard in apple**

| Density category | Standard variety | Spur variety |
|---|---|---|
| Very low | 226/seedling | - |
| Low | 270/M26 | - |
| Medium low | 670/M26 | 1000/M26 |
| Medium | 670/MM106 | 1000/MM106 |
| Medium high | 1000/M9 | 2000/M26 |
| High | 1000/M26 | 3000/M9 |
| Very high | 2000/M26 | 5000/M9 |
| Bed system | - | 10000/M9 |
| Meadow orchard | - | 40000/MM106 |

Six type of planting systems *viz.*, low density, moderate density, high density, ultra-high density and super high-density plantings or meadow orcharding were categorised. The exact limits of plant density in high density planting are based on several factors and for many crops it is not yet well defined. It varies with the region, land fertility, species, variety, rootstock, economic feasibility of the farmer to adopt the system, farmer's knowledge about HDP and crop, labour and the likely return from the orchard and natural resource to support HDP. The brief about different HDPs are as under-

## Low Density Planting (LDP)

It is the traditional method of planting in which trees are planted at wider spacing and accommodates the minimum number of plants (100 plants/ha). There is no use of dwarfing rootstocks, minimum training and pruning, favours maximum development of canopy availing the scarce water and nutrients. The trees come in to commercial production potential after 10 to 15 years of planting and the output of orchard during early years found to be very low.

## Medium Density Planting (MDP)

The medium density planting accommodates 250 to 500 plants /ha. Proper pruning is undertaken to restrict the tree for desired shape. It requires intensive care and labour since the return found to be sustainable. It gives more yield and quality fruit in crops and better water and nutrient managements are also required for maintaining the orchards for a long productive life.

## High Density Planting (HDP)

In HDP, the tree density is more than 1000 plants/ha depending on fruit crops. It requires heavy or rigorous training and pruning to restrict the growth. Dwarfing rootstock and chemicals are normally required to maintain the optimum growth. Both yield and expenses are high under this planting density. Establishment and maintenance of such a high density orchard could be demanding and require technical backup.

## Ultra-High Density Planting (UHDP)

Ultra-High-Density Planting can accommodate more than 2000-5000 trees per hectare and requires severe pruning and training after planting. Ultra-high density in apple accommodated 1111 plants /ha at a spacing of 3 × 3 m (444 plants /acre). For sustainable yield and quality under ultra-high density, proper canopy management, chemical assistance and nutrient management are required.

## Meadow orcharding

"Meadow - grassland system" is also known as "Super High Density Planting", widely popular in temperate crops. It can accommodate more than 10,000-40000 or more plants per hectare in which severe top pruning is practiced similar to mowing of grass land. The tree yields after one or two years of planting. Heavy uses of growth regulators as well as judicial canopy managements are required for fruiting. In apple, 10000 plants /ha can be accommodated at a spacing of 1 × 1m (4000 plants /acre). For high density planting, compact varieties are used whereas in ultra-high density and super high-density planting, dwarfing rootstocks are used to limit the natural growth of the plant besides the constant use of growth retardants every now and then to have a control over the growth.

# Factors affecting HDP

## Tree quality

Tree quality is an essential components that must be considered on priority. Several studies have shown that greater the number of lateral branches or feathers the greater the yield in the second and third year. Modern high-density orchards can have significant 2$^{nd}$ and 3$^{rd}$ year yield if highly-feathered trees are planted and if trees grow well the first year. If growers use whips or small calliper trees these do not produce significant quantities of fruit until year 4 or 5. We recommend that nursery trees must have 10-15 well positioned feathers with a maximum length of 1 ft (30cm) and starting at a minimum height to 30" (80cm) on the tree.

## Scion varieties

**Table 2. Compact growth habit type temperate fruit varieties**

| Crop | Genetically dwarf cultivars | Desirable features |
|---|---|---|
| Apple | Spur bearing varieties Wijik Mc Intosh and Nugget Golden Delicious | Bear on short stems, spurs; grow to 60-70% the standard cultivar in vigour and bear more spurs and yield more |
| Peach | Redheaven | Dwarfing and high yielding |
| Cherry | Sweetheart, Hedelfinger, Compact Lambert, Meteor and North star | High yielding, self-fruitful, dwarf |

## Dwarfing rootstocks/interstocks

**Table 3. Dwarfing rootstocks and interstocks recommended for temperate fruits**

| Crop | Dwarfing rootstocks |
|---|---|
| Apple | M-9, M-26, M27, MM-111, MM-106, Bud.9, Bud.146, P-22 and Ottawa 3 |
| Pear | EM Quince A & C |
| Peach | Siberian C, St. Julien x *P. besseyi* and Rubira |
| Plum | Pixy |
| Cherry | Growth is 35 to 50% of standard (Gisela 5, Edabriz, Weiroot 72) to semi-dwarfing (Gisela 12, Ma x Ma 14, Weiroot 158) to 90 to 110% of standard (Gisela 6, M x M 60, Colt); Gisella-5, Edabriz, Ma x Ma 14, Weiroot 158, Colt, Charger and Rubira |

Rootstocks, dwarfed than M.9, like P.16 and P.22 are able to increase yield efficiency by 5 to 8 % but loose total production by up to 30% on a hectare basis.

## Tree establishment

In high density planting, the allotted space should be filled rapidly to ensure that the ceiling production level is quickly reached. Fill planting holes with fresh, organic soil mixtures and providing the trees with adequate water by localised irrigation. Enriching the irrigation water with nutrients adds improved flower-bud formation, indispensable for a crop in the second leaf. As fertigation may have more consistent effects on growth than soil sterilisation, it might become an environment-friendly alternative to overcome replant problems in apple. The quantity of water required for a good start depend on tree size, soil and climatic conditions. More insight in the water requirement of a newly planted trees are needed to get a good balance between growth and early cropping on the one hand and to avoid waste of costly water on the other. Obviously, more water is needed for heavily feathered trees to maintain a certain water status in and around the planting hole. Although it is clear that shoot growth, fruit set, yield, and fruit weight are reduce by low quantities of water, but it is not known what quantities are optimal.

## Growth retardants

Various growth retardant have been used to restrict the vegetative growth of the plant. Among them the commercially adopted are AMO 1618, CCC, Ancymidal, Paclobutrazol, B-9 (Phosphon D) and chloramquat.

## Induction of viral infection

Though still not popular and emerged as commercially but tree size can be reduced by inducing viral infection e.g. in apple, virus free rootstocks series East Malling Long Ashton (EMLA) are vigorous than their infected counterparts.

## Use of incompatible rootstock

The use of graft incompatible scion and rootstocks also induces dwarfness in the composite plant e.g. in pear Bartlett and Quince.

## Branch angle manipulation

An important method of shifting the balance between vegetative growth and cropping in young trees is tying down of the scaffold branches below horizontal to induce cropping. In the vertical position, a shoot grows more vigorously than in a horizontal or pendant position and tends to remain non-fruitful. A horizontal or pendant limb grows less vigorously and then crop heavily the next year and bend under the weight of the fruit. The fruits are also strong competitors for resources and moreover it limit the growth of the branches. If a vertical limb is manually bent horizontal, lateral buds are released from dormancy. If the vigour of the branch is excessive, these buds may grow into vigorous upright shoots by themselves and remain unfruitful. However if the branch has more moderate vigour, the lateral buds grow into short shoots which become fruitful.

After the initial tying or weighting down of feathers at planting, new lateral branches that arise along the leader do not need to be tied down. In most climates, if moderate vigour shoots are not pruned, often crop load in the third year will bend branches down below horizontal and a natural balance between vigour and cropping will be established without additional limb positioning. Thus with the tall spindle, no additional limb tying is needed after the initial tying or weighting down of the feathers at planting. However, in vigorous and/or warmer climates where winter chilling is insufficient, often limbs become too large before they set sufficient crop loads to bend the branches down. In these climates, tying down of all vigorous limbs must be done annually for the first 3-5 years until the tree settles down and begins to crop heavily. However, in most traditional apple growing areas, growers often invest too much money in limb tying which should be limited to only the feathers at planting. Thereafter, the precocity of the rootstock induces heavy cropping and a natural balance is established.

## Irrigation and fertigation

Irrigation is crucial for viability of the production system. Large, highly-feathered trees often undergo water stress shortly after planting despite adequate soil moisture levels. This is due to the damaged root system of a transplanted tree which cannot adequately support the large top without frequent irrigation. Large, highly feathered trees produce much more leaf area shortly after planting than unfeathered trees which creates a high water demand before the root system can re-grow sufficiently to support the trees. In-addition during many years, dry weather following planting results in water stress of newly planted trees which can limit tree growth. Frequent and early trickle irrigation can help these trees to produce good growth in the first year. In humid areas growers are unaccustomed to installing irrigation immediately after planting and delay its installation until mid-summer. It is recommend to the growers to install trickle irrigation soon after planting with high density orchards that use feathered trees to prevent water stress and maximize first year tree growth. Frequent low doses of nitrogen fertilizer delivered at least twice in a week through the trickle system (fertigation) for the first 12 weeks of the season will greatly improve tree growth during the first 2 years to speed up the development of the canopy. Ground fertilization with dry fertilizers is often ineffective since too much time elapses between applying the fertilizer and uptake by the tree. With high tree densities as with the Tall Spindle system and highly-feathered trees, almost no lateral tree growth is required and only vertical extension growth is needed. In this case fertigation for the first 2 years is essential both for the water to avoid water stress and for the nitrogen which is rapidly moved to the root zone and is readily available to the tree as soon as it starts growing. Adoption of fertigation considerably improves tree growth and yield potential in the 2nd and 3rd year. For moderate tree densities such as with the vertical axis, slender pyramid or Y-trellis, trees must be grown vigorously for several years to fill the allotted space with canopy and relatively high nitrogen fertilization is desirable for 2-3 years after planting. However, excessive fertilization, especially nitrogen, can cause too much growth which results in delayed flowering, reduced yields, poor fruit quality and greater pruning.

After the first few years, low nitrogen fertilization is desirable to keep the trees calm with a balance between fruiting and cropping. Many mature high density orchards receive excessive nitrogen fertilizer rates which cause severe canopy management problems. Soil or fertility must be considered when calculating the amount of nitrogen to apply to mature high-density orchards especially with vigorous and poor colouring varieties. Soils having excess fertility often results in excessive vegetative growth, delayed cropping and soft and poorly coloured, unmarketable fruit.

## Weed control

Weed competition can drastically reduce tree growth during the first few years and can cause failure of the orchard to fill its allotted space which always results in diminished yield and profitability. Good weed control during the first 3-4 months of a growing season is the most critical time period. In later summer months if weed control is poorer it is not detrimental to the trees. Thus, It has been recommended that growers provide excellent weed control for the first 4 months of the season for the development years of the orchard. In India, weed control in often carried out manually which involves maximum production cost. Therefore, chemical weed control offers fruitful results to increase the net returns per unit area.

## Crop load management

Management of cropping with high-density orchards during the first 4 years to avoid biennial bearing is critical in maintaining a proper balance between vegetative growths and cropping as the trees begin to bear. With precocious dwarfing rootstocks, young apple trees can often overset in the 2nd or 3rd year resulting in biennial bearing as early as the 4th year. These then results in increased vigour in the 4th year just when the trees have filled their allotted space and when reduced vigour is needed. Varieties differ in their biennial bearing tendency and this must be incorporated into the crop load allowed on young trees. For annual cropping varieties like Gala, a crop load of 6 fruits/cm2 TCA (25-40 apples/tree in the second year, 40-60 apples/tree in the third year, and 100-120 apples/tree in the fourth year) is recommended. For slow growing and biennial bearing varieties like Honeycrisp we recommend crop load of 4 fruits/cm2 TCA (15-20 apples/tree in the second year, 25-40 apples/tree in the third year, and 50-70 apples per tree in the fourth year). Within each year, the low end of the range should be used for low vigour trees and the high end of the range for high vigour trees.

## Pruning

During the developmental years of an orchards life the trees have a juvenile character and the balance between vegetative growth and cropping is shifted toward vegetative growth. With high density systems such as the tall spindle the goal is to get the trees into cropping as soon as possible. This is best accomplished by minimizing pruning during the first 4 years. No heading cuts should be done to the leader or lateral branches at planting or for the next 4 years since the maximum growth and earliest cropping is achieved with no pruning. For the first 4 years pruning should be limited to the complete removal of unsuitable branches such as those lateral branches that are larger than 2/3 the diameter of the leader. Of much greater importance during the first 4 years is limb angle manipulation to change a young and vigorous tree from a vegetative state into a reproductive state. With the Tall Spindle and Vertical Axis systems the artificial limb bending is limited to the 1st tier of branches while with the Solaxe even upper branches tied down. With the Tall Spindle and the Vertical Axis systems the bending of upper branches is achieved naturally by crop load.

As the orchard reaches maturity, containment pruning of the canopy is essential in maintaining trees within the allotted space and to improve the light distribution to the lower portion of the tree. Good light distribution and good fruit quality can be maintained as trees age if the top of the tree is kept narrower than the bottom of the tree and if there is a good balance between vegetative growth and cropping. Pruning strategies based on shortening or stubbing back permanent branches that outgrow their allotted space generally are not as successful as limb renewal pruning strategies. This is partially because the most productive fruiting wood is cut off when a branch is shortened. In addition, stubbing cuts stimulate localized vigour on the affected branches which results in shading of the lower canopy. In our studies on how to manage the canopies of high-density systems, treatments where branches were shortened to maintain the conic shape to the tree, resulted in unacceptable yield reductions, a dense canopy resulting in interior canopy shading, excessive vigour compared to an unpruned control and poor fruit quality.

## Use columnar apple

Columnar habit apple trees have short internodes and axillary buds grow into spurs rather than lateral branches, so the tree naturally grows like a sturdy cordon.

## Optimum use of light resource

High density plantings achieve earlier cropping than wider plantings, but they show a decline in fruit quality, fruit bud formation and yield. These are mainly attributed owing to its influence on light interception and distribution. Therefore, high light interception is a prerequisite for higher yield, better fruit quality, especially colour. Light interception increases as the height increases for trees of triangular in shape.

## Commercial use of HDP

High density production technology in various fruit and nut crops has already made a dent in the several countries which showed significant results in terms of productivity and net returns. The salient details of the high density production technologies are given in the tables includes the high yielding varieties, compatible rootstock combination, tree training system, appropriate spacing, tree density per hectare and yield per hectare.

## Apple

In modern high-density orchards in-row spacing range from 3-5 ft. and between row spacing from 10-14 ft. (780-1450 trees/acre). Within this range, the decision of how close to plant a new orchard should be made with consideration of the vigour of the variety, vigour of the rootstock, and soil strength. With vigorous scion cultivars, growers should use a more dwarfing clone of M.9 or B.9 rootstock and/or greater planting distances. With weak scion cultivars, a more vigorous clone of M.9 rootstock should be used and/or closer planting distances. Despite some latitude in planting distances, growers should remember that to obtain high early yields high tree densities are essential. In New Zealand in the 1970s semi-intensive system of apple growing was found most profitable. The results from Lincoln University in apple grown on 'Malling 9' rootstock with tree density of 1905 trees/ha gave higher yields as compared to those grown on 'Malling-Merton-106' rootstock at 635 trees/ha.

In apple, most productive system recorded in 'Empire' grafted on dwarfing rootstock M-26 and trees trained on Y-trellis system at a closer spacing (2.1 x 3.7 m) accommodated 1283 trees per hectare gave maximum yield (275 t/ha). High density in apple around the world entirely depends upon few varieties (Red Delicious, Empire, Granny Smith, Fuji, Gala, McIntosh, Golden Delicious, Oregon Spur and Red Chief) and rootstocks (M-9, M-26, MM-106 and MM-111). Open canopy and vertical type training systems like Y-trellis, slender spindle, central leader as well as modified leader systems are highly efficient for photosynthesis results in to greater yields. Closer tree spacing along with open canopy were found more productive. However, higher densities gave significantly more yield during initial years (Table 4).

**Table 4. Assessment of high densing Planting in India and abroad on apple**

| Variety | Root-stock | Training system | Spacing (m) | Density (trees/ ha) | Yield (t/ ha) | Reference (Country) |
|---|---|---|---|---|---|---|
| Empire | M-26 | Y-trellis | 2.1 x 3.7 | 1283 | 275.0 | Robinson *et al.* 1991 |
| Granny Smith | EM-9 | V- shape | 0.4 x 3.5 | 7140 | 210.0 | - |
| COE Red Fuji | M.26 | Vertical axis | 0.75x1.5 | 8889 | 135.30 | CITH, 2013-14 |
| Red Delicious | M-26 | Y-trellis | - | 960 | 130.0 | - |
| Mollies Delicious | M-9 | Head and spread | 0.75 x 1.5 | 8889 | 126.33 ($5^{th}$ Yr) | CITH, 2012-13 |
| Red Delicious | M-26 | Y-trellis | 2.4 x 4.3 | 961 | 122.0 | Robinson *et al.* 1991 |
| Mollies Delicious | M-9 | Modified leader | 0.75 x 1.5 | 8889 | 108.33 ($5^{th}$ Yr) | CITH, 2012-13 (India) |
| COE Red Fuji | M-9 | Espalier | 1.5 x 3.0 | 2222 | 95.36 ($5^{th}$ Yr) | CITH, 2012-13 (India) |
| Red Delicious | M-9 | Slender spindle | - | 1495 | 95.0 | - |
| COE Red Fuji | M-9 | Vertical Axis | 0.75 x 1.5 | 8889 | 93.82 ($5^{th}$ Yr) | CITH, 2012-13 (India) |
| Fuji | M-26 | Spindle bush | 2.0 x 4.0 | 1250 | 88.00 | |
| Royal Delicious | MM 106 | Modified leader | 2.5 x 3.5 | 1143 | 67.73 ($10^{th}$ Yr) | CITH, 2012-13 (India) |
| Marshal McIntosh | M-9 | Slender spindle | 1.7 x 4.0 | 1429 | 60.0 | Wunsche and Lasko 2000 (USA) |
| Golden Delicious | M-9 | Central leader | 1.5 x 3.0 | 2222 | 47.4 ($8^{th}$ year) | CITH, 2010 (India) |
| Red Spur | Standard | Modified Leader | 4.0 x 4.0 | | 52.5 | CITH, 2012-13 (India) |
| Vista Bella | M-9 | Central leader | 1.5 x 3.0 | 2222 | 42.5 ($8^{th}$ year) | CITH Annual Report, 2010 (India) |
| Golden Delicious | MM-106 | Central leader | 2.5 x 3.5 | 1143 | 39.5 ($8^{th}$ year) | CITH, 2010 (India) |
| Mollies Delicious | M-9 | Central leader | 1.5 x 3.0 | 2222 | 35.5 ($8^{th}$ year) | CITH, 2010 (India) |
| Gala Redlum | M-9 | Tall Spindle | 3.0 x 1.5 | 2222 | 86.5 ($7^{th}$ year) | CITH Annual Rep. 2023 |
| Super Chief | M-9 | Tall Spindle | 3.0 x 1.5 | 2222 | 53.0 ($7^{th}$ year) | CITH Annual Rep. 2023 |
| Red Jona Prince | M-9 | Tall Spindle | 3.0 x 1.5 | 2222 | 60.0 ($7^{th}$ year) | CITH Annual Rep. 2023 |

| Variety | Root-stock | Training system | Spacing (m) | Density (trees/ha) | Yield (t/ha) | Reference (Country) |
|---|---|---|---|---|---|---|
| Golden Delicious Reindeers | M-9 | Tall Spindle | 3.0 x 1.5 | 2222 | 48.5 (7th year) | CITH Annual Rep. 2023 |
| Red Velox | M-9 | Tall Spindle | 3.0 x 1.5 | 2222 | 46.0 (7th year) | CITH Annual Rep. 2023 |
| Pinnova | M-9 | Tall Spindle | 3.0 x 1.5 | 2222 | 52.0 (7th year) | CITH Annual Rep. 2023 |
| Elstar | M-9 | Tall Spindle | 3.0 x 1.5 | 2222 | 46.0 (7th year) | CITH Annual Rep. 2023 |

**Figure 1. High density orchard of apple at CITH, Srinagar**

**Table 5. Spacing and planting density for different scion-stock combinations**

| Scion/ cultivar | Rootstock | Tree size (Vigour) | Spacing (mxm) | Density (tree/ha) |
|---|---|---|---|---|
| Standard | MM 109 | Semi-vigorous | 6.0 x 6.0 | 278 |
| | MM 111 | Semi-vigorous | 6.0 x 6.0 | 278 |
| | M 7 | Semi-dwarf | 4.5 x 4.5 | 494 |
| | MM 106 | Semi-dwarf | 4.5 x 4.5 | 494 |
| | M 9 | Dwarf | 1.5 x 1.5 | 4444 |
| Spur type | Seedling | Semi-vigorous | 5.0 x 5.0 | 400 |
| | MM 109 | Semi-dwarf | 3.5 x 3.5 | 816 |
| | MM 111 | Semi-dwarf | 3.5 x 3.5 | 816 |
| | M 7 | Dwarf | 3.0 x 3.0 | 1111 |
| | MM 106 | Dwarf | 3.0 x 3.0 | 1111 |

In the management of an ultra-high density (20000 trees/ha) apple orchard, control of tree size is an important objective to obtain optimal yield of quality fruit. If horticultural practices are applied to supplement the effect of rootstock in vigour management then these practices should be economical, effective, and environmentally sound. Root pruning was effective in reducing vegetative growth in the 2nd year after treatment.

In India, Red Spur varieties found productive on PSM-1, an indigenous rootstock (*Malus baccata* bio-type Shillong) and maximum yield (62.79 t/ha) was recorded at 4444 density with lowest leaf nutrient content. However, tree density 1111 (3 x 3m) was found ideal with moderate yield (25.13 t/ha) for sustainable orchards life under Himachal Pradesh conditions (Pramanick *et al.* 2012).

## Potential HDP systems in apple

The seven leading systems in the world are: *Tall Spindle, Super Spindle, Vertical or V-trellis, Solaxe, Bi-Axis and Fruiting Wall.* All seven systems use high tree densities (900-2,200 trees/acre) and depend on early production to repay the initial investment.

- The Tall Spindle is the most preferred system with most growers using 1,000-1,300 trees/ha. It is simple to learn and requires moderate initial investments, has high early yield and is estimated to provide the highest profitability over 20 years.
- The Super Spindle orchards are planted at 2,200 trees/acre but these are limited to growers who produce their own trees. The super spindle has a simplified pruning recipe and high fruit quality.
- The Organized trellis systems (V-trellis and Vertical trellis) are the two systems which are designed for precision pruning where bud numbers are reduced through pruning to a pre-calculated number. These systems also have less sunburn to fruits.
- The Solaxe system uses extensive limb bending and manual bud extinction to achieve a balance of vegetative growth and cropping even when vigour is excessive. The bud extinction also helps to control biennial bearing.
- The bi-axis system uses a two-stem tree to achieve a very high number of leaders per acre (1,800) with only a moderate number of trees per acre (900). It also distributes tree vigour to two leaders giving more moderate growth in each leader. The system has its greatest potential when tree vigour is difficult to manage with traditional Tall Spindle training. This system is well adapted to mechanical pruning.
- The "Mur Frutiere" or Fruiting Wall system uses mechanized sidewall shearing during the early summer to reduce pruning costs and achieve high yields and good fruit quality. It is interesting that some successful apple growers manage apple trees with complete precision of bud load as with the organized trellis systems while other successful apple growers use a completely mechanized system with no precision of bud management.

## Pear

In pear, high density planting systems gave very high yields in varieties like Abbe Fetel, Conference and Doyenne du Comice. These all varieties found productive on dwarfing rootstock Quince and its various clones. Higher yields were mostly recorded at very high densities (5555 to 7936 trees/ha) which indicates that the planting in hedge-row pattern and trained with open centre systems like V-shape and Y-trellis facilitates maximum sunlight interception and results into maximum yields (Table 6).

**Table 6. High density orcharding systems in pear**

| Variety | Rootstock | Training system | Spacing (m) | Density (Trees/ ha) | Yield (t/ha) | Reference |
|---|---|---|---|---|---|---|
| Abbe Fetel | Quince clone (MC) | V-shape | 3.6 x 0.50 | 5555 | 262.0 After 7th year cumm. | Musacchi *et al.* 2005 |
| Abbe Fetel | Quince clone (MC) | Vertical axis | 3.6 x 0.50 | 5555 | 257.0 After 7th year cumm. | Musacchi *et al.* 2005 |
| Conference | Quince clone (MC) | V-shape | 3.6 x 0.50 | 5555 | 181.0 After 7th year cumm. | Musacchi *et al.* 2005 |
| Doyenné du Comice | Quince clone (MC) | Vertical axis | 3.6 x 0.35 | 7936 | 132.0 After 7th year cumm. | Musacchi *et al.* 2005 |
| Conference | Quince-A | Y-trellis | 4.0 x 1.50 | 1666 | 47.5 | - |

## Peach

Peaches are the very important fruit crops adapted from temperate to subtropical climate. In this group, both peaches and nectarines are very productive. Genetically, dwarf rootstocks are not common in stone fruits. Therefore, these fruit crops could not be planted as closer as pome fruits. Results indicated that few varieties like Royal Giant, Flavour Crest, Baby Gold-5, Stark Red Gold, Early Grande, Red Globe, Cresthaven, Glohaven and Fantasia are found highly productive under high density (900 to 1600 trees/ha). Open canopy type training system always favours the higher productivity in peaches and nectarines as in case of pome fruits (Table 7).

In peach, the main tools for tree size control are rootstocks, summer pruning, deficit irrigation, genetic modification of tree growth habit, root restriction, growth regulators. Around the world, the most important training systems for HDP are free and sprint palmette, fusetto, delayed vasette, Tatura trellis and related training systems. In medium or low fertile soil the fusetto facilitates uniform distribution of leaf area, early entry into production, good light penetration and good fruit quality; fusetto allows maximum planting density of 1200-1500 trees ha. In contrast, on extremely fertile soil it is necessary to adopt training systems that allow the canopy to develop as a single or double wall of limited thickness. The palmette allows maximum planting densities of 700-900 trees/ha. For orchards with planting densities greater than 1000-1500 trees/ha, the Tatura trellis and its variants (KAC V, perpendicular V) appear the most appropriate. Overall, the right choice of training system and orchard density will have to be made on the basis of careful economic analysis, taking into account the greater costs involved for the establishment and often also for the management of HDP.

**Table 7. High density orcharding systems assessed at various places in India and abroad in peach**

| Variety | Root stock | Training system | Spacing (m) | Density (Trees/ha) | Yield (t/ha) | Reference (Country) |
|---|---|---|---|---|---|---|
| Royal Giant | Nema guard | V-shape | 2.0 x 5.5 | 909 | 353.0 | - |
| Flavour Crest | Nema guard | V-shape | 2.0 x 5.5 | 909 | 206.0 | - |
| Baby Gold-5 | Nema guard | V-shape | 2.0 x 5.5 | 909 | 169.0 | - |
| Cresthaven | Nema guard | Tatura trellis | 2.5 x 2.5 | 1600 | 47.26 (6th Yr) | CITH, 2013-14 |
| Glohaven | Nem guard | Tatura trellis | 2.5 x 2.5 | 1600 | 42.99 (6th Yr) | CITH, 2013-14 |
| Red Globe | Nema guard | Tatura trellis | 2.5 x 2.5 | 1600 | 41.12 | CITH, 2013-14 |
| Glowhaven | Standard | V-Shape | 3.0 x 3.0 | 1111 | 40.08 (5th Yr) | CITH, 2012-13 |
| Fantasia | Balle | Central leader | 2.4 x 4.5 | 900 | 38.0 | - |
| Stark Red Gold | GF-655-2 | Free Spindle | 2.0 x 4.0 | 1250 | 34.3 | - |
| Early Grande | - | Meadow system | - | - | 29.6 | - |
| Fantasia | Standard | Modified leader | 2.5 x 2.5 | 1600 | 20.6 (3rd Year) | CITH, Annual Report, 2010 (India) |
| Red Globe | Standard | Modified leader | 3.0 x 3.0 | 1111 | 16.86 (3rd Year) | CITH, Annual Report, 2010 (India) |
| Red Globe | Standard | Modified leader | 2.5 x 2.5 | 1600 | 15.77 (3rd Year) | CITH, Annual Report, 2010 (India) |
| Cresthaven | Standard | Modified leader | 2.5 x 2.5 | 1600 | 13.8 (3rd Year) | CITH, Annual Report, 2010 (India) |

| Variety | Root stock | Training system | Spacing (m) | Density (Trees/ha) | Yield (t/ ha) | Reference (Country) |
|---|---|---|---|---|---|---|
| Glohaven | Standard | Modified leader | 3.0 x 3.0 | 1111 | 13.22 (3rd Year) | CITH, Annual Report, 2010 (India) |
| Fantasia | Standard | Modified leader | 3.0 x 3.0 | 1111 | ….. (3rd Year) | CITH, Annual Report, 2010 (India) |
| Cresthaven | Standard | Modified leader | 3.0 x 3.0 | 1111 | 10.89 (3rd Year) | CITH, Annual Report, 2010 (India) |

## Plum

Growing of plum trees under high density is also a successful venture around the world. But a very few varieties were reported to grow under high densities. Simca, Mallard, Opal and Elda grafted on Nemaguard and St. Julian rootstock were found compatible as well as productive combinations (Table 8).

**Table 8. High density orcharding systems assessed at various places in India and abroad in plum**

| Variety | Rootstock | Training system | Spacing (m) | Density (Trees/ha) | Yield (t/ ha) | Reference (Country) |
|---|---|---|---|---|---|---|
| Simca | Nemaguard | Central Leader | 2.0 x 5.5 | 909 | 169.8 | - |
| Mallard | St. Julian A | Y-trellis | 0.5 x 4.0 | 5000 | 95.1 | Meland, 2005 (Norway) |
| Opal | St. Julian A | Y-trellis | 0.5 x 4.0 | 5000 | 77.6 | Meland, 2005 (Norway) |
| Elda | St. Julian A | Y-trellis | 0.5 x 4.0 | 5000 | 53.0 | Meland, 2005 (Norway) |
| Mariposa | Standard | Modified leader | 5. 0 x 5.0 | 625 | 27.26 (9th leaf) | CITH, 2014-15 |
| AU-Cherry | Standard | Modified leader | 5. 0 x 5.0 | 625 | 23.99 (9th leaf) | CITH, 2014-15 |
| Tarrol | Standard | Modified leader | 5. 0 x 5.0 | 625 | 23.40 (9th leaf) | CITH, 2014-15 |

## Apricot

High density apricot cultivation is not common due to non-availability of the dwarfing rootstocks. Therefore, plantations were made in medium density. In 10th leaf stage, the maximum productivity was recorded in cultivar 'CITH-Apricot-3' as compared to exotic cultivar viz. Harcot and Rival (Table 9)

**Table 9. High density orcharding systems assessed at various places in India and abroad in plum**

| Variety | Rootstock | Training system | Spacing (m) | Density (Trees/ha) | Yield (t/ ha) | Reference (Country) |
|---|---|---|---|---|---|---|
| CITH-AP-3 | Standard | Modified leader | 5 x 5 | 625 | 23.64 ($10^{th}$ Yr) | CITH, 2014-15 |
| CITH-AP-2 | Standard | Modified leader | 5 x 5 | 625 | 23.02 ($10^{th}$ Yr) | CITH, 2014-15 |
| CITH-AP-1 | Standard | Modified leader | 5 x 5 | 625 | 19.06 ($10^{th}$ Yr) | CITH, 2014-15 |
| Harcot | Standard | Modified leader | 5 x 5 | 625 | 18.20 ($10^{th}$ Yr) | CITH, 2014-15 |
| Rival | Standard | Modified leader | 5 x 5 | 625 | 17.51 ($7^{th}$ Yr) | CITH, 2014-15 |

## Sweet cherries

Production of sweet cherries (*Prunus avium* L.) in high density systems can be accomplished in a number of ways, but underlying principles exist across all systems that are critical to their success. Inherent vigour and productivity factors - scion, soil type, climate, and rootstock - must be taken into account before deciding upon a particular training strategy to impose. To develop high density sweet cherry orchards, for which growers have the choice of traditional rootstocks (Mazzard and mahaleb seedlings) that tend to promote excessively vigorous growth or relatively new dwarfing rootstocks (Gisela, Edabriz) that, thus far, tend to promote excessive cropping. Poor management of a high density orchard system that is prone to either excessive vigour or excessive cropping can provide disastrous results for yields, fruit quality, and even tree longevity.

For sweet cherry, increased tree density translates into more manageable cropping targets (yield and fruit quality) than traditional systems. Calculation of the number of fruit/tree/ton of yield of desired fruit size readily creates manageable production goals on a tree-by-tree basis. For example, at the relatively low density of 300 trees/ha and a target fruit size of 10 g, for each metric ton (per ha) of marketable yield, a tree must bear ~366 fruit (333 marketable fruit plus 33 additional fruit to account for a 10% factor representing cullage or fruit missed at harvest). In an orchard target of 15 MT/ha, each tree must produce 5,490 fruit. At a higher density of 900 trees/ha, each tree must only produce 1,830 fruit. In view of the production and management potential of various training systems such as central leader, slender spindle, Vogel spindle, Zahn spindle, HYTEC, Solaxe, Tatura trellis, Open vase, Spanish Bush, Steep Leader, Brunner, Palmette, etc.

Sweet cherries are the choicest fruit among stone fruits. It has been successfully growing in selected pockets of Kashmir valley. Bing and Summit grafted on Gisela-6 rootstock planted at closer spacing found very productive. New Cherry clones selected by CITH, Srinagar also showed promising results under high density. Trees trained with vertical as well as open centre canopies have positive effect on yield and fruit quality (Table 10).

**Table 10. High density orcharding systems assessed at various places in India and abroad in sweet cherry**

| Variety | Rootstock | Training system | Spacing (m) | Density (Trees/ha) | Yield (t/ha) | Reference (Country) |
|---|---|---|---|---|---|---|
| Bing | Gisela-6 | Palmette | 2.6 x 4.9 | 864 | 27.1 (9 yr old) | Whiting *et al.* 2005 (USA) |
| Bing | Gisela-6 | Central leader | 2.6 x 4.9 | 864 | 25.8 (9 yr old) | Whiting *et al.* 2005 (USA) |
| Bing | Gisela-6 | Y-trellis | 2.6 x 4.9 | 864 | 23.9 (9 yr old) | Whiting *et al.* 2005 (USA) |
| Van | Damil | Y-trellis | 0.5 x 4.0 | 5000 | 14.5 | Meland, 1998 (Norway) |
| Summit | Tabel Edabriz | V-shape | 0.7 x 3.8 | 3759 | 19.3 ($6^{th}$ year) | Radunic *et al.* 2011 (Croatia) |
| CITH Cherry-1 | Colt | Modified centre | 2.5 x 2.5 | 1600 | 7.93 ($8^{th}$ year) | CITH Annual Report 2010 (India) |
| Bigarreau Noir Grossa | Colt | Modified centre | 2.5 x 2.5 | 1600 | 5.2 ($8^{th}$ year) | CITH Annual Report 2010 (India) |
| CITH -05 | Colt | Modified centre | 2.5 x 2.5 | 1600 | 8.3 ($8^{th}$ year) | CITH Annual Report 2010 (India) |
| CITH-05 | Colt | Modified centre | 3.0 x 3.0 | 1111 | 8.3 ($8^{th}$ year) | CITH Annual Report 2010 (India) |
| CITH-01 | Colt | Modified centre | 3.0 x 3.0 | 1111 | 8.3 ($8^{th}$ year) | CITH Annual Report 2010 (India) |
| CITH-02 | Colt | Modified centre | 3.0 x 3.0 | 1111 | 8.3 ($8^{th}$ year) | CITH Annual Report 2010 (India) |

## Almond

Almond is one of the most important nut crop grown only in the monsoon free areas of Kashmir valley. Domestic demand of almond is very high in India. But the productivity is very low. So, India is annually importing about more than 60% of the world almond imports. This demand could be compensated by resorting to high density orcharding in almonds using Non Pariel, Waris, Waris, Pranyaj and Shalimar (Table 11). Lack of dwarfing rootstocks for almond is a primary hurdle in high density orcharding. Therefore, high density orchards are more productive during early 5 years of tree bearing, but later they are at par with medium density orchards for nut yield.

In India, there is a lack of data from commercial temperate fruit plantings in relation to high density plantings. In apple, few experimental orchards have been established in apple growing areas of India including Jammu & Kashmir, Himachal Pradesh and Uttarakhand. The trials carried out in India have compared different rootstocks, densities and cultivars.

The effect of higher early yield and increased yield at full production with increased tree is density evident in several high density experiments around the world. However, it is the labour intensive system that involves higher expenses. A study conducted at Lincoln University showed that the more labour is required per hectare basis in semi-intensive orchards than the high density orchards during first 6 years for training and pruning. Because dwarf trees (0.6-2.4 m height) require half cost than the semi-intensive vigorous trees (3.0-3.7 m height). Although establishment costs for high density plantings are higher than any other semi-intensive systems and total establishment costs are not proportional to tree numbers.

**Table 11. High density orcharding systems assessed at various places in India and abroad in almond**

| Variety | Rootstock | Training system | Spacing (m) | Density (trees/ha) | Yield (t/ha) | Reference (Country) |
|---|---|---|---|---|---|---|
| Supernova | Seedling | V-shape | 2.5 x 5.0 | 800 | 24.5 (cummu. Yield 10 yrs) | Monastra *et al.* 1998 |
| Ferragnus | Seedling | V-shape | 2.5 x 5.0 | 800 | 23.5(cummu. Yield 10 yrs) | Monastra *et al.* 1998 |
| Waris | Seedling | Modified leader | 3.5 x 3.5 | | 3.24 (12th Yr) | CITH, 2012-13 |
| Non Pariel | Nemaguard | Standard | 3.0 x 6.6 | 500 | 3.28 | Duncan R., Univ. California (USA) |
| Non Pariel | Hensin | Standard | 3.0 x 6.0 | 500 | 3.63 | Duncan R., Univ. California (USA) |
| Pranyaj | Seedling | Modified leader | 4.0 x 4.0 | | 3.07 (12th Yr) | CITH, 2012-13 |
| Non Pariel | Price | Hedge row | 2.2 x 6.7 | 667 | 2.45 | Duncan R., Univ. California (USA) |
| Pranyaj | Seedling | Modified Leader | 4.0 x 4.0 | 625 | 2.94 (10th yr) | CITH Annual Report 2010 (India) |
| Waris | Seedling | Modified Leader | 4.0 x 4.0 | 625 | 2.78 (10th yr) | CITH Annual Report 2010 (India) |

| Variety | Rootstock | Training system | Spacing (m) | Density (trees/ha) | Yield (t/ha) | Reference (Country) |
|---|---|---|---|---|---|---|
| IXL | Seedling | Modified Leader | 4.0 x 4.0 | 625 | 2.58 ($10^{th}$ yr) | CITH Annual Report 2010 (India) |
| Shalimar | Seedling | Modified Leader | 4.0 x 4.0 | 625 | 2.37 ($10^{th}$ yr) | CITH Annual Report 2010 (India) |

## Advantages of HDP

- Early production and higher returns i.e. many higher density orchards are breaking even in 6-7 years compared to 10-12 years for traditional systems.
- Efficient use of labour to harvest and prune from the ground or from a short stool.
- It has potential to produce high quality fruit for a longer period of time by maintaining light interception in the smaller trees of higher density orchards.
- Higher pesticide application efficiency.

## Disadvantages of HDP

- It cannot be adopted in very steep, unfertile, shallow and drought-prone areas.
- The high cost of orchard establishment.
- Require highly skilled manpower.
- Require more training and minimal pruning during the first 6 years than traditional systems, especially during the summer. A lack of attention in the early life of this type of orchard creates a very high probability that the orchard will never be profitable given the high costs of establishment.

## Orchards of future

Tall spindle: This system was developed to revive the New York apple industry economically. In 1993, Robinson and his cohorts at Cornell gave the birth of the high-density tall spindle system. "It contained concepts from slender spindle, vertical axe, Solaxe and super spindle. Adoption of the tall spindle system is yielding more and showing its economic superiority. The basic elements of tall spindle are:

- *High density*- The optimum tall spindle spacing for an average vigour variety and soil is 3 feet by 11 feet, or 1,289 trees per acre.
- *Full dwarfing rootstocks*- The most successful tall spindle orchards established to date have been on M.9 and B.9. Precocious dwarfing stocks are important since early cropping is essential.
- *Highly feathered nursery trees*- Nursery stock ideally have from 10 to 15 feathers per tree.

- *Minimal pruning at planting*- No heading.
- *Support system*- To carry the crop and support potentially brittle rootstocks.
- *Branch devigoration*- Upright scaffold branches are devigorated by bending them below the horizontal. This early investment of labour pays off in early cropping and less pruning later.
- *Limb renewal*- All scaffolds are removed as they become too large for the available space and become out of balance within the tree.
- *Early fruiting*- Fruiting in the second and third leaf is essential to keep a low tree vigour level and provide income from early fruit sales.
- This system gave cumulative production of more than 81.65 t/acre in the first five years, resulting in a 40 percent higher crop value than the slender vertical axis and Sol Axis planting systems.

## Futuristic needs for HDP

The planting of a new high density orchard is a long time (more than 20 year) commitment. Therefore, before planting a grower should consider the possible changes in orchard management that will occur in the next 20 years and try to plant the new orchard so he can benefit from advances in orchard management that are likely to occur. The 5 basic principles or orchard design outlined above (1. high light interception, 2. good light distribution throughout the canopy, 3. high early yields, 4. simple canopies are more adaptable to partial mechanization, and 5. planting density is governed by the law of diminishing returns) will be a part of any future orchard system. Based on these five principal the design of a HDP should be

- Tall (10-11 feet) enough due to the need of intercepting 70-75% of available light.
- It will have narrow simple canopies no wider than 4-6 feet due to the need to have good light distribution in all regions of the canopy.
- Use of highly branched trees for high early yields. The only exception to this rule will be those growers who choose to plant very high tree densities (>1,500 trees/acre) where the cost of feathered trees is too high and the value of feathers is less due to very small space between trees.
- Simple canopies will offer significant benefits in mechanization with simple platforms for pruning, thinning and harvesting.
- The optimum planting density will remain close to 1,000 trees/acre.

In addition to high, early yields, a key goal is to increase the percentage of high quality fruit. That means having narrow canopies that allow good exposure to light from bottom to top. This can be achieved by narrowing canopies with mechanical hedging to form tree walls at suitable time. The next economic challenge is labour reduction. We need to grow temperate fruits with half the labour (180 hours/acre) we now (360 hours/acre) use. Key ingredients are machines platforms to thin, prune, and serve as carriers for harvest crews, but also hedgers to do much of the pruning now done by hand. Use of chemicals can reduce the thinning labour by half of the physical lines.

## Conclusion

The change in planting systems over the last 50 years has been dramatic. High yield and high fruit quality can be achieved with a high-density orchard when the orchard has good light distribution throughout the tree canopy and there is a balance between vegetative growth and cropping ("calm trees"). This can best be accomplished by maintaining a narrow canopy shape and through regular limb renewal pruning and the development of pendant fruiting branches. Successful growers will maintain a balance between vegetative growth and cropping by regularly renewing the fruiting wood on the tree, moderate levels of nitrogen and proper cropload management. Growers should try to obtain 50cm of leader shoot growth in the first year, 75-100cm of leader shoot growth in the second and third years and 50cm of leader shoot growth in the fourth year. If this is combined with minimal pruning and a precocious rootstock, significant production should be obtained in the second-fourth years which will limit vegetative growth in future years resulting in a "calm" tree.

## Acknowledgement

This book chapter is a compilation of information collected from journals, annual reports, bulletins and author work. Therefore, author is grateful to Central Institute of Temperate Horticulture for use of the information available in public domain. Special thanks are given to Dr(s) Nazeer Ahmed, A.A. Sofi, and scientists working on temperate fruits. I also thanks to the Research Associates, Technical and Supporting Staff for their Help and cooperation that led to development of high density plantings.

## References

Bhatia H. S. and Kumar J. 2009. Performance of new apple cultivars on different rootstocks under high density plantation. *Agricultural Science Digest*, 29(4):303-305.

CITH Annual Report. 2010, 11, 12, 13, 14. ICAR- Central Institute of Temperate Horticulture, Srinagar, J&K.

CITH Annual Report 2022. ICAR- Central Institute of Temperate Horticulture, Srinagar, J&K.

Corelli, L and Sansavini, S. 1989. Light interception and photosynthesis related to planting density and canopy management in apple. *Acta Hort*. 243: 159-174.

Kumar Dinesh, Ahmed N., Verma M.K. and Dar T.A. 2013. Growth, yield, quality and leaf nutrient status as influenced by planting densities and varieties of apricot. *Indian Journal of Horticulture*, 70(2): 195-199.

Kumar Dinesh, Ahmed N. and Verma, M.K. 2012. Studies on high density planting in almond in Kashmir valley. *Indian Journal of Horticulture*, 69(3): 328-332.

Meland M. 1998. Yield and fruit quality of 'Van' sweet cherry in four high density production systems over seven years. *Acta Hort*. 468:425-432

Meland M. 2005. High density planting systems of European plums- the effect of growth and productivity of three cultivars after nine years. *Acta Agriculturae Scandinavica Section B-Soil and Plant*. 55: 51-57.

Monastra F., Martelli G. and Chiarotti A. 1998. Economic advantages of high-density planting of almond orchards. In . *X GREMPA p. 81-85 Meknès* (Morocco)

Musacchi, S., Ancarani, V., Gamberini, A., Gaddoni, M., Grandi, M. and Sansavini, S. 2005. Response of training system, planting density and cultivar in pear. *Acta Hort*., 671:463-469.

Pramanick, K.K., Kishore, D.K., Singh, R. and Kumar, J. 2012. Performance of apple (*Malus x domestica* Borkh.) cv. Red Spur on a new apple rootstock in high density planting. *Scientia Horticulturae*, 133: 37–39.

Raduni M, Jazbec A., Pecina M., osi T. and Pavi i N. 2011. Growth and yield of the sweet cherry (*Prunus avium* L.) as affected by training system. *African Journal of Biotechnology*, 10(24): 4901-4906. DOI: 10.5897/AJB10.278.

Robinson T. L., Lasko A. N. and Carpenter S. C. 1991. Canopy development, yield, and fruit quality of 'Empire' and 'Delicious' apple trees grown in four orchard production systems for ten years. *J. American Soc. Hort. Sci.* 116(2):179-187.

Robinson, T., Hoying, S., Sazo, M.M., DeMarree, A. and Dominguez, L. 2013. A Vision for Apple Orchard Systems of the Future. *New York Fruit Quarterly*. 21(3): 11-16.

Webster T. 2006. Control of Growth and Cropping of Temperate Fruit Trees. *Chronica Horticulturae*, 46(3): 20-26.

Whiting M. D., Lang G. and Ophardt D. 2005. Rootstocks and training systems affect sweet cherry growth, yield and fruit quality. *HortScience*, 40(3): 582-586.

Wunsche, J. N and Alan N. Lasko 2000. The relationship between leaf area and light interception by spur and extension shoot leaves and apple orchard productivity. *HortScience*, 35(7):1202-1206.

## CHAPTER 2

# POLLINATION AND BUNCH MANAGEMENT IN DATE PALM

**Akath Singh[1], P.R. Meghwal[2] and Anshuman Singh[1]**

*[1]ICAR-Central Arid Zone Research Institute, Jodhpur 342003, India*
*[2]ICAR-Central Institute for Subtropical Horticulture, Lucknow, India*
*Corresponding author: akath2005@yahoo.co.in*

Date palm (*Phoenix dactylifera* L.) is one of the most potential fruit crop in dry arid zone of the country. It is grown for its sweet edible fruits and is considered to be the oldest amongst the cultivated tree fruits (Alotaibi *et al.*, 2023; Madani *et al.*, 2021). It is one of the economically important perennial plant in the Arabian desert of the Middle East and the Sahara Desert in North Africa. Egypt, Saudi Arabia, Iran, Algeria, and Iraq are currently the world's top producers of date palm (FAOSTAT, 2023). With an export value of over 330.09 million US dollars in 2022, Israel was the world's top exporter of both fresh and dried dates, followed by Tunisia, the United Arab Emirates, Iran, and Saudi Arabia (SATISTA, 2023). India is the world's biggest date importer with imports estimated to be worth around 230 million US dollars annually (SATISTA, 2023). Due to its adaptability to different regions, its cultivation is becoming popular in India and now it is cultivated in Gujarat, Rajasthan, Telangana, Andhra Pradesh, Haryana, Punjab, Maharashtra, Tamil Nadu, West Bengal and Madhya Pradesh.

In addition to being an abundant source of carbohydrates, the date palm fruits are also reasonably rich in vitamins (A, B-2 and B-7), minerals (potassium, calcium, copper, manganese, phosphorus and iron), phenolic compounds and antioxidants (Hussain *et al.*, 2016). A wide variety of products like, fresh and pitted dates, date powder, (date crush), date syrup, date juice, date confectionery, date jam and date pickle are produced from date fruits.

Date palm requires very specific climatic conditions for its successful cultivation i.e. prolonged hot dry summer, moderate winters and almost rain free period at fruit ripening (July-Aug). The requirement of heat summation unit (HSU) for date palm varies from 1950 to 3650 depending upon cultivars. The HSU in the Indian desert during March to August has been worked out to be 2000 to 2400, above a base of 180C and from 3500 to 4000 above a base of 100C (Meghwal, 2011). This suggests that the climatic conditions in the Indian desert region are congenial for the commercial production of date palms (Meghwal, 2011; Kumar and Naqvi, 2022).

## Popular varieties

Date palm maturation passes through four important stages, namely Kimri (raw fruit), Khalal/ Doka (fresh fruit), Rutab/ Dang (semi-ripened) and Tamar/ Pind (fully ripened). Early maturing cultivars are non-astringent during the doka stage such as Halawy, Barhee, and Khunezi are suitable for fresh consumption, whereas Medjool and Khadrawy are most appropriate for producing dry dates (Meghwal and Singh, 2017; Singh *et al.*, 1997). Rainfall during ripening, however, still poses a challenge to commercial date palm farming in India's dry regions because none of the aforementioned varieties can be harvested prior to the onset of the monsoon season. The date palm variety ADP-1 developed by the Anand Agricultural University Anand, Gujarat, India is an early ripening variety and its doka stage fruits can be harvested before the onset of rainfall in the arid regions of Rajasthan.

**Figure 1. Fruits of Popular date varieties**

## Floral characters

Date palm is a dioecious, meaning it has separate female and male trees. The number of inflorescences borne on each palm varies from none to more than 25 that contains 150 to 200 spikelets each, depending on the age and vigor of the tree (Zaid et al., 2002). The flower stalks are produced from the axils of the leaves in similar positions to those in which offshoots are produced. Inflorescences of both female and male trees are enclosed in a hard and fibrous cover (the spathe) that protects the flowers from heat and sunlight during the early stage of flower development. The male palm produces the pollen and the female palm produces the fruit. The male flowers are borne single and are waxy white, while the female flowers are borne in clusters of three and are yellowish green in colour. Flowering periods of male and female palms should be synchronised in order to have enough pollen when the female spathes open. It is preferable if the male spathe opens 2 to 4 days earlier than the female spathe. Favourable temperature for spath emergence is 18°C, and for fruit setting and development may be more than 25°C (Zaid *et al.*, 2002). The inflorescence consists of many unbranched rachillae, known as "strands," arranged in spirals on the rachis. Later the flowers are borne on theses strands in the female trees. Male date palm flowers usually have three sepals, three petals, six stamens, and three carpels (which rarely develop parthenocarpically). In contrast, female flowers have three sepals, three petals, six staminodes, and three separate carpels. The fruit normally develops after fertilization from one of the three carpels within each pistillate flower.

**Spathe emergence**

**Inflorescence emergence after spath cracking**

**Inflorescence**

**Figure 2. Characteristics and emergence of date palm inflorescence**

## Pollination

Being a dioecious plant, date palm requires artificial pollination, which is significantly impacted by even minor variations in climatic conditions (Salomon-Torres *et. al.*, 2021). In general, female trees open their spathes (inflorescences) earlier than male plants do, which prevents viable pollen from being available at the proper stage of pollination (Mesnoua *et al.*, 2018). Inadequate or improper pollination resulted in an excessive number of fruits on individual strands within the same bunch, which eventually leads to the production of small and low-quality fruits (Abu-Zahra and Shatnawi, 2019). Understanding the ideal pollination timing in relation to pistil receptivity is essential for achieving an acceptable level of fruit set and yield (Kadri *et al.*, 2022). Several studies documented the role of pollen source in fruit set and yield of different date palm cultivars (Omar and El-Ashry, 2015). Metaxenia phenomenon, which was defined as a direct effect of pollen from a male clone on the morphology and other characters of seed and fruit tissues surrounding the embryo and endosperm was proposed first time in date palm (Swingle, 1928) and has been evaluated in different date cultivars. The physical and chemical characteristics of fruits have also been considerably variable in many date cultivars under influence of pollen sources (Rezazadeh *et al.*, 2013). Male date palm plants play equally an important role in the fertilisation process since pollen's viability and pollination methods influence the quality, development and yield of fruits. Superior clonal male genotypes having early flowering (flowering before female flowers), presence of more fertile pollens, and larger quantity of pollens in each male flower is essential for commercial cultivation of date palm. Madsari, Ghanami, Succary, Sallag, Dikhiny, Menify are popular male cultivars.

Pollination time interval varies according to the date cultivar and climate, ranging from the day of spathe opening to 10 days later (Hammami et. al., 2024; Kadri et al., 2022). The standard procedure currently in use is manual pollination, which involves dusting the pollen with a cotton swab at specific times throughout flowering. Pollens are sprayed on the inflorescence two or three times, depending on the spathe opening. If not done at the right time, this does not ensure a high fruit set (Salomon-Torres et. al., 2021). Moreover, this process necessitates two to three times as many pollen grains in addition to increased labor costs. It is, therefore, imperative to determine the appropriate stage of pollination for optimal fruit set and economically viable date farming (Hammami et. al., 2024). The findings of the study conducted at CAZRI Jodhpur revealed that pollination in between 12 hours and 36 hours after spathe splitting was found most appropriate to harvest potential yield with superior quality date fruits.

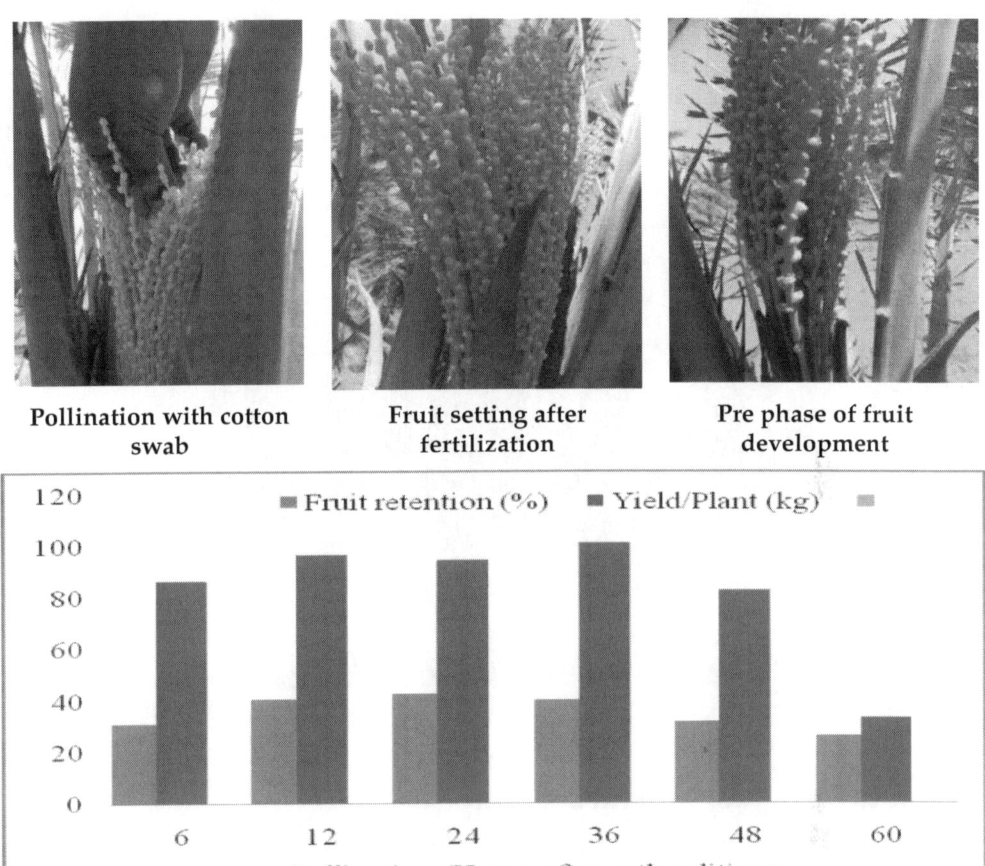


**Figure 3. Yield and fruit retention of date palm cv. ADP-1 as influenced by pollination time**

## Bunch management

Fruit thinning is another major cultural practice in date palm, which frequently helps to increase the fruit size, as well as the physical and chemical quality and reduces compactness within the bunch of fruits without having a negative impact on overall fruit production (Marzouk *et al.*, 2007). Reducing the number of fruits per bunch, the number of strands per bunch, or the number of bunches per palm could all improve the fruit size and quality (Soliman *et. al.*, 2017). Manual pollination often results in excessive fruit setting; hence fruit thinning during both the pollination and developmental stages is necessary to produce high-quality dates (Madani *et al.*, 2021). It has been demonstrated in many date palm cultivars that increased bunch and fruit weight can be achieved by removing 20–30% of the bunch strand by either thinning out or cutting back before pollination (Salomon-Torres *et. al.*, 2021). For instance, removal of up to 30% of the strands from the bunch center after 8 weeks of pollination resulted in the maximum productivity and fruit quality in date cultivar Khadrawi (Moustafa *et al.*, 2019). Similarly, complete removal of some strands and/or fruits from each strands had a favorable impact on overall yield as well as the physical and chemical attributes of the fruits (Ghazzawy *et. al.*, 2023). In an experiment conducted at CAZRI suggest that in order to get a greater commercial output with the best physico-chemical properties of fruits for table or processing, the ideal retention of fruit load may be 12 bunches per tree coupled with 25% strand thinning. Excess bunch should be removed completely coupled with around 25 % strand thinning just after fruit setting.

**Figure 4. Bunch management in date plam**

## Bunch covering for enhancing fruit quality

The date palm crop is also frequently adversely impacted by the abrupt changes in weather conditions including early rains, high temperatures, and desiccating winds at the time of fruit maturity (Alotaibi *et al.*, 2023; Dinar *et al.*, 2023). In contrast to other fruit crops, date palm fruits are commonly covered before harvest in order to enhance fruit quality by reducing the negative effects of weather fluctuations. Unlike other fruit crops, preharvest covering of date palm fruits is also common practice to improve overall fruit quality by avoiding weather vagaries. The materials used for covering, the stage at which the fruit is covered, and cultivar responses all determine the influence of covering on overall fruit quality (Harhash *et al.*, 2020; Madani *et al.*, 2021; Singh *et, al.*, 2023). The ripening rate of 'Helali' dates was enhanced by bunch bagging with various materials and colors after one month from pollination to harvest. Similarly, bunch bagging with various materials and colors after four weeks of pollination till the bisir stage, boosted fruit and bunch weight and enhanced the quality of both 'Succary' and 'Khalas' dates (Harhash and Al-Obeed, 2010).

Hence the cultural practice like bagging of individual bunch can improve fruit quality with increasing ripening process, enabling to harvest greater proportion of *pind* fruits. Our results of study undertaken at CAZRI, Jodhpur showed that bunch covering using perforated LDPE bag or nonwoven fabric bag delivered the highest marketable yield of unblemished, highly colored date fruits by creating a favorable internal microclimate. In addition, bagging material can also influence the fruit quality characteristics including fruit color, avoid fruits damage from potential threat like dust storm, attack of some insects and bird.

| Nonwoven bags | Polypropylene bags | LDPE bags |
|---|---|---|

**Figure 5. Bunch covering in date palm**

## References

Abu-Zahra, T.R., Shatnawi, M.A., 2019. New pollination technique in date palm (Phoenix dactylifera L.) Cv. "Barhee" and "Medjool" under Jordan valley conditions. Am-Euras. J. Agric. Environ. Sci. 19, 37–42.

Dinar, Hassan Ali, Munir, M., Mohammed, M., 2023. Drought-tolerance screening of date palm cultivars under water stress conditions in arid regions. *Agronomy 13*(11), 2811; https://doi.org/10.3390/agronomy13112811

Ghazzawy, H.S., Alqahtani, N., Munir, M., Alghanim, N.S., Mohammed, M., 2023. Combined impact of irrigation, potassium fertilizer, and thinning treatments on yield, skin separation, and physicochemical properties of date palm fruits. Plants, 12(5), 1003-1031

Hammami Z., Mahmoudi H., Al Janaahi, A., Singh, R.K., 2024. Evaluation of date palm fruits quality under different irrigation water salinity levels compared to the fruit available in the market. Front. Sustain. Food Syst. 7:1322350. doi: 10.3389/fsufs.2023.1322350

Harhash, M.M., Al-Obeed, R.S., 2010. Effect of bunch bagging colour and fruit quality of date palm. American-Eurasian J. Agric. And Environ. Sci. 7(3), 312-319

Harhash, M.M., Mosa, W.F.A., El- Nawam, S.M., Gattas, H.R.H., 2020. Effect of bunch covering on yield and fruit quality of "Barhee" date palm cultivar. Middle East J. Agric. Res. 9(1), 46-51

Hussain, I., Ahmad, S., Amjad, M., Ahmed, R., 2016. Execution of strands thinning improves the phytochemical and sugars profiling in date palm (Phoenix dactylifera L.) fruits. Pakistan J. Pharma. Sci. 4, 1209-1215

Kadri, K., Elsafy, M., Makhlouf, S., and Awad, M.A., 2022. Effect of pollination time, the hour of daytime, pollen storage temperature and duration on pollen viability, germinability, and fruit set of date palm (*Phoenix dactylifera* L.) *cv* "Deglet Nour". Saudi J. Biol. Scie. 29(2), 1085-1091

Kumar, R., Naqvi, A. R. 2022. Susceptibility of different date palm varieties against *Parlatoria blanchardii* in hyper arid region of Rajasthan. J. Krishi Vigyan. 10(2), 306-310. 10.5958/2349-4433.2022.00054.X

Madani, B., Dastjerdy, A.M., Shahriyar, A., 2021. Improving ' Piyarom' date palm fruit quality with fruit thinning and bunch covering treatments. Adv. Hort. Sci. 35(1), 11-19

Marzouk, H.M., El-Salhy, A.M., Abdel-Galil, H.A., Mahmoud, A.E., 2007. Yield and fruit quality of some date palm cultivars in response to some flower thinning rates. The Fourth Symposium on date palm in Saudi Arabia. King Faisal Univ. Al-Hassa. 5-8 May, p. 110.

Meghwal, P. R., Singh, A., Kumar, P., Khapte, P. S., Saxena, A., 2018. Upliftment of arid zone economy through horticultural and protected cultivation. Indian Farming 68(09), 46–51

Meghwal, P.R., 2011. Date palm Production in India- Prospects and Problems. Conference: National Symposium on Resource Utilization through Integrated Farming System and Biodiversity conservation in Drylands At: CAZRI, RRS, Bhuj, Gujarat, India

Mesnoua, M., Roumani, M., Salem, A.,2018. The effect of pollen storage temperatures on pollen viability, fruit set and fruit quality of six date palm cultivars. Sci. Hortic. 236, 279–283

Moustafa, A.R., Abdel-Hamid, N., El-Hamid A.A., El- Sonbaty, M.R., El-Naby, S.A., 2019. Strand thinning of Khadrawi date palm cultivar in relation to yield and fruit quality. Bull. Natl. Res. Cen. 1, 204-209

Salomon-Torres, R., Krueger, R., Garcia-Vazquez, J.P., Villa-Angulo, R., Villa-Angulo, C., Noe Ortiz-Uribe , Jesus Arturo Sol-Uribe and Laura, S.S., 2021. Date Palm pollen: features, production, extraction and pollination methods. Agronomy 11(3), 504-524.

SATISTA. 2023, Leading exporters and importers of dates worldwide in 2022, Statista Research Department, Aug 3, 2023 (https://www.statista.com/statistics/961359/global-leading-exporters-of-dates/).

Singh, P., Singh, H.K., Singh, R., Roy, M.K., Kumar, A. and Kumar A., 2023. Effect of bunch cover on production, fruit quality and damage by scarring beetle of banana (*Musa paradisiaca* L.). The Pharma Innovation Journal 12(9), 331-334

Soliman, S.S., Alebidi, A.I., Al-Saif, A.M., Al-Obeed, R.S., Al-Bahelly, A.N., 2017. Impact of pollination by pollen-grain-water sus-pension spray on yield and fruit quality of segae date palm cultivar (Phoenix dactylifera L.). Pak. J. Bot. 49, 119–123.

Zaid, A.; Arias-Jimenez, E.J. *Date Palm Cultivation*; Food and Agricultural Organization: Rome, Italy, 2002; ISBN 92-5-104863-0.

# CHAPTER 3

# OFF-SEASON VEGETABLE FARMING: CHALLENGES TO REWARDS

**Hare Krishna[1*], Manjunatha T. Gowda[1], V. Karuppaiah[2]**
**Vinay Kumar Patel[1], Shekhar Singh[1], Manoj Kumar Singh[1]**
**Shubham Kumar Tiwari[1] and Anant Bahadur[1]**

*[1] ICAR-Indian Institute of Vegetable Research, Shahanshahpur Varanasi (U.P.)*
*[2] ICAR-Directorate of Onion and Garlic, Rajgurunagar, Pune, India*
**Corresponding author; Email: kishun@rediffmail.com*

---

The increasing recognition of the vital role nutrition plays in our daily diet has underscored the significance of vegetables in our overall well-being. Serving as rich sources of essential nutrients such as vitamins and minerals, vegetables are globally acknowledged as "protective foods," valued for their abundance of nutrients, vibrant color pigments, antioxidants and their indispensable role in ensuring food security, nutrition, immunity enhancement and livelihood sustainability (Krishna *et al.*, 2022). In India, the demand for vegetables continues to grow steadily, driven by heightened consumer awareness of health and the pressure of rapidly expanding population. India ranks as the world's second-largest vegetable producer, closely following China. Despite a substantial increase in total vegetable production from 58.5 million tonnes in 1991-92 to 212.91 million tonnes in 2022-23, traditional cultivation methods remain constrained by seasonal and regional requirements (Anonymous, 2023). This increased awareness has spurred a higher demand for vegetable production. However, conventional open-field farming techniques prove inadequate to meet the rising needs of the global population. Continued reliance on traditional farming practices has resulted in low yields, diminished productivity and inconsistent produce quality (Navnath *et al.*, 2019). As a result, there has been a surge in off-season cultivation practices, where crops are grown beyond the usual cropping seasons to mitigate supply shortages and capitalize on favorable market conditions. Various methods are employed for off-season vegetable cultivation, including leveraging a country's agro-climatic diversity, utilizing diverse crop varieties, cultivating in diara lands or riverbeds and adopting protected cultivation. However, this chapter has specifically focused on diara/riverbed cultivation.

## Harnessing nature for off-season cultivation

The three hill states of India Jammu and Kashmir, Uttarakhand and Himachal Pradesh situated in the North-Western Himalayan region, have earned the title of the "Natural Glass House" of the country. This accolade stems from their varied climate, ranging from subtropical to dry temperate, rendering them well-suited for off-season vegetable farming. The geographical characteristics of these hills, encompassing altitude, latitude and slopes, give rise to valleys that are conducive to growing off-season vegetables. Vegetables cultivated in these hills are renowned for their unique flavor, freshness, crispness, sweetness, exceptional quality and lack of pesticide residues. Winter vegetables, typically grown during the summer months in the hills, find a ready market in the plains, offering farmers lucrative returns during their off-peak seasons. Numerous vegetables have emerged as key players in off-season farming in the hills, including tomatoes, green peas, beans, cabbage, cauliflower, green capsicum, colored capsicum, summer squash, cucumber, broccoli, lettuce, Chinese cabbage, Brussels sprouts, European carrots and snow peas (Sharma *et al.*, 2023).

## Varieties for continuous harvest throughout the year

Through the dedicated efforts of breeders, a variety of vegetable cultivars have been developed, enabling year-round cultivation and availability of specific vegetables. In recent years, there has been widespread adoption of off-season varieties, particularly in radish, tomato, onion, cabbage and carrot, leading to a significant increase in vegetable production (Krishna *et al.*, 2024). For instance, tomato cultivars such as Pusa Sheetal and Pusa Hybrid 1 have been specifically bred for fruit setting at both low and high temperatures, respectively. Similarly, the development of cultivars like Pusa Chetki and Pusa Desi has extended the radish growing season throughout the year. Traditionally a winter crop, onions can now be grown during the monsoon season as well, due to the development of cultivars like N-53, Agrifound Dark Red, Arka Kalyan and Baswant 780. Additionally, adjusting planting schedules has further prolonged the availability of carrots. For example, cultivars such as Pusa Vrishti and Pusa Meghali can be sown in July-August for harvest between October and December. Conversely, Pusa Yamdagini can be planted from December to February to yield roots from March to May, while Pusa Nayanjyoti can be sown in March-April for harvesting in June-July (Aditika *et al.*, 2017).

## Utilizing *Diara* Lands for Sustainable and Off-Season Cultivation

Diara lands, situated amidst natural levees within active flood plains, undergo an annual cycle of flooding, erosion and sediment deposition. This natural process enriches the soil with fresh silt and clay deposits during the rainy season, enhancing its fertility and suitability for crop cultivation. The agricultural productivity of Diara lands is thus sustained through this cyclic rejuvenation, making them valuable areas for cultivation. Diara land or riverbed farming presents an effective approach for off-season vegetable production, offering a cost-effective and easily adaptable technology (Maharjan, 2017). This method empowers landless households to utilize marginal lands effectively, particularly benefiting small and marginal farmers. Riverbed farming enables these farmers and their families to engage in cultivation of cucurbits and other vegetables (Wadhwani *et al.*, 2007). Approximately 65% of the total cucurbits cropped area in the country is located in riverbeds (Choudhary *et al.*, 2019). During the summer season, a significant proportion (70-75%) of cucurbits production occurs in riverbeds or Diara land areas (Kumar and Baraik, 2023). These vegetables are available in the market from February to June and October to January. Various riverbeds, including those of the Yamuna, Ganges, Gomati, Saryu and their tributaries in Haryana, Uttar Pradesh and Bihar, as well as regions like the Tonk district in Rajasthan, Madhya Pradesh, Maharashtra, Gujarat Andhra Pradesh and Kerala, are important for extensive cucurbits cultivation (Choudhary *et al.*, 2019). Cucumbers, bitter gourds, long melons, sponge gourds, ridge gourds, watermelons, muskmelons, bottle gourds and pumpkins are among the cucurbits grown in these riverbeds (Panda, 1996). In Kerala, for example, bitter gourds and snake gourds thrive in the riverbeds of Pamba and Manimala. The seepage of subterranean moisture from adjacent rivers and streams enriches the upper layers of Diara land, creating favorable conditions for early vegetable cultivation. This form of farming not only provides an additional source of income for farmers but also meets the market demand for off-season vegetables. *Diara* lands are predominantly found in states like Uttar Pradesh, Bihar, Assam and Orissa. In Uttar Pradesh, the estimated area of flood-prone *Diara* lands covers approximately 1.5 million hectares across various river systems, while in Bihar, it is about 0.9 million hectares. In Assam, Diara lands may exceed 2.3 million hectares, considering that 30% of the state's geographical area is flood-prone. Although, precise data on Orissa's *Diara* land extent is not available, it is estimated to be around 1.0 million hectares, primarily in the flood-prone areas of the Mahanadi delta (Panda, 1996). The prevalence of *Diara* lands in these states underscores their importance for various agricultural activities, despite their vulnerability to flooding due to their proximity to rivers and plains.

## Optimizing Cropping Patterns for Enhanced Harvest

Mixed cropping is a prevalent practice in riverbed cultivation, where different crops are grown together to maximize space and resource utilization. For instance, watermelon and muskmelon are often cultivated simultaneously in riverbeds. Additionally, various other cucurbits are intercropped, such as summer squash, bottle gourd, round melon, cucumber, sponge gourd, bitter gourd and long melon in North India, ridge gourd in Rajasthan, Madhya Pradesh and Uttar Pradesh and pointed gourd in Bihar (Singh, 2012). It is advisable to harvest crops when fruits are tender and ready for consumption. To maintain quality, fruits should be harvested every 2-3 days, as they tend to harden due to seed maturity if left unharvested. Regular harvesting intervals can be maintained from late June to late October, with the potential for off-season harvesting as early as February-March, resulting in early yields and higher returns. After harvesting, crops are transported to local market centers for sale. This cropping pattern and harvesting strategy are designed to optimize yield and crop quality in riverbed cultivation, ensuring a continuous and profitable agricultural cycle (Kumari *et al.*, 2018; Krishna *et al.*, 2024). Despite its strengths and constraints (see Table 1), many small and marginal farmers engage in riverbed cultivation, underscoring the need for solutions to address these challenges and ensure equitable distribution of benefits (Choudhary *et al.*, 2019). Diara land, the riverbed between two or more streams, becomes dry after the monsoon season, offering opportunities for income generation and natural resource management. Proper management of these areas could contribute to income generation, natural resource utilization and better disaster/risk management. Despite facing challenges, riverbed cultivation offers livelihood prospects, warranting further exploration to enhance its relevance (Kumari *et al.*, 2018; Choudhary *et al.*, 2019; Kumar and Baraik, 2023). The advantages of riverbed cultivation highlight its potential as a sustainable and beneficial agricultural practice, especially for small-scale and resource-limited farmers (Ghimire *et al.*, 2022).

**Table 1. Prospects and challenges of river bed cultivation**

| Prospects | Challenges |
|---|---|
| **High Net Return per Unit Area**: Riverbed cultivation offers a high net return, making it economically favorable for farmers. | **Non-availability of Quality Seeds**: Riverbed cultivation faces challenges in accessing quality seeds, leading farmers to rely on seeds from private seed companies with high cort.. |
| **Early and High Yield**: Riverbed crops often yield early, providing farmers with a quicker and higher return on their investment. | **Cross-pollination Issues**: Many vegetables in riverbed cultivation are produced through cross-pollination, resulting in unreliable quality, especially in sweetness and flesh color. |
| **Ease in Irrigation**: Despite challenges, riverbeds generally have a high-water table, easing the process of irrigation for cultivated crops. | **Seed Purity Concerns**: Each farmer preserving his/her own seeds raises concerns about cultivar purity, with no guarantee of maintaining purity for more than one season. |
| **Low Cost of Cultivation**: Riverbed cultivation is cost-effective, requiring fewer resources and inputs compared to other forms of cultivation. | **Uncertainity in Land Availability**: Riverbeds may remain underwater for extended periods, making land availability uncertain for cultivation. |

| Prospects | Challenges |
|---|---|
| **Less Nutrients Requirement Due to High Fertility**: The naturally high fertility of riverbed soil reduces the need for additional nutrients inputs, contributing to cost savings. | **Ownership Challenges**: Vegetable growers often do not own the land they cultivate, making it difficult to secure for cultivation. |
| **Limited Weed Growth**: Riverbeds tend to have limited weed growth, reducing competition with cultivated crops. | **River Course Changes**: Changes in the river's course pose problems, requiring growers to cultivate new areas. |
| **Easy Control of Pest and Disease by Cultural Means**: Cultural practices can effectively control pests and diseases in riverbed cultivation, minimizing the need for chemical interventions. | **Leaching Issues**: Heavy leaching of soils in riverbeds results in low fertility, necessitating additional efforts for manures and fertilizers. |
| **Low-Cost Labor Facilities**: The nature of riverbed cultivation allows for the utilization of low-cost labor, contributing to overall cost-effectiveness. | **Continuous Soil Fertility Impact**: Continuous leaching affects the fertility status of the soil in riverbeds. |
| **No Land Ownership Required**: Riverbed cultivation does not demand land ownership, making it accessible to those without substantial property. | **Irrigation Challenges**: Diara lands are primarily rainfed and assured irrigation facilities are not available, posing challenges for water management. |
| **High Economic Returns**: Despite challenges, riverbed farming can yield high economic returns, contributing to the financial well-being of farmers. | **High Water Table**: Despite a high-water table, effective irrigation remains a challenge in riverbed cultivation. |
| **Income and Food Security for Landless and Marginal Farmers**: Riverbed cultivation plays a crucial role in ensuring income and food security for landless and marginal farmers who may not have access to conventional agricultural resources. | **Water Lifting Constraints**: Small-scale water lifting mechanisms or bamboo tube wells could help address irrigation challenges if provided on a subsidized basis. |
| **Local Adaptation to Climate Change**: The local adaptation of riverbed cultivation practices help farmers cope with the impacts of climate change, making it a resilient option. | **Crop Rotation Challenges**: Riverbed cultivation does not align well with conventional crop rotation practices, especially with cucurbits. |
| **Utilization of Underexploited Resource**: Riverbed farming utilizes an underexploited resource and enhances smallholders' skills on marginal soils. It provides small farmers with sustainable options for coping with environmental shocks like floods. | **Profit Distribution Disparities**: The majority of profits often go to businessmen and middlemen who purchase cucurbits from farmers and sell them in the market, leaving small and marginal farmers deprived of a significant share of profits. |

| Prospects | Challenges |
|---|---|
| | **Ecological Constraints**: During the monsoon season, rivers like the Ganges and Brahmaputra cause catastrophic flooding in their plains, inundating large areas from June to September (Hofer, 1993). This flooding can both create fertile alluvial soil and cause significant damage to crops and infrastructure, posing ecological challenges for riverbed cultivation. |

## Human Resource Development and Management in *Diara* Land Cultivation

Krishi Vigyan Kendra (KVKs) and extension agencies play a pivotal role in disseminating essential information on soil, water and crop management, as well as promoting farm technology adoption. They achieve this through various means such as demonstrations, fairs and training programs aimed at empowering farmers with improved skills. Extension agencies also support the formation of farmers' associations and provide training to cultivate a self-reliant workforce. Education on the impact of drainage, encouraging farmers participation and raising awareness about available financial support which are essential components for ensuring economic sustainability in Diara land cultivation. In challenging terrains, a holistic approach that addresses agriculture, environment and ecology is crucial. Sustainable measures require scientific study and active community participation in policy-making, ensuring the success of agricultural practices (Choudhary *et al.*, 2019).

## Balancing Challenges and Possibilities of Off-Season Vegetable Farming through Protected Cultivation

To meet the growing demand for off-season and high-quality vegetables, farmers are transitioning from traditional cultivation methods to contemporary agri-business models (Chauhan *et al.*, 2016). One such approach gaining popularity in India is "protected cultivation," traditionally associated with affluent Western nations facing adverse environmental conditions. This method aims to enhance overall productivity and vegetable quality, addressing issues encountered in open-field cultivation, especially during the rainy season (Li *et al.*, 2022; Krishna *et al.*, 2024). Protected cultivation involves controlling the microclimate around plant bodies, either partially or entirely, to meet the specific needs of different plant species during their growth stages. Structures like poly houses, poly tunnels, trench greenhouses, shade nets, net-houses and walk-in tunnels have proven beneficial for cultivating off-season vegetables. Forward-thinking farmers are now embracing commercial protected cultivation of high-value vegetables (Maitra *et al.*, 2020). Through the use of these structures and techniques, a variety of vegetables—such as tomatoes, brinjals, chilies, capsicums, broccoli, cabbages, cauliflowers, radishes, carrots, spinach, coriander, lettuce, cucumbers, bottle gourds and bitter gourds—have captured the market by being available in different seasons.

Regulating climatic factors like temperature, light, humidity and $CO_2$, coupled with the use of soilless media, has made year-round vegetable availability. Techniques such as geoponics, soil-less culture, hydroponics and aeroponics facilitate crop growth and allow for the establishment of nurseries (Li *et al.*, 2022).

Off-season farming of summer vegetables, has shifted towards cultivation in winter with the help of greenhouses providing controlled temperature conditions. Polyhouses have become a specialized production technique for ensuring year-round production of slow-season vegetables (Sahu *et al.*, 2020). Greenhouses also support the early production of seedlings for summer-season vegetables, further enhancing productivity.

For winter vegetables, the use of shade nets and enclosed structures helped to extend their lifespan and improve yields. Shade houses, composed of frames and cladding materials in various colors, create a conducive microenvironment for vegetable cultivation, particularly at higher altitudes.

During the rainy season, when vegetable prices soar due to low production caused by high soil moisture and heavy rainfall, protected structures such as rain shelters and poly tunnels mitigate crop damage and promote growth. Rain shelters, roofed with plastic film or waterproof materials, shield crops from rain, ensuring stable production of vegetables like cabbage, cauliflower, French beans, radishes, spinach and coriander (Sahu *et al.*, 2020).

## Yield and quality of greenhouse produce

Off-season cultivation under protective structures has demonstrated remarkable success in enhancing both yield and quality of greenhouse produce (Krishna *et al.*, 2024):

**Hybrid Cucumber in Greenhouses**: Phookan and Barua (2016) found that growing hybrid cucumber in naturally ventilated greenhouses during the off-season resulted in higher quantities and increased ascorbic acid content.

**Capsicum Production in Polyhouses**: Kumar *et al.* (2016) highlighted the considerable impact of planting capsicum in polyhouses, leading to increased yield. Bhatnagar *et al.* (1990) reported a significant increase in capsicum yield under greenhouse conditions compared to open fields, reaching up to 50 quintals per hectare.

**Enhanced Microclimate in Polyhouses**: Cheema *et al.* (2004) noted that the improved microclimate provided by polyhouses significantly contributed to the enhanced output of various vegetable crops, including capsicum and cauliflower.

**Increased Crop Longevity and Yield**: Protected technologies, such as polyhouses, have been shown to increase crop longevity and yield compared to open field conditions. Polyhouse demonstrated higher fruit weight and yield compared to open fields, with the highest net return and Benefit-Cost ratio (B: C) observed in polyhouses (Singh *et al.*, 2011).

**Influence of Greenhouse Cover Materials**: The choice of greenhouse cover materials, such as films or screens, significantly influences the microclimate and affects vegetable growth and quality. Studies by Yamaura *et al.* (2022) and Wen *et al.* (2022) examined the impact of different materials on crops like tomatoes and Chinese flowering cabbages, respectively.

**Utilization of Color Shade Nets**: Color shade nets have emerged as an effective method to shield crops from adverse environmental conditions, enhance vegetable quality and prolong post-harvest freshness (Ili *et al.*, 2018). Ramasamy *et al.* (2021) noted the consistent improvement in yield of amaranths and water spinach under shade nets in different seasons.

**Successful Off-Season Cultivation**: Off-season cultivation of various vegetables, including bitter gourd and spinach, in protective environments like trench greenhouses and shade nets, has been successful (Asati *et al.*, 2018; Angmo *et al.*, 2017).

**Profitability of Polyhouse Cultivation**: Although growing cucumbers in polyhouses incurs a higher cost per hectare compared to open-field cultivation, the production and net return per hectare are significantly higher, contributing to increased yields and income for growers (Kumar *et al.*, 2017; Malik *et al.*, 2017).

**High Benefit-Cost Ratio in Protected Environments**: In Kerala, cucumbers cultivated in cost-effective polyhouses with natural ventilation achieve the highest Benefit-Cost ratio when nutrients are applied at recommended doses (Chand *et al.*, 2014).

Overall, off-season cultivation under protective structures has proven to be a lucrative and sustainable approach for increasing yield and improving the quality of greenhouse produce, offering significant benefits to farmers.

## Positive Outlook and Challenges in Greenhouse Cultivation

India's diverse agricultural landscape presents a promising backdrop for vegetable cultivation, with protected cultivation techniques offering significant advantages. Encouragingly, these methods have the potential to engage youth and women in agriculture, fostering pride in the profession. However, the current extent of protected cultivation in India, at only 2,75,000 hectares, pales in comparison to China's 2 million hectares (Dixit *et al.*, 2023). To capitalize on this potential, a substantial increase in protected farming over the next few years is imperative.

Protected cultivation brings a multitude of benefits, including optimized resource utilization leading to a substantial increase in yields, sometimes up to 3-5 times more than traditional open-field methods (Jayasurya *et al.*, 2021). A range of techniques, from plastic mulch to soilless farming, addresses specific agricultural needs such as early crop protection and off-season cultivation (Chandra *et al.*, 2023; Krishna *et al.*, 2024). Adoption of these methods is crucial for propelling India's vegetable cultivation sector forward. However, several challenges impede the widespread adoption of protected cultivation techniques. These include a lack of specialized knowledge among farmers, insufficient funds for infrastructure, high production costs, limited access to skilled labor, elevated seed costs and issues related to market accessibility and cold storage facilities (Sethi *et al.*, 2009; Jayasurya *et al.*, 2021; Chandra, 2023; Dixit *et al.*, 2023). Moreover, the heterogeneity of farms complicates the scaling of protected cultivation technologies, particularly for smallholders. Addressing these challenges requires targeted interventions such as capacity building, financial support for infrastructure development and improving market linkages and cold storage facilities (Table 2). Policy interventions, similar to China's successful vegetable basket plan, can play a pivotal role in promoting protected cultivation and driving its widespread adoption (Nordey *et al.*, 2017). By targeting resource-endowed producers, promoting knowledge and skills and addressing information gaps, India can significantly enhance its vegetable cultivation sector and achieve food security goals.

**Table 2. Prospects and challenges of protected cultivation**

| Prospects | Challenges |
|---|---|
| **Increased Production**: The use of low-cost protected structures can boost vegetable production by approximately four times, increasing productivity per unit of land. | **Availability of Materials**: Construction materials like GI pipes, polyethylene sheets, etc., are not consistently found in local markets and often need to be imported at high costs, including freight and custom duties. |
| **High-Quality Crop**: Protected cultivation ensures the production of high-quality and clean vegetables. | **Skilled Labor**: The installation, monitoring and supervision of protected cultivation may require skilled workers, which are not readily available in village areas. |
| **Intensive Crop Production**: Greenhouses enable multifold intensification of vegetable production compared to traditional open-field cultivation. | **Regional Variation**: The package of practices for year-round cultivation of different vegetable crops in protected environments has not been standardized. Recommendations for one region may not be applicable to others. |
| **Healthy Seedling Production**: It supports the production of high-quality and healthy vegetable seedlings, resulting in strong and resistant crop stands. | **Lack of Awareness**: Limited awareness among farmers about the potential benefits of protected vegetable production hampers large-scale cultivation. |
| **High Purity Hybrid Seed Production**: Greenhouse cultivation eliminates cross-pollination, ensuring high purity in hybrid seed production. | **Design**: The current use of conventional screen net houses for off-season vegetable cultivation in North India's composite climate faces limitations such as poor structural design, high constructional costs, lack of a greenhouse effect in winter and elevated plant temperatures in summer. Similarly, polyethylene sheet-covered greenhouses also encounter issues, including significantly higher constructional and operational costs and elevated indoor air temperatures during summers. |

| | |
|---|---|
| **Disease-Free Seed Production**: Protected cultivation facilitates disease-free seed production, ensuring the quality of seeds for future crops. | **R&D Shortcomings**: There is a scarcity of comprehensive research on protected vegetable farming for various commercial crops, limiting the availability of region-specific guidelines and best practices. AICRP on PEASEM is inadequate for the diverse needs of the country in developing plasticulture technologies. Higher education programs are needed to support greenhouse-related research and development. |
| **Weather and Pest Protection**: Protective structures shield high-value crops from unfavorable weather conditions and protect them from pests and diseases. | **Structural Integrity and Skills**: Recurrent greenhouse structural failures due to winds highlight the need for climate-specific designs. Lack of operational skills is a serious constraint that requires addressing before providing subsidies. |
| **Off-Season Horticultural Production**: Controlled microclimates in greenhouses allow year-round production, addressing off-season demand and enhancing profitability. | **Skilled Workforce**: Ensuring greenhouse managers and workers receive adequate training is crucial for successful production. Establishing state-level skill development centers for protected cultivation is recommended. |
| **Expanding Growing Seasons**: In regions like Ladakh, greenhouses extend the growing season from 4-5 months to around 10 months. | **Education-Work Ecosystem**: Flexibility for students to learn and earn in greenhouses should be promoted. |
| **High-Quality Planting Material**: Favorable microclimates in greenhouses facilitate intensive preparation of high-quality planting material. | **Biodegradable Glazing Materials**: Efforts are needed to accelerate the availability of biodegradable materials for greenhouse glazing. |
| **Enhancing Horticultural Produce Availability**: Greenhouse technology can address gaps and enhance the availability of horticultural produce in India. | **Marketing Infrastructure**: Establishing a well-defined marketing infrastructure, including cold-chain and traceability, is essential as in its absence protected cultivation is still limited. |
| **Conservation of Input Resources**: Greenhouse technology improves resource utilization efficiency, conserving water, nutrients and agro-chemicals. | **Comprehensive Insurance**: Comprehensive and user-friendly insurance portfolios are crucial for reassuring greenhouse growers in emergencies, which is lacking currently. |
| **Job Creation**: Greenhouses offer employment for about five persons per hectare, potentially generating direct and indirect employment opportunities for millions. | |
| **Income Enhancement for Greenhouse Farmers**: Greenhouse cultivation significantly increases farmers' income, contributing to the country's GDP. | |
| **Meeting Urban Demand**: Greenhouses play a vital role in meeting the fresh produce demands of the growing urban population. | |

| | |
|---|---|
| **Expanding Food Production in Non-Terrestrial Spaces**: Greenhouse activities can be expanded to unconventional spaces like underground, seas, ocean floors and high mountains, addressing land limitations. | |
| **Bridging Supply-Demand Gaps**: Greenhouse cultivation helps overcome short-term shortages in open-field cultivation, ensuring a consistent supply of targeted commodities. | |

## Conclusion

Cultivating vegetables during off-season provides the opportunities for higher income and increased employment to farmers, which require specialized techniques and consistent supervision. Despite the inherent risk of pests and diseases, the benefits outweigh the costs when modern production technologies are utilized. Many farmers, particularly those in hilly regions, have found success in growing off-season vegetables under polyhouse using organic cultivation methods, proving to be profitable and providing employment, especially for female family members. Prioritizing and promoting off-season vegetable cultivation under polyhouses is recommended (AERC, 2017). However, ensuring consistent technical support, training and a reliable supply of inputs such as seeds and fertilizers, along with access to market information, is crucial for the sustained adoption of agricultural technologies. Long-term success hinges on accessible extension services and the combined impact of various technologies contributes to improved off-season agricultural practices. This underscores the importance of farmer-centric policies in guiding enduring technology adoption. Socioeconomic characteristics and institutional effectiveness significantly influence off-season vegetable production (Kunwar *et al.*, 2018). Policies should leverage positive factors and address negative influences. Factors like farm size, gross return, access to extension services and credit availability significantly affect farmers' decisions to adopt off-season vegetables production. Engaging village youth in marketing, promoting contract farming (Dhanta *et al.*, 2022), implementing motivational initiatives, establishing regular extension contracts, ensuring credit accessibility, providing training and offering input support are recommended to enhance farmers' knowledge of off-season vegetable production practices (Maruf *et al.*, 2022).

## References

Adarsh A, Kumari A and Gupta S.K. 2021. Pointed Gourd cultivation in Diara land for nutritional security and economic prosperity. Food and Scientific Reports 2(9): 30-32.

Maharjan, S.K. 2017. Riverbed farming as a source of income, family nutrition and food security for landless and poor farmers in Terai Region of Nepal. Innovative Techniques in Agriculture 2(1): 316-319.

Aditika, Kanwar, H.S., Priyanka and Sharma A. 2017. Vegetable Improvement in India; Recent Past, Present and Future: A Review. International Journal of Current Microbiology and Applied Science 6: 3246-3255.

AERC 2017. Economic Analysis of Cost and Return of Off-Season Vegetables with Focus on Poly House Effect in Sikkim; Study No.-185, Agro-Economic Research Centre (For the States of West Bengal, Sikkim and Andaman and Nicobar Islands), Visva-Bharati, Santiniketan, West Bengal, pp.-xiv+105.

Angmo S, Angmo P, Dolkar D, Norbu T, Paljor E, Kumar B and Stobdan T. 2017. All year round vegetable cultivation in trenches in cold arid trans Himalayan Ladakh. Defence Life Science Journal 2(1): 54-58.

Anonymous 2023. Second advance estimate for the year 2022-23 released for area and production of horticultural crops. [Online] Available: https://pib.gov.in/PressReleasePage.aspx?PRID=1968772#:~:text=The%20production%20of%20vegetables%20is,in%20the%20year%202021%2D22.&text=Production%20of%20plantation%20crops%20is,an%20increase%20of%20about%201.78%25.

Asati K.P., Barche S. and Kirad K.S. 2018. Varietal hybrid performance of bitter gourd under polyhouse conditions. International Journal of Current Microbiology and Applied Science 7(2): 3581-3584.

Bhatnagar R.R., Ved Prakash R.C. and Srivastav R.C. 1990. Production of vegetables in polyhouse greenhouse during winters in mid hills of Uttar Pradesh. Progressive Horticulture 22: 97-100.

Chand Jisha A.R. 2014. Nutrient use efficiency and economics of salad cucumber using drip fertigation in naturally ventilated poly house. Journal of Agricultural and Veterinary Science 7(12): 22-25.

Chauhan V.B.S., Mukherjee A., Pati K., Hanume Gowda K. and Bansode V.V. 2016. Low-cost protected cultivation of vegetable crops for sustainable farm income. Training manual of Model training course on "Root and tuber crops based integrated farming system: A way forward to address climate change and livelihood improvement" conducted in between 19-26th September, conducted by ICAR- Central Tuber Crop Research Institute (CTCRI), 41-46.

Cheema D.S., Kaur P. and Kaur S. 2004. Off-season cultivation of tomato under net house conditions. ISHS Acta Horticulture 659: 177-181.

Dhaliwal MS, Sharma SP, Jindal SK, Dhaliwal LK and Gaikwad AK. 2017. Growth and yield of bell pepper as influenced by growing environment, mulch and planting date. Journal of Crop Improvement 31(6): 830–846.

Dhanta R, Tewari SC and Yadav A. 2020. From Farm to Market: Analyzing production cost and marketing management of off-season vegetables in Himachal Pradesh. Journal of Xi'an University of Architecture and Technology 12(3): 5746-5760.

Dixit PS, Singh AK, Tripathi CM, Singh R and Kumar P. 2023. A Review on Protected Cultivation of Vegetables: Opportunities and Challenges. International Journal of Environment and Climate Change 13(11): 3371-3380.

Ghimire R, Suvedi M and Kaplowitz M. 2022. Adoption of improved agricultural practices: Learning from off-season vegetable production in Nepal. Journal of International Agricultural and Extension Education 29(3): 32-47.

Ghosal MK and Das RK. 2012. A study on the cultivation of capsicum in a greenhouse during the off-season in the warm and humid climate of India. International Journal of Agricultural Sciences 8(1): 220-223.

Hofer T. 1993. Himalayan deforestation, changing river discharge and increasing floods: Myth or Reality. Mountain Research and Development 13(3): 213-233.

Ili ZS, Milenkovi L, Šuni L and Manojlovi M. 2018. Color shade nets improve vegetables' quality at harvest and maintain quality during storage. Contemporary Agriculture 67: 9–19. doi: 10.2478/contagri-2018-0002

Jayasurya P, Chatterjee S, Biswas A, Viswanath M and Roy A. 2021. Low-cost protected structures for off-season vegetable production: a review. The Pharma Innovation Journal 10(7): 778-783.

Krishna H, Janakiram T, Singh MK, Karuppaiah V, Yadava RB, Prasad RN, Singh J and Behera TK. 2022. Immunomodulatory potential of vegetables vis-à-vis human health. The Journal of Horticultural Science and Biotechnology 97(5): 560-579.

Krishna H, Hebbar S, Kumar P, Sharma S, Kumar R, Tiwari S K, Maurya S, Srivastava K, Pal G, Bahadur A and Behera T K. 2024. Navigating Challenges and Prospects in Off-Season Vegetable Production. Vegetable Science 51: 97-105.

Kumar P, Chauhan RS and Grover RK 2017. An economic analysis of cucumber (Cucumis sativus L.) cultivation in the eastern zone of Haryana (India) under poly house and open-field condition. Journal of Applied and Natural Science 9(1): 402-405.

Kumar P, Chauhan RS and Grover RK. 2016. Economic analysis of capsicum cultivation under poly house and open-field conditions in Haryana. International Journal of Farm Sciences 6(1): 96-100.

Kumar M and Baraik VK. 2023. Assessing Livelihood Vulnerability in Diara with Special Reference to Flood: A Case Study of West Champaran District of Bihar. Annals of the National Association of Geographers, India 43(1): 93-109.

Kumari R, Sharma A, Bhagta S and Kumar R. 2018. Riverbed cultivation: a kind of vegetable forcing for remunerative returns. International Journal of Current Microbiology and Applied Sciences 7(04): 359-365.

Kunwar B, Dhakal D and Panta HK. 2018. Determinants of smallholders' adoption of off-season vegetable production technology in Okhaldhunga District of Nepal. Journal of the Institute of Agriculture and Animal Science 33: 221–228.

Lal G, Mehta RS, Singh R, Kakani RK, Meena NK and Maheria SP. 2016. Effects of nitrogen levels on plant growth and leaf yield of off-season coriander under shade nets. International Journal of Seed Spices 6(2): 46-49.

Li X, Hu X, Song S and Sun D. 2022. Greenhouse Management for Better Vegetable Quality, Higher Nutrient Use Efficiency and Healthier Soil. Horticulturae 8: 1192.

Maitra S, Mitra B, Sarkar R, Chatterjee R, Khalko S and Chaturvedi OP. 2020. Small Scale Protected Cultivation for Off-season Vegetables: A Promising Option for Terai Farmers towards Income Generation. International Journal of Current Microbiology and Applied Sciences 9(04): 972-980.

Malik K. 2017. Economic viability of cucumber cultivation in greenhouses. International Journal for Innovative Research in Multidisciplinary Field 3(6): 366-368.

Maruf SA, Ahmed JU and Khan JA. 2022. Prospect of off-seasonal vegetable production in Bangladesh: a socioeconomic diagnosis. Qual Quant 56: 3441–3463.

Nagalakshmi S, Nandakumar N, Palanisamy D and Sree Narayanan VV. 2001. Naturally ventilated poly house for vegetable cultivation. South Indian Horticulture 49: 345-346.

Navnath I, Kannuji P, Kale S.J, Panda N and AN M. 2019. Offseason vegetable production under protected cultivation. International Journal of Engineering and Science Research 9(5): 54-57.

Nordey T, Basset-Mens C, De Bon H, Martin T, Déletré E, Simon S, *et al.* 2017. Protected cultivation of vegetable crops in sub-Saharan Africa: limits and prospects for smallholders. A review. Agronomy for Sustainable Development 37: 53. doi: 10.1007/s13593-017-0460-8

Panda BC. 1996. Monitoring of the Diara lands of Eastern India with respect to vegetation, wasteland, soil erosion and inundation through remote sensing techniques. Division of Agricultural Physics, Indian Agricultural Research Institute (IARI) New Delhi-110, 12.

Pandey S, Dubey R.K, Singh S, Kumar S, Singh S and Behera TK. 2023. Diara land cultivation of cucurbitaceous crops. Indian Horticulture 68(2): 77-81.

Phookan DB and Barua S. 2016. Performance of hybrid cucumber varieties during the off-season under naturally ventilated polyhouse conditions. Journal of Eco-friendly Agriculture 12(1): 95-96.

Ramasamy S, Lin MY, Wu WJ, Wang HI and Sotelo-Cardona P. 2021. Evaluating the potential of protected cultivation for off-season leafy vegetable production: prospects for crop productivity and nutritional improvement. Frontiers in Sustainable Food Systems 5: 731181.

Sahu B, Dalai S and Mallikarjunarao K. 2020. Cultivation of Off-season Vegetables under Protected Conditions. In: Maitra S, Gaikwad D.J and Shankar T. (eds.) Protected Cultivation and Smart Agriculture. New Delhi Publishers, New Delhi: pp. 161-166.

Sethi VP, Dubey R.K and Dhath AS. 2009. Design and evaluation of a modified screen net house for off-season vegetable raising in a composite climate. Energy Conversion and Management 50(12): 3112-3128.

Sharma A, Lata H, Sood P, Thakur A, Sharma KC, Sharma P. 2023. Off-Season Vegetable Growing for Nutrition and Entrepreneurship. In: Singh B, Kalia P. (eds) Vegetables for Nutrition and Entrepreneurship. Springer, Singapore.

Singh AK, Singh B and Gupta R. 2011. Performance of sweet pepper (Capsicum annuum) varieties and economics under protected and open-field conditions in Uttarakhand. Indian Journal of Agricultural Sciences 81(10): 973-975.

Singh PK. 2012. "Cucurbits Cultivation under Diara-Land", Asian Journal of Agriculture and Rural Development, Vol. 2, No. 2, pp. 243-247.

Wen J, Ning S, Wei X, Guo W, Sun W, Zhang T and Wang L. 2022. The Impact of Insect-Proof Screen on Microclimate, Reference Evapotranspiration and Growth of Chinese Flowering Cabbage in Arid and Semi-Arid Region. Horticulturae 8: 704.

Yamaura H, Furuyama S, Takano N, Nakano Y, Kanno KT, Amasaki I, Watanabe Y, Iwasaki Y and Isozaki M. 2022. Effects of NIR Reflective Film as a High Tunnel-Covering Material on Fruit Cracking and Biomass Production of Tomatoes. Horticulturae 8: 51.

# CHAPTER 4

# AUTOMATION IN PROTECTED CULTIVATION

**T. Thomson, N.V. Gowtham Deekshithilu, P. Ashok and T. Janakiram**

*Dr. YSR Horticultural University, Venkataramanna gudem,*
*West Godavari Dt., Andhra Pradesh*
*Corresponding author: thomsonmonu@gmail.com*

Protected cultivation is one of the most accepted approach for production of horticultural crops and ensures judicious use of water, land and other inputs. It also provides opportunities for climate-resilient production of quality planting material and has vast potential to provide near optimal conditions by controlling the climate using different protected structures/methods/devices. Improving the microclimate in polyhouses is crucial for enhancing crop growth and yield. It is feasible to establish ideal growing conditions for a variety of crops by employing automatic climate control systems and other methods to adjust the temperature, humidity, and other environmental factors. In order to improve crop growth and productivity in polyhouses, it is necessary to take into account the numerous variables that have an impact on the microclimate and to put in place the necessary controls. This chapter automation, need for automation, monitoring and control systems in protected cultivation using IoT, including the use of sensors for environmental monitoring, irrigation and fertilizer management, pest and disease management, and data management and analysis.

Protected cultivation technology is an integral component of precision farming which has the potential to enhance input use efficiency to a significant level to achieve sustainability in food production. It uses artificial techniques to climatic change factors such as natural light, temperature and humidity to create environmental conditions suitable for the growth of plants, we describe protected cropping environments to include Hoop houses, Tunnel houses, Canopy protection. The protected cultivation environment is a relatively easy system for automated machinery due to its structured nature. The development of systems is easier under the protected cultivation environment. The automation systems for protected cultivation deal with **micro-climate control, seedling production, spraying, harvesting, irrigation** *etc.*

## Automation

It is a buzz term in the agriculture technology industry, and it can refer to any technology that reduces operator workload.

### Why Automation?

Many of the farmers have migrated from fully outdoors to protected cropping and have seen great improvements in yield and quality of their crops. Some studies had shown that you can get an extra four to six weeks of production with protected cropping techniques versus open field cropping.

### Advantages

- Lower operating costs.
- Robots can perform the work of three to five people, depending on the task.
- Improved workers safety.
- Reduced factory lead times.
- Ability to be more competitive.
- Increased production output.
- Consistent and improved production and quality.
- Smaller environmental footprint.

## What should be automated in growing environment

There are things that can be automated and take the hassle out of some of the more time-consuming tasks. Depending on the size, type and structure, the following things can be monitored or controlled:

- Irrigation
- Fertigation
- Run-Off
- Root Zone

### Need for Automation

With additional external factors to consider, plants in these environments need to be closely monitored to ensure the conditions which are ideal for crop growth. The climate can change rapidly over the day requiring different watering schedules to be established. Automation will take care of all the variables and then trigger the system to make changes.

## Benefits of Automation

- **Decreases labor costs** - Automation can either decrease labor costs or allow to free up workers to focus on other areas of importance.
- **Be accurate Know** – To know exactly what's happening in greenhouse and make data driven decisions based on facts and not assumptions.
- **Increase quality and yield** - If it is not measured it can't be improved. Automation will give the knowledge to increase quality and yield, learn from previous crop cycles, and give better return on investment (ROI)

## What we can automate

**Climate control** - It controls Manage factors like temperature, humidity, lighting and $CO_2$ with a single device through climate control solutions.

**Hydroponics and substrates** - It is a cost-effective and sustainable approach to grow crops like microgreens, leafy greens and medicinal cannabis without the need for soil.

**Irrigation and fertigation** - It monitors climate, irrigation and fertigation with comprehensive controllers to ensure energy efficiency.

**Multi-crop solutions** - There are various tools which offer the customization and flexibility to ensure all crops are neatly managed when growing multiple crops on the same site.

**Multi-area solutions** - Automate growing empire as it expands to stay on top of day-to-day management and progress – remotely.

**Water right in quantity every time** - If watering is done manually it can be easy to overwater or underwater plants. To be accurate, it would need to stick to a good schedule to make sure there were no inconsistencies. Ensuring accuracy by setting a schedule which can be adjusted based on type of crop, quality, yield and local weather parameters.

**Self-management** - The key to automation is being able to rely on the technology to do the work. Once parameters are set can later be focused on other important tasks. If a problem comes up, the system will then alert (via email or text message) and can log in to check the data.

**Timeless growth** - Because indoor farming removes much of the dependencies on outside influences like rain and sunshine. The seasonality and space can effectively be controlled with automated tools.

**Precision and adaptability** - Seasonality is always a factor, even with indoor farming, but automated environmental controllers will adapt and adjust the distribution of crops as needed. Room temperature and humidity control also gives visibility into the ongoing data, giving the knowledge to be able to plan and strategize effectively for future yields.

**Increased Efficiency** - Effective irrigation and fertigation is a direct contributor to a successful yield, so it's vitally important that not to leave it up to chance.

## Internet of Things

The Internet of Things (IoT) is a system of interrelated computing devices, mechanical and digital machines, objects, that are provided with unique identifiers and the ability to transfer data over a network without requiring human-to-human or human-to-computer interaction. In recent years, the agriculture industry has undergone a digital transformation, with the integration of emerging technologies such as the IoT to optimize crop production, reduce costs and increase sustainability. Among the various applications of IoT in agriculture, the use of IoT technology in protected cultivation has gained significant attention due to its potential to improve crop yields, quality and resource efficiency. IoT has been defined as a network of physical objects embedded with sensors, software and connectivity that enables the exchange of data between the objects and with other devices and systems. In the context of agriculture, IoT can be defined as the integration of sensors, controllers and other smart devices to collect and analyze data on various aspects of crop production and management. IoT technology has been increasingly adopted in the agriculture sector due to its potential to revolutionize farming practices, improve efficiency, and increase sustainability. In protected cultivation, IoT can help farmers to optimize the growing environment by collecting real-time data on environmental factors such as temperature, humidity, light, soil moisture and nutrient levels. The use of IoT in protected cultivation has gained significant attention due to the growing demand for food production, increasing awareness for sustainability and the need to maximize crop yields while minimizing resource usage. IoT devices and sensors can help farmers to manage and monitor their crops more effectively, reduce waste and increase profitability.

One of the primary benefits of using IoT in protected cultivation is the ability to monitor and adjust the growing environment in real-time. This can help farmers to optimize crop growth by providing the ideal conditions for the plants, which can increase yields and improve the quality of the produce. In addition, the use of IoT can help to reduce waste by ensuring that resources such as water and fertilizer are used efficiently. IoT can also improve crop protection by enabling farmers to monitor pest and disease outbreaks in real-time and take proactive measures to prevent or minimize their impact. This can lead to reduced use of pesticides and fungicides, which can lower costs and decrease the environmental impact of farming. Another important benefit of using IoT in protected cultivation is the ability to collect and analyze data over time. This data can be used to identify patterns, trends, and correlations that can help farmers to make more informed decisions about their crop management practices.

### IoT devices and sensors

IoT devices and sensors play a critical role in enabling the use of IoT in protected cultivation. By collecting and transmitting data about the growing environment, these devices and sensors can provide farmers with valuable insights that can help to optimize crop growth and improve efficiency.

## Types of IoT devices and sensors

There are several types of IoT devices and sensors that are commonly used in protected cultivation, including:

- **Temperature and humidity sensors**: These sensors measure the temperature and humidity levels in the growing environment, providing farmers with critical data on the ideal conditions for plant growth.
- **Light sensors:** Light sensors measure the intensity of light in the growing environment, which can help farmers to optimize the amount of light that plants receive.
- **Soil moisture sensors:** These sensors measure the moisture content in the soil, which can help farmers to optimize irrigation and prevent over or under-watering.
- **$CO_2$ sensors:** $CO_2$ sensors measure the levels of carbon dioxide in the growing environment, which can help farmers to optimize plant growth by ensuring that the plants receive the right amount of $CO_2$.
- **Motion sensors:** Motion sensors can detect movement in the growing environment and can be used to detect the presence of pests or animals that are damaging crops .

The use of these IoT devices and sensors in protected cultivation enables farmers to collect and analyze data on a real-time basis, which can help to optimize crop growth and reduce costs.

## Applications of IoT devices and sensors

IoT devices and sensors have numerous applications in protected cultivation, including:

- **Monitoring environmental conditions:** IoT devices and sensors can be used to monitor temperature, humidity, light, $CO_2$ levels, and soil moisture in the growing environment. Their data can be used to optimize growing conditions, reduce water usage, and prevent plant diseases.
- **Crop management:** These can be used to monitor crop growth, detect plant diseases and pests, and optimize nutrient requirement. This information can be used to make decisions about when to fertilize, irrigate, or apply pesticides.
- **Energy management:** These can be used to monitor energy usage in protected cultivation, such as lighting and heating. These data can be analyzed to optimize energy usage and reduce costs.
- **Supply chain management:** These can be used to track and monitor the movement of produce from the farm to the consumer. These data can be used to improve logistics, reduce food waste, and enhance food safety.

By providing real-time data on the growing environment, IoT devices and sensors can help farmers to reduce costs, increase yields, and improve the quality of their produce.

## Examples of IoT devices and sensors

There are several examples of IoT devices and sensors that are commonly used in protected cultivation. Some of these examples include:

- **Wireless Sensor Networks (WSNs):** WSNs consist of multiple sensor nodes that are wirelessly connected to a central gateway. They are used to monitor environmental conditions, such as temperature and humidity, in real-time. WSNs are highly scalable, cost-effective, and can be easily deployed in large-scale cultivation facilities.
- **Smart Irrigation Systems:** Smart irrigation systems use soil moisture sensors to monitor soil moisture levels in the root zone. This data is used to control irrigation, ensuring that crops receive the optimal amount of water. Smart irrigation systems can reduce water usage by up to 50% and improve crop yields.
- **Automated Pest Management Systems:** Automated pest management systems use IoT sensors to detect pests in the growing environment. This data is used to control pests through the release of natural predators, parasitoids and thus reducing the need for chemical pesticides.
- **Light Control Systems:** Light control systems use IoT sensors to monitor light levels in the growing environment. This data is used to optimize the use of artificial lighting, ensuring that crops receive the optimal amount of light for growth. Light control system can improve crop yields and reduce energy usage.

## Monitoring and control systems

Protected cultivation using IoT involves the integration of various monitoring and control systems to manage and optimize the growing environment. These systems provide real-time data on environmental conditions and enable farmers to remotely control and adjust parameters such as temperature, humidity, lighting, and irrigation. The use of IoT monitoring and control systems can improve crop yields, reduce resource usage, and increase operational efficiency in protected cultivation.

### i. Environmental monitoring

Environmental monitoring using IoT sensors is one of the key aspects of controlling and optimizing the growing environment in protected cultivation. IoT sensors are used to measure various environmental parameters such as temperature, humidity, light intensity, CO2 concentration, and soil moisture. These sensors can provide real-time data on environmental conditions, allowing farmers to remotely monitor and adjust the growing environment as needed.

### ii. Irrigation and fertilization management

IoT can be used for irrigation and fertilization management in protected cultivation. IoT sensors and devices can provide real-time data on soil moisture levels, nutrient content, and other relevant parameters, allowing for precise control of irrigation and fertilization. This can lead to more efficient use of water and fertilizers, as well as better crop growth and yield.

### iii. Pest and disease management

IoT sensors can detect early signs of infestations or diseases, allowing for prompt and targeted interventions. Moreover, IoT devices can automate the release of biological agents or pesticides, reducing the need for manual labor and minimizing the risk of exposure to harmful chemicals. The system used sensors to monitor temperature and humidity levels and provide real-time alerts to farmers when the conditions are conducive to disease development. The results showed that the IoT-based system reduced the incidence of powdery mildew by 67% compared to traditional disease management methods.

### iv. Data management and analysis

Monitoring and control systems in protected cultivation using IoT are not only capable of collecting data from the field but also provide a platform for data management and analysis. With the help of IoT, large amount of data can be collected and processed to make better-informed decisions. This technology provides real-time data on environmental conditions, plant growth, and the performance of the irrigation and fertilization systems, which can be used to optimize crop production. One example of data management and analysis using IoT in protected cultivation is the use of machine learning algorithms to predict crop yield based on environmental conditions and other factors. The system provides real-time data on environmental conditions, crop growth, and irrigation and fertilization management, which can be used to optimize crop production.

## Advancements and challenges in IoT technology

Protected cultivation using IoT technology is a relatively new area of research that has gained significant attention in recent years. As the field continues to evolve, advancements are being made to improve the efficiency and effectiveness of IoT systems in protected cultivation.

## Advancements in IoT technology and their impact

Advancements in IoT technology has led to significant improvements in the efficiency and effectiveness of protected cultivation systems. For instance, the use of wireless sensor networks, cloud computing, and machine learning algorithms has enhanced the ability to monitor and control environmental factors, irrigation, fertilization, pest and disease management, and crop quality in protected cultivation. These advancements have the potential to increase yields, reduce resource usage, and improve the quality of crops.

## Challenges and limitations of using IoT technology

One of the major challenges of using IoT technology in protected cultivation is the cost of the devices and sensors. IoT devices and sensors can be expensive, which can be a major barrier to adoption, especially for small-scale farmers. Another challenge is the complexity of the technology. IoT devices and sensors require a certain level of technical expertise to set up and maintain, which can be a challenge for some farmers.

## Solutions and strategies to overcome challenges and limitations

Solutions and strategies are being developed to overcome the challenges and limitations of using IoT technology in protected cultivation. One approach is to improve the connectivity of IoT devices and sensors using advanced communication technologies such as 5G networks. Another strategy is to enhance the accuracy and reliability of IoT sensors by incorporating artificial intelligence and machine learning algorithms. Moreover, the integration of IoT technology with other emerging technologies such as blockchain and cloud computing can provide a more secure and efficient data management system for protected cultivation. Furthermore, the adoption of a systematic approach for IoT implementation, including proper planning, testing, and maintenance, can lead to better performance and sustainability in protected cultivation. Several studies have suggested that the adoption of IoT technology in protected cultivation can result in significant benefits, including higher yields, better resource management, and reduced labor costs. Therefore, it is essential to address the challenges and limitations of IoT technology to fully realize its potential in protected cultivation.

## Future prospects and recommendations for IoT

Future prospects of IoT in protected cultivation are immense, and many researchers have already begun exploring new ideas and technologies to improve the performance of IoT devices and systems. One such area of focus is the development of low-cost, energy-efficient, and reliable sensors and devices that can collect and transmit data accurately and consistently. Another area of interest is the use of machine learning and artificial intelligence algorithms to analyze data collected from IoT devices to provide predictive insights into plant growth and health. Moreover, the use of blockchain technology has also been proposed as a potential solution to the challenge of data security and privacy in IoT systems in protected cultivation. Implementing blockchain-based systems can help to secure data transmission and storage, provide traceability, and ensure transparency in the management of protected cultivation systems. To ensure the widespread adoption of IoT technology in protected cultivation, there is a need for greater collaboration among stakeholders, including farmers, technology providers, and policymakers. Efforts must be made to raise awareness about the benefits of IoT technology and provide training to farmers to enable them to use these systems effectively. Additionally, there is a need to address the issue of data standardization and interoperability among different IoT devices and systems to ensure seamless integration and compatibility.

## Optimizing Microclimate Control in Polyhouses

The microclimate within a polyhouse is influenced by factors such as light, temperature, humidity, and carbon dioxide concentration. Maintaining proper microclimate control in polyhouse is crucial for successful crop growth, good genetics and soil quality, incorrect temperature and humidity management and inadequate carbon dioxide levels can result in reduced productivity or even crop loss. It is possible to improve crop productivity in a polyhouse by carefully controlling the microclimate. This can involve cultivating multiple crops per year using controlled conditions and efficiently utilizing resources such as water, fertilizers, seeds, and plant protection chemicals. With the aid of technology, it is also possible to automate irrigation and precisely control other inputs and the environment, enabling the successful growth of tissue culture plants and high value crops in polyhouses.

### Polyhouse Micro Climate Parameters

Plants need specific climatic conditions that support growth through photosynthesis. Some of the important parameters include temperature, relative humidity, light, and carbon dioxide.

## Micro Climate Control Methods

All polyhouse cultivation systems, have fundamental climate control components that provide varying degrees of climate control and plant growth and productivity based on their design and complexity. It is important to maintain optimal ambient control to achieve these goals, including reducing emissions and production costs. The microclimate control system includes the climatic parameters inside the polyhouse (such as soil and air temperature, relative humidity, CO2 concentrations, electrical conductivity, and soil moisture), outside the polyhouse (such as temperature, relative humidity, solar radiation, wind speed, wind direction, and rainfall rate), and equipment (such as pipe temperature, vents, and curtain positions). Each microclimate parameter should be kept at the optimal level determined by the type and state of the crop. Some of the important micro climate control methods include ventilation (Natural ventilation, forced ventilation), cooling and shading.

### Naturally Ventilation

Natural ventilation in a polyhouse is the movement of air through openings in the structure due to differences in pressure caused by wind and temperature variations between the inside and outside. This process helps to regulate humidity and temperature within the polyhouse and ensures adequate air exchange. It is a key factor in the design of a polyhouse, as a well-ventilated system can improve climate control and optimize energy usage.

### Forced Ventilation

The use of forced ventilation using fans is an effective way to ventilate a polyhouse, but it consumes electricity. The fans are installed to maintain uniform temperature and humidity inside the polyhouse. Fresh air enters through one side and hot, stale air is forced out through the opposite side. The ventilation fans should be located on the windward side of the polyhouse and the distance between two fans should not exceed 8-10 meters.

### Shade Covers

The main objective of a polyhouse is to reduce air temperature, especially when high temperatures occur. Solar radiation is the primary source of heat gain in polyhouse, which can be controlled through shading or reflection. There are various methods for shading, such as internal and external shade screens, paints, and nets. However, shading can negatively impact plant development and photosynthesis by reducing light and potentially affecting ventilation rates and gas exchange. Therefore, it is important to consider the type of shading and associated control strategies. Shade nets, when properly chosen and installed, can improve crop production in adverse climatic conditions.

### Evaporative Cooling

One solution for mitigating the adverse effects of the polyhouse climate is to use evaporative cooling systems, which convert sensible heat into latent heat through the evaporation of water. This can be done through direct application of water into the polyhouse atmosphere using methods such as misting or fogging, or through the use of evaporative pads.

### Internal air circulation system

To promote optimal plant growth through effective gas exchange, it is recommended that the internal air velocities in a polyhouse be kept between 0.5 to 0.7 m/s. To achieve this, fans are installed above the crops at a rate of 0.01 m3 /s per m2 and positioned along the ridge. The distance between the fans should not exceed 30 times the diameter of the fans.

### Automatic Polyhouse Climate Control Systems

Automatic polyhouse climate control systems are computer-based systems that helps to regulate and optimize the internal microclimate of a polyhouse. These systems are designed to improve crop growth and productivity by maintaining optimal temperature, humidity, light intensity, and other environmental conditions. One of the main benefits of automatic polyhouse climate control systems is the ability to accurately monitor and control the microclimate in real-time. These systems use sensors and actuators to measure and adjust various parameters, such as temperature, humidity, light intensity, and CO2 concentration. By precisely regulating the temperature and humidity, these systems can minimize the use of energy-intensive heating and cooling systems. They can also optimize the use of natural light and ventilation, further reducing energy consumption.

There are several different types of automatic polyhouse climate control systems available, each with its own set of features and capabilities. Some systems are more basic and only control temperature and humidity, while others are more advanced and can control a wide range of environmental parameters. Some systems are designed to work with specific types of polyhouses, such as hydroponic or traditional soil-based polyhouses, while others are more flexible and can be used in any type of polyhouse.

### Automation at CERTPC, Dr YSR Horticultural University

With the advent of protected cultivation technology, Dr. Y.S.R. Horticultural University has persistently endeavored to serve the farming community. Befitting this virtue, this university has established "Centre of Excellence for Research and Training in Protected Cultivation" in the year 2016 at Dr. Y.S.R. Horticultural University, Venkataramannagudem. The project was aimed to demonstrate different technologies for intensive and commercial cultivation of horticultural crops for improved quality and productivity. It was established to act as a nodal centre of R and D work, training and transfer of technology and to demonstrate protected cultivation technologies in horticultural crops and hence aptly named as "Centre of Excellence for Research and Training in Protected Cultivation".

The centre has a Hi-Tech nursery with a capacity to produce 10 million vegetable seedlings per year and providing services to the farming community. This facility includes a climate-controlled structure used for growing high value crops with precision technology. By the use of polyhouse farming, crops can be protected from adverse environment such as high humidity or high temperatures. Temperature, humidity and ventilation can be controlled by the automated equipment fixed in the polyhouse. The polyhouse is equipped with automation and fertigation unit through which requires quantity of water along with desired dosage of fertilizers that can be applied to the plants.

## Infrastructure at CERTPC

### Automation system for plug type seedling production

Sowing seeds manually is time consuming and the automation ensures sowing in less time, accuracy in sowing and efficient use of inputs. The components of automation include diffuser, mixer, level control mechanism, trays dispenser machine, auto seeder, topping machine, watering station and conveying belt were used for raising seedlings. The capacity of the seed sowing machine is approximately 70,000 seed/hour.

### Hi-tech nursery

Hi-Tech nursery greenhouse frame is made of galvanized with zinc cover. Polycarbonated structure with fan and pad cooling was constructed in an area of 1075 m2 with high durability in gothic shaped design for plug type nursery production. Motorized aluminum net was installed to cut down the radiation system. Irrigation and fertigation and spraying of pesticides are performed by boomer irrigation system. The seedling production capacity is approximately 4,70,000 seedlings /

cycle. Computerized climate control system with weather station and internet-based software program (Priva office) are used to sense, control and monitor the light, temperature, relative humidity, roof screen and boomer etc.

### Seedling hardening facility

It has been installed in an area of 2112 m2 with an approximate capacity of 9,00,000 seedlings/cycle. Shade cladding on roof by 200-micron UV stabilized standard poly film having 85% light transmission, multi-layer with roof screen and all sides covered by 50:50 shade net UV stabilized shade net.

### Fan and pad

Fan and pad systems consist of exhaust fans at one end of the greenhouse and a pump circulating water through and over a porous pad installed at the opposite end of the greenhouse. Three climate-controlled fan and pad polyhouses of 580 m2 each with motorized side ventilation operating system were established to raise off-season vegetables. One of the structures established is ear-marked for production of chrysanthemum flowers throughout the year by providing artificial light. The structures can be used for raising high value vegetables like lettuce, chives, broccoli, brussels sprout, red cabbage, cucumber *etc.*

### Automated fertigation unit

In fertigation, nutrients are supplied to the crop together with the irrigation water as a nutrient solution. For the preparation of this solution, water soluble fertilizers and micro nutrients are to be injected.

## Conclusion

Protected Cultivation Technology with smart farming, automation and IoT-driven agriculture pave the way for what can be called a "Third Green Revolution" in agriculture. After the plant breeding and genetics revolutions, the Third Green Revolution is taking over the agriculture based upon the combined application of Information and Communication Technologies (ICT) such as precision equipment, the Internet of Things, sensors and actuators, geo-positioning systems, Big Data, Unmanned Aerial Vehicles (UAVs, drones), robotics, etc. In the future depicted by this revolution, pesticide and fertilizer use will drop while overall efficiency will be optimized. IoT technologies will enable better traceability of food which in turn will lead to increased food safety. It will also be beneficial for the environment, through more efficient use of water, or optimization of treatments and inputs. Therefore, Protected Cultivation Technology has a real potential to deliver a more productive and sustainable agricultural production, based on a more precise and resource- efficient approach.

## References

Aardra, K., Ray, S.K., Shili, K. and Jinu, A. (2019). Development and evaluation of greenhouse automation system using microcontroller. Department of Soil and Water Conservation Engineering.

Ahmad, Latief and Nabi, Firasath 2020 Agriculture 5.0: Artificial Intelligence, IoT and Machine learning. CRC Press Taylor and Francis Group.

Al-Helal, A.M. and Abdel-Ghany, I.M. (2010). Responses of plastic shading nets to global and diffuse PAR transfer: Optical properties and evaluation. NJAS - Wageningen Journal of Life Sciences, 57:125-132.

Altieri, M.A., Nicholls, C.I., Henao, A. and Lana, M.A. (2015). Agroecology and the design of climate change-resilient farming systems. Agronomy for Sustainable Development, 35(3): 869–890.

Alzoubi, Y. I., Al-Ahmad, A. and Kahtan, H. 2022. Blockchain technology as a Fog computing security and privacy solution: An overview. Computer Communications, 182, 129-152.

ASAE, (2000). American Society for Agricultural Engineers. ANSI/ASAE EP406.3 MAR98, Heating, venting and cooling greenhouses, 675: 682.

Atzori, L., Iera, A., and Morabito, G. 2010. The internet of things: A survey. Computer networks, 54(15), 2787-2805.

Badgery-Parker, J., James, L., Jarvis, J. and Parks, S. (2015). Commercial greenhouse cucumber production: 2010 Edition. NSW Agriculture.

Bailey, B.J. (2000). Constraints, limitations and achievements in greenhouse natural ventilation. Acta Horticulturae, 534:21–30.

Bhuvaneswari, A. 2017. A Survey on Internet of Things [IoT]. International Journal of Advanced Research in Computer Science, 8(1).

Brewster, C., Roussaki, I., Kalatzis, N., Doolin, K., and Ellis, K. 2017. IoT in agriculture: Designing a Europe-wide largescale pilot. IEEE communications magazine, 55(9), 26-33.

Briassoulis, D., Mistriotis, A. and Eleftherakis, D. (2007). Mechanical behaviour and properties of agricultural nets - Part I: Testing methods for agricultural nets. Science Direct. Polymer Testing, 26: 822–832.

Contreras-Medina LM, Torres-Pacheco I, Guevara-Gonzal-ez RG, Romero-Troncoso RJ, Terol Villalobos IR, De Pascale, S.T. and Maggio, A. (2008). Plant stress management in semiarid greenhouses. Acta Horticulturae, 797:205–215.

Engler, N., and Krarti, M. 2021. Review of energy efficiency in controlled environment agriculture. Renewable and Sustainable Energy Reviews, 141, 110786.

García-Sánchez, F., Valencia-García, R., Martínez-Béjar, R., and Fernandez-Breis, J. T. 2009. An ontology, intelligent agent-based framework for the provision of semantic web services. Expert Systems with Applications, 36(2), 3167-3187.

Ioslovich, I., Gutman, P. and Linker, R. (2009). Hamilton-Jacobi-Bellman formalism for optimal climate control of green-house crop. Automatica, 45(5): 1227-1231.

Kamp, P.G.H., Timmerman, G.J. (1996). Computerized environmental control in greenhouses. Ed. IPC Plant, The Netherlands.

Katsoulas, N. Kittas, C. Tsirogiannis, H., Kitta E, Savvas, D. (2007). Greenhouse microclimate and soilless pepper crop production and quality as affected fog evaporative cooling system. Transaction of ASABE. 50:1831–1840.

Kelly, S. D. T., Suryadevara, N. K., and Mukhopadhyay, S. C. 2013. Towards the implementation of IoT for environmental condition monitoring in homes. IEEE sensors journal, 13(10), 3846-3853.

Kumar, S., Bairwa, D., Kumar, K., Yadav, R., and Yadav, L. (2022). Climate regulation in protected structures: A review. Journal of Agriculture and Ecology, 13, 20–34.

Lakhiar, I. A., Jianmin, G., Syed, T. N., Chandio, F. A., Buttar, N. A., and Qureshi, W. A. 2018. Monitoring and control systems in agriculture using intelligent sensor techniques: A review of the aeroponic system. Journal of sensors, 2018, 1-18.

Mahmud, M. S. A., Abidin, M. S. Z., Emmanuel, A. A., and Hasan, H. S. 2020. Robotics and automation in agriculture: present and future applications. Applications of Modelling and Simulation, 4, 130-140.

Maitra, S., Gaikwad, DJ, Shankar, T., Eds., New Delhi Publishers, pp. 300–312. Huber, L. and Gillespie, T. (1992). Modeling leaf wetness in relation to plant disease epidemiology. Annual Review of Phytopathology, 30(1): 553–577.

Nagarajan, S. M., Deverajan, G. G., Chatterjee, P., Alnumay, W., and Muthukumaran, V. 2022. Integration of IoT based routing process for food supply chain management in sustainable smart cities. Sustainable Cities and Society, 76, 103448.

OuldKhaoua, S.A., Bournet, P.E., Migeon, C., Boulard, T., Chasseriaux, G. (2006). Analysis of green-house ventilation efficiency based on computational fluid dynamics (CFD). Biosystems Engineering, 95(1):83–88.

Palai, J.B., Purohit, S., Dalai, S. and Maitra, S. (2020). Control of abiotic factors in greenhouse, In: Protected Cultivation and Smart Agriculture, New Delhi Publishers, India, pp. 18-31.

Rawat, P., Singh, K. D., Chaouchi, H., and Bonnin, J. M. 2014. Wireless sensor networks: a survey on recent developments and potential synergies. The Journal of supercomputing, 68, 1-48.

Roy, S. K., Misra, S., Raghuwanshi, N. S., and Das, S. K. 2020. AgriSens: IoT-based dynamic irrigation scheduling system for water management of irrigated crops. IEEE Internet of Things Journal, 8(6), 5023-5030.

Sagheer, A., Mohammed, M., Riad, K., and Alhajhoj, M. 2020. A cloud-based IoT platform for precision control of soilless greenhouse cultivation. Sensors, 21(1), 223.

Santosh, D.T. and Maitra, S. (2021). Estimation of irrigation water requirement of zucchini squash (cucurbita pepo l.) Under protected cultivation structures and in open field conditions. Indian Journal of Natural Sciences, 12(69): 37380–37385.

Santosh, D.T., Raju, G. and Nataraj, K. Internet of things in protected cultivation: Advancements, applications and challenges. Recent advances in agricultural sciences and technology.

Santosh, D.T., Nataraj, K., Sagar, K.K. and Kumar, K.P. (2023). Optimizing micro climate control in polyhouses for enriched crop growth and productivity. Advances in agricultural technology, 1-15.

Singh, A. (2006). Flower crops: cultivation and management. New India Publishing, India.

Singh, M.C., Singh, J., Pandey, S., Mahay, D. and Srivastava, V. (2017). Factors affecting the performance of greenhouse cucumber cultivation-a review. International Journal of Current Microbiology and Applied Sciences, 6(10), 2304–2323.

Tap, F., van Willigenburg, L.G. and van Straten, G. (1996). Receding horizon optimal control of greenhouse climate based on the lazy man weather prediction, 13th IFAC World Congress; San Francisco; USA; pp. 387–392.

Wang, L. and Wang, B. (2020). Greenhouse microclimate environment adaptive control based on a wireless sensor network. International Journal of Agricultural and Biological Engineering, 13(3): 64–69.

Zhang, S., Mahrer, Y. and Margolin, M. (1997). Predicting the microclimate inside a greenhouse: an application of a one-dimensional numerical model in an unheated greenhouse, Agricultural and Forest Meteorology, 86: 291–297.

# CHAPTER 5

# PRECISION FARMING TECHNIQUES IN FRUIT PRODUCTION

**Uday Raj Patial[1*], Anirudh Thakur[1], Antima Sharma[1], Nav Prem Singh[1] and Prabhat Kumar[2]**

*[1] Department of Fruit Science, Punjab Agricultural University, Ludhiana, India;*
*[2] Ministry of Agriculture and Farmer Welfare, Krishi Bhawan, New Delhi*
**Corresponding author: uday-20101004@pau.edu*

---

Precision farming is an emerging technology which works on the principle of precise application of inputs on the right time at right place in right amount and in right manner, with the aim of efficient input use and improved productivity. Today, a global shift has been witnessed towards fruit culture owing to its lucrative returns and their nutritional significance. However, there are certain barriers in improved fruit production stemming from field variability and changing climatic scenarios. The integrated application of precision farming tools (GPS, GIS & remote sensing) in the orchards has significantly augmented better decision making, consequently hiking the returns. Leveraging computerized and sensor-based technologies (hyper-multispectral imaging, deep learning, VS-NIR spectroscopy) enables the creation of comprehensive crop inventories, real time assessment of soil-plant dynamics, maturity status and yield projections for framing better management programs. Remarkable strides have been made in mapping crop specific area-production estimates, species identification, automated harvesting, forecasting biotic and abiotic stress in fruit crops. This provides an overview of the application of cutting-edge technologies in fruit cultivation, elucidating their potential for optimizing pomiculture practices amidst current horticultural landscapes.

Today, the world is witnessing a paradigm shift of the growers towards the field of horticulture, since they found it a more remunerative investment as compared to the field crops. In a country like India, where 55 percent of the total population is dependent on the agriculture as a source of livelihood, farmers are generating more income per unit area by growing fruit and vegetables. The agriculture sector of India contributes 15 percent to the national GDP and of which, horticulture sector shares 30.4 percent to the total agriculture GDP. The proportion of horticulture crops relative to the total value of agricultural crops has exhibited an upward trajectory, rising from 39 percent in the fiscal year 2011-12 to 43.37 percent by 2023-24. A sudden hike has been observed in the area blanketed under the horticulture crops from 18.5 to 28.70 million hectares, since 2001 to 2024, whereas the land cover under the field crops had shrunk from 198 to 124 million hectares. Moreover, the production of horticulture crops has raised 55.5 percent (355.25 million tonnes) in 2024 compared to 2001, while in case of field crops it increased up to 25 percent only (Anonymous, 2024). Six decades ago, in 1960's during the period of the green revolution, each farmer had a responsibility of feeding 155 people on his shoulders. Today human population has increased many folds and 9.6 billion population has been expected by 2050, resulting each farmer should be able to feed 265 mouths with same acreage. Likewise, the current market need for fruit crops stands at approximately 112 million metric tons, poised to surge to 244 million metric tons by 2047, marking a growth of 2.18 times. This population explosion followed with poor economy has led to horticultural intensification and extensification into marginal soils with an increased risks of crop failure in near future. So, it's necessary to follow improved production techniques for obtaining higher productivity per unit area.

Precision agriculture is an emerging technology which deals with the following principles and raising technologies for the better management of the spatial and temporal variability with in a field associated with the agriculture production. The precision farming runs on the 5R principle, wherein agricultural inputs are meticulously applied in accordance with the specific requirements of soil, prevailing weather conditions and crop exigencies, thereby aiming to enhance crop quality, productivity and profitability (Fig.1). Professor Pierre C. Robert (father of precision farming) defines precision farming as more than the mere integration of novel technologies; rather, it signifies an information revolution facilitated by these technologies, leading to the development of a refined and precise agricultural management system. The concept of precision farming in some ways was applied in past, as the farm size were small, making it easy to observe the variation within the field. Today, the increased farm size and with the mechanization of farming, it becomes difficult to observe the variation with in the field. Previously, the concept of precision agriculture was confined primarily to field crops and was only introduced to horticultural crops in the 1990s. On one side, precision agriculture is a kind of management system (considering variability) to increase productivity and economic returns with a reduced impact on the environment. Whereas, conservation agriculture is a principle aimed at maximizing crop yields along with environmental and economic returns. There is a minimum mechanical soil disturbance, permanent organic soil cover is kept and diversified crop rotations are followed. The combination of both these precision and conservation farming is the need of hour as it's a kind of long term ecological & economic sustainable farming.

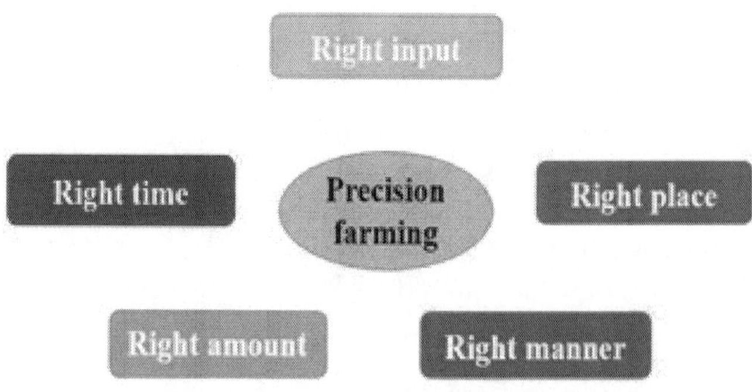


**Fig. 1. R principle of precision farming**

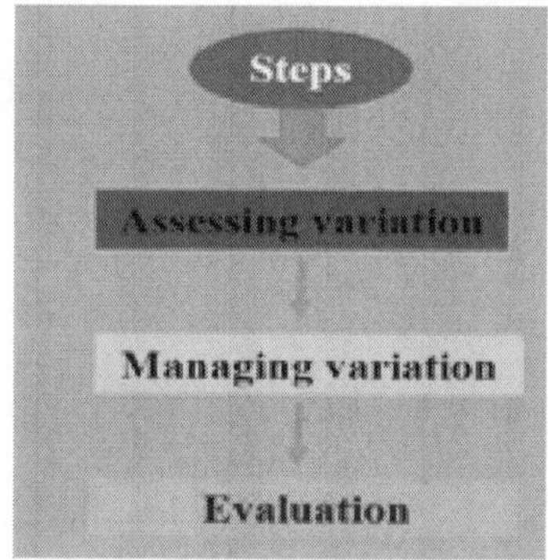


**Fig. 2. Protocol of precision farming**

## Steps in precision farming

The concepts which lead to the precision farming is the presence of the variability in theThe precision technique works in three steps; a) assessing variability b) managing variability and finally c) evaluating it.

**a) Assessing variability:** One cannot manage the problem which is not known to him, so this is the first and foremost step in a precision farming programme. The plant performance and the final yield depends upon various processes which vary both spatially and temporally. Though, there are various techniques and technologies for assessing the temporal or spatial variability individually by modelling, sensors, surveying or mapping through high resolution satellite data but, assessing them both simultaneously is very hard.

**b) Managing variability:** Once the variability has been assessed in the field, the next step is to make management procedures which are specific to the variability present in the field. The success of any precision farming approach depends on how efficiently management procedures will be raised as per different biotic-abiotic variables and how accurately corrective measures will be taken with respect to the variability noticed in the field. The principles like site specific management and variable rate technology are generally followed.

**c) Evaluation:** The evaluation is done in three categories (i) economic, (ii) environmental and (iii) feasibility of technology. Economic evaluation stresses that weather the noted horticultural traits have given any marketable benefit or not. Environment evaluation checks weather precision farming has improved soil health of saved microflora. Lastly, we check if the technology seems beneficial, how it can be transferred to other orchards.

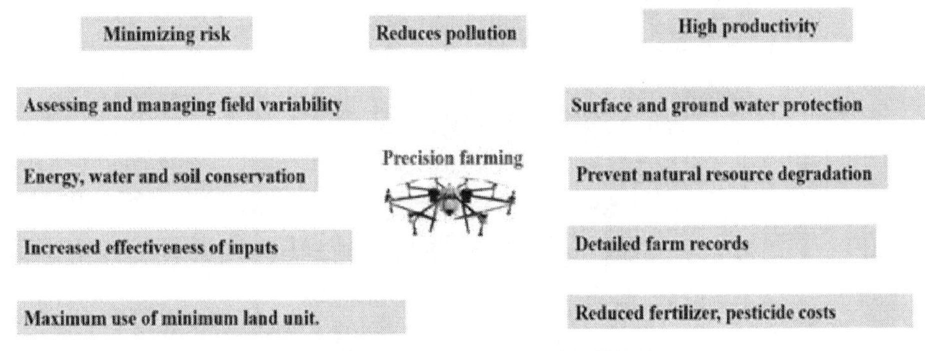


**Fig. 3. Advantages of following precision farming approach.**

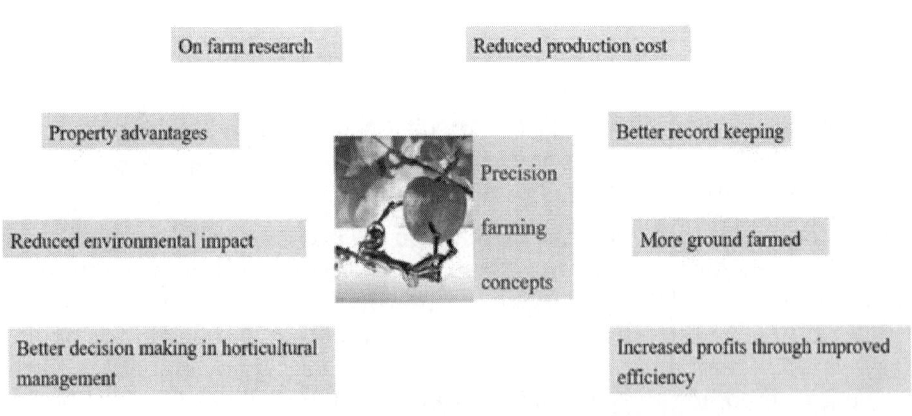


**Fig. 4. Concepts of precision farming.**

# Technologies employed in precision farming

## Global Positioning System (GPS)

The history of the precision farming does not date back to eons ago, it actually started around 30 years ago, when GPS sensors were made available in the public domain. The GPS system was initially developed by United States Department of Defence (US DoD), with the name NAVSTAR GPS for its military use but later, it was made available in the public domain with certain restrictions.

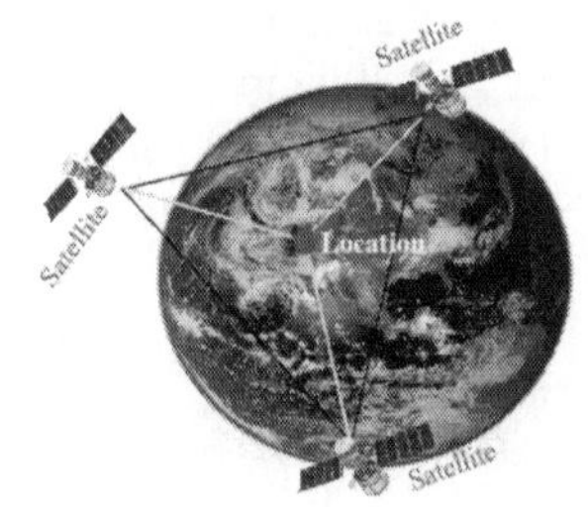


**Fig. 5. Trilateration principle of GPS**

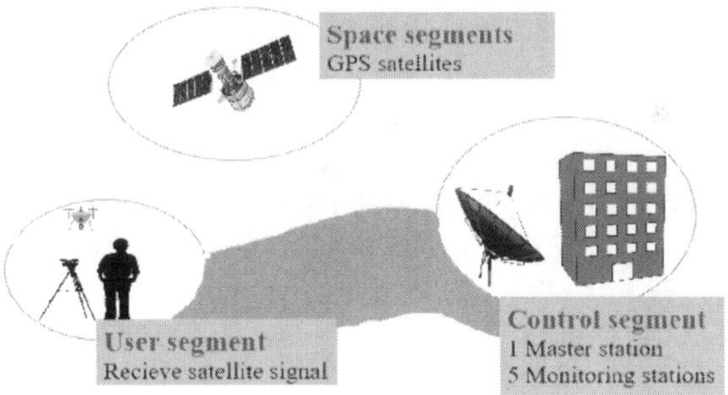


**Fig. 6. Components of GPS**

The GPS is basically a satellite-based system which provides navigation all around the world, 24 hours a day. It works on the trilateration principle i.e., the ground location navigated at some point of time is the mean value of area lying between three satellites (Fig.6.). 24 satellites orbiting around the earth enables GPS to mark precise location on three-dimensional basis (elevation, longitude and latitude). Alike USA, today many countries have developed their own positioning systems such as, GLONASS (Russia), GNSS and Galileo of Europe. Recently, in last few years India has also released a series of satellites like IRNSS (NavIC), Resourcesat-2, Resourcesat-2A, RISAT-1, Cartosat-1, Kalpana-1, INSAT-3D, INSAT-3DR which offers standard positioning services for the improvisation of agriculture (ISRO, 2022). A GPS system comprises three components namely a) space segment

b) control segment and c) user segment. The space segment transmits signal at a defined time variables with respect to the position of the satellite (Fig.6.). These dual frequency data signals from the space segment are uploaded every second in the control segment. Each user segment is governed by four different satellites present anywhere globally and they navigate the real time position by solving the range difference equations (elevation, longitude, latitude and time) of all satellites.

The GPS has gained its popularity due to its free of cost, precise, reliable, unlimited user capacity, all weather information or can be used anytime and anywhere. Different GPS mounted tools like variable rate technology, site specific farming, field mapping etc. are being used in precision farming which has boosted the productivity. The GPS tool is installed in a central vehicle (UAV's, tractor etc.) and all the data is recorded and distributed from a single system. The real time kinematic-differential GPS (RTK-DGPS) works on the double differencing technique and is used for topographic mapping. It can give accurate elevation data up to 5 cm range since, the data is offered after examining of the carrier wave of GPS signal.

## Geographic Information system (GIS)

The GIS tool comprises of a hardware, software and set of procedures which helps in collecting, storing, retrieving, manipulating and analysing the geographical data for better decision making. The GIS differentiates a geographical location into a multilayer map (base map) on the basis of certain parameters like elevation, soil type, nutrient status, air temperature, humidity etc. These base maps are integrated with the help of different GIS software. Since, most of the software are costly and a technical know-how is required for their working, they are less used in the farmer's field. Many software packages are introduced by Farms Works TM, AGRIS Corporation, John Deere Precision Farming Group, Agri Logic Inc., Rockwell International, Case Corporation. Different web-based GIS software available are as MapServer, GISNET Server, Planet GIS, Field-level GIS (FIS), Alta Map Server, Geo Server, ESRI ArcGIS StoryMaps. A few advances made in recent past, has improved the usability and functioning of the Web-GIS interfaces (e.g., ArcGIS online) by offering multiple geoprocessing tools as compared to the desktop GIS. These Web-GIS interfaces (e.g., Leaflet) allow the person to use GIS without any accessibility to the proprietary software.

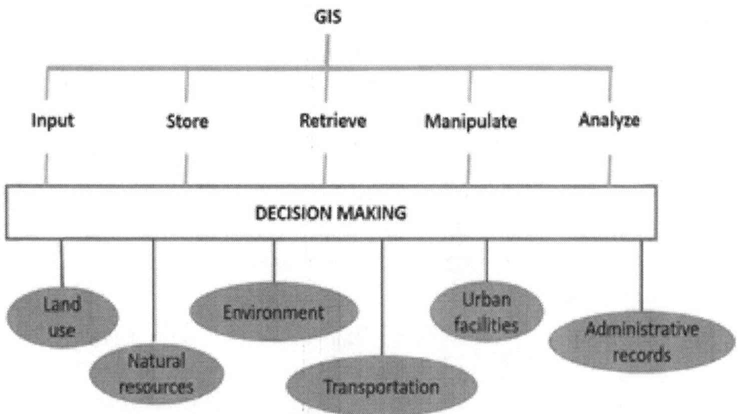


**Fig. 7. GIS in decision making.**

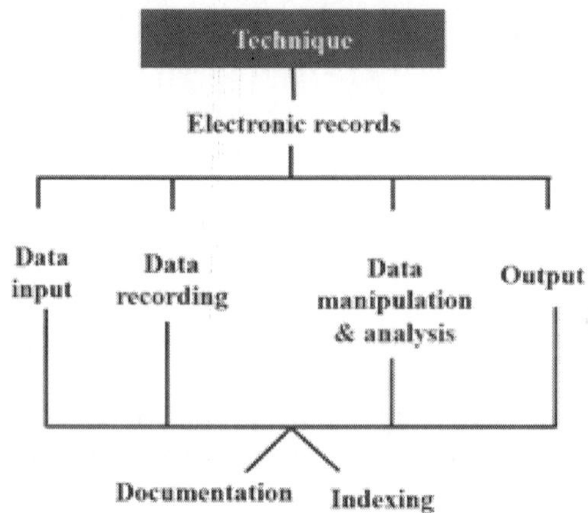


**Fig. 8. Working mechanism of GIS**

The climate change has disturbed the cropping pattern all over the world and the selection of suitable crops as per locality is necessary for ensuring fruit production in future. Analytical hierarchy process (AHP) using GIS has proved to be helpful in the multicriteria spatial analysis for the cultivation of strawberry, pistachio, walnut and almond trees in Portugal conditions. Yedage *et al.*, (2013) conducted multi-criteria land assessment for pomegranate cultivation in Maharashtra using AHP based fuzzy logic method in ArcGIS 10 software. Zabihi *et al.*, (2015) determined the certain critical factors leading to decreased citrus productivity under Iran conditions using MCDM (multi criteria decision making) including AHP and ANP (Analytical network process) in ArcMap software. Under APDMP (Andhra Pradesh Drought Mitigation Project), Srinivasan *et al.*, (2020) classified some arid crops (Aonla, Ber, Custard Apple, Jamun and Mango) into 5 suitable classes (highly, moderately, marginally, currently not and permanently not) after land suitability evaluation in Arc GIS 10.2 environment for the mitigation of the drought in South Deccan Plateau. GIS has been found to be a better tool for the land assessment since it allows managing and analysing the temporal or multi-spatial data at one time.

GIS not only helps in identification of suitable niche for a crop but at varietal level also. In an investigation at CISH, Lucknow Rajan *et al.*, (2004) located suitable zones for the better performance of different cultivars of mango and later, he mapped diversity of guava genotypes in different parts of U.P. using Flora Map. Similarly, crop combinations of different vegetable, fruit and plantation crops were studied in Tamil Nadu using MapInfo and GIS.

Bioenergy systems are in trend worldwide, in these days and the extraction of biofuels from the agricultural commodities/remains is gaining momentum. The discarded fruits, plant wastes or foliage of some fruit plants (citrus, banana etc.) has shown a marked potential for methane and ethanol production. Guerrero *et al.*, (2016) estimated available banana biomass potential in El Oro province of Ecuador for ethanol production using ArcGIS 10 software with network analyst extension approach. Many investigations are conducted on the growing conditions, land assessment, irrigation requirement, yield forecasting, diversity mapping of citrus fruits around the world.

## Remote sensing

It is a science where, information about a plant or any object on the earth surface is gathered without coming in any physical contact with it. This technology monitors things on the earth surface with greater accuracy and precisely through different platforms (Fig.9.). Remote sensing works on the principle of electromagnetic radiations. Every element on the earth when came in contact with light radiations of different wavelength, they emit back radiations of specific wavelength which is detected by the different sensors. This differed response to different wavelengths helps in differentiating in bare soil, vegetation and other features. Sensing of plant canopies provides various insights into plant vegetative parameters. The electromagnetic radiations received by the sensors are converted into a signal (spectral signature) and presented either an image or numeral data. Remote sensing presents a cost-effective strategy for data collection across extensive geographic regions and this information can be used in determining phenological data, crop yield estimation, biotic-abiotic stress forecasting, plant identification, decision making and precision agriculture.

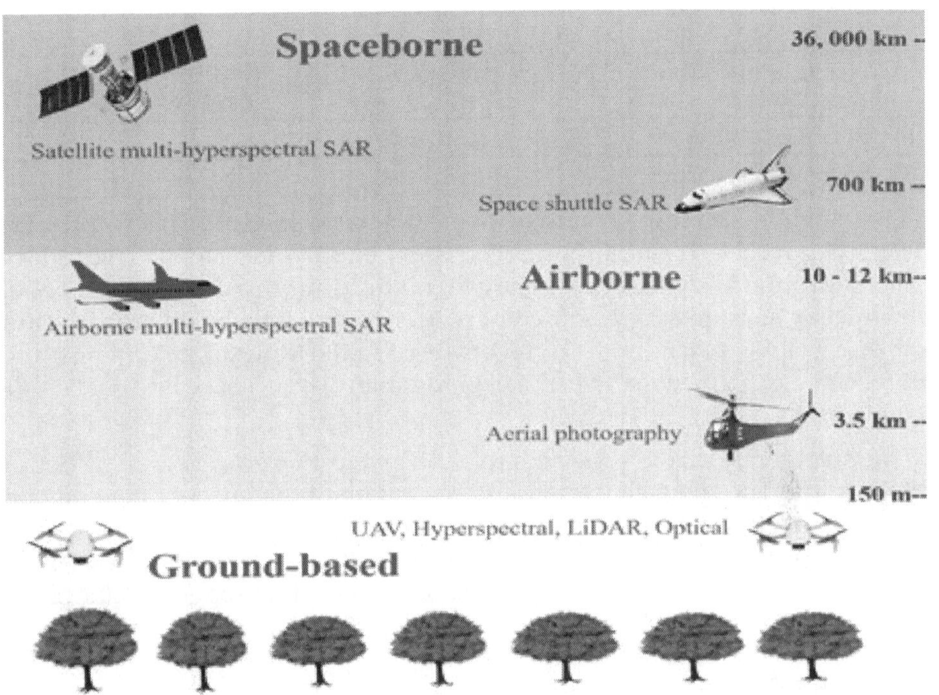


**Fig.9. Platforms of Remote sensing**

**Table 1. Sensing indicators for vegetation indices**

| Parameter | Equation |
|---|---|
| CI. (Chlorophyll. based. Difference Index) | $R_{850}$· - $R_{710}/R_{850}$ - $R_{680}$ |
| CT.VI (Corrected. Transformed Vegetation. Index) | ND VI+0.5 /a bs*(N DV I+0.5) *[√ab s (ND VI + 0.5)] |
| MN.DW.I (Modified Normalized. Difference Water. Index) | ρ G – ρ SW IR1/ ρ G + ρ SW IR1 |
| MS.R (Modified. Red Edge. Simple Ratio.) | $(R_{800}/R_{670})$ -1. / $(\sqrt{R_{800}}/R_{670})$ + 1 |
| ND.VI (Normal. ised Difference. Vege.tation Indices.) | $R_{800}$ - $R_{670}$ / $R_{800}$ + $R_{670}$ |
| OS.AV. I (Optimised. Soil. Adjusted Vegetation. Index) | (1+0.16) * $(R_{800} - R_{670})/R_{800} + R_{670}$ + 0.16) |
| PI (Photosynthesis Index) | $R_{531}$ - $R_{570}/R_{531}$ + $R_{570}$ |
| PRI (Photochemical Reflectance Index) | $R_{570}$ - $R_{531}/R_{570}$ + $R_{531}$ |
| R N (Nitrogen. Index) | $R_{550}$ - $R_{600}/R_{800}$ - $R_{900}$ |
| RV SI (Red .Edge Vegetation. Stress. Index) | (R714 + $R_{752}$/2) - R733 |
| SA VI (Soil Adjusted Vegetation Index) | (ρ NIR – ρ R) * (1 + L)./ (ρ NI.R – ρ R + L) |
| SI PI (Structure. Insensitive Pigment Index) | $R_{800}$ - $R_{445}/R_{800}$ - $R_{680}$ |

| SR (Simple. Ratio) | $R_{800}/R_{670}$ |
|---|---|
| TC AR I (Transformed Chlorophyll. Absorption in Reflection Index) | $3 * [(R_{700} - R_{670}) - O.2 * (R700 - R_{550}) * R_{700}/R_{670})]$ |
| W BI (Water Band Index) | $R_{900}/R_{970}$ |
| W MI (Water Moisture Index) | $R_{1600}/R_{820}$ |

Gravimeter, solid scanner, microwave radiometer, spectrometer, laser distance or water depth meter, optical mechanical scanner, camera, satellite and radar are some of the common sensors used. The geographical information or the spatial data is obtained by sensors at three different platforms divided on the basis of altitude such as a) ground based: upto 50 meters (e.g., Aerial vehicle), b) airplane based: upto 50 km (e.g., Parachute, helicopter, aeroplane) and, c) space based: 100 – 3600 km (e.g., rockets, satellite, shuttle).

The spatial resolution is the sum total of smallest objects that can be detected on the surface. The spatial coverage and resolution of the imagery depends upon both the quantity of images contributing to individual pixels on the sensor and the altitude of the sensor relative to the surface. The image footprint and pixel-size depend on the number of images forming pixels in sensor and distance of the sensor from the ground. The sensors vary in their capacity to detect smaller objects as Sentinel-2 (10 m/pixel), GSD SPOT-6 imagery (1.5 m) and 30 cm in DigitalGlobe (Rabbi *et al.*, 2020).

Plants reflect majority of incident sunlight in NIR (Near Infrared Wavelengths) since it is sensitive to the cell structure. The physiological stage or any disease, stress, nutritional deficiency, injury in the plant can be identified with the help of VIR (visible and infrared) and fluorescence spectroscopy. The reflectance values of the different wavelengths coming from the object are inserted in the equation and the information is obtained for a particular parameter (Table 1).

Remote sensing is divided into different classes (Fig.10.) on the basis of energy source and the range of electromagnetic spectrum.

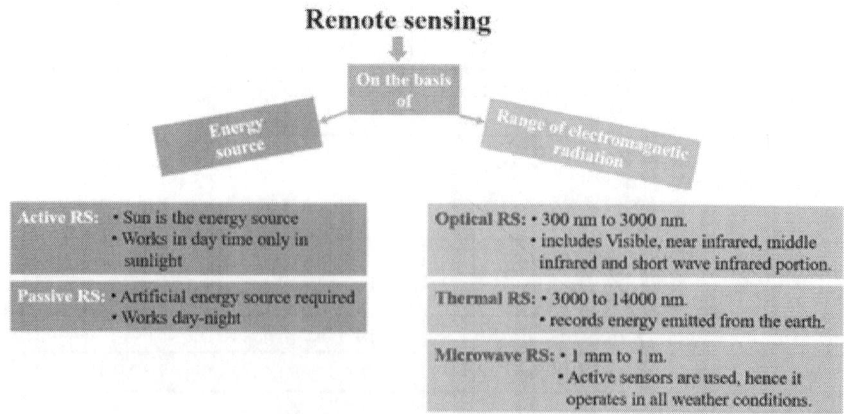


**Fig.10. Types of remote sensing**

## Application of precision farming tools

### Land cover estimation

There is always a great fluctuation in the trends of horticultural crops. Hence, regular statistics of area and production are required for the efficient market planning. The remote sensing applications facilitates synoptic, regular, multi temporal and multi spectral coverage of a land area. Mushtaq and Asima (2014) observed the apple orchards in the major fruit growing belts of the Kashmir, recorded some terrain parameters and signatures were generated using AWIFS and Landsat data. A Digital Elevation Model was created on the basis of the obtained data which characterised the orchards on the basis of slope, aspect and elevation. Sharma and Panigrahy (2010) used high resolution IRS satellite P6 data for mapping apple orchards in Shimla and a plantation database was generated for different blocks of the district. Mango has blanketed a major area in the country and various reports of the mapping of potential mango belts using high resolution remote sensed data have been published. Verma *et al.*, (2020) used multitemporal Landsat 8 OLI images for the area estimation under mango orchards in Lucknow and a decision tree classification (DTC) system has been developed. They found images collected in the February month as best for the orchard classification. Pena and Brenning (2015) used NDVI and NDWI profiles with SITS (Satellite Image Time Series) for classifying area under crops like grape, walnut, plum and almond in central Chile. Area estimation of grapes, citrus, apple, plantation crops had been done in India using IRS LISS III satellite.

### Age and stage identification

The aging and old senile orchards are a major reasons for the reduction in yield and productivity. Remote sensed data can be used to extract the age of orchards using; a) remote sensing images, b) LiDAR point cloud data and c) combination of both. The variation in the spectral or textural characteristics are the basis for the determination of plant age. Chemura *et al.*, (2015) estimated the age of plantation and area occupied by oil palm in Ghana with a regression model using WorldView-2 satellite images. Bignell and Mulo (2018) used LiDAR data and determined the height and age of macadamia orchards in Queensland. Rizeei *et al.*, (2018) did regression analysis using Worldview-3 satellite and LiDAR airborne imaging for counting and estimating age of oil palm plantation in Malaysia. Since, the cultivation and management strategies differ from orchard to orchard, effective determination of plant age is not possible on the basis of textural and structural characters. A higher accuracy (98.14 %) was recorded in determining age of apple orchards in China using Sentinel-2 NDVI data and phenological information.

### Crop inventory

Chaudhari *et al.* (2019) conducted pixel-based hybrid classification for the mapping of mango, banana and citrus orchards. Appropriate bio-window of all three crops were selected in 20 districts Bihar, Uttar Pradesh, Madhya Pradesh and Gujarat.

The analysed data on comparison with state departmental statistics was found to be 80 % accurate and the maps generated were uploaded on BHUVAN, SAC, NRSC and VEDAS portals of Indian space research organisation. However, there are certain limitations in the classification of horticultural crops using conventional per pixel classifiers as a result of huge spatial variability. The spectral signatures of crops are mixed that confuse with similar crops like mango and sapota. In this context, the object-oriented techniques are found to be more promising for improved classification. Hebbar *et al.* (2014) used semi-automated object orientation feature for delineation of banana, mango, citrus, coconut and coffee plantations in different agro-climatic zones using Erdas Objective Imagine software.

## Canopy measurements

The canopy size has a major role in the horticultural crops, since all the cultural operations are based on the canopy volume. Application of normalized difference vegetative indices is common in relation to canopy volume in horticultural plantations of different planting configurations at different growth stages. Efficient role of airborne remote sensing has been reported in precision viticulture for determining vines canopy parameters. The use of ultrasonic and laser sensors has been reported in observing canopy volume, leaf area and foliage density of fruit crops.

## Estimating fruit maturity

Almost all the fruit and vegetable crops are perishable in nature, their developmental stages are highly influenced by environment and the market is regulated by the time and quality of harvest. It is necessary to harvest the fruit at optimum stage to have good shelf life. Remote sensing can identify the maturity non-destructively using techniques like colour mapping, chlorophyll fluorescence, hyper or multi-spectral imaging. Many reports of pre harvest fruit maturity estimation have been published for sweet orange, grapes, apple, canola, mango, banana, peach and strawberry using hyper or multi-spectral imaging. Zhang *et al.*, (2022) reported the use of a laser induced fluorescence based handy sensor for characterising the tropical fruits.

## Yield prediction

Predicting crops yield has a major role in orchard management, market forecasting and accessing the variation in the field. The traditional methods of manual yield estimation were time consuming and did not provide yield variability estimates within an orchard. Automated yield prediction can be done by following certain direct or indirect methods (Fig.11.). Prediction models with yield related traits are developed in indirect methods whereas, direct methods predicts by estimating the statistics of quantity. Different computer vision technologies are there for automated yield mapping of fruit and vegetable orchards. Multispectral imaging-based VI's which are combination of NIR and visible wavebands are well known for yield estimation. Since certain limitations are posed by satellite or aircraft sensing, hence the use of proximal sensors like depth cameras, thermal imaging and certain optical methodologies are more efficient for the yield estimation. The use of vision-based mapping algorithms has been reported in pineapple (Moonrinta *et al.*, 2010) and grapefruit with a success rate of 70 – 80%. Acoustic ultrasonic sensors are long been used from last 3 decades for yield mapping of apple, blueberry, citrus any many other crops.

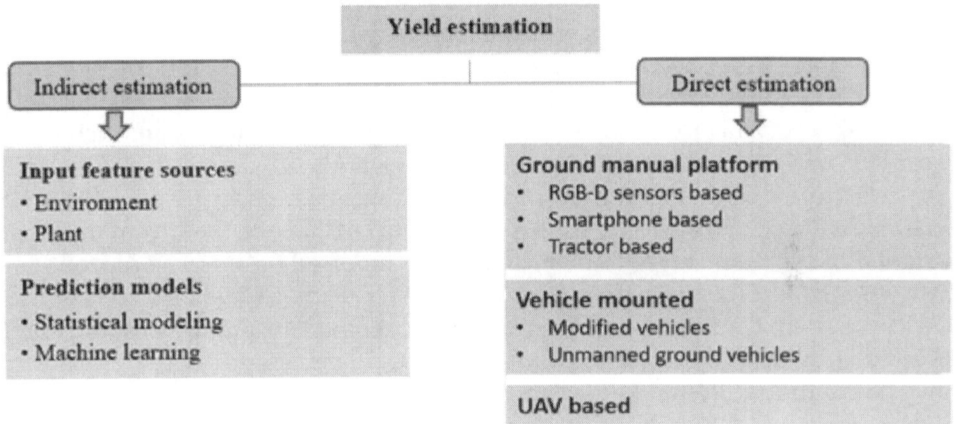


**Fig. 11. Modes of automated yield estimation**

Digital imaging techniques simulate a bas-relief effect which analyses texture on the basis of image enhancement, border detection, spatial fusion and filtering which are reported to be used for yield estimation in various orchards. Image processing-based machine learning vision algorithms yield prediction models have been reported to estimate yield in grapes, mango, citrus, apple. However, more accurate results are obtained with machine learning as compared to image processing (Gongal *et al.*, 2015).

## Robotic harvesting

Most of the horticultural crops require selective harvesting as the fruits mature at different time on the same plant. The harvesting operation is done manually which is labour intensive and time consuming. Since, past three decades around 50 robots have been developed for the harvesting of high value crops. A dual arm robot was developed by Ling *et al.*, (2019) for the detection and localization of tomato fruits with a binocular vision sensor. The success rate of 87.5% was achieved and a harvesting cycle completes in less than 30 s. Mechanical tree shakers are available for the crops in which harvesting of fruits is done at once, but selective harvesting is one of the difficult tasks. However, detection of fruits using deep neural network technology has been reported to be quite successful. The development of crop specific harvesting robots has been going on and several investigations have been reported in litchi, grape, coconut, pear, apple, cherry and kiwifruit.

## Variable rate technology (VRT)

The variable rate or site-specific application is the most important approach of precision farming. The whole field is divided into small zones and the inputs are supplied as per the variation present in the field, so as to reduce the wastage and make efficient use of resources. In conventional spraying practices only 30-40% of chemical hits the target and rest is lost in surrounding and harmful to the biodiversity. Hence, VRT protect the excess wastage and will ensure precision fruit cultivation. It can either be map or sensor based. In map-based technique, the present or past year data is collected through grid sampling, GIS prescription maps for the preparation of irrigation and nutrient or pesticide application schedules. Wang *et al.*, (2021) conducted variable rate fertilization in a pomegranate orchard and recorded 26:54:34 % reduction in N:P:K usage with 41.52% increase in fruit yield as compared to conventional farmers practices. However, in case of sensor based VRT, operations are scheduled according to the real time data obtained from the sensors installed in the field. Vorotnikova *et al.* (2014) reported an improved profitability with reduced fertilizer application and crop losses using a sensor based strawberry advisory system than calendar-based spray management in anthracnose and botrytis affected strawberry farm. Several ultrasonic and laser sensor-based technologies have been reported for advanced spraying in fruit orchards.

**Table 2. Disease diagnosis using certain models in fruit crops.**

| Fruit | Disease | Model | Parameter | Prediction | Reference |
|---|---|---|---|---|---|
| Apple | Scab | WSN | T, H, LT | Disease incidence | Bhargava et al., 2014 |
| Banana | Panama disease | BLR | Multispectral imaging | Disease incidence | Ye et al., 2020 |
| Cherry | Blossom blight | LDA, QDA, PLD, CCT | T, H, W | Disease occurrence | Ilic et al., 2018 |
| Citrus | Gummosis | SVR, MLR | T, H, ST, SM, LAI | Disease severity | Badnakhe et al., 2018 |
| Grape | Downy mildew | RF | MT, RA | Disease severity and incidence | Chen et al., 2020 |
| | Powdery mildew | BN | Weather, cultivar, pathogenic stage | Disease risk | Lu et al., 2020 |
| Grapefruit | Canker | PLSR | Fluorescence spectra | Disease incidence | Saleem et al., 2020 |
| Mango | Anthracnose | RF, SVM | T, H | Disease incidence | Ardila et al., 2020 |
| Orange | Greening, scab, canker | DLNN | T, R, H, imaging | Disease prediction | Kaur and Kaur 2018 |

| Fruit | Disease | Model | Parameter | Prediction | Reference |
|---|---|---|---|---|---|
| Olive | Verticillium wilt | LDA, SVM | Hyperspectral imaging | Severity index (1-5) | Calderón et al., 2015 |
| Oil palm | Ganoderma | ANN | Spectral analysis | Disease severity | Ahmadi et al., 2017 |
| Pomegranate | Bacterial blight, wilt | SVN | T, H | Disease prediction | Nilkanth et al., 2021 |
| Strawberry | Anthracnose | SAM, SDA, CM | Hyperspectral imaging | Disease stage | Yeh et al., 2016 |

ANN: Artificial Neural Network; BLR: Binary Logistic Regression; BN: Bayesian Network; CCT: Compact Classification Tree; CM: Correlation Measure; DLNN: Deep Learning Neural Network; H: Humidity; KNN: K-Nearest Neighbor; LAI: Leaf Area Index; LDA: Linear Discriminant Analysis; LT: Leaf Temperature; MLR: Multi-Linear Regression; MT: Monthly Temperature; PLD: Pseudo Linear Discriminant; PLSR: Partial Least Squares Regression; QDA: Quadratic Discriminant Analysis; RA: Rainfall; RF: Random Forest; SAM: Spectral Angle Mapper; SDA: Stepwise Discriminant Analysis ; SM: Soil Moisture; ST: Soil Temperature; SVR: Support Vector Regression; T: Temperature; W: Wind; WSN: Wireless Sensor Network

**Table 3. Non-destructive spectroscopic determination of pest incidence in fruit crops**

| Fruit | Pest | Wavelength (nm) | Reference |
|---|---|---|---|
| Citrus | Red scale | VIS (410 – 774 nm) | García-Guerra et al., 2021 |
| Cherry | Plum curculio | Vis/SWNIR (550–980 nm) | Xing & Guyer, 2008 |
| Chestnut | Weevil & Totrices | NIR (1100–2300 nm) | Moscetti et al., 2014 |
| Jujube | Leaf roller | Vis/SWNIR (400–1000 nm) | Wang et al., 2010 |
| Mango | Fruit fly | N.I.R (4.00 – 10.00 nm) | Saranwong et al., 2011 |
| Pomegranate | Carob. moth | V.is/SW.NI.R (500–1000 nm) | Jamshidi et al., 2019 |
| Wild. blueberry | Maggot. fly | V.is/SW.NIR (6.50–11.00 nm | Xing & Guyer, 2008 |

## Disease forecasting

The occurrence of any disease in the orchards is highly influenced by the climatic conditions. The quantification of disease incidence, prevalence, intensity and severity is done by the estimation of plant symptoms by different ways. The data for this quantification can be obtained from weather stations, UAVs, remote sensing and through online web services. Several crop specific disease prediction models (ANN, CNN, SVM, LDA, SVR, DT etc.) have been developed differing on the basis of input parameters as; a) weather data b) image processing and c) data from heterogenous resources. A brief detail of some of the models used for fruit disease diagnosis has been discussed in Table 2.

## Pest forecasting

Insect-pests pose a major threat to all the fruit crops and are responsible for serious decline in the production all around the globe. Targeted pest monitoring is essential to check the pest population and to make a making a sustainable pesticide application. The sensor-based technologies identify the insect population by observing the changes in plant biomass, plant pigments and skeletonizing of the leaves in the susceptible areas. Several colour based automated counting devices are available for determining the morphological features of some of the large sized insects (Arthropods) but, unavailable for small sized insects. However, application of spectroscopic techniques has found to be efficient in the automatic determination of pests even in non-destructive stages (Table 3). The use of acoustics and hyperspectral imaging has been reported for detecting the codling moth infestation in apple orchards. Visible light sensing (computer vision) techniques has been investigated for assessing thrips, scales, psyllid and medfly infestation in citrus orchards. Vanegas *et al* (2018) demonstrated the use of RGB, multispectral and hyperspectral sensors with UAVs for pest surveillance in the vineyards. Since, identification of minute insects was difficult, Wang *et al*. (2022) proposed a deep learning approach (MPest-RCNN), offering efficient recognition, counting and reorganization features with success rate of 99.11%–99.50%.

## Status of precision horticulture in India

Recently, Ministry of Agriculture, Government of India has established Mahalanobis National Crop Forecast Centre for operationalizing the space and AI technologies in agriculture sector. A project C.H.A.M.A.N. (Coordinated. Horticulture Assessment & Management using geoinformatics) had been initiated under Mission for Integrated Development of Horticulture (M.I.D.H.) with the objectives of developing crop inventories and geospatial studies. It is running in collaboration with ISRO, NHRDF, ICAR, IMD, state remote sensing centres and 12 state horticulture departments. The project initiated in two phases, first starting from 2014-2017, in which area assessment of 7 major horticultural crops was done in some selected districts. The developmental and management plans were prepared using geo-spatial techniques in the 2$^{nd}$ phase (2018-2020). The area and production data of selected crops were mapped on the Bhuvan Geo-portal (Fig.12.) and is freely available in public domain.

Though many advancements have been made around the world, but there are still many limitations in India in proper adoption of precision farming such as uneven land holdings, lack of local expertise, area coverage, data management, scale biasness and high investment. Therefore, there is a need of developing multidisciplinary committees involving scientists, engineers or economists of different field and awareness programs should be started to make farmers aware from emerging technologies by conducting pilot studies.

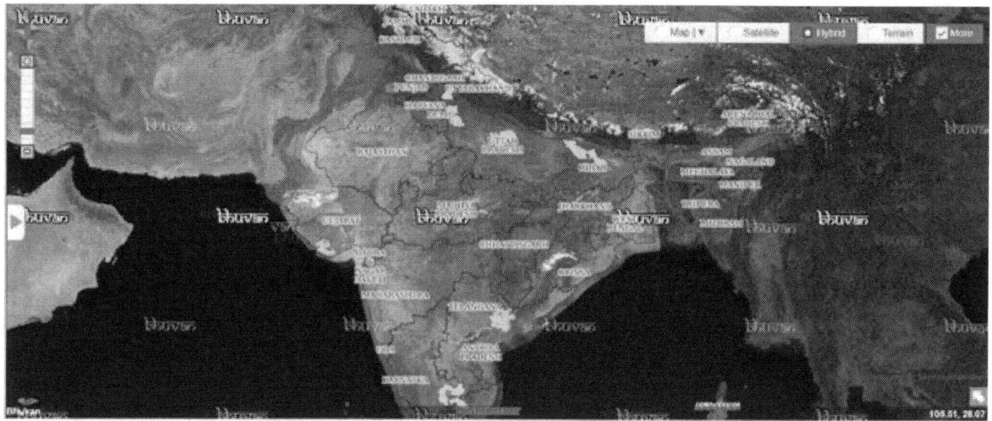

**Fig.12. Prominent mango, citrus and banana belts mapped on Bhuvan geo-portal (https://bhuvan-app1.nrsc.gov.in/chaman/)**

## Conclusion

The continuous use of conventional or some faulty practices in the orchards has resulted in a significant land degradation, nutrient losses and lower production of horticultural crops than their marked potential. Non-availability of yield or inventory records pose various restrictions in framing efficient farm plans. The concept of mechanization, artificial intelligence and sensors has revolutionized the horticulture sector and made it easy for growers to predict and manage crops from various biotic & abiotic stresses. The precision approach is emerging with new technologies every day but is still at nascent stage in many developing countries like India. The approach is beneficial in minimizing production costs, environmental impacts with improved input use efficiency, crop production and is the need of the hour.

## References

Ahmadi, P., Muharam, F.M., Ahmad, K., Mansor, S., Seman, I.A. (2017). Early Detection of Ganoderma Basal Stem Rot of Oil Palms Using Artificial Neural Network Spectral Analysis. *Plant Dis*. 2017, 101, 1009–1016. https://doi.org/10.1094/PDIS-12-16-1699-RE

Anonymous. (2024). Area and Production of Horticulture crops for 2023-24 (1st Advance Estimates). *Ministry of Agriculture & Farmers Welfare. https://agriwelfare.gov.in/en/StatHortEst*

Ardila, C. E. C., Ramirez, L. A., & Ortiz, F. A. P. (2020). Spectral analysis for the early detection of anthracnose in fruits of Sugar Mango (*Mangifera indica*). *Computers and Electronics in Agriculture, 173*, 105357. https://doi.org/10.1016/j.compag.2020.105357

Badnakhe, M. R., Durbha, S. S., Jagarlapudi, A., & Gade, R. M. (2018). Evaluation of Citrus Gummosis disease dynamics and predictions with weather and inversion based leaf optical model. *Computers and Electronics in Agriculture, 155*, 130-141. https://doi.org/10.1016/j.compag.2018/10.009

Bhargava, K., Kashyap, A., & Gonsalves, T. A. (2014, February). Wireless sensor network based advisory system for Apple Scab prevention. In *2014 Twentieth national conference on communications (NCC)* (pp. 1-6). IEEE. https://doi.org/10.1109/NCC.2014.6811263

Bignell, G., and Mulo, S. (2018). Using LIDAR to measure the impact of tree height on yield and kernel recovery in mature macadamia orchards. y Grant Bignell and Shane Mulo of Horticulture and Forestry Science, Agri-Science Queensland, Department of Agriculture and Fisheries.

Calderón, R., Navas-Cortés, J. A., & Zarco-Tejada, P. J. (2015). Early detection and quantification of Verticillium wilt in olive using hyperspectral and thermal imagery over large areas. *Remote Sensing, 7*(5), 5584-5610. https://doi.org/10.3390/rs70505584

Chaudhari, K., Nishant, N., Upadhyay, G., More, R., Singh, N., Vyas, S. P., & Bhattacharya, B. K. (2019). Crop inventory of orchard crops in India using remotely sensed data. *International Archives of the Photogrammetry, Remote Sensing & Spatial Information Sciences*. https://doi.org/10.5194/isprs-archives-XLII-3-W6-269-2019

Chemura, A., van Duren, I., & van Leeuwen, L. M. (2015). Determination of the age of oil palm from crown projection area detected from WorldView-2 multispectral remote sensing data: The case of Ejisu-Juaben district, Ghana. *ISPRS journal of photogrammetry and remote sensing, 100*, 118-127. https://doi.org/10.1016/j.isprsjprs.2014.07.013

Chen, M., Brun, F., Raynal, M., & Makowski, D. (2020). Forecasting severe grape downy mildew attacks using machine learning. *PLoS One, 15*(3), e0230254. https://doi.org/10.1371/journal.pone.0230254

García-Guerra, C. E., Burgos-Fernández, F. J., Canals, E., Díaz-Doutón, F., Zaragoza, A., Virgili, A., & Vilaseca, M. (2021, November). Use of spectral information for red scale pest control. In *Color and Imaging Conference* (Vol. 2021, No. 29, pp. 253-257). Society for Imaging Science and Technology. https://doi.org/10.2352/issn.2169-2629.2021.29.253

Gongal, A., Amatya, S., Karkee, M., Zhang, Q., & Lewis, K. (2015). Sensors and systems for fruit detection and localization: A review. *Computers and Electronics in Agriculture, 116*, 8-19. https://doi.org/10.1016/j.compag.2015.05.021

Guerrero, A. B., Aguado, P. L., Sánchez, J., & Curt, M. D. (2016). GIS-based assessment of banana residual biomass potential for ethanol production and power generation: a case study. *Waste and biomass valorization, 7*(2), 405-415. https://doi.org/0.1007/s12649-015-9455-3

Hebbar, R., Ravishankar, H. M., Subramoniam, S. R., Uday, R., & Dadhwal, V. K. (2014). Object oriented classification of high resolution data for inventory of horticultural crops. *The International Archives of Photogrammetry, Remote Sensing and Spatial Information Sciences, 40*(8), 745. https://doi.org/0.5194/isprsarchives-XL-8-745-2014

Ilic, M., Ilic, S., Jovic, S., & Panic, S. (2018). Early cherry fruit pathogen disease detection based on data mining prediction. *Computers and electronics in agriculture, 150*, 418-425. https://doi.org/10.1016/j.compag.2018.05.008

ISRO, 2022. Indian Space Research Organisation, Indian Regional Navigation Satellite System (IRNSS): NavIC. 24 March 2022. https://www.isro.gov.in/irnss-programme

Jamshidi, B., Mohajerani, E., Farazmand, H., Mahmoudi, A., & Hemmati, A. (2019). Pattern recognition-based optical technique for non-destructive detection of Ectomyelois ceratoniae infestation in pomegranates during hidden activity of the larvae. *Spectrochimica Acta Part A: Molecular and Biomolecular Spectroscopy, 206*, 552-557. https://doi.org/10.1016/j.saa.2018.08.059

Kaur, K., & Kaur, M. (2018). Prediction of plant disease from weather forecasting using data mining. *Int. J. Future Revolut. Comput. Sci. Commun. Eng, 4*, 685-688.

Ling, X., Zhao, Y., Gong, L., Liu, C., & Wang, T. (2019). Dual-arm cooperation and implementing for robotic harvesting tomato using binocular vision. *Robotics and Autonomous Systems, 114*, 134-143. https://doi.org/10.1016/j.robot.2019.01.019

Lu, W., Newlands, N. K., Carisse, O., Atkinson, D. E., & Cannon, A. J. (2020). Disease risk forecasting with Bayesian learning networks: Application to grape powdery mildew (*Erysiphe necator*) in Vineyards. *Agronomy, 10*(5), 622. https://doi.org/10.3390/agronomy10050622

Moonrinta, J., Chaivivatrakul, S., Dailey, M. N., & Ekpanyapong, M. (2010). Fruit detection, tracking, and 3D reconstruction for crop mapping and yield estimation. In *Proceedings of the International Conference on Intelligent Robots and Systems* (pp. 1181–1186). Singapore: IEEE. https://doi.org/10.1109/ICARCV.2010.$_{570}$7436

Moscetti, R., Haff, R. P., Saranwong, S., Monarca, D., Cecchini, M., & Massantini, R. (2014). Nondestructive detection of insect infested chestnuts based on NIR spectroscopy. *Postharvest Biology and Technology, 87*, 88-94. https://doi.org/10.1016/j.postharvbio.2013.08.010

Mushtaq, G., & Asima, N. (2014). Estimation of apple orchard using remote sensing and agro-meteorology land based observation in Pulwama district of Kashmir valley. *Int. J. Remote sensing and Geo. Sci, 3*(6), 2319-3484.

Nilkanth, S., Dange, Y., Naikwade, S., Sayyad, Z., & Chandgude, A. (2021). Disease Prediction of Pomegranate using Data Mining. *International Journal of Research in Engineering and Science.* 9 (6):42-46

Pena, M. A., & Brenning, A. (2015). Assessing fruit-tree crop classification from Landsat-8 time series for the Maipo Valley, Chile. *Remote Sensing of Environment, 171*, 234-244. https://doi.org/10.1016/j.rse.2015.10.029

Rabbi, J., Ray, N., Schubert, M., Chowdhury, S., & Chao, D. (2020). Small-object detection in remote sensing images with end-to-end edge-enhanced GAN and object detector network. *Remote Sensing, 12*(9), 1432. https://doi.org/10.3390/rs12091432

Rajan, Sh., Mathur, P.N., Kumar, R., Pathak, R.K. & Yadava, L.P. (2004). GIS application for managing mango genetic resources. In: Proceedings of the First Indian Horticulture Congress on "Improving productivity, quality, post-harvest management and trade in horticultural crops " 6-9 November, New Delhi. pp.69.

Rizeei, H. M., Shafri, H. Z., Mohamoud, M. A., Pradhan, B., & Kalantar, B. (2018). Oil palm counting and age estimation from WorldView-3 imagery and LiDAR data using an integrated OBIA height model and regression analysis. *Journal of Sensors, 2018*. https://doi.org/10.1155/2018/2536327

Saleem, M., Atta, B. M., Ali, Z., & Bilal, M. (2020). Laser-induced fluorescence spectroscopy for early disease detection in grapefruit plants. *Photochemical & Photobiological Sciences, 19*(5), 713-721. https://doi.org/10.1039/c9pp00368a

Saranwong, S., Haff, R. P., Thanapase, W., Janhiran, A., Kasemsumran, S., & Kawano, S. (2011). A feasibility study using simplified near infrared imaging to detect fruit fly larvae in intact fruit. *Journal of near infrared spectroscopy, 19*(1), 55-60. https://doi.org/10.1255/jnirs.915

Sharma, A., & Panigrahy, S. (2010). Apple orchard characterization using remote sensing and GIS in Shimla district of Himachal Pradesh. National Horticulture Technology Mission Programme.

Srinivasan, R., Hegde, R., Srinivas, S., Niranjana, K. V., & Maddileti, N. (2020). Arid fruit crops suitability evaluation for drought mitigation using remote sensing and GIS techniques in South Deccan Plateau, Andhra Pradesh, India. *Climate Change and Environmental Sustainability, 8*(1), 13-23. https://doi.org/10.5958/2320-642X.2020.00002.2

Vanegas, F., Bratanov, D., Powell, K., Weiss, J., & Gonzalez, F. (2018). A novel methodology for improving plant pest surveillance in vineyards and crops using UAV-based hyperspectral and spatial data. *Sensors, 18*(1), 260. https://doi.org/10.3390/s18010260

Verma, H. C., Ahmed, T., & Rajan, S. (2020). Mapping and Area Estimation of Mango Orchards of Lucknow Region by Applying Knowledge Based Decision Tree to Landsat 8 OLI Satellite Images. *Int J Innov Technol Explor Eng, 9*, 3627-35. https://doi.org/10.35940/ijitee.B8109.019320

Vorotnikova, E., Borisova, T., & VanSickle, J. J. (2014). Evaluation of the profitability of a new precision fungicide application system for strawberry production. *Agricultural Systems, 130,* 77-88. https://doi.org/10.1016/j.agsy.2014.06.006

Wang, F., Zhao, C., Yang, H., Jiang, H., Li, L., & Yang, G. (2022). Non-destructive and in-site estimation of apple quality and maturity by hyperspectral imaging. *Computers and Electronics in Agriculture, 195,* 106843. https://doi.org/10.1016/j.compag.2022.106843

Wang, J., Nakano, K., Ohashi, S., Takizawa, K., & He, J. G. (2010). Comparison of different modes of visible and near-infrared spectroscopy for detecting internal insect infestation in jujubes. *Journal of Food Engineering, 101*(1), 78-84. https://doi.org/10.1016/j.jfoodeng.2010.06.011

Wang, W., Cai, L., Peng, P., Gong, Y., & Yang, X. (2021). Soil Sampling Spacing Based on Precision Agriculture Variable Rate Fertilization of Pomegranate Orchard. *Communications in Soil Science and Plant Analysis, 52*(20), 2445-2461. https://doi.org/10.1080/00103624.2021.1928175

Xing, J., & Guyer, D. (2008). Comparison of transmittance and reflectance to detect insect infestation in Montmorency tart cherry. *Computers and Electronics in Agriculture, 64*(2), 194-201. https://doi.org/10.1016/j.compag.2008.04.012

Ye, H., Huang, W., Huang, S., Cui, B., Dong, Y., Guo, A. & Jin, Y. (2020). Recognition of banana fusarium wilt based on UAV remote sensing. *Remote Sensing, 12*(6), 938. https://doi.org/10.3390/rs12060938

Yedage, A. S., Gavali, R. S., & Jarag, A. P. (2013). Land Assessment for Horticulture (Pomegranate) Crop Using GIS and Fuzzy Decision Analysis in the Sangola Taluka of Solapur District. *Int J Remote Sens GIS, 2*(3), 104-113.

Yeh, Y. H., Chung, W. C., Liao, J. Y., Chung, C. L., Kuo, Y. F., & Lin, T. T. (2016). Strawberry foliar anthracnose assessment by hyperspectral imaging. *Computers and Electronics in Agriculture, 122,* 1-9. https://doi.org/10.1016/j.compag.2016.01.012

Zhang, Q., Sun, Y., & Svanberg, S. (2022, February). Compact fluorosensor for close-range remote-sensing characterization of fruits. In *International Conference on Optoelectronic Materials and Devices (ICOMD 2021)* (Vol. 12164, pp. 16-19). SPIE. https://doi.org/10.1117/12.2628584

CHAPTER 6

# AGRIVOLTAICS, CURRENT STATUS AND ITS FUTURE PROSPECTS

**J. Suresh Kumar[1], R. Sarvanan[1], S. Sunitha[1]**
**Himani B. Patel[2], Adarsh A[3]**

*[1] ICAR-Central Tuber Crops Research Insitute, Sreekariyam, Kerala, India*
*[2] ASPEE College of Horticulture, Navsari Agricultural University, Navsari, Gujarat, India*
*[3] CSIR-National Institute for Interdisciplinary Science and Technology, Pappanamcode, Thiruvananthaurama, Kerala, India*
*Corresponding author: suresh.kumar.jagu@gmail.com*

Fossil fuel burning and nuclear reactors are the major sources for power generation in the world. Bio-fuels are claimed to be a promising substitute to fossil fuel, but growing biomass crops on arable land raises serious concerns about food security. Hence production of power by wind mills was initiated, but it couldn't meet the supply and demand gap. Hence other renewable energy like solar power generation was focused. Development of huge ground mounted solar panels in arable land lead to strong protests by farmers in developed countries. Hence an idea of Agrivoltaics was started in 1980s.

Agrivoltaics is the concept of producing crop and solar power in the same piece of land. The concurrent land is used for dual purpose namely agriculture and photovoltaic power generation which offers a prospective solution to the competition for land resources between agriculture and renewable energy. This concept is very useful for semi-arid and arid regions where sufficient lighting is available for tapping solar energy, and the partial shade by the solar panels will help in reducing the excess light, heat and drought stress on the crop plants, increases water and nutrient use efficiency. The research on the agrivoltaics started few decades ago and is still at nascent stage with various pilot scale studies. Apart from production of crops and solar energy, agrivoltaics is effective in meeting different sustainable developmental goals (SDGs) of United Nations (UN).

### SDG 1: No Poverty

Agrivoltaics promotes stable and supplementary income sources by power generation and crop yields. It stabilizes financial stability of the farmers and industry.

### SDG 2: Zero Hunger

Apart from solar energy, increase in agricultural (crop and animal) production will enhance food security.

### SDG 7: Affordable and Clean Energy

Agrivoltaics will promote the production of clean and affordable energy (green energy) through solar power. It will reduce the dependence on fossil fuel.

### SDG 8: Decent Work and Economic Growth

This approach creates employment opportunities, stimulates economic growth and offers decent work prospects to different communities. It reduces dependency on grid electricity which is often unreliable and lacks connectivity in many rural areas, allowing rural population to use power generated on-site to meet various needs at low or no cost.

### SDG 9: Industry, Innovation, and Infrastructure

It integrates agriculture and solar energy production in the same piece of land, fostering innovation and sustainable development in both sectors.

### SDG 11: Sustainable Cities and Communities

Agrivoltaics promotion on urban landscape by roof top gardening under the solar panels will facilitates the creation of resilient and sustainable communities. Its decentralized nature enable energy production closer to consumption centers, reduces transmission losses and enhance energy security, reduces food miles.

### SDG 13: Climate Action by actively reducing greenhouse gas emissions

It is a carbon credit technology. Unlike the power production by fossil fuels (thermal energy, nuclear energy), agrivoltaics based solar power generation will not create any green house gases. It mitigates climate change and building resilience to its impacts.

## Agrivoltaics *vs.* Traditional Farming

Traditional farming is practiced since centuries ago to produce the crop and live stock. The implements used are from primitive form of hand tools, animal drawn implements to advanced mechanized equipments aiming to maximize crop and animal yields and ensure food security.

Agrivoltaics represents a symbiotic approach combining agriculture with solar energy production. By co-locating solar panels and crops/animal, agrivoltaic systems offer multiple benefits for food and energy production.

**Potential Benefits and Risks of Agrivoltaics:** The pros and cons of agrivoltaics are highly contextual, depending on various factors like choice of crop and animal species, climatic conditions, design and configuration of the system, etc. Agrivoltaics is at nascent stage, not all aspects are fully studied, many of the impacts either theoretical or contested without concrete scientific evidences. Summarizing the intended pros and cons of agrivoltaics based on the available information is given in Table 1.

**Table1. Pros and Cons of agrivoltaics**

| Pros | Cons |
|---|---|
| **1. Increased land productivity:** | |
| The main benefit of agrivoltaics is an increase in land-use efficiency, leading to higher total output from a given piece of land compared to farming or energy production separately (Trommsdorff et al., 2022). If the crop and power were produced separately, ~1.86 times the total land would be required. | Initial investment is very high becouse; it requires a significant upfront investment for solar panel installation, system setup, and infrastructure development. Multiple studies reported higher LER values (greater than one), indicating better land-use efficiency (Guerin, 2019; Kostik et al., 2020; Trommsdorff et al., 2022). But LER assigns equal value to agriculture and power production yields. An increase in one may compensate for a reduction in the other. For instance, setting up agrivoltaics on fertile land may significantly reduce crop yield, but the LER can exceed one through optimized power production. Productivity benefits need to be assessed relative to the bigger-picture needs for food and energy production, prioritizing food above energy as essential for human survival. |
| **2. Symbiotic relationship** | |
| Shading may improve the yields of some crops. | But it is not same for all crops. It depends on various factors namely the species of crops selected for the study, panel density and climatic conditions. The symbiotic and mixed results were reported by Gonocruz et al., 2021; Homma et al., 2016; Trommsdorff, 2021; Weselek et al., 2019. Hence there is a need of optics based approach for uniform light spreading under the panels for cultivation of shade intolerant crops. |
| Congenial microenvironment created by crops may improve the panel efficiency | The impact of crop on the efficiency of PV panel performance/efficiency is not well studied. The reduced panel temperatures were reported by Santra et al., 2017; Barron-Gafford et al., 2019; CIAL, 2021; Johansson et al., 2022; Williams et al., 2023. It may not apply to all climatic regions and needs more scientific investigation. |
| Agricultural activities in agrivoltaic land can reduce the incidence of weed menace and helps in reduction of land maintenance.<br>Crop can protect PV panels from weather hazards like hailstorms and decreases dust erosion and deposits on the panel. | There may be benefits for land maintenance; there are associated risks from electric wiring on farmers' safety and the impact on day-to-day operations in agriculture (SolarPower Europe, 2021), but, these have not been fully studied. |

## Agrovoltaics studies

### Agrivoltaics in arable land

**Viticulture agrivoltaic system in France**: The approach of dynamic agrivoltaic (DAV) systems over grape orchard with overhead PV was studied at the Vaucluse department of southeastern France in 2019-2020. The study reported that the enhanced plant growth and water productivity, increased water use efficiency and grape quality. Under DAV in hot weather controlled shading reduced the temperature of vines up to 5°C. Apart from heat, DAV thermal blanket protects the plants against harmful spring frost during bud burst. About 40% reduction in potential evapo-transpiration (PE) reduced the amount of water used up to 34%. The grapes produced had higher anthocyanins (up by 10% to 15%), fruits with 15% more acidic than control (Catherene, 2020).

**Dynamic agrivoltaic systems over grape orchard in France**
***Source*: Gape-based agrivoltaics italy - Google Search**

**Viticulture agrivoltaic system in India:** A study was modelled to assess the energy aspects and expected output per acre of farmland with the dual use of land for crop and power in Nasik district of Maharashtra. PV panels were installed in the area available between the trellis on a grape farm. The electricity generation potential was determined per unit area and economic benefits for the cultivators were also quantified. The study shows that using the agrivoltaic system model, the economic value of the farms may increase more than 15 times compared to conventional farming alone and could maintain approximately the same grape production per unit area (Malu et al., 2017).

## Agrivoltaic system in industry areas

**Vegetable Agrivoltaic system in CIAL, Kochi:** Successful agrivoltaic farming started in Cochin international airport (CIAL) in an area of 20 acres in 2020-2022. The water used to clean PV panels is utilized to irrigate the crops between the PV arrays. This cropping modifies the micro-climates underneath PV modules by reducing the temperature, which results in increasing efficiency in power generation. Being a high rainfall area, the crop coverage in between PV arrays will check the erosion of soil and thus reduce the dust load on the PV module. Another advantage is that it reduces the weed population underneath the PV panel mounts. Apart from the crop yield, CIAL PV plants produce 1.6 lakh units of power a day, which is above the daily consumption of 1.3 lakh units.

**Vegetable Agrivoltaic system in CIAL, Kochi in India.**
**Source: cial agrivoltaics kochi - Google Search.**

## Agrivoltaics in dry tracts

**ICAR-CAZRI, Jodhpur:** A study was conducted at arid regions of Rajasthan at ICAR-Central Arid Zone Research Institute, Jodhpur in India. It was observed that about 49% land area of a solar PV installation system can be used to cultivate crops, which are otherwise left as fallow (Santra et al., 2020). The crops grown in interspaces area between PV arrays were moong bean, moth bean and cluster bean. It has been found that 1400 litres of rainwater could be harvested per kW of agrivoltaic system. The stored water is used for irrigation in the *Rabi* season. When compared to the income from PV-generated electricity, income from the agricultural components were comparatively less, but has several environmental and societal benefits namely: (i) It will help in effective utilization of precious resource of scarce rainwater of arid regions, (ii) It will reduce and control soil erosion through wind action and thus reduce the dust accumulation on PV modules, (iii) It will improve microclimate surrounding the PV module and thus helps in optimum generation of electricity from PV module, (iv) It will improve land equivalent ratio (LER) etc. The study confirmed that the annual income from PVgenerated electricity has been estimated Rs 6000/- per kW installation, whereas the cost involved for installation was Rs 49,840/-, and thus, the breakeven period of the system is about 9–10 years. This breakeven period can be reduced if cash crops are used for cultivation.

### Agrivoltaics in urban space

**Roof top agrivoltaics:** Scientific studies on rooftop agrivoltaics are at conceptual stage only when compared to open field agrivoltaics. Rooftop agrivoltaics have the potential to revolutionize urban farming across the globe under the present fast growing urban scape. It is like the multifunctionality of agroforestry, which perhaps revolutionized rural farming in the 1980s and 1990s. The multifunctionality of rooftop agrivoltaics can provide parallel co-benefits in urban spaces. Agroforestry has the co-benefits of growing food, sequestering carbon, and providing other ecosystem services in the same space – rooftop agrivoltaics is an urban analog where plants evaporatively cool solar panels, and solar panels partially shade plants in the high temperature, water-limiting space on green roofs. The protection from solar panels slows the water-use rate of the plants below them, reducing the drought and heat stress of food crops.

**A model of Hinren Agri-PV Rooftop (APVRT) System, Bangalore.**
**Source: Agro-PV -Hinren Engineering, Bangalore**

## Different Aspects of Agrivoltaics

### Technological Innovations in Agrivoltaics

The upcoming field of agrivoltaics is a testament to how technological innovation can drive sustainable coexistence between renewable energy production and agriculture. This interdisciplinary approach, blending photovoltaic (PV) systems with crop/animal component, is not just about placing solar panels over farmland. It's about leveraging cutting-edge technologies to create synergies between sunlight capture and agricultural productivity, thereby transforming the way we think about and utilize land resources. The technological advancements that are propelling agrivoltaics into the spotlight of sustainable development are discussed under.

### Semi-Transparent and Bifacial Panels

Semi-transparent solar panels are at the forefront of agrivoltaic innovation. These panels allow sunlight to pass through them, which is essential for the photosynthesis process in plants, while simultaneously generating electricity. This translucency is achieved through the use of selective light filtration technologies, which can target specific wavelengths of light that are most beneficial for plant growth and allow them to pass through to the crops below.

Bifacial solar panels, which capture sunlight from both their top and bottom surfaces, offer another layer of efficiency. By collecting reflected light from the ground and the direct sunlight from above, these panels can produce significantly more energy than traditional single-faced panels, making them especially suitable for agrivoltaic systems where ground reflection can be enhanced by the underlying vegetation.

### Dynamic Solar Tracking Systems

Dynamic solar tracking system is the advanced method of capturing solar light. In this system it adjust the orientation of solar panels throughout the day to follow the sun's path, maximizing energy capture. In agrivoltaics, tracking systems can be designed to also consider the light requirements of the crops beneath, optimizing the balance between shade and sunlight exposure to enhance plant growth without compromising solar efficiency.

### Smart Farming Integration

### IoT and Sensor Networks

The integration of Internet of Things (IoT) technology and sensor networks is crucial for the real-time monitoring and management of agrivoltaic systems. Sensors can measure a wide range of environmental conditions, including soil and atmospheric moisture and temperature, sunlight intensity, and more. This data, when processed through advanced algorithms, can inform precise irrigation, fertilization, and crop management practices, ensuring optimal growth conditions while conserving resources with more profits.

### AI and Predictive Analytics

Artificial Intelligence (AI) and machine learning algorithms are explored to predict and optimize both crop yields and energy production. By analyzing data from sensor networks, weather forecasts, and historical crop performance, AI can provide actionable insights for farmers, such as the best planting schedules and crop varieties for specific agrivoltaic configurations. Furthermore, AI can optimize energy production by predicting solar output and adjusting panel angles in real-time through dynamic tracking systems.

### Energy Storage and Grid Integration

As agrivoltaics scale, the question of how to store and manage the generated energy becomes increasingly important. Advanced energy storage solutions, including next-generation batteries and super capacitors, are being developed to store excess energy produced during peak sunlight hours. This stored energy can then be used to power farm operations after sunset or be fed into the grid, contributing to the overall resilience and stability of local energy networks.

Moreover, smart grid technologies enable the seamless integration of agrivoltaic systems into existing energy infrastructures. These technologies allow for two-way communication between energy producers and consumers, facilitating efficient energy distribution and consumption based on real-time demand and supply dynamics.

## Economic Analysis of Agrivoltaic Systems

Agrivoltaics presents a compelling solution to the dual challenges of energy sustainability and agricultural efficiency. But the in-depth economic analysis and different factors effecting market potential, and strategic considerations that shape the financial landscape of agrivoltaic solutions are summarized here under.

### Capital Expenditure and Operational Expenditure

The upfront costs for agrivoltaic systems include solar PV installation, modifications to agricultural infrastructure, and potential land adjustments. Economies of scale and technological advancements are gradually reducing these costs, enhancing investment appeal to implementation of more such units in different agro ecological regions.

Operational costs are often lower in agrivoltaic systems due to reduced water usage, lower energy costs for irrigation, and potential decreases in pest management expenses due to the microclimate alterations under PV panels. But it is depends on the model followed and crop chosen.

### Payback Period

The early period to reach break even point in agrivoltaic projects is influenced by several factors including electricity generation, crop/animal yield increase, and savings on water and other inputs. The payback period can vary widely but it is generally becoming more favorable with increased efficiency and declining solar costs. Detailed case studies, such as those conducted in regions with high insolation and agricultural output, have shown payback periods becoming increasingly competitive when compared to traditional farming or standalone solar projects as mentioned in the previous models.

### Risk mitigation, Economic Resilience

Agrivoltaics offer an economic risk reduction by providing a dual income stream by energy sales and agricultural produce. This diversification is particularly valuable in the face of volatile agricultural markets and changing climatic scenarios.

The shading provided by solar panels can protect crops from extreme weather conditions, potentially reducing crop loss and insurance costs, thereby adding an element of resilience to farming operations.

**Strategic Economic Implications:**

a. **Land Use Efficiency and Zoning Regulations**: By increasing the productivity of a given land area, agrivoltaics can alleviate pressures on land resources, which is a critical factor in densely populated or ecologically sensitive regions. However, zoning regulations and land use policies must evolve to accommodate and encourage the dual use of agricultural lands for energy production.

b. **Technological Innovation and System Design**: The net income from agrivoltaic systems are closely dependant on technological advancements in solar PV and agricultural practices. Innovations such as dynamic solar tracking, transparent solar panels, and optics based light management under panels, and precision agriculture techniques can enhance both energy output and yield from crop and animal components, further improving economic returns.

c. **Supply Chain and Infrastructure Development**: The growth of agrivoltaics can spur developments in related supply chains and infrastructure, from solar panel manufacturing and agricultural equipment to energy storage solutions. This can lead to job creation, rural development, and enhanced energy security.

## Social and Community Impacts

Worldwide there were many conflicts with locals and solar industry due to conversion of fertile land to solar energy hubs. The agrivoltaics projects pave the way to avoid these conflicts with locals and industry. Before implementation of agrivoltaics community acceptance is very crucial. It is possible by educating the people on the benefits of agrivoltaics.

Social barriers are still there in implementation of photovoltaics and agrivoltaics due to the physical structures and pv panels, which may not permit the easy farm operations unlike traditional farming. Hence the technological innovations in agri industry are the key to overcome this drawback and there is need for more pilot scale studies to take up the agrivoltaics in huge areas.

## Policy and Regulatory Environment

### Policy Frameworks and Incentives

The economic viability of agrivoltaics is significantly influenced by policy frameworks, including subsidies for renewable energy, agricultural grants, and land use regulations. In regions where such policies are favorable, agrivoltaics can see a substantial boost in economic attractiveness. Feed-in tariffs and net metering policies for solar energy can provide stable and predictable revenue streams, enhancing the financial models of agrivoltaic projects.

### Solar Incentives beneficiaries (Farmers)

Several countries from tropical regions to sub tropical and temperate conditions are promoting solar energy and agrivoltaics. Several states inside the US provide incentives for farmers to encourage the adoption of solar electricity, consisting of agrivoltaics. These incentives can encompass tax credits, rebates, and offers. Farmers can check with their state's electricity branch to see what incentives they will receive. Similarly different schemes like PM_KUSUM (Pradhan Mantri Kisan Urja Suraksha evam Utthaan Mahabhiyan Yojana) started in India to encourage farmers to start solar energy production on farm.

### Funding for Agrivoltaic Research

The US Department of Energy (DOE) has announced investment for agrivoltaic studies tasks throughout six states and the District of Columbia. Likewise government of India funded few pilot research projects with successful projects like ICAR-CAZRI, Jodhpur agrivoltaics project.

## Environmental and Ecological Impact

### Environmental Credits and Carbon Trading

Agrivoltaic projects contribute to carbon sequestration and can qualify for environmental credits under various carbon trading schemes. These credits represent an additional revenue stream that can be monetized, further improving the economic case for agrivoltaics.

The economic viability of agrivoltaics extends beyond the apparent benefits, touching upon deeper economic, strategic, and policy-driven factors. As the world gravitates towards sustainable solutions, agrivoltaics stands out not just for its environmental benefits, but for its potential to redefine agricultural and energy economies. Clear understanding of its economic dynamics, coupled with supportive policies and continuous innovation, will be a key to unlocking the full potential of agrivoltaics in the global pursuit of sustainable development.

## Future Directions and Challenges

### The areas of research needed in agrivoltaics:

- So far a limited variety of crops and agricultural landscape were studied under agrivoltaics, no studies on staple foods like paddy and wheat, and scarce research on live stock production.
- Use of bifacial solar panels, semi-transparent solar panels and sun tracking system may be further explored.
- Studies are limited to small scale production only, and involvement of commercial scale studies is meager for generating more scientific results.

- Location based studies are lacking for similar systems for technology up scaling.
- No or very minimal studies are available on economical analysis and social impacts of agrovoltaics.

## The Road Ahead

The technological advancements in agrivoltaics are not just about enhancing efficiency and productivity; they embody a holistic approach to sustainable development that respects the delicate balance of our ecosystems. As research continues and these technologies evolve, agrivoltaics stands poised to redefine the landscapes of agriculture and energy, promising a future where they thrive in tandem, powered by the sun.

In embracing these innovations, stakeholders across sectors farmers, energy producers, policymakers, and communities can work together to harness the full potential of agrivoltaics, turning the dual challenge of food security and energy sustainability into an opportunity for growth, resilience, and environmental stewardship.

## References

Barron-Gafford, G. A., Pavao-Zuckerman, M. A., Minor, R. L., Sutter, L. F., Barnett-Moreno, I., Blackett, D. T., Thompson, M., Dimond, K., Gerlak, A. K., Nabhan, G. P. and Macknick, J. E. 2019. Agrivoltaics provide mutual benefits across the food–energy–water nexus in drylands. Nature Sustainability, 2(9): 9. https://doi.org/10.1038/s41893-019-0364-5

Catherine R. 2020. A good year for solar: Agrivoltaics in vineyards. Pv magazine, 31st March, 2020.

CIAL 2021. New Indian Express. Cochin Airport scales up agri-voltaic farming with joint production of food and energy (newindianexpress.com) 13 Dec 2021. Source: agrovotaics cial kochi - Google Search.

Gonocruz, R. A., Nakamura, R., Yoshino, K., Homma, M., Doi, T., Yoshida, Y. and Tani, A. 2021. Analysis of the rice yield under an agrivoltaic system: A case study in Japan. Environments, 8(7): 7. https://doi.org/10.3390/environments8070065

Guerin, T. F. 2019. Impacts and opportunities from large-scale solar photovoltaic (PV) electricity generation on agricultural production. Environmental Quality Management, 28(4): 7–14. https://doi.org/10.1002/tqem.21629

Homma, M., Doi, T. and Yoshida, Y. 2016. A field experiment and the simulation on agrivoltaic-systems regarding to rice in a paddy field. Journal of Japan Society of Energy and Resources, 37(6): 23–31. https://doi.org/10.24778/jjser.37.6_23

Johansson, F., Gustafsson, B. E., Stridh, B. and Campana, P. E. 2022. 3D-thermal modelling of a bifacial agrivoltaic system: A photovoltaic module perspective. Energy Nexus, 5, 100052. https://doi.org/10.1016/j.nexus.2022.100052.

Kostik, N., Bobyl, A., Rud, V. and Salamov, I. 2020. The potential of agrivoltaic systems in the conditions of southern regions of Russian Federation. IOP Conference Series: Earth and Environmental Science, 578(1), 012047. https://doi.org/10.1088/1755-1315/578/1/012047

Malu, P. R., Sharma, U. S. and Pearce, J. M. 2017. Agrivoltaic potential on grape farms in India. Sustainable Energy Technologies and Assessments, 23: 104-110. https://doi.org/10.1016/j.seta.2017.08.004.

Santra, P., Singh, R.K., Meena, H.M., Kumawat, R.N., Mishra, D., Machiwal, D., Devi Dayal, Jain, D. and Yadav, O.P. 2017. Agri-voltaic system for crop production and electricity generation from a single land unit. In: Advances in Energy Research, Vol 1 (eds. Suneet Singh, Venkatasailanathan Ramadesigan), Springer Proceedings in Energy. Springer, Singapore, pp. 45-56.

Trommsdorff, M. 2021. Agrivoltaics for arid and semi-arid climatic zones: Technology transfer and lessons learned from Japan and Germany. International Webinar Series on Agrivoltaics in Africa, Fraunhofer Institute for Solar Energy Systems ISE, 13 May 2021..

Trommsdorff, M., Gruber, S., Keinath, T., Hopf, M., Hermann, C., Schönberger, F., Högy, P., Zikeli, S., Ehmann, A., Weselek, A., Bodmer, U., Rösch, C., Ketzer, D., Weinberger, N. and Vollprecht, J. 2022. Agrivoltaics: Opportunities for agriculture and the energy transition. Fraunhofer Institute for Solar Energy Systems ISE. https://www.ise.fraunhofer.de/content/dam/ise/en/documents/publications/studies/APV-Guideline.pdf.

Uchanski, M., Hickey, T., Bousselot, J., Barth, K.L. 2023. Characterization of Agrivoltaic Crop Environment Conditions Using Opaque and Thin-Film Semi-Transparent Modules. *Energies 16:* 1-11. https://doi.org/10.3390/en16073012

Weselek, A., Ehmann, A., Zikeli, S., Lewandowski, I., Schindele, S. and Högy, P. 2019. Agrophotovoltaic systems: Applications, challenges, and opportunities. A review. Agronomy for Sustainable Development, 39(4), 35. https://doi.org/10.1007/s13593-019-0581-3

Williams, H. J., Hashad, K., Wang, H. and Max Zhang, K. 2023. The potential for agrivoltaics to enhance solar farm cooling. Applied Energy, 332, Article 120478. https://doi. org/10.1016/j.apenergy.2022.120478.

# CHAPTER 7

# TROPICAL TUBER CROPS FOR CLIMATE RESILIENT AND ENTREPRENEURSHIP DEVELOPMENT

**M. Nedunchezhiyan*[1], K. Pati[1], V.B.S. Chauhan[1], K.H. Gowda[1], R. Arutselvan[1], and G. Byju[2]**

[1] *Regional Centre of ICAR-Central Tuber Crops Research Institute, Bhubaneswar 751019, Odisha*

[2] *ICAR-Central Tuber Crops Research Institute, Thiruvananthapuram 695017, Kerala*

**Corresponding author: m.nedunchezhiyan@icar.gov.in*

---

Tropical tuber crops are the most important food crop after cereals and grain legumes. They are used for food, medicine, animal feed and raw material for starch-based industries. Tropical tuber crops are having varied growth habit, drought and flood resistance and crop duration. Though tropical tuber crops are perennial in nature but domesticated as seasonal/ annual. This provides an opportunity for staggered harvesting as per household and market needs. They have great flexibility in planting and can fit into any cropping/ farming system. This is possible because the propagating material is asexual stem or vine or tuber cuttings. As the economic part is swollen roots or modified stem, photoperiod has no significant effect on yield forming factors. Thus, these are both thermo and photo insensitive. However, extreme high and low temperature affects the growth and yield. These crops grow well in marginal soil with fewer inputs where other crops usually fail to grow. It also produces high amount of dry matter per unit area per unit time compared to cereals. These crops have great flexibility in mixed cropping systems to generate additional employment and income. Crops like yam and elephant foot yam grow as intercrops in horticultural and plantation crops. Tuber crops are capable of utilizing available resources more efficiently especially in partial sunlight and residual moisture. Generally, tubers are consumed directly by boiling and baking or by and preparation of curries. Value addition of tropical tuber crops to the acceptable food preference form increases income and create opportunities for entrepreneurship development. Fried chips, sauce, jam and jelly are prepared from fresh cassava and sweet potato tubers. Glutein free noodles, pasta, pappad, extruded products, bread, cookies and snack foods are prepared from cassava and sweet potato flours. From cassava starch, sago, wafers, liquid glucose, hydrogel, addehives, biodegradable plastics, bioethanol etc are produced. Bio-pesticides are produced from leaves and seeds of cassava and yam bean. Thus, tropical tuber crops are climate resilient as well as provide great opportunities in entrepreneurship development.

The green revolution helped to solve the problem to a great extent. But the exploitative tendency of modern agriculture aimed at higher production per unit area, through specialized monocropping by using chemical fertilizers and pesticides has in fact, rapidly degraded the basic production capacity of the ecosystem. Some of the problems like depletion and degradation of natural resources, increase in effects of biotic and abiotic stresses, increased draining of soil nutrients without replacement, indiscriminate use of systemic chemical pesticides, adverse change in soil microbial dynamism, poor seed replacement rate due to unavailability of quality germplasm and planting materials timely and overall non availability of technical inputs like knowledge for holistic approach pose alarming threat to Indian Agriculture. The repeated natural calamities as flood and drought, poor cropping intensity, lower return and profitability, poor and uneconomical land holding pattern and existing socio-politico-economic penury further aggravate the above problems. Along with this there is an increase in unemployment rate, migration of youth and work force to other region in search of livelihood is alarmingly high and lack of sustained effort to develop agriculture sector is being witnessed. Due to poor profitability of agricultural and allied activities, the youth have started migrating to other sector like services etc. Agriculture is being adopted as a livelihood avenue by compulsion by the youths who are unable to get any other avenues.

Poverty and paddy cultivation has become nearly synonymous worldwide where crop diversification can be a saviour from the situation. It is proved beyond doubt statistically that agriculture growth reduces poverty but it will be negligible if growth is driven by large farms. But real growth in food production, nutrition security and poverty alleviation can occur only when advanced production technologies are made success with the marginal farmers who have less than a hectare.

Diversification with different crops and enterprises bringing in stability, meeting the diversified needs of the farm family, offer of an insurance against crop/ market risk and overall sustainability of soil have proven advantages over the monocropped situations. Further the crop /enterprise diversification, the farmer and labour engaged all the year round in different activities, less risk to market price of the product, the by products of this farm can utilize properly as cattle, poultry, birds, etc. are reared with crop production and regular and quicker return is obtained from various enterprises.

Conservation agriculture is a farming system that can prevent losses of arable land while regenerating degraded lands. Crop diversification is one of the three principles of conservation agriculture. The other two principles are maintenance of a permanent soil cover and minimum soil disturbance.

## Tropical tuber crops

The commonly cultivated tropical tuber crops are: Cassava (*Manihot esculenta*), Sweet potato (*Ipomoea batatas*), Elephant foot yam (*Amorphophallus paeoniifolius*), Taro (*Colocasia esculenta* var. *anticorum*), Bunda (*Colocasia esculenta* var. *esculenta*), Swamp taro (*Colocasia esculenta* var. *stoloniferum*), Tannia (*Xanthosoma sagittifolium*), Lesser yam (*Dioscorea esculenta*), Greater yam (*Dioscorea alata*), White yam (*Dioscorea rotundata*), Aerial yam (*Dioscorea bulbifera*), Yam bean (*Pachyrhizus erosus*), Chinese potato (*Solenostemon rotundifolius*), Arrowroot (*Maranta arundinacea*) etc.

These crops are rich source of energy and carbohydrates although each of them also provides other important nutrients as well. These are rich in minerals and vitamins too. As per FAOSTAT (2012), cassava is produced in 20.4 million ha with production of 262.6 million t and productivity of 12.8 t/ha. Sweet potato is produced in 8.1 million ha with production of 104.5 million t and productivity of 12.9 t/ha. Yams are produced in 5 million ha with production of 58.8 million t and productivity of 10.7 t/ha. Taro is produced in 1.3 million ha with production of 9.6 million t and productivity of 7.8 t/ha. Cocoyam is produced in 0.04 million ha with production of 0.4 million t and productivity of 8.4 t/ha. In terms of percentage contribution of World production as follows: cassava (60.2%), sweet potato (24.0%), yams (13.5%), taro (2.2%) and cocoyams (0.1%). Cassava is mostly produced in Africa, Asia and South America while sweet potatoes are in Asia. Africa dominates the production of yams and taro.

This is possible because the propagating material is asexual stem or vine or tuber cuttings. As the economic part is swollen roots or modified stem, photoperiod has no significant effect on yield forming factors. Thus, tropical tuber crops are both thermo and photo insensitive. However, extreme high and low temperature affects the growth and yield. Tropical tuber crops grow well in marginal soil with fewer inputs where other crops usually fail to grow. They are tolerant to drought and some of them grow fast and provide a wide soil cover to prevent erosion. It also produces high amount of dry matter per unit area per unit time compared to cereals. They are most efficient in converting solar energy, for example cassava producing 250x103 kcal/ ha and sweet potato 240x103 kcal/ha as compared to 76x103 kcal/ha for rice, 110x103 kcal/ha by wheat and 200x103 kcal/ha for maize. Thus, tropical tuber crops are suitable candidature to include in crop diversification programme and conservation agriculture. (Nedunchezhiyan and Laxminarayana, 2006).

## Sweet potato

Sweet potato (*Ipomoea batatas* L. Lam.) is cultivated through out the tropics, subtropics and warmer temperate regions. In India, sweet potato is cultivated in 0.112 million ha with the production of 1.09 million tonnes (NHB, 2013). Odisha ranks first in area (43,460 ha) and production (4,10,100 tonnes) (NHB, 2013). Sweet potato roots and tops are highly nutritious, which can be used to combate nutritional deficiencies in many parts of the developing world. The roots are primarily used as human food after boiling, frying, steaming and baking. Value added products like noodle, liquid glucose, sorbitol, manitol, yoghurt, wine, ethanol etc are also prepared from sweet potato. Apart from source of energy, roots also contain significant quantities of water soluble vitamins *i.e.,* ascorbic acid, thiamin, riboflavin and niacin. The contents of pyridoxine, folic acid and pantothenic acid may be relatively high. Raw leaves and tender tips are also excellent sources of ascorbic acid and B-vitamins especially riboflavin which is deficient in many Asian diets. Orange flesh sweet potatoes contain beta carotene as high as 14 mg/100 g fresh tuber (Nedunchezhian and Ray, 2010). Regional Centre of ICAR-CTCRI, Bhubaneswar has released purple colour anthocyanin (85 mg/100 g fresh tuber) rich variety Bhu Krishna. Green tops and culled tubers are used as animal feed. Foliage is good for hay making.

Sweet potato is a short duration (100-120 days) crop that can be grown during rainy season in hilly and plateau regions and post rainy season in plains with supplemental irrigation. Sweet potato can be intercropped with cereals and pulses (Nedunchezhiyan et al., 2011). Nedunchezhiyan et al. (2010a) and Nedunchezhiyan (2011) reported that strip intercropping of sweet potato and pigeonpea (1:1 ratio) increased net income and improved the soil health. Nedunchezhiyan et al. (2010b) also reported that incidence of weevil reduced when sweet potato was strip intercropped with maize. Nedunchezhiyan et al. (2004) also recommended sweet potato as an intercrop in upland rice for a contingent plan against the crop failure as well as for soil and water conservation. In Odisha, rice-fallow-sweet potato is a popular three-year rotation in uplands which minimises weed infestation and sweet potato weevil incidence, rebuilding of soil fertility status and greater rice productivity (Nedunchezhiyan et al., 2004).

In rice-based cropping system in the coastal plains of Odisha, West Bengal, Karnataka, Tamil Nadu and Andhra Pradesh, sweet potato is planted after harvest of rice to take advantage of the residual soil moisture and fertilizers (Nedunchezhiyan et al., 2004). Sweet potato planted with minimum tillage in rice fallow required lower irrigation water than conventional method of planting (Nedunchezhiyan et al., 2012, 2013).

## Cassava

Cassava (*Manihot esculenta* Crantz) is a starchy root tuber crop. It is popularly known as tapioca, produces more calories/unit area. It is grown in Asia, Africa, South America and Latin America. In India, cassava is being cultivated on an area of 0.24 m ha with a production of 6.7 m t (FAOSTAT, 2005). Cassava is an important crop in South India (Kerala, Tamil Nadu and Andhra Pradesh) and is slowly spreading to Western (Maharastra) as well as Eastern (Odisha) and Northeastern (Assam, Meghalaya and Tripura) India.

Cassava tubers are utilized primarily as human food after boiling, frying, baking or steaming. The other uses of cassava are in industry where it is used for production of starch, liquid glucose, dextrin, high fructose syrup, monosodium glutamate, alcohol and preparation of biodegradable plastics. The starch is also used in textile industries in warping and to produce ply wood. Cassava produces huge foliages. The foliage can produce up to 5 t of crude protein per hectare per year (Moore, 1976). Leaves and tubers also form an important component of animal feed and are extensively used in feed for cattle, poultry and swine. Green foliage is used for silage making. Culled tubers and thippi is used as animal feed as well as fuel in starch industries.

Cassava is a long duration crop. In Odisha, it is grown in Koraput, Rayagada and Gajapati districts where monsoon rain is for 4 months (June-September) and residual moisture is available for 2 months (October-November) due to cyclonic depression. Usually it is grown as sole crop. Maize and pigeonpea can also be intercropped with cassava for higher returns and soil health.

## Greater yam

Greater yam (*Dioscorea alata*) is grown throughout the tropics and sub-tropics, where rainfall is sufficient for their growth. In India, though yam is cultivated in allmost all the state in homestead gardens. It is a most common crop in tribal belts. Greater yam is commercially cultivated in certain specific locations in Andhra Pradesh, Bihar, Gujarat, Kerala, Odisha, Madhya Pradesh, Tamil Nadu and Rajasthan.

Normally greater yam tubers are consumed as boiled, baked or fried vegetables. In West Africa it is mostly eaten as stiff glutinous dough called 'fufu'. They contain 18-20% starch with mucilaginous substance. Extraction of starch is difficult in yams but on cooking the sticky character is lost. Greater yam also contains alkaloids, tannins and steroids which have pharmaceutical value.

Greater yam is a long duration crop and it requires staking for trailing. Maize is a suitable intercrop in greater yam (Nedunchezhiyan et al., 2006). After harvest of maize cobs, the haulm will be supporting greater yam for trailing. Greater yam is intercropped in coconut gardens.

## Elephant-foot yam

Elephant-foot yam (*Amorphophallus paeoniifolius* (Dennst.) Nicolson) is cultivated for its edible corms in India, Philippines and Malaysia. Elephant foot yam or Suran or Jimmikand is gaining popularity because of its yield potential and culinary properties. In India, it is cultivated in Andhra Pradesh, Bihar, Gujarat, Kerala, Maharashtra, North-eastern states, Odisha, Tamil Nadu, Uttar Pradesh and West

Bengal. The starchy tuberous corm is rich in dietary fibre and ash. It is also rich in potassium, calcium, phosphorus, zinc and vitamins. The corm is consumed as vegetable after boiling, baking and frying. Young leaves are also used as vegetable after chopping and boiling. Elephant foot yam corms are also used as medicine in many preparations. They are recommended in case of piles, dysentery, asthma, swelling of lungs, vomiting, abdominal pain and as blood purifier. The paste of corms is applied externally to reduce pain in arthritis.

In wide spaced elephant foot yam, close growing pulses, turmeric and ginger were recommended as intercrops (Nedunchezhiyan and Byju, 2005; Nedunchezhiyan, 2014). Elephant foot yam also grown as intercrop in coconut gardens, mango, litchi and cashew orchards.

## Taro

Taro (*Colocasia esculenta* Schott.) is most important and one of the oldest crops. Taro otherwise known as cocoyam is grown throughout the tropical and sub-tropical countries. Two types of taro – eddoe (*Colocasia esculenta* var. *antiquorum*) and dasheen (*Colocasia esculenta* var. *esculenta*) are commonly cultivated throughout India. The eddoe type is commonly called arvi and dasheen as bunda. Eddoe type is most prevalent as vegetable. In India, it is grown in allmost all the states, but commercially cultivated in Andhra Pradesh, Tamil Nadu, Kerala, Orissa, Uttar Pradesh, Maharastra and Gujarat.

Taro is a tuberous vegetable crop consumed after boiling and seasoning. Tubers are used for chips production and starch making. Taro starches are used in talc along with kaoline. Young leaves and petioles are also consumed as green vegetable. Culled tubers are used as animal feed especially for pigs.

In upland ecosystem, taro is grown as sole crop as well as intercropped with maize. In low land ecosystem, it is grown as sole crop.

## Yam bean

Yam bean (*Pachyrrhizus erosus* (L) Urban) popularly known as potato bean. In India it is called 'Mishrikkand' in Bihar, 'Kesaru' in Eastern Uttar Pradesh and 'Sank alu' in Orissa, West Bengal and Assam.

The starchy conical or turnip shaped fleshy tubers are eaten. High sugar content in tubers imparts sweet taste when eaten raw. The fresh tubers are used as salad and can also be made into chips. The young tubers have crisp juicy and refreshing flesh. The over matured tubers become fibrous, hence unsuitable for consumption. In China mature dried roots are reported to be used as a cooling agent for people suffering from high fever. In many countries young immature pods are used as a vegetable. The stem is tough and fibrous and is used for making fish nets in Fiji. The mature seeds have high content of alkaloids and insecticidal properties. In many developed countries, the tubers are processed, canned and used for making many sweet preparations.

Yam bean is propagated through sexually developed seed. Generally it is cultivated as sole crop. Maize is intercropped in yam bean to provide staking support apart from additional yield. Yam bean also intercropped in sweet potato fields to reduce sweet potato weevil incidence.

## Tuber crops based integrated farming system

During the year 2014-15, the participatory research on Tuber crops based integrated farming system (0.4 ha model) under upland ecosystem was conducted in Khanjuguda (village), Chakapada (Block), Kandhamal (District), Odisha state by the Regional Centre of ICAR-CTCRI, Bhubaneswar under AICRPTC-Tribal Sub Plan. Tuber crops based integrated farming system (0.4 ha) was laid out in 52 tribal researcher fields. The components of tuber crops based integrated farming system and their area of cultivation are given in the Table 1. Sole rice cultivation in 0.4 ha was laid out in four tribal researcher fields as a check. Land preparation for sowing/ planting crop components was initiated as soon as pre-monsoon rain received. Farm yard manure of 10 t/ha was applied at last plough and incorporated. Sowing/ planting were carrying out during June month onwards. Rice was transplanted with 22-25 days old seedling raised in nursery. The spacing adopted was 25 x 15 cm. Ragi was also raised in nursery and transplanted 20 days old seedling at 25 x 15 cm spacing. Maize and red gram were sown at 60 x 30 cm spacing. Tuber crops were planted in ridge and furrow system. Sweet potato, yam bean, greater yam, elephant foot yam, taro and cassava were planted at 60 x 20, 60 x 30, 90 x 90, 90 x 90, 60 x 30, 75 x 75 cm spacing respectively. Weeding was carried out as and when required. Greater yam, elephant foot yam and taro were mulched with dried leaves and paddy straw immediately after planting. 'Handi Khata' was sprayed 30 days after planting of tuber and vegetable crops as a prophylactic measure against pests and diseases. It act as repellent against pests. It reduces root and tuber rot diseases and nematode infection. It also acts as growth enhancer.

Broken rice, maize kernels and chopped culled tubers were fed to poultry birds .However, during day time poultry birds were scavenging in backyards. The poultry manures collected from the sheds were composted and then applied to crops.

The crops as soon as matured were harvested regularly. Fruits of vegetables were plucked as soon it was matured. Eggs of poultry bird collected regularly. Chicken yield of poultry birds recorded at the end of a year. Yield of crops and poultry for the 52 tuber crops based integrated farming system were collected independently; averaged it and presented in the Table 1. Similarly the sole rice crop cultivation data also collected from 4 tribal farmers field independently and averaged it. In tuber crops based integrated farming system, among the crops, sweet potato recorded tuber yield of 516 kg, yam bean tuber yield of 514 kg, rice 381 kg and greater yam tuber yield of 376 kg. The Vanaraja poultry bird attained live weight 35 kg and produced 400 eggs. The tuber crops based integrated farming system registered gross and net returns of Rs 52170 and Rs 34770, respectively. The B: C ratio of 3.00 was realised in the tuber crops based integrated farming system. Lenka et al. (2013) also reported greater income when tuber crops were included in the cropping system.

**Table 1. Integrated farming system components yield and economics (0.4 ha)**

| Crop/animal | Area (ha) | Yield (kg) | Gross income (₹) | Expenditure (Rs) | Net income (₹) | B: C ratio | Employment generation (man days) |
|---|---|---|---|---|---|---|---|
| Rice | 0.20 | 381 | 11430 | 5500 | 5930 | 2.08 | 44 |
| Maize | 0.03 | 62 | 930 | 350 | 580 | 2.66 | 3 |
| Ragi | 0.02 | 25 | 500 | 250 | 250 | 2.00 | 2 |
| Redgram | 0.02 | 14 | 700 | 250 | 450 | 2.80 | 2 |
| Sweet potato | 0.04 | 516 | 5160 | 1400 | 3760 | 3.69 | 6 |
| Yam bean | 0.03 | 514 | 7710 | 1200 | 6510 | 6.43 | 6 |
| Greater yam | 0.02 | 376 | 7520 | 2000 | 5520 | 3.76 | 10 |
| Taro | 0.02 | 305 | 6100 | 1400 | 4700 | 4.36 | 6 |
| Elephant foot yam | 0.008 | 115 | 2300 | 700 | 1600 | 3.29 | 3 |
| Cassava | 0.002 | 38 | 380 | 250 | 130 | 1.52 | 2 |
| Vegetable (*Amaranthus*, Bhendi, bitter gourd, ridge gourd etc.) | 0.01 | 237 | 4740 | 1500 | 3240 | 3.16 | 12 |
| Backyard poultry | 20 (nos.) | 35 | 3500 | 1200 | 2300 | 2.92 | 8 |
| | Egg | 400 | 1200 | 500 | 700 | 2.40 | 2 |
| | 0.4 | 2630 | 52170 | 17400 | 34770 | 3.00 | 106 |

**Check/Control**

| Sl. No. | Crop/ animal | Area (ha) | Yield (kg) | Gross Income (₹) | Expenditure (₹) | Net income (₹) | Employment generation (man-days) |
|---|---|---|---|---|---|---|---|
| 1 | Rice | 0.4 | 800 | 24000 | 11000 | 13000 | 88 |

**Rice ₹ 30/kg; Maize ₹ 15/kg; Ragi ₹ 20/kg; Redgram ₹ 50/kg; Sweet potato ₹ 10/kg; Yam bean ₹ 15/kg; Greater yam ₹ 20/kg; Taro ₹ 20/kg; Elephant foot yam ₹ 20/kg; Cassava ₹ 10/kg, Vegetables ₹ 20/kg; Poultry live bird ₹ 100/kg; egg ₹ 3**

The gross and net return from sole rice cultivation was ₹ 24000 and ₹ 13000, respectively. The employment generated in tuber crops based integrated farming system was 106 man-days (Table 1), whereas sole rice cultivation recorded 88 man-days. Inclusion of tuber crops in the integrated farming system spread the employment opportunities through out the year. Pali et al. (2012) also reported generation of additional 12 days employment in 0.4 ha pond based integrated farming system. Inclusion of long duration tuber crops like grater yam, elephant foot yam and taro spread the labour requirement for a longer period. Thus dependency on out-side labours was much less in tuber crops based integrated farming system. Whereas in rice, ragi and maize cultivation the total employment generated was spread within seven months (June-December) only.

Part of the yields of grains, tuber crops, vegetables, egg and meat were sold for cash income. By marketing of vegetables, tubers, egg and meat a farmer is able to earn sufficient money to meet out daily needs. Further, the availability of tubers for household consumption for long period due to high storability along with other vegetables, rice, egg and meat enhanced food and nutritional security of the household. The cash income improved the livelihoods of the farm families.

## Value -addition tuber crops

Generally, tubers are consumed directly by boiling and baking, and preparations of curries. Shelf-life of some of the tuber crops only last too few days. Hence, the availability of tuber crops is in the specific period only. Further, farmers also fetches lower price due to market glut during harvesting. Cassava tubers shelf-life is 2 days and sweet potato shelf-life is 20-30 days only. Within this period, these tubers are to be consumed. Minimal processing and value addition will increase the availability of these tubers for more periods. Value addition of tuber crops to the acceptable food preference form will increases income and the demand for the tuber crops.

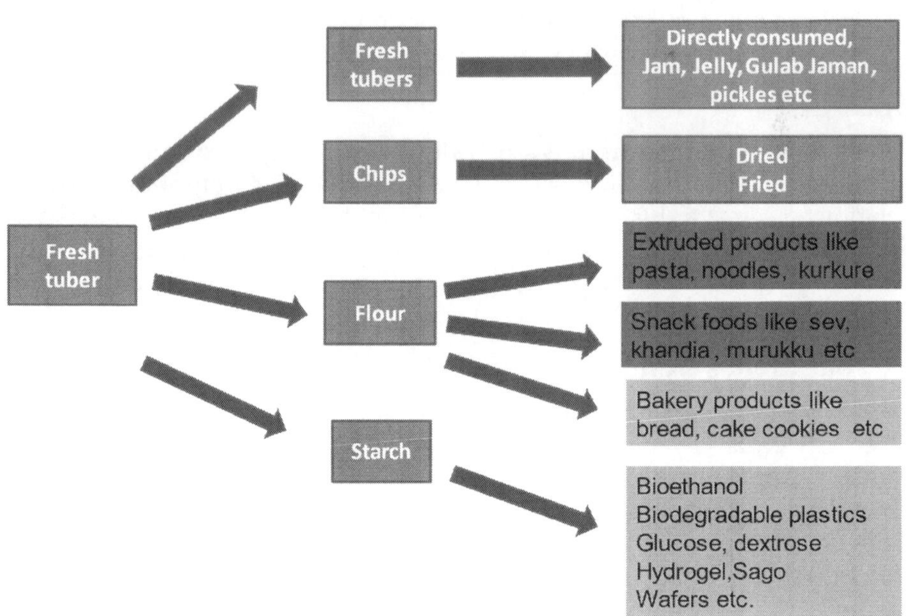


**Fig.1. Value-added products from tuber crops**

## Minimal processing

### Dried chips

Cassava and sweet potato tubers are sliced/chipped after washing. The slices/chips are sundried or tunnel dried for 48-72 hours and then packed in air tight containers. The dried slices/chips can be stored up to one year.

### Par-boiled chips

Cassava and sweet potato tubers are sliced/chipped after washing. The slices/chips are put in to boiling water and cooked for 5-10 minutes. The cooked slices/chips are sundried or tunnel dried for 48-72 hours and then packed in air tight containers. The dried par-boiled slices/chips can be stored up to one year.

## Value added products from cassava

### Fried chips

Fresh cassava tubers are washed and peeled. The tubers are sliced/chipped 1.5-2.0 mm thickness. Then the slices/chips are soaked in acetic acid-brine solution for 1 h, parboiling for 5 minutes followed by surface drying and deep frying in oil.

### Pasta

Besides, pasta getting wide popularity among the young Indians and in the metros as a convenient food, transformation to new health and wellness foods is essential to add value to cassava and sustain their cultivation in India.

### Pappad

Papads are prepared from cassava flour by adding fibre sources like wheat bran, oat meal, rice bran and cassava fibrous residue. The fibre sources are added to gelatinized cassava slurry and mixed thoroughly. The spicy condiments are also added and spread on plastic sheets which are then dried in the sun for 36 h. The papads are peeled off from the sheets and packed. The deep fried products are soft with crisp texture.

### Fried snack foods from

The cost of making value added snack foods from cassava could be considerably reduced, if wet cassava paste is used instead of cassava flour. Such an innovation was made in making a highly acceptable crisp snack food viz., chitchore from cassava. The wet cassava tuber paste is mixed with ingredients like maida, cheese, salt, sugar, baking powder and white pepper. The dough after proofing for 1 h is spread into sheets and cut into small discs of 1 cm diameter.

**Pakkavada**: This is a hot snack food having good texture and taste made out of cassava flour. The other ingredients include maida, bengal gram flour, salt, chilli powder, asafoetida, baking soda and oil. The ingredients are thoroughly mixed and made into dough with hot water (50 C), proofed for 1 h and then extruded through hand extruder having flat rectangular holes, into hot oil.

**Sweet fries**: This is a sweet snack food made out of cassava flour, maida, baking soda and oil. The ingredients are mixed well and made into dough with hot water (50 C). The dough after proofing for 1 h is hand extruded through die having round holes, into hot oil. The fried product is then coated with sugar by dipping for a few minutes in sugar syrup having thick consistency.

**Nutria-chips**: This is a high protein snack food made out of cassava flour by mixing with other ingredients like maida, groundnut paste, egg, salt, sugar, sesame, coconut milk, baking soda and oil. After mixing the ingredients, hot water is added and mixed to form smooth dough. The dough after proofing is made into small balls which are then spread into sheets of 2 mm thickness. This is then cut into diamond shape using a sharp knife and deep fried in oil.

**Crisps**: This is a soft and good textured crispy snack food made from cassava flour, maida, rice flour, bengal gram flour, salt, baking soda, turmeric powder and oil. The dough made with hot water is proofed for 1 h and then extruded through the small pore size die having round holes. The deep fried material is mixed with fried nuts, curry leaves etc. before packing.

## Value-added products from sweet potato

At present it is consumed after boiling or baking as secondary staple. It is also used as vegetable in 'Curry' and 'Dalmah' preparation. These roots are rich in starch, sugars, vitamins and minerals. Some sweet potato varieties contain coloured pigments anthocyanin, β-carotene and unidentified flavonoids. These pigments are regarded as beneficial antioxidants having several physiological attributes such as anti-oxidation, anticancer and protection against liver injury. Recently World Health Food Organization (WHFO) labelled sweet potato as an "antidiabetic" food

Some studies conducted in animals showed that sweet potato diet helped to stabilize blood sugar levels and lowered insulin resistance. The ICAR-Central Tuber Crops Research Institute, Regional Centre, Bhubaneswar has released two nutritionally important sweet potato varieties i.e. Bhu Sona (Orange fleshed-β-carotene rich) and Bhu Krishna (Purple fleshed anthocyanin rich). Sweet potato is a rich source of dietary fibres. The importance of dietary fibre in food is well recognized. Dietary fibre reduces serum cholesterol, prevents colon cancers, and maintains good intestinal health as well as prophylactic action on cardiovascular diseases, diabetes and obesity.

Though the Regional Centre of ICAR-CTCRI, Bhubaneswar has developed many value added technologies in sweet potato, no value-added product is available in the market. Yet, the Regional Centre of ICAR-CTCRI, Bhubaneswar has demonstrated some of the simple value-added products development to the tribal farmers and entrepreneuring youth through TSP.

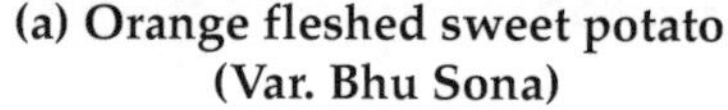

**(a) Orange fleshed sweet potato (Var. Bhu Sona)**

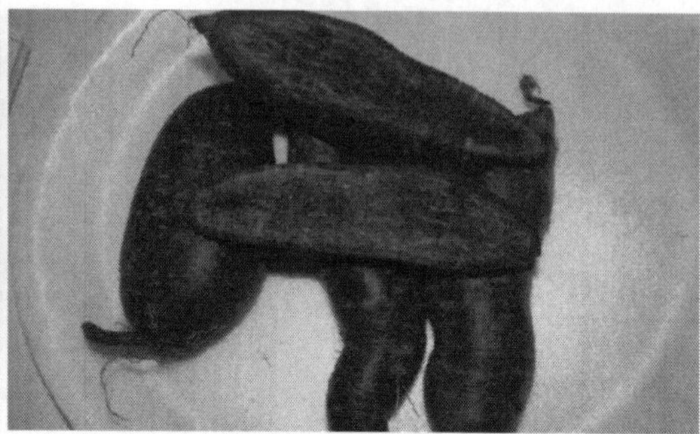

**(b) Purple fleshed sweet potato (Var. Bhu Krishna)**

**Biofortified varieties of sweet potato**

## Sweet potato fried chips

Excellent quality fried chips can be made from sweet potato tubers, by subjecting to mild acid treatment. This can prevent the discoloration of sweet potato slices, with the result that light yellow crispy chips can be obtained, having soft mouthfeel and good texture. About 4 months old tubers are chosen for making chips. After slicing into round pieces with thickness of about 2 mm, soaked for 15 minutes in 0.5% citric acid-0.5% brine (salt) solution. After soaking, wash with clean water and keep slices in shade drying for 30 minutes to remove surface moisture. Fry the slices at 150°C for 5 minutes, mix the seasonings and pack it.

## Fried snacks from sweet potato flour

Sweet potato flour is gluten free and having low protein content. Hence fortification is done by mixing with nutritionally superior flours to enhance the nutritional and textural value of food products. Various fried snack foods have been developed. They are sweet potato nutri sticks, crisps etc.

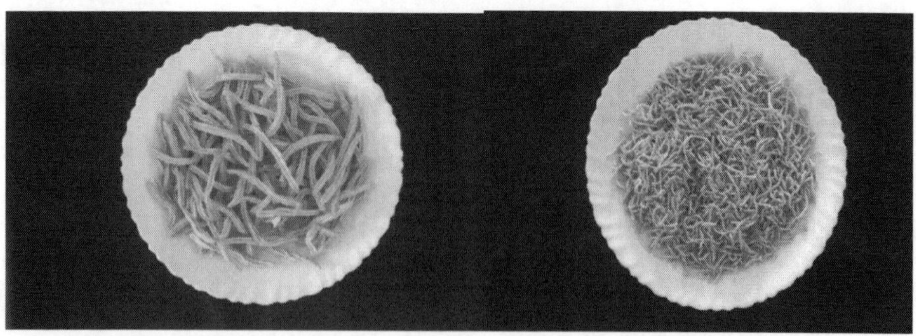

**Nutri-sticks from sweet potato** **Nutri-crisp from sweet potato**

**Sweet potato chips**
**Snack food from sweet potato**

## Chappati, *puri* and *parota* from sweet potato flour

Chappati, ***puri*** and ***parota*** can be prepared from sweet potato flour. As sweet potato flour is devoid of gluten, some binding agent should be added for the preparation of the above products.

## Sweet potato cake

Anthocyanin rich, gluten free cake could be prepared from purple fleshed sweet potato (variety BhuKrishna) flour, sugar, eggs, butter and baking powder. The preparation method includes steps like; beating of butter/vanspati ghee, adding ground sugar and make the cream, separately beat the eggs and add into the cream. Add sieved sweet potato flour and baking powder in to the cream-egg mixture, thorough mixing of batter, pour batter into previously greased baking pan. Bake at 160 C for 25-30 minutes; cool the cake at room temperature. The prepared cake is healthy for consumers.

## Sweet potato pasta

Use of purple fleshed sweet potato flour gives dark purple coloured pasta which has low starch digestibility and high resistant starch content besides high anthocynin and protein content. The preparation steps include: mixing of ingredients (sweet potato flour, whole wheat flour, semolina and water) for 15 minutes. Transfer and knead in pasta machine for 5 minutes. Extrude with single screw, dicing to desirable size, then drying for 55°C for 4 hours, packing and storage.

## Sweet potato jelly

Purple coloured, attractive and transparent jelly could be prepared from purple fleshed sweet potato extract with the help of sugar, pectin and citric acid. The preparation method includes; Selection of good quality tubers, washing, peeling and cutting into slices. Add water into slices (1:1), boil and collect the strained sweet potato extract for jelly preparation. Add sugar (60%), pectin (1%) and citric acid (0.5%). Check the TSS (Total soluble solids) by refractometer to maintain the TSS (67.5 Brix). Fill the hot product into glass bottles followed by sealing, cooling and storage.

**Anthocyanin-rich cake (Variety, Bhu Krishna)**

**Anthocyanin-rich pasta (Variety, Bhu Krishna)**

**Anthocyanin-rich jelly**

**Value-added products from Bhu Krishna variety**

## Value-added product from arrowroot

At present in Odisha only arrowroots are used for processing into starch. In Odisha, three types of arrowroots are cultivated for starch purposes. They are (1) West Indian arrowroot (*Maranta arundinacea*), (2) East Indian arrowroot (*Curcuma* spp.) and (3) Queensland arrowroot (*Canna edulis*). The starch is extracted by traditional methods from the above three arrowroot crops by the farmers as an off-seasonal activity and marketed locally. The starch is mainly used as a functional food. Local farmers consume arrowroot starch by dissolving in milk or warm water for instant energy during diarrhoea and other stomach ailment.

Ninety percent of the starch produced is sold in the local market and 10% is kept for household consumption. Arrowroot starch extraction starts from February month and continue up to June month. It is a main off-season activity of the farmers. Local traders' visits villages during this period and collect the starch. The local traders sell the starch to the exporters. Some time, local traders keep middlemen (brokers) in the villages for collecting the starch. Local traders or middlemen on behalf of local traders negotiate the rate for the starch. KASAM, an NGO based in Odisha exports 100-200 tons of arrowroot starch to Germany every year.

The ICAR-Central Tuber Crops Research Institute, Thiruvananthapuram has developed multi crop mobile starch extraction machine, in which *Canna, Curcuma, Maranata,* cassava and sweet potato tubers can be crushed for starch extraction. The efficiency of the machine is 20kg/hour, whereas in traditional method it is 1.0 kg per hour (Table 34). The starch recovery is 20%, whereas in traditional method, it is 10-12%.

The Regional Centre of ICAR-CTCRI during the year 2013, a mobile starch extraction machine was installed at Lentrimaliguda (Village), Koraput Block, Koraput disrict, Odisha under TSP. All the farmers from Lentrimaliguda were allowed to use the machine. During the year 2013, approximately 1250 kg starch was produced in this village from canna using the mobile starch extraction machine. During the year 2014, a mobile starch extraction machine was introduced in a Pottamaliguda village, Koraput Block, Koraput District, Odisha under CIP-IFAD and TSP. Through this machine 750 kg starch was extracted from 3700 kg canna tubers. Farmers from Koraput district realized the benefits of starch machine. They modified the machine as per their economical conditions in such a way that machine is mainly used for pulping (crushing) the canna tubers. Rest of the operations were carried out by traditional way. Pulping (crushing) is the biggest drudgery in the starch extraction process. At present (the year 2022), approximately 250 starch extraction machines are under operation in Koraput district.

**Table 34. Efficiency of mobile starch extraction machine**

| Particulars | Traditional method | Starch extraction machine |
|---|---|---|
| Efficiency | 1 kg/hour | 20 kg/hour |
| Starch recovery from Canna (*palua*) | 12% | 20% |
| Electricity charges | - | 1.1 unit/20 kg |
| Extraction charges | Own labour | Rs 1.00/kg<br>(Rs 0.75 labour charge + 0.25 sustainability fund) |

## Conclusion

Tropical tuber crops are most suitable for crop diversification in both upland and low land ecosystem as well as conservation agriculture. Its great flexibility in planting and harvesting is the most spectacular feature and fitting into any cropping/ farming system. Tropical tuber crops give assured returns in spite of natural calamities like drought and floods because the economic part is asexually formed and remains under ground. The productivity of tuber crops is quite high compared to cereals and pulses and doubles the farmers income. Many of the research findings concluded that tuber crops based integrated farming system increased farmers income 2-3 times. Tropical tuber crops are rich in vitamins and minerals apart from carbohydrates. Thereby it provides food and nutritional security. Further, tuber crops provide opportunities for developing value added products and entrepreneurships. Thus, in changing climate, tuber crops find special niche in crop diversification programmes as well as value addition for doubling farmers income.

## References

FAO STAT, 2005. Area, production and productivity of root and tuber crops. FAO Statistics 2005. Available HTTP: http://apps.fao.org.

FAOSTAT. 2012. Food and Agriculture Organization of the United Nations. faostat. fao.org/site/567/default/aspx.

Lenka, A., Jata, S.K. and Nedunchezhiyan, M. 2013. Tuber crops for improving livelihood security in the disadvantaged districts of Odisha. *Journal of Root Crops*, 39 (2): 207-210.

Moore, C.P. 1976. So del torraje de yucca en la alimentacion de ruminantes. In: memoria del seminario Internacional de Ganaderia Tropical, Acapulco, Guerreso Mexico 8-12, March 1978. Pp. 47-63.

Nedunchezhiyan, M. 2011. Evaluation of sweet potato (*Ipomoea batatas*) based strip intercropping systems for yield, competition indices and nutrient uptake. *Indian Journal of Agronomy*, 56 (2): 98-103.

Nedunchezhiyan, M. 2014. Production potential of intercropping spices in elephant foot yam (*Amorphophallus paeoniifolius*). *Indian Journal of Agronomy*, 59 (4): 596-601.

Nedunchezhiyan, M. and Byju, G. 2005. Productivity potential and economics of elephant foot yam based cropping system. *Journal of Root Crops*, 31 (1): 34-39.

Nedunchezhiyan, M. and Laxminarayana, K. 2006. Tuber crops based cropping system in eastern India. Kisan World, 45-46.

Nedunchezhiyan, M. and Ray. R.C. 2010. Sweet potato growth, development production and utilization: overview. In: Sweet potato: post harvest aspects in food, Ray, R.C. and Tomlins, K.I. (Eds.). Nova science Publishers Inc., New York, pp. 1-26.

Nedunchezhiyan, M., Byju, G. and Naskar, S.K. 2006. Effect of intercrops and planting pattern on incidence of anthracnose, productivity potential and economics of greater yam (*Dioscorea alata*). *The Indian Journal of Agricultural Sciences*, 76 (2): 132-134.

Nedunchezhiyan, M., Byju, G. and Ray, R.C. 2012. Effect of tillage, irrigation and nutrient levels on growth and yield of sweet potato in rice fallow. *ISRN Agronomy 2012*: 1-13, Article ID 291285, doi: 10.5402/2012/291285.

Nedunchezhiyan, M., Laxminarayana, K., Rajasekhara Rao, K. and Satapathy, B.S. 2011. Sweet potato (*Ipomoea batatas* L.)–based strip intercropping: I. Interspecific interactions and yield advantage. *Acta Agronomica Hungarica*, 59 (2): 137-147.

Nedunchezhiyan, M., Rajasekhara Rao, K. and Satapathy, B.S. 2010a. Productivity potential, biological efficiency and economics of sweet potato (*Ipomoea batatas*) – based strip intercropping systems in rainfed Alfisols. *Indian Journal of Agricultural Sciences*, 80 (4): 327-330.

Nedunchezhiyan, M., Rajasekhara Rao, K., Laxminarayana, K. and Satapathy, B.S. 2010b. Effect of strip cropping involving sweet potato (*Ipomoea batatas* L.) on soil moisture conservation, weevil infestation and crop productivity. *Journal of Root Crops*, 36 (1): 53-58.

Nedunchezhiyan, M., Sinhababu, D.P. and Sahu, P.K. 2013. Effect of soil amendments and irrigation regimes on minimum tillage planted sweet potato (*Ipomoea batatas*) in rice (*Oryza sativa*) fallows under lowland conditions. *Indian Journal of Agricultural Sciences*, 84 (3): 371–375.

Nedunchezhiyan, M., Sivakumar, P.S., Naskar, S.K. and Misra, R.S. 2004. Growth potential of sweet potato in rice based cropping system. In: *National Symposium on Recent Advances in Rice-based Farming Systems*, 17-19 November 2004, Abstracts, pp. 86.

NHB. 2013. National Horticultural Board. www.nhb.org.

Pali, G.P., Sahu B., Savu, R.M., Chaudhary, J.L., Sharma, G.K. and Thakur, A.K. 2012. Integrated farming system models for sustainable productivity and income of small and marginal rainfed rice farmers of Chhattisgarh. Extended Summaries Vol 3, 3rd International

CHAPTER 8

# IMPACT OF CLIMATE CHANGE ON VEGETABLE PRODUCTION AND ITS MANAGEMENT PRACTICES

**Arpita Srivastava and Bhoopal Singh Tomar**

*Division of Vegetable Science, ICAR-IARI, New Delhi*
*Corresponding author: bst_spu_iari@rediffmail.com*

Climate change significantly impacts agriculture, particularly vegetable production. This chapter examines the effects of rising temperatures, altered precipitation, and extreme weather events on vegetable cultivation, leading to decreased yields, compromised quality, and increased vulnerability to pests and diseases. Various management practices have been developed to mitigate these effects, including shifting planting schedules, adopting drought-tolerant varieties, implementing efficient irrigation systems, practicing agroforestry, and employing precision farming techniques. Sustainable practices like organic farming and integrated pest management are crucial for enhancing resilience and minimizing environmental degradation. Genetic improvement of vegetable crops is vital for developing resilient varieties. Grafting, which utilizes wild and underutilized genetic resources, has successfully developed stress-tolerant vegetables, broadening genetic diversity. Despite controversies, transgenic technology and genome editing hold promise for creating climate-resilient vegetable varieties. This chapter emphasizes the need to address the interplay between climate change and vegetable production. By implementing appropriate management strategies, the agricultural sector can safeguard food security, promote sustainable farming practices, and foster resilience in the face of a changing climate.

The global average surface temperature has increased by approximately 1.62 F (0.9 C) since the late 19$^{th}$ century, primarily due to elevated carbon dioxide levels and other anthropogenic emissions into the atmosphere (IPCC, 2021). The heat-trapping nature of carbon dioxide and other gases was confirmed in the mid-19th century (IPCC, 2021).

Vegetables play a crucial role in human nutrition, providing essential nutrients, vitamins, and minerals. However, like other crops, vegetables are facing the repercussions of climate change, including global warming, alterations in seasonal and monsoon patterns, and impacts from both living (biotic) and non-living (abiotic) factors (IPCC, 2021). Under changing climatic conditions, common issues in vegetable cultivation, such as crop failures, reduced yields, lower quality and increased pest and disease problems, render vegetable cultivation unprofitable (IPCC, 2021). As many physiological processes and enzymatic activities in vegetables are temperature-dependent, they are particularly susceptible to environmental extremes. High temperatures and limited soil moisture are major causes of low yields, affecting processes like reduced photosynthetic activity, altered metabolism and enzymatic activity, thermal injury to tissues, reduced pollination and fruit set, which will be further magnified by climate change (Hatfield et al., 2011).

Drought and salinity are two important consequences of an increase in temperature, worsening vegetable cultivation (IPCC, 2021). Anthropogenic air pollutants, such as CO2, CH4 and CFCs, contribute to global warming (IPCC, 2021). These climate change impacts also influence pest and disease occurrences, host-pathogen interactions, distribution and ecology of insects, time of appearance, migration to new places, and their overwintering capacity, thereby becoming a major setback to vegetable cultivation (IPCC, 2021).

Despite directly affecting rainfed vegetable cultivation, climate change also impacts water storage and availability for irrigation. With limited water availability, drought becomes a major stress factor for vegetable production, further stressing farming systems. Many vegetable crops like cole crops, onions, and root crops are pollinated by insects, and changes in insect species distribution will affect pollination. (Vermeulen et al., 2010).

## Vegetable reactions to climatic shifts

### Response to elevated $CO_2$ levels in the atmosphere

Alterations in atmospheric $CO_2$ concentration can have significant impacts on plant tissues, influencing their growth and physiological behavior. Summarily, these investigations reveal that elevated atmospheric $CO_2$ levels result in changes to net photosynthesis, biomass production, protein, sugar, and organic acid contents, stomatal conductance, firmness, seed yield, light utilization, water use efficiency, nutrient utilization, and plant water potential.

#### Effect of elevated $CO_2$ on total plant yield in vegetables

Tomato fruit yield exhibits an increase ranging from 7% to 125% under elevated $CO_2$ concentrations (Doddrell et al. 2023). Pepper crops similarly demonstrate enhanced yields, with reported increase ranging from 12.9% to 370.2% when grown at elevated $CO_2$ concentrations Most studies indicate yield enhancements in the range of 12.9% to 47.4%, excluding other influencing factors. However, controlled environments reveal that a 4°C temperature increase negatively impacts yield, even under elevated $CO_2$ conditions (750 ppm). This suggests that carbon enrichment alone may not suffice to counteract reduced yield in settings with elevated temperatures, especially in extreme climate change scenarios. Notably, carbon-enriched growth contributes to increased capsaicinoid content in chili fruits, leading to higher Scoville Heat Units (SHU). Cucumber, among cucurbitaceae fruit crops, has been extensively studied concerning growth in carbon-enriched atmospheres. Improved carbon assimilation rates of up to 99% and 112% have been reported for cucumber and melon, respectively, under elevated $CO_2$ conditions. In instances of high nitrogen supplementation, fruit yield reaches as high as 106% under elevated $CO_2$ of 800 ppm, indicating the potential for combining increased nitrogen fertilization with $CO_2$ enrichment to achieve optimal yield increases in cucumber (Doddrell et al. 2023).

## Effect of Elevated $CO_2$ on nutritional quality of vegetables

Numerous studies, predominantly focusing on leafy vegetables, have explored the effects of enriched $CO_2$ on product quality, suggesting a potential nutritional enhancement characterized by increased levels of sugars, ascorbic acid, phenols, flavonoids, and antioxidants (Bisbis et al., 2018). For instance, red leaf lettuce cultivated at 1000 ppm $CO_2$ exhibited elevated sugars, flavonoids and caffeic acid derivatives (Becker, 2016). Similarly, higher vitamin C content has been observed in lettuce, celery and Chinese cabbage when grown under $CO_2$ concentrations ranging from 800 to 1000 ppm (Jin et al., 2009). Effect of elevated $CO_2$ on various nutritional quality of vegetables has been summarized in Table 1.

**Table 1: Effect of Elevated $CO_2$ on nutritional quality of vegetables**

| Vegetable | Nutritional Traits | | | | | | | | | | Reference |
|---|---|---|---|---|---|---|---|---|---|---|---|
| | Pr | AA | FA | MaN | MiN | Vit C | AO | Ni | Su | TPhe | Rosales *et al.* 2011 |
| Tomato | ↓ | - | - | ↓ | ↓ | ↑↓ | - | - | - | - | Saure, 1998 |
| Cabbage | - | - | - | - | - | ↑ | - | - | - | - | Jin *et al.* (2009) |
| Carrot | ↓ | ↓ | ↓ | ↓ | ↓ | ↓ | - | - | | - | Ibrahim *et al.* 2006 |
| Celery | - | - | - | - | - | ↑ | - | ↓ | - | - | Jin *et al.* 2009 |
| Lettuce | - | - | - | ↓ | ↓ | ↑ | ↑ | ↓ | - | - | Oh *et al.* 2009 |
| Potato | ↓ | ↓ | - | ↓ | ↓ | ↑ | | ↑ | ↑↓ | - | Kumari and Agrawal (2014) |
| Radish | ↓ | ↓ | ↓ | ↓ | ↓ | ↓ | - | - | - | - | Azam *et al.* 2013 |
| Spinach | - | ↑ | - | ↓ | ↓ | - | - | - | - | - | Giri *et al.*, 2016 |
| Turnip | ↓ | ↓ | ↓ | ↓ | ↓ | ↓ | - | - | - | | Azam *et al.*, 2013 |

Pr: Protein; AA:Amino acids; FA: Fatty Acids; MaN: Macronutrients, MiN: Micronutrients; Vit C: Vitamin C; AO: Antioxidants; Ni:Nitrates; Su:Sugars; TPhe: Total Phenols,

## Response to Elevated temperatures

### Effect of elevated temperature on total yield and physiology

The temperature plays a crucial role in determining the harvest index, with higher temperatures during the growing season accelerating crop maturity. Exposure to elevated levels of carbon dioxide and ozone, along with high temperatures, can directly and indirectly impact the production and quality of vegetable crops (Mattos et al., 2014). Research has highlighted that lettuce, celery, and cauliflower subjected to higher temperatures exhibited earlier maturation compared to the same crops grown under lower temperatures. Many physiological, biochemical and metabolic activities crucial for plant development are temperature-dependent. Various physiological disorders in vegetable crops induced by high temperatures are summarized in Table 2. High temperatures during development can influence essential processes such as photosynthesis, respiration, aqueous relations, membrane stability and levels of plant hormones and metabolites. Seed germination may be hindered or inhibited by elevated temperatures, contingent upon the species and stress level (Motsa et al., 2015). The ratio between photosynthesis and respiration is critical for achieving a high yield, with the optimal ratio typically exceeding ten around temperatures of 15 C (Went, 1953). Higher temperatures also indirectly impact photosynthesis by elevating leaf temperatures and influencing stomatal conductance (Moore et al., 2021).

**Table 2: Physiological disorders of various vegetable crops caused by high/fluctuating temperature**

| Crop | Disorders appearing due to high/fluctuating temperatures | Refernces |
|---|---|---|
| Asparagus | High fiber in stalks and spheres, Feathering Lateral branch growth | Spaldon *et al.*, 2015 |
| Bean | High fibre in pods | Spaldon *et al.*, 2015 |
| Carrot | Low carotene | Spaldon *et al.*, 2015 |
| Cauliflower & Broccoli | Hollow Stem, Leafy heads, No heads<br>Branching | Spaldon *et al.*, 2015 |
| Cole crops, Lettuce | Tip burning | Spaldon *et al.*, 2015 |
| watermelon | Delayed ripening of fruits Reduction in the sweetness of melons | Spaldon *et al.*, 2015 |
| Cucurbits | Blossom end rot<br>Increased vegetative growth<br>Poor production of female flowers | Singh, 2010 |

| Crop | Disorders appearing due to high/fluctuating temperatures | Refernces |
|---|---|---|
| Tomato | Decreased fruit set as well as lower quality fruits, bud drop, abnormal flower development, viability and reduced carbohydrate availability sunburn, disruption of lycopene<br>Synthesis, appearance of yellow areas in the affected<br>tissues, poor fruit set, delay in ripening, yellow shouldered fruit, white core and blossom-end rot | Vijayakumar et al. 2021 |
| Pepper/ Chillies | Inhibits fruit set<br>Affects capsanthin content in ripened chilli fruits<br>Flower drop,<br>Ovule abortion,<br>Poor fruit set,<br>Fruit drop in chili<br>Reduced size of fruits, average fruit weight and number of seeds per fruit | Thuy and Kenji, 2015 |
| Potato | Severe reduction in the potato tuber yield at temperatures above 210C while complete inhibition of tuber formation at 30 C occurs. | Sandhu *et al.*, 2018 |
| Cucumber | Decreased yield, early flower drop<br>Bitterness of fruits | Kumar *et al.*, 2011 |

## Effect of elevated temperature on quality of vegetables

### Perceived Quality

Sunburn stands out as the most prevalent temperature-induced disorder documented in vegetable crops. For common beans, temperatures surpassing 27/22°C (day/night) during seed development led to the production of smaller seeds compared to beans grown at 21/16°C (Abdus and Goodwin, 1980). Pea plants subjected to short periods of excessive temperature experienced accelerated maturity and reduced seed size (Bisbis et al., 2018). In the case of head lettuce, temperatures beyond 17-28/3-12°C (day/night) increased the incidence of loose and puffy heads, along with tipburn and leaf chlorosis, accompanied by an accumulation of bitter compounds. Onions exhibited reduced size when grown at 32°C instead of 27°C (Coolong and Randle, 2003). Leafy and brassica vegetables reported an increase in the tipburn disorder at high night temperatures (Saure, 1998). For tomato plants, elevated temperatures were shown to decrease macronutrients, carotene, lycopene, and antioxidant content (Rosales et al. 2011). In asparagus, heightened air temperatures negatively impacted quality, leading to rapid head opening, purple discoloration, and undesirable fibers in asparagus spears (Peet and Wolfe, 2000).

## Nutritional Quality

Rising temperatures linked to climate change may potentially enhance both the perceived and nutritional quality of fruits and vegetables. This improvement could stem from the heightened activity of metabolic pathways, such as photosynthesis and secondary metabolism, resulting in positive effects. For instance, in carrots, high temperatures lead to the accumulation of antioxidants, including monoterpenes and sesquiterpenes, as well as essential oils associated with flavor attributes (Ibrahim et al., 2006). In tomatoes, increased temperatures have been found to elevate carotene and antioxidant content, including phenols, flavonoids, anthocyanins, and ascorbic acid, compared to harvests during lower temperature periods (Rosales et al., 2011). Similarly, lettuce subjected to heat stress demonstrated enhanced tocopherol and antioxidant content (Oh et al., 2009).

## Effect of drought and salinity on vegetable production

The growth, photosynthesis, biochemistry, and overall quality of cucumber fruits were significantly constrained by exposure to elevated salinity and drought (Ouzounidou et al. in 2016). The application of 150 mM NaCl led to noteworthy reductions in both plant biomass and total pigment concentration in cucumber cultivars and broad beans (Ouzounidou et al. in 2014). Additionally, the impact of drought on vegetable storage quality varies among different crops. For instance, tomatoes subjected to water stress exhibited extended shelf-life and reduced weight loss (Conesa et al. in 2014). Conversely, carrots experiencing drought stress during growth demonstrated increased water loss during storage and heightened susceptibility to chilling damage during cooling (Toivonen and Hodges in 2011). Table 3 summarises the effect of drought at critical stages of growth of various vegetables.

**Table 3: Impact of water stress at critical stages of growth of vegetables crops**

<table>
<tr><th>Vegetable</th><th>Impact of Water Stress</th><th>Critical Period for Irrigation</th></tr>
<tr><td colspan="3">Solanaceus Vegetables</td></tr>
<tr><td>Tomato</td><td>1. Flower shedding<br>2. Lack of fertilization<br>3. Reduced fruit size<br>4. Fruit splitting<br>5. Poor seed viability<br>6. Development of blossom end rot (BER)</td><td rowspan="2">(i) Flowering<br>(ii) Fruit set<br>(iii)Fruit development</td></tr>
<tr><td>Brinjal</td><td>1. Reduced yield<br>2. Poor color development in fruits<br>3. Poor seed viability</td></tr>
</table>

| Vegetable | Impact of Water Stress | Critical Period for Irrigation |
|---|---|---|
| Chilli and Capsicum | 1. Flower and young fruit shedding<br>2. Reduced dry matter production<br>3. Poor nutrient uptake<br>4. Poor seed viability | |
| Potato | 1. Poor tuber growth and yield<br>2. Splitting<br>3. Internal brown spot | Tuberization and enlargement |
| | **Malvaceae** | |
| Okra | 1. Considerable yield loss<br>2. fiber development<br>3. high mite infestation<br>4. poor seed viability | Flowering and pod development |
| | **Brassicaceae/Cole Crops** | |
| Cauliflower, cabbage, and broccoli | 1. Tip burning<br>2. Head splitting (cabbage)<br>3. browning, and buttoning (cauliflower) | Head/curd formation & Curd enlargement |
| Root crops | | |
| Carrot, radish, and turnip | 1. Distorted<br>2. Rough roots<br>3. Strong odour in carrot<br>4. Accumulation of harmful nitrates | Root enlargement |
| | **Allium Vegetables** | |
| Onion | 1. Splitting<br>2. Doubling of bulb<br>3. Poor storage life | Bulb formation and enlargement |
| | **Cucurbitaceous vegetables** | |
| Cucumber | 1. Deformed, non-viable pollen grains<br>2. Bitterness<br>3. Deformity in fruits<br>4. Poor seed viability | Flowering and fruit development |

| Vegetable | Impact of Water Stress | Critical Period for Irrigation |
|---|---|---|
| Melons | 1. Poor fruit quality (muskmelon)<br>2. increased nitrate content (watermelon)<br>3. poor seed viability | Flowering and fruit development |
| Summer squash | 1. Deformed<br>2. non-viable pollen grains<br>3. misshapen fruits | Bud development and flowering |
| **Leafy Vegetables** | | |
| Leafy vegetables | 1. Tough leaves<br>2. Poor foliage growth<br>3. Nitrate accumulation | Throughout growth and development |
| Asparagus | 1. Reduced spear size<br>2. Increased fiber content<br>3. Tougher<br>4. Lower grade spears | Spear production and fern development |
| Lettuce | 1. Tough leaves<br>2. Poor growth<br>3. Tip burning | Throughout development |
| **Legume Vegetables** | | |
| Vegetable pea | 1. Reduced root nodulation<br>2. Poor pod filling<br>3. Poor seed viability | Flowering and pod filling |
| Lima bean | 1. Greyish leaf color<br>2. Blossom drop<br>3. Poor seed viability | Pollination and pod development |
| Snap bean | 1. Blossom drop<br>2. Inadequate pod filling<br>3. Poor seed viability | Flowering and pod enlargement |
| Sweet potato | 1. Reduced root enlargement<br>2. Poor yield<br>3. Growth cracks | Root enlargement |

## Effect of climate change on vegetable seed production

The increase in temperature has a detrimental impact on the activity of pollinating agents, leading to a decrease in seed yield. Many bee species employ physiological and behavioral mechanisms to regulate the temperatures in their flight muscles before, during, and after flight (Willmer and Stone, 2004). Concerning potential repercussions of future global warming, the behavioral responses of pollinators to avoid extreme temperatures could significantly reduce pollination services (Corbet et al. 1993). Examples of behavioral strategies for thermal regulation include extended periods of sunbathing to warm up and seeking shade or returning to the nest to cool down. These behaviors reduce the time pollinators spend visiting flowers, subsequently impacting pollination and the setting of fruit or seeds (Willmer and Stone, 2004).

## Effect of climate change on diseases and pests of vegetables

Beyond the direct effects on plants, climate change alters pest and disease dynamics, affecting factors such as incidence, host-pathogen interactions, insect distribution, and overwintering capacity. Elevated temperatures, for instance, down-regulate genes involved in host defense responses, increasing disease severity in plants (Sun et al., 2011). Crop plants exposed to stress conditions are generally more vulnerable to pathogens compared to those growing under normal circumstances (Desprez-Loustau et al., 2006). Increased rainfall coupled with higher temperatures creates favorable conditions for the development of bacterial and fungal pathogens. Air pollution, especially ozone and UV-B radiation, along with nutrient availability, also play crucial roles in shaping plant diseases.

Longer surface wetness on plant green foliage, combined with elevated temperatures, intensifies fungal diseases like anthracnose in peppers (Than et al. 2008; Park et al. 2009). Warmer temperatures promote late blight caused by *Phytophthora capsici* in peppers, with a more virulent A2 type observed, exhibiting a higher optimum growth temperature than the A1 type (Sheu et al., 2009). Climate change aids the adaptability and global spread of invasive species with high fertility and dispersal capabilities. Examples include the swift global dissemination of the tomato pinworm, *Tuta absoluta* (Desneux et al. 2011). Elevated CO2 conditions enhance feeding behavior in common armyworms (*Spodoptera litura*) on mungbean plants due to increased sugar levels (Srivastava et al., 2002). Green peach aphid (*Myzus persicae*) reproduction accelerates on *Brassica oleracea* plants under elevated CO2 (Bezemer et al., 1999), while rising temperatures facilitate the rapid multiplication of whiteflies (*Bemisia tabaci*), posing a threat to solanaceous crops as vectors for virus diseases like tomato leaf curl disease (Hanson et al., 2011). Temperature variations also impact the development, fertility and mortality of pests such as *Aphis gossypii* on cucumber (Satar et al., 2005). The increasing incidence of viroids, a recent concern for solanaceous crop cultivation, is favored by rising temperatures, especially above 25 C, posing risks to major vegetables like tomatoes and peppers. Sustained temperatures of 35 C proved lethal to immature stages of *A. gossypii* (Satar et al. 2005). Similarly, heat-shock treatments within the range of 40-47 C inflicted damage on pod borers (*Helicoverpa armigera*) (Mironidis and Savopoulou-Soultani, 2010). Table 4 summarises the expected impact of climate change on certain insect pests affecting vegetables

**Table 4: Projected Impact of Climate Change on Certain Insect Pests Affecting Vegetables**

| Common name | Scientific name | Current habitat | Climate change impact | References |
|---|---|---|---|---|
| Melon fly | *Bactrocera cucurbitae* | Located in south of the Japanese archipelago | Climate change expected to facilitate invasion into Japanese archipelago | Gutierrez et al., 2021 |
| Sweet potato weevil | *Cylas formicarius* | Found in tropical and subtropical regions worldwide | Likely expansion into temperate zones due to rising temperatures | Jackai et al., 2006 |
| Aphids | *Aphis gossypi, Aulacorthum solani* | Widely distributed globally | Accelerated generational cycles and increased infestation potential driven by lower developmental thresholds and reduced thermal requirements | Skend i et al., 2021 |
| Fruit and pod borer | *Helicoverpa armigera; Spodoptera litura* | Globally distributed | Heightened infestation rates attributed to augmented generational frequency | Kocmánková et al., 2010 |
| Colorado potato beetle | *Leptinotarsa decemlineata* | Present in North America, Southern Europe, Asia, and much of the Pacific ocean | Potential northward range expansion by 400 km, elevating risk levels in 99% of potato cultivation areas | Wang et al., 2017 |
| Diamond back moth | *Plutella xylostella* | Originating in Europe, cosmopolitan in distribution | Increased overwintering occurrences and heightened pest activity levels | Chen et al., 2021 |
| Cabbage butter fly | *Pieris brassicae* | Found in Europe, NW Africa, and Asia; Introduced to America and Australia | Range expansion and increased abundance and variety; Decreased biodiversity | Ryan et al., 2019 |

## Impact of climate change on vegetables

Efforts are needed to counteract climate change's impact on vegetable crops. Identifying heat and drought-tolerant germplasm and advancing breeding lines are key focuses. Research aims to enhance water-use efficiency in production systems to combat hot and dry conditions. Strategies include optimizing fertilizer application, precise water delivery, grafting for disease tolerance, and soil amendments. These measures collectively contribute to resilient and productive vegetable cultivation under challenging climatic conditions.

### Innovative irrigation management for water conservation

To address heat stress in open-field vegetable crops, advanced irrigation systems have proven effective (Liu and Kang, 2006). These systems improve the microclimate and crop production by controlling canopy-level climate. Sustainable smart irrigation is crucial for food security and sustainable development goals. It ensures precise water application, reducing losses from evaporation and seepage, saving water, energy and labor. Smart irrigation methods include micro-irrigation, sprinkler irrigation, and surface irrigation, making it a key climate change adaptation strategy.

### Grafting

Grafting has primarily been employed to manage soilborne diseases that impact the production of fruit vegetables, especially solanaceous crops and cucurbits. Beyond disease control, grafting can confer tolerance to various soil-related environmental stressors such as drought, salinity, and low soil temperatures when appropriate, tolerant rootstocks are utilized. Grafted plants exhibit enhanced tolerance to low soil temperatures, exemplified by *Solanum lycopersicum* *S.habrochaites* rootstocks providing resilience to temperatures ranging from 10 C to 13 C for their grafted tomato scions. Similarly, eggplants grafted onto *S. integrifolium* x *S. melongena* rootstocks demonstrate improved growth at lower temperatures (18 C to 21 C) compared to non-grafted plants (Okimura et al., 1986). The application of grafting extends to watermelon plants, where grafted specimens display relatively higher tolerance to sub-optimal temperatures compared to their ungrafted counterparts, offering the potential for production under stressful conditions (Mohamed et al., 2018).

### Harnessing genetic resources for growing of climate-resilient vegetable varieties

Climate change necessitates developing vegetable varieties with enhanced resistance to abiotic and biotic stresses. Breeding programs benefit from local and traditional varieties and the genetic diversity of wild relatives of cultivated vegetables. Collecting, conserving, and characterizing landraces and wild relatives is essential to mitigate climate change challenges (Lane and Jarvis, 2007). Wild relatives provide plant breeders with vital genes for developing stress-resilient plants. Agrobiodiversity in farmers' fields and bioreserves also offers valuable gene pools. Global gene banks house diverse collections, including improved varieties, landraces, and wild species.

**Table 5: List of abiotic stress tolerant genetic resources identified in some vegetable crops**

| Vegetable | Tolerant sources | Tolerance to abiotic stress | Reference |
|---|---|---|---|
| Tomato | *S. habrochaites, S. pennellii, S. pimpinellifolium, S. esculentum var. cerasiforme, S. hirsutum, S. cheesmanii, S. chilense, S. habrochaites, S. sitiens L. pennellii, L. chilense, S. sitiens,* | Drought | Rai et al. (2011) |
| | LA3317 (Campbell), CL5915-93D4-1-0-3 and LA2661 ("Nagcarlang"); BJ01, BJ02; WAC1, ATS020 | High Temperature | Ayenan et al. 2021 |
| | *L. chilense* (LA 1969 and LA 1971), and *Solanum lycopersicoides* Dun. (LA 1964)] | Cold Tolerance | Wolf et al. 1986 |
| Brinjal | *S. macrocarpon, S. gilo S. macrosperma, S. integrifolium, S. sodomaeum (syn. S. linneanum), S. insanum* | Drought | Rai et al., 2011; Cebeci et al. 2023 |
| | *Solanum insanum, Solanum melongena* (Guhala Chatua Local, DBL-21 and DBL-08) | High temperature | Cebeci et al. 2023; Santhiya et al. 2019 |
| Chilli | *C. chinense, C. baccatum var. pendulum, C. eximium,* Arka Lohit, IIHR-Sel.-132 | Drought | Kopta et al. 2020 |
| | *C. annuum (DLS-161-1, DLS-152-1, DLS-20-11)* | High temperature | Srivastava et al. 2020; Aruna et al 2023 |
| | *C. annuum* (DLS-CT-IR2-1, DLS-P-1R-3 , DLS-L-2R-5 , DLS-CT-IR2-2) | Cold Tolerance | Mangal et al. 2022 |
| Potato | *C. chinense, C. baccatum var. pendulum, C. eximium,* Arka Lohit, IIHR-Sel.-132 | Drought | Arvin and Donnelly (2008) |
| | *Solanum tuberosum* (FA, D73, and C132) | High temperature | Zhang et al., 2024 |
| Okra | *A. caillei, A. rugosus, A. tuberosus* | Drought | Charrler (1984) |
| Onion | *Allium fistulosum, A. munzii,* | Drought | Singh (2010) |
| French bean | *P. acutifolius* | Drought | Kavar et al. (2011) |
| Water melon | *Citrullus colocynthis* (L.) Schrad. | Drought | Dane et al. (2007) |
| Winter Squash | *Cucurbita maxima* | Drought | Chigumira and Grubben (2004) |
| Cucumis spp. | *Cucumis melo* var. *momordica, Cucumis pubescens, Cucumis melo* var. *callosus, Cucumis melo* | Drought | Kusvuran (2012), Pandey et al. (2011) |
| Carrot | *D. carota* L. subsp. *carota, D. carota* L. subsp. *capillifolius,* and *D. carota* L. subsp. *gummifer* | High Temperature & Drought | Simon et al. 2021 |

## Harnessing modern biotechnological and genomic tools

Advancements in biotechnology have equipped plant breeders with an array of tools to augment phenotypic screening, including *in vitro* screening, molecular markers, marker-assisted selection, and genetic engineering. The ongoing research utilizes these tools to develop plants with enhanced stress tolerance to effectively combat the challenges posed by climate change (Mtui, 2011). Traditional field performance assessment of genotypes under stress requires significant resources. In-vitro screening is a valuable alternative, as demonstrated in evaluating 30 potato genotypes for heat stress. This method identified genotypes with superior growth and microtuberization under heat stress conditions, providing valuable genetic material for breeding heat-adapted elite potato genotypes (Mohamed et al., 2016). The rise of genomics, facilitated by whole genome sequencing and the advent of high-throughput genetic and molecular technologies, enables the genetic manipulation of genes associated with environmental stress tolerance. Plant molecular genetics research complements traditional plant breeding, contributing to the increase and sustainability of crop productivity. The integration of novel insights from genomic research with traditional breeding methods enhances the capacity to improve crop plants effectively.

## Discovery of quantitative trait loci (QTLs) and Genes

Lin et al. (2006) identified random amplified polymorphic DNA (RAPD) markers linked to heat tolerance in the tomato line CL5915. Additionally, QTLs were pinpointed for various traits crucial to reproductive success, such as pollen viability under continuous mild heat conditions (Xu et al., 2017). These findings offer valuable insights for the development of more heat-tolerant tomato varieties. An association mapping approach, utilizing a high-throughput genomic array, was employed to enhance heat tolerance in tomatoes, revealing 15 common markers associated with the studied traits. These promising candidate genes can potentially be transferred to cultivated tomatoes to enhance their performance under high temperatures (Ruggieri et al., 2019). The integration of QTL analysis with gene discovery facilitates a comprehensive understanding of stress tolerance, aids in developing effective markers for marker-assisted selection, and identifies candidate genes for genetic engineering.

## Transgenics incorporating genes for climate change resilience

While stress response genes, particularly in Arabidopsis thaliana, have been identified, few have improved tolerance when overexpressed in vegetables (Zhang et al., 2004). For example, expressing AVP1, a vacuolar H+ pyrophosphatase from *A. thaliana*, in tomatoes enhanced performance under water deficit conditions by developing a stronger root system (Park et al., 2005). Similarly, CBF/DREB1 genes have engineered drought tolerance in tomatoes (Hsieh et al., 2002), and ROB5 from bromegrass improved transgenic potatoes' heat tolerance (Gusta, 2012). Integrating genetic engineering with conventional breeding accelerates the development of climate-adapted vegetable cultivars (Ortiz et al., 2014).

## Adaptive Agriculture Practices to deal with Climate Change

### Harvesting Rainwater

The collection of rainfall provides an economical water supply method, particularly valuable in drought-prone regions. While it conserves water resources, the operation of rainwater storages requires additional inputs. However, this approach is not without challenges, as the collection of rainwater may lead to declines in groundwater levels, impacting the balance of ecosystems.

### Precision Agriculture

Precision farming represents a significant advancement in addressing global climate change and agricultural challenges. It involves site-specific farm management, utilizing technologies such as drone observations, satellite data, and online farming software for data processing. This approach allows farmers to focus on critical areas instead of treating the entire field, saving resources and reducing environmental pollution.

### Cover Crops

Integrating cover crops into farming practices proves to be an effective method for preventing soil erosion, enhancing water retention, and facilitating nitrogen fixation. Legumes, known for nitrogen production, play a crucial role in converting atmospheric nitrogen into plant-accessible forms. Cover crops also serve as organic manure and fodder for grazing cattle.

### No-Tillage or Minimum Tillage

No-till farming, characterized by minimal or no soil disturbance, helps prevent soil erosion and promotes carbon sequestration. This method contributes to climate change mitigation and improves the overall health of the soil and natural environment. Additionally, it reduces the need for soil depletion and minimizes inputs required for tillage activities and equipment maintenance.

### Crop Diversification and Rotation

Crop rotation, an established agricultural practice, has proven effective in weed and pest management while minimizing the use of chemicals. Crop diversification contributes to ecosystem biodiversity and enriches the culinary experiences of humanity.

## Conclusion and Future Prospects

Climate change, driven partly by human activities like industrialization, is increasingly affecting agriculture and crop production. Temperature changes and $CO_2$ levels play significant roles, impacting crop growth and susceptibility to pests and diseases, thus threatening global food security. Key effects include drought, salinity, and flooding from rising sea levels. Addressing these challenges requires a holistic approach, emphasizing improved water use efficiency and protective agronomic practices, especially in developing countries. Vegetable breeders are crucial in this effort, using genetic diversity from landraces and wild species to develop stress-tolerant crops. Sustainable strategies must be adopted to mitigate climate change's impact on agriculture and ensure future food security.

## References

Abdus Siddique, M. and P. Goodwin (1980). Seed vigour in bean (*Phaseolus vulgaris* L. cv. Apollo) as influenced by temperature and water regime during development and maturation. Journal of Experimental Botany, 31: 313-323.

Aruna TS, Srivastava A, Tomar BS, Behera TK, Harikrishna, Jain PK, Pandey R, Singh B, Gupta R, Mangal M (2023) Genetic analysis of heat tolerance in hot pepper: insights from comprehensive phenotyping and QTL mapping. Front Plant Sci. 2023;14:1232800

Arvin MJ and Donnelly DJ (2008) Screening potato cultivars and wild species to abiotic, stresses using an electrolyte leakage bioassay. J. Agric. Sci. Technol. 10: 33-42.

Ayenan, M.A.T., Danquah, A., Hanson, P. *et al.* (2021) Identification of new sources of heat tolerance in cultivated and wild tomatoes. *Euphytica* 217, 33.

Azam, A., I. Khan, A. Mahmood and A. Hameed (2013).Yield, chemical composition and nutritional quality responses of carrot, radish and turnip to elevated atmospheric carbon dioxide. Journal of the Science of Food and Agriculture, 93: 3237-3244.

Becker, C. and H. P. Klaring (2016). CO2 enrichment can produce high red leaf lettuce yield while increasing most flavonoid glycoside and some caffeic acid derivative Concentrations. Food Chemistry, 199: 736-745.

Bezemer, T. M., K. Knight, E. J. Newington and T. H. Jones (1999). How general are aphid responses to elevated atmospheric CO2? Annals of the Entomological Society of America, 92: 724-730.

Bisbis, M. B., N. Gruda and M. Blanke (2018). Potential impacts of climate change on vegetable production and product quality-a review. Journal of Cleaner Production, 170: 1602-1620

Cebeci E., Boyacı HF., Do an Y., Toppino L., Rotino GL., (2023) Determination of Heat and Drought Tolerant Lines in Segregating Populations Produced by Interspecific Crosses in Eggplant. Ekin J. 9(2):81-90.

Charrler A (1984) Genetic resources of Abelmoschus (Okra) IBPGR Secretariat, Rome. 5pp.

Chen Y, Liu Z, Régnière J, Vasseur L, Lin J, Huang S, Ke F, Chen S, Li J, Huang J, Gurr GM, You M, You S. (2021) Large-scale genome-wide study reveals climate adaptive variability in a cosmopolitan pest. Nat Commun. 12(1):7206.

Chigumira NF and Grubben GJH (2004) Cucurbita maxima Duchesne. In : Grubben GJH and Denton OA (Eds.) PROTA2: Vegetables/ Legumes, PROTA, Wageningen, The Neitherlands.

Conesa, M. A., J. Galmes, J. M. Ochogavia, J. March, J. Jaume, A. Martorell, D. M. Francis, H. Medrano, J. K. Rose and J. Cifre (2014). The postharvest tomato fruit quality of long shelf-life Mediterranean landraces is substantially influenced by irrigation regimes. Postharvest Biology and Technology, 93: 114-121.

Coolong, T. W. and W. M. Randle (2003). Temperature influences flavor intensity and quality in Granex 33'onion. Journal of the American Society for Horticultural Science, 128: 176-181.

Corbet, S. A., M. Fussell, R. Ake, A. Fraser, C. Gunson, A. Savage and K. Smith (1993). Temperature and the pollinating activity of social bees. Ecol. Entomol., 18: 17-30.

Dane F and Liu J (2007) Diversity and origin of cultivated and citron type watermelon (Citrullus lanatus). Genet. Resour. Crop. Eol. 54:1255-1265

Desneux, N., M. G. Luna, T. Guillemaud and A. Urbaneja (2011). The invasive South American tomato pinworm, *Tuta absoluta*, continues to spread in Afro-Eurasia and beyond: The new threat to tomato world production. Journal of Pest Science, 84: 403-408.

Desprez-Loustau, M., B. Marcais, L. Nageleisen, D. Piou and A. Vannini (2006). Interactive effects of drought and pathogens in forest trees. Annals of Forest Science, 63: 597-612.

Doddrell NH, Lawson T, Raines CA, Wagstaff C, Simkin AJ. Feeding the world: impacts of elevated [CO2] on nutrient content of greenhouse grown fruit crops and options for future yield gains. Hortic Res. 2023 Feb 21;10(4):uhad026. doi: 10.1093/hr/uhad026. PMID: 37090096; PMCID: PMC10116952.

Giri, A., B. Armstrong and C. B. Rajashekar (2016). Elevated carbon dioxide level suppresses nutritional quality of lettuce and spinach, American Journal of Plant Sciences, 7: 246- 258.

Gusta, L. (2012). Abiotic stresses and agricultural sustainability. Journal of Crop Improvement, 26: 415-427.

Gutierrez, A.P., Ponti, L., Neteler, M. et al. (2021) Invasive potential of tropical fruit flies in temperate regions under climate change. Commun Biol 4, 1141.

Hanson, P., P. A. Gniffke, J. Shieh and C.-W. Tan (2011). Solanaceous vegetable breeding at AVRDC. The World Vegetable Center to meet the challenges of climate change in the tropics. In: D.-H. Wu, M.-T.Lu, T.-H.Tseng,

Hatfield, J. L.; Boote, K. J.; Kimball, B. A.; Ziska, L. H.; Izaurralde, R. C.; Ort, D. R,; Thomson, A. M.; and Wolfe, D., "Climate Impacts on Agriculture: Implications for Crop Production" (2011). Publications from USDA-ARS / UNL Faculty. 1350.

Hsieh, T. H., J. T. Lee, Y. Y. Charng and M. T. Chan (2002). Tomato plants ectopically expressing Arabidopsis CBF1 show enhanced resistance to water deficit stress. Plant Physiol., 130: 618-626.

Ibrahim, M. A., A. Nissinen, N. Prozherina, E. Oksanen and J. Holopainen (2006). The influence of exogenous monoterpene treatment and elevated temperature on growth, physiology, chemical content and headspace volatiles of two carrot cultivars (*Daucus carota* L.). Environmental and Experimental Botany, 56: 95-107

IPCC, 2021: Climate Change 2021: The Physical Science Basis. Contribution of Working Group I to the Sixth Assessment Report of the Intergovernmental Panel on Climate Change [Masson-Delmotte, V., P. Zhai, A. Pirani, S.L. Connors, C. Péan, S. Berger, N. Caud, Y. Chen, L. Goldfarb, M.I. Gomis, M. Huang, K. Leitzell, E. Lonnoy, J.B.R. Matthews, T.K. Maycock, T. Waterfield, O. Yelekçi, R. Yu, and B. Zhou (eds.)]. Cambridge University Press, Cambridge, United Kingdom and New York, NY, USA, 2391 pp. doi:10.1017/9781009157896

Jackai L. E., B. Sosinski, D. M. Jackson et al., (2006) "Occurrence and intra-specific variation of sweetpotato weevil (Brentidae: Coleoptera) in relation to its potential spread in southern United States of America and the Caribbean," in ISHS Acta Horticulturae 703: II International Symposium on Sweetpotato and Cassava: Innovative Technologies for Commercialization, pp. 197–204, International Society for Horticultural Science, Leuven, Belgium,.

Jin, C., S. Du, Y. Wang, J. Condon, X. Lin and Y. Zhang (2009). Carbon dioxide enrichment by composting in greenhouses and its effect on vegetable production. Journal of Plant Nutrition and Soil Science, 172: 418-424.

Kavar T, Maras M, Kidriè M, Šuštar-Vozliè J and Megliè Vladimir (2011) The expression profiles of selected genes in different bean species (Phaseolus spp.) as response to water deficit. J. Central European Agri.12(4): 557-576

Kocmánková E., Trnka M., Juroch J., Dubrovsk M., Semerádová D., Mo n M., alud Z., Pokorn R., Lebeda A. (2010) Impact of climate change on the occurrence and activity of harmful organisms. Plant Prot. Sci. 45:S48–S52. doi: 10.17221/2835-PPS

Kopta, Tomas, Agnieszka Sekara, Robert Pokluda, Vojtech Ferby, and Gianluca Caruso. 2020. "Screening of Chilli Pepper Genotypes as a Source of Capsaicinoids and Antioxidants under Conditions of Simulated Drought Stress" Plants 9, no. 3: 364. https://doi.org/10.3390/plants9030364

Kumari, S. and M. Agrawal (2014). Growth, yield and quality attributes of a tropical potato variety (*Solanum tuberosum* L. cv Kufri Chandramukhi) under ambient and elevated carbon dioxide and ozone and their interactions. Ecotoxicology and Environmental Safety, 101: 146-156.

Kusvuran S (2012). Effects of drought and salt stresses on growth, stomatal conductance, leaf water and osmotic potentials of melon genotypes (Cucumis melo L.). Afr. J. Agril. Res. 7(5): 775-781

Lane, A. and A. Jarvis (2007). Changes in climate will modify the geography of crop suitability: agricultural biodiversity can help with adaptation. Journal of SAT Agricultural Research 4(1): 1-12.

Lin, K. H., H. F. Lo, S. P. Lee, C. G. Kuo and J. T. Chen (2006). RAPD markers for the identification of yield traits in tomatoes under heat stress via bulk segregant analysis. Hereditas, 143:142-154.

Mangal M, Srivastava A, Singh K, Kumari S and Tomar B S (2022) Biochemical and molecular responses of hot pepper (*Capsicum annuum*) to cold stress. Indian Journal of Agricultural Sciences 92 (7): 871–5.

Mattos, L. M., C. L. Moretti, S. Jan, S. A. Sargent, C. E. P. Lima and M. R. Fontenelle (2014). Climate changes and potential impacts on quality of fruit and vegetable crops. In: Emerging technologies and management of crop stress tolerance (pp. 467-486). Academic Press

Mironidis GK, Savopoulou-Soultani M. Effects of heat shock on survival and reproduction of Helicoverpa armigera (Lepidoptera: Noctuidae) adults. J Therm Biol. 2010 Feb;35(2):59-69. doi: 10.1016/j.jtherbio.2009.11.001.

Mironidis, G. K. and M. Savopoulou-Soultani (2010). Effects of heat shock on survival and reproduction of *Helicoverpa armigera* (Lepidoptera:Noctuidae) adults. Journal of Thermal Biology, 35(2): 59-69.

Mohamed, F. H., M. A. N. Hussein, K. E. Abd-El-Hamid and M. W. M. Elwan (2018). Response of watermelon plants grafted onto different cucurbit rootstocks to sub-optimal growing temperature. Hortscience Journal

Mohamed, F., K. El-Hamed, M. Elwan, N. El-Magawry and A. El-Salam (2016). *In Vitro* screening of different potato genotypes for heat stress tolerance. Catrina: The International Journal of Environmental Sciences, 15(1): 77-93.

Moore, C. E., K. Meacham-Hensold, P. Lemonnier, R. A. Slattery, C. Benjamin, C. J. Bernacchi, T. Lawson and A. P. Cavanagh (2021). The effect of increasing temperature on crop photosynthesis: from enzymes to ecosystems. Journal of experimental botany, 72(8), 2822-2844.

Motsa, M. M., M. M. Slabbert, W. Van Averbeke and L. Morey (2015). Effect of light and temperature on seed germination of selected African leafy vegetables. South African Journal of Botany, 99: 29-35.

Mtui, G. Y. (2011). Involvement of biotechnology in climate change adaptation and mitigation: Improving agricultural yield and food security. International Journal for Biotechnology and Molecular Biology

Oh, M. M., E. E. Carey and C. Rajashekar (2009). Environmental Stresses Induce healthpromoting Phytochemicals in Lettuce, Plant Physiology and Biochemistry, 47: 578-583.

Okimura, M., S. Matsou, K. Arai and S. Okitso (1986). Influence of soil temperature on the growth of fruit vegetable grafted on different stocks. Bull. Veg. Ornam. Crops Res. Stn. Japan, C9:3-58

Ortiz, R., A. Jarvis, P. Fox, P. K. Aggarwal and B. M. Campbell (2014). Plant genetic engineering, climate change and food security.CCAFS working paper no.

72.CGIAR Research Program on Climate Change, Agriculture and Food Security (CCAFS). Copenhagen, Denmark.

Ouzounidou, G., A. Giannakoula, I. Ilias and P. Zamanidis (2016). Alleviation of drought and salinity stresses on growth, physiology, biochemistry and quality of two *Cucumis sativus* L. cultivars by Si application. Brazilian Journal of Botany, 39: 531-539.

Ouzounidou, G., I. Ilias, A. Giannakoula and I. Theoharidou (2014). Effect of water stress and NaCl triggered changes on yield, physiology, biochemistry of broad bean (*Vicia faba*) plants and on quality of harvested pods. Biologia, 69: 1010-1017.

Pandey S, Ansari WA, Jha A, Bhatt KV and Singh B (2011) Evaluations of melons and indigenous Cucumis spp. genotypes for drought tolerance, 2nd Internatioanal symposium on underutilized plant species,27th June - 1st July, The Royal Chaulan Kuala Lumpur, Malaysis, (A61). 95 pp

Park, S. K., S. H. Kim, H. G. Park and J. B. Yoon (2009). *Capsicum* germplasm resistant to pepper anthracnose differentially interact with *Colletotrichum* isolates. Horticulture Environment and Biotechnology‹ 50(1): 17-

Park, S., Li, J., Pittman, J.K., Berkowitz, G.A., Yang, H., Undurraga, S., Morris, J., Hirschi, K.D. and Gaxiola, R.A. (2005) Up-regulation of a H+ -pyrophosphatase (H+-PPase) as a strategy to engineer drought-resistant crop plants. Proc. Natl Acad. Sci. USA, 102, 18830–18835.

Peet, M. M. and D. W. Wolfe (2000). Crop Ecosystem Responses to Climate Change: Vegetable Crops, New York, and Wallingford, UK, CABI Publishing

Rai N, Tiwari SK, Kumar R, Singh M and Bharadwaj DR (2011) Genetic Resources of Solanaceous Vegetables in India. National symposium on vegetable biodiversity. Jawaharlal Nehru Krishi Vishwa Vidyalaya, Jabalpur, M.P. April, 4-5. 91-103pp

Rosales, M. A., L. M. Cervilla, E.Sanchez Rodriguez, M. D. M.Rubio Wilhelmi, B. Blasco, J. J. Rios, T Soriano, N. Castilla, L. Romero and J. M. Ruiz (2011). The effect of environmental conditions on nutritional quality of cherry tomato fruits: evaluation of two experimental Mediterranean greenhouses. Journal of the Science of Food and Agriculture, 91: 152-162.

Ruggieri, V., R. Calafiore, C. Schettini, M. M. Rigano, F. Olivieri, L. Frusciante and A. Barone (2019). Exploiting genetic and genomic resources to enhance heat tolerance in tomatoes. Agronomy, 9(1): 22.

Ryan SF, Lombaert E, Espeset A,Vila R, Talavera G, Dinc V, Doellman MM, Renshaw MA, Eng MW, Hornett EA, Li Y, Pfrender ME, Shoemaker deW (2019) Global invasion history of the agricultural pest butterfly (*Pieris rapae*) revealed with genomics and citizen science. Proceedings of the National Academy of Sciences. 116:40, 20015-20024. doi: 10.1073/pnas.1907492116}.

Sandhu SK, Kingra PK & Kaur, S (2018) Effect of Climate Change on Productivity and Disease Scenario of Potato - A Review Journal of Agricultural Physics 18(2): 141-157.http://www.agrophysics.in

Santhiya S, Saha P, Tomar BS, Ghoshal C (2019) Heat stress tolerance study in eggplant based on morphological and yield trait. Indian Journal of Horticulture 76(4): 691.

Satar, S., U. Kersting and N. Uygun (2005). Effect of temperature on development and fecundity of *Aphis gossypii* Glover (Homoptera:Aphididae) on cucumber. Journal of Pest Science, 78(3): 133-137.

Saure, M. (1998). Causes of the tipburn disorder in leaves of vegetables. Scientia Horticulturae, 76: 131-147.

Sheu, Z. M., J. R. Chen and T. C. Wang (2009). First report of the A2 mating type of *Phytophthora capsici* infecting peppers (*Capsicum annuum*) in Taiwan. Plant Disease, 93(5): 548.

Simon, P. W. (2020). Carrot (*Daucus carota* L.) breeding. In J. M. Al-Khayri (Ed.), Advances in plant breeding strategies: Vegetable crops. Cham, Switzerland: Springer. doi:10.1073/pnas.1907492116,

Singh HP (2010). Ongoing research in abiotic stress due to climate change in horticulture, Curtain Raiser Meet on Research Needs Arising due to Abiotic Stresses in Agriculture Management in India under Global Climate Change Scenario, Baramati, Maharashtra, October 29-30. 1-23pp.

Singh, A. K. (2010). Climate change sensitivity of Indian horticulture- Role of technological interventions. pp: 85-95. Souvenir of Fourth Indian Horticultural Congress, HSI, New Delhi

Skend i S, Zovko M, ivkovi IP, Leši V, Lemi D. (2021) The Impact of Climate Change on Agricultural Insect Pests. Insects. 12(5):440. doi: 10.3390/insects12050440. PMID: 34066138; PMCID: PMC8150874.

Spaldon, S., R. K. Samnotra and S. Chopra (2015). Climate resilient technologies to meet the challenges in vegetable production. International Journal of Current Research and Academic Review, 3(2): 28-47.

Srivastava A, Mangal M, Parihar BR, 2020, Development and characterization of heat tolerant genotypes in chilli (Capsicum annuum L.) Annals of Horticulture. 13(1): 34-38

Srivastava, A. C., L. D. Tiwari, M. Pal and U. K. Sengupta (2002). CO2-mediated changes in mungbean chemistry: Impact on plant herbivore interactions. Current Science, 82: 1148-1150

Sun, Y., J. Yin, H. Cao, C. Li, L. Kang and F. Ge (2011). Elevated CO2 influences nematode induced defense responses of tomato genotypes differing in the JA pathway. PLoS ONE, 6, e19751

Than, P. P., R. Jeewon, K. D. Hyde, S. Pongsupasamit, O. Mongkolporn and P. W. J. Taylor (2008). haracterization and pathogenicity of *Colletotrichum* species associated with anthracnose on chili (*Capsicum spp.*) in Thailand. Plant

Thuy TL and Kenji M (2015) Effect of High Temperature on Fruit Productivity and Seed-Set of Sweet Pepper (*Capsicum annuum* L.) in the Field Condition. Journal of Agricultural Science and Technology A and B & Hue University Journal of Science 5 (2015) 515-520. doi: 10.17265/2161-6256/2015.12.010

Toivonen, P. M. and D. M. Hodges (2011). Abiotic stress in harvested fruits and vegetables. In: Abiotic Stress in Plants-Mechanisms and Adaptations. A. Shanker and B. Venkateswarlu (eds.), pp. 39-58. Intech Open Access Publishers.

Vermeulen SJ, Aggarwal PK, Ainslie A, Angelone C, Campbell BM, Challinor AJ, Hansen J, Ingram JSI, Jarvis A, Kristjanson P, Lau C, Thornton PK, Wollenberg E (2010). Agriculture, Food Security and Climate Change: Outlook for Knowledge, Tools and Action. CCAFS Report 3. Copenhagen, Denmark: CGIARESSP Program on Climate Change, Agriculture and Food Security. Downloaded from http://ccafs.cgiar.org/sites/default/ files/pdf/ccafs_report_3-low-res_final.pdf

Vijayakumar A, Shaji S, Beena R, Sarada S, Sajitha Rani T, Stephen R, Manju RV, Viji MM. High temperature induced changes in quality and yield parameters of tomato (*Solanum lycopersicum* L.) and similarity coefficients among genotypes using SSR markers. Heliyon. 2021 Feb 3;7(2):e05988. doi: 10.1016/j.heliyon.2021.e05988.

Wang, C., Hawthorne, D., Qin, Y. et al. (2017) Impact of climate and host availability on future distribution of Colorado potato beetle. Sci Rep 7, 4489. https://doi.org/10.1038/s41598-017-04607-7

Went, F. W. (1953). The effect of temperature on plant growth. Annual Review of Plant Physiology, 4: 347-362.

Willmer, P.G. and G. N. Stone (2004). Behavioral, ecological, and physiological determinants of the activity patterns of bees. In: Advances in the Study of Behavior, San Diego, CA, Elsevier Academic Press Inc.,34: 347-466.

Wolf, S., Yakir, D., Stevens, M.A., & Rudich, J. (1986). Cold Temperature Tolerance of Wild Tomato Species. Journal of the American Society for Horticultural Science, 111(6), 960-964.

Xu, J., N. Driedonks, M. J. Rutten, W. H. Vriezen, G. J. de Boer and I. Rieu (2017). Mapping quantitative trait loci for heat tolerance of reproductive traits in tomato (*Solanum lycopersicum*). Molecular Breeding, 37(5): 58.

Zhang, J. Z., R. A. Creelman and J. K. Zhu (2004). From laboratory to field.Using information from Arabidopsis to engineer salt, cold, and drought tolerance in crops. Plant Physiol., 135:615-621.

Zhang, Sujie, Han Ye, Lingshuang Kong, Xiaoyu Li, Yeqing Chen, Shipeng Wang, and Bailin Liu (2024) "Multivariate Analysis Compares and Evaluates Heat Tolerance of Potato Germplasm" *Plants* 13, no. 1: 142.

# CHAPTER 9

# FLORICULTURE–A PROFITABLE AND ECO-FRIENDLY AGRI-BUSINESS IN BANGLADESH

**Farjana Nasrin Khan**

*Head, Floriculture Division, Horticulture Research Centre,*
*Bangladesh Agricultural Research Institute, Bangladesh*
*Corresponding author: farjanakhan2015@gmail.com*

---

Flowers are one of the most liked and loved things on the earth and it symbolize beauty and purity. Floriculture could be considered as one of the most colourful sector of horticulture, which includes cut and loose flowers, foliage, potted plants, ornamentals and cut greens. With urbanization and increase in disposable income level, the demand for floriculture products has increased significantly. The world consumption of cut flowers and plants is increasing and there is a steady annual increase of 10–15 per cent in all importing countries. Floriculture has emerged as a lucrative agribusiness venture in the world, particularly in developing countries including Bangladesh which ensures higher potential to return compared to other crops.

As a component of floriculture industry, cut flower development has brought a dramatic increase of farmers' income in Bangladesh. Flower trading worth of near about 2 million $ has been recorded at a local flower market, Jashore district simply on the occasion of spring festival and valentine day (2023). This is the testimony of booming flower market in Bangladesh. As flower cultivation is emerging as a profitable business due to higher return compared to other crops, farmers are very much interested to shift their business to floriculture sector.

Apart from the profitability, floricultural components have significant health benefits to people living inside the house and outside. There are so many plants that can purify the air of indoors and outdoors. Multiple studies have proven certain plants act as effective natural air purifiers and are able to absorb polluting organic compounds like Volatile Organic Compounds (VOCs), formaldehyde, and benzene through their leaves and roots. A NASA study even highlighted several plants that excelled in cleaning the air around them. If the pot plants are happy and well maintained in indoors they can help cleanse indoor air of VOC pollutants and improve almost every aspect of indoor environmental quality. The pollution from VOCs, even at very low levels, can cause 'sick building syndrome' and building-related illnesses with symptoms such as loss of concentration, headaches, dry eyes, nose, throat, woozy head and nausea (Carrer, 1999).

Urban-dwellers generally spend about 90% of their time indoors, where the air is likely to be significantly more polluted than outdoors (Wood *et al.*, 2006). Poor indoor air quality is responsible for 6.7 million premature deaths globally (WHO, 2000). Therefore, indoor air pollution's potential consequences on human health are a global concern. As we spend a long time of our lives in indoors so, house plants can be an easy way to bring nature into our home. Therefore, floriculture has the potential to expand more in Bangladesh regarding area and production. Along with playing a vital role in employment generation and poverty reduction to make a profitable agribusiness, floriculture can also play a key role in environmental protection by reducing air pollution.

## Floriculture in Bangladesh

The present floriculture industry in Bangladesh is a new dimension. But some innovative farmers in late seventies started flower cultivation with tuberose on a small scale. Large scale commercial production started from early eighties in Jhikargacha upazila of Jashore district which later on spreaded in the other region of the country. Flower cultivation has adopted commercially in 26 districts among 64 of our country (Khan, 2013). Although cultivation of flowers is prevalent among marginal and small farmers but the involvement of large farmers' in this industry is gradually increasing. About 15–20 lakhs people are directly or indirectly involved in this business and passing their livelihood. The flower market in the country started to expand significantly from 2,000 onward with rapid urbanization and cultural proliferation with new festivals like Valentine's Day, 1st Falgun *etc.* The annual turnover in flower market is near about 150 million $. A wide variety of flowers, especially gerbera, rose, gladiolus, tuberose, orchids, lilium *etc.* are now being commercially cultivated in Bangladesh. Moreover, aglaonema, aralia, diffenbanchia, dracaena, ferns, coleus, croton, *etc.* are also being added in commercial floriculture.

## Why floriculture in Bangladesh?

- Flower cultivation is more profitable than rice and vegetables (3-5 times and 1.5-2 times more returns than rice and vegetable cultivation, respectively (Dadlani, 2003).
- Increasing domestic demand.
- Shorter production cycle.

- Suitable agro ecology.
- Less recurring cost of production materials.
- Employment opportunity is high specially for women.

## Flower export scenario of Bangladesh.

Recently flower is added into the export item in Bangladesh though it is not remarkable. According to Export Promotion Bureau (EPB), the country's export earnings from cut flowers and foliage for July-January period of 2018-19 touched $3.98 million, which was $0.02 million in 2017-18. Though Bangladesh exports very poor portion of the world flower demand, it produces most attractive flower items of the world in a lower price than the most flower exporting country. There is a huge opportunity for Bangladesh to grab the international flower market due to its very much cheap production cost. In spite of various limitations, flower industry is rising and Bangladesh is trying to increase the quantity of exportable flowers to international market mainly ethnic flower market.

## Import of flowers

To meet the market demand, chrysanthemum, lilium, carnation, orchids, rose, anthurium are imported from China, India, Vietnam and Thailand every year. Retailers prefer imported flowers because of their freshness (they are kept in a cold chain) and breadth of product selection. Flower imports still represent somewhere between 20 and 30% of the market in Bangladesh (Khan, 2013). For reducing the flower import custom duty has been increased from 12 to 25% in 2013-14, which may give additional protection to local growers (Kallol, 2013).

## Potential and prospects of floriculture in Bangladesh

Bangladesh has advantage due to its favourable climate, topography, lower labour cost and relatively low production cost for growing many flower and ornamental plants which can help development of flower industry. For quick returns and good market prospects, a vast area of agricultural land has been converted into flower growing areas (Wittstock and Quinto, 2008). As flower cultivation is emerging as a profitable business due to its higher return compared to other crops, farmers are very much interested to shift their business to the floriculture sector. It is also very positive that a good number of large farmers are coming up to invest money in this area. Bangladesh enjoys a very conducive environment for flower production and business due to following reasons:

- Increasing labour cost that is putting pressure on the cost of cultivation of major flower producing countries like the Netherlands, Japan, Taiwan and Israel.
- Environment degradation and cost of land which impede further expansion of cut-flowers production in EU countries.

- Increasing demand in nearer markets of West Asia and South East Asia, where the rising standards of living are pushing-up demand for floriculture products.

Besides cut flower, foliage and ornamental plants also may be a good item for floriculture trade. At present, use of ornamental plants for outdoor and indoor decorations is gaining importance in Bangladesh. The tropical climatic conditions available in the country are ideal for multiplication of many leading ornamental plants of the global trade. The high humidity levels help the plants to grow quickly and give them a vibrant, healthy appearance. Another business is going to start to supply indoor and outdoor plants to various offices, restaurants *etc.* These leaves of ferns, palms, dracaenas, and philodendrons which are perfect as fillers to add a touch of beauty to floral arrangements. Bangladesh has very good potential to become an important supplier of ornamental plants for Asia, the Middle East and Europe (Campo, 2001).

Currently, dried flowers and plant materials provide a great deal of potential as an alternative to fresh flowers and foliage for a wide range of commercial and aesthetic applications, including interior decoration which is very initial stage in Bangladesh. Out of all floricultural goods, dried ornamentals industry is the most lucrative in terms of overall sales and unit value. The Indian export basket comprises 71 per cent of dry flowers which are exported to USA, Europe, Japan, Australia, Far East and Russia. Dry flowers constitute more than two-thirds of total floriculture exports in India (Dilta *et al.*, 2011) which could be a potential area for Bangladesh flower industry. After a successful training from Dr Y.S.R. Horticultural University, Andhra Pradesh, some innovative entrepreneur started to make processed products in Bangladesh.

The oil extraction industry from different fragrant flowers has a bright future as flower oil is widely used in various industries such as perfumes, cosmetics, food and beverage *etc.* The demand for floral oil is rising along with increasing demand for natural and organic products. Bangladesh has the potential in that area as a large number of fragrant indigenous flowers are available that could be used in perfume and oil industry.

Various potting media such as cocopeat, compost, vermicompost and tricho-compost are also selling in market and business is expanding day-by-day. Apart from that business, so many services like repotting of indoor plants, pick-up and delivery services, consultations for selection and management of indoor plants and change the plants after specific time of intervals in home/office/hospital/restaurant *etc.*, are also observing in Bangladesh.

An online business has been started in Bangladesh mainly with potted plants for indoor and roof top gardening. During covid period when people were stacked in their homes, there were tremendous development of home and roof top gardening which later on turned into online business. In addition to producing more revenue, floriculture industry has demonstrated a great deal of promise for creating jobs for rural people, especially women and younger generations. So, commercial floriculture is a profitable venture which can contribute to boost up our national economy. There is hardly any doubt about the potential of flower industry in Bangladesh.

## Profitable agri-business

In Bangladesh, floriculture has become a profitable agri-business endeavor with a larger potential for profit which depends on many factors like area, season, special occasions, market stability, own land or hired land *etc.* If gladiolus is considered as example, following calculation (draft) may be given for taking an assumption about the profitability of gladiolus production in Bangladesh (Table 1).

Area: 1 Bigha (33 decimal), Study area: Savar

Variety: BARI Gladiolus 3

Time of sowing/planting: October-November

Harvesting: January-March

Corm rate per bigha: 28,000 corms/*bigha* (>2.5 cm sized corm should be used)

Yield (flower sticks/ha): 1,80,000-2,00000

Cost of production (for winter season)

**Table 1. Cost benefit ratio for gladiolus production in Bangladesh**

| Item (broad head) | Total cost (BDT) |
|---|---|
| Hiring land | 12,000/- |
| Land preparation (applying Cow dung, other fertilizer, tractor, labour *etc.*) | 11,200/- |
| Corm/seed (28000 corm @ Tk. 2.85) with carrying cost | 79800/- |
| Fertilizers | 6000/- |
| Insecticides and pesticides | 4000/- |
| Irrigation (8 irrigation @ Tk 300 from BADC deep irrigation) | 2400/- |
| Labour for Intercultural operations (25 labour @ Tk 300 | 7500/- |
| Flower harvesting | Self |
| Flower grading and making bundle | 5000/- |
| **Total cost** | **1,27,900/-** |

Return

| Product | Amount | Selling price (BDT) |
|---|---|---|
| Flower sticks | 26,000 (@ Tk. 8.0 each flower sticks) | 2,08,000/- |
| Corm | 25,000 (@ Tk. 2.5 each corm) | 62,500/- |
| Cormel | Due to soil condition of Savar cormel production is negligible | - |
| **Total income** | | **2,70,500/-** |

**Net profit:** 1,42,600/ for winter season only

**Benefit Cost Ratio:** 2.11

Two times gladiolus production is possible /unit area/year. In summer season (planting time May-June), flower price is comparatively high (BDT. Av.10-12) and profit is also high but corm production is little bit lower due to heavy rain. If storage facility is available year round production is possible but not in the same field.

## Eco- friendly nature

Floricultural elements are not only earth-friendly but also help protect environment in various ways and significantly improve the health of both indoor and outdoor residents. Not only do indoor plants make for beautiful home decor, they may also boost our health and well-being. Air purification refers to removal of various impurities like Volatile Organic Compounds (VOCs), Sulphur oxides ($SO_2$), Carbon oxides ($CO_2$) *etc.* from the air by which a person can easily breathe clean air. To reduce the air pollution, plantation is very much needed as outdoor/roof top/ indoor/balcony/window gardening which have so many benefits:

- **Provide shade:** There are so many large tree whose primary role is to provide shade in the surrounding environment due to it's spreading canopy and crown.
- **Improve aesthetics:** Ornamental plants can add beauty and charm with their unique forms and colour both in indoor and outdoor. Flowers are a sign of purity, love, and serenity and have aesthetic value in human life.
- **Improve air quality:** Plants can clean the air and improve air quality of both outdoor and indoor. Plants also replace carbon dioxide with fresh oxygen. Multiple studies have proven certain plants are able to absorb polluting organic compounds like formaldehyde and benzene through their leaves and roots. That absorption purifies the air around the plant. According to NASA study, houseplants can help improve air quality by removing cancer-causing chemicals like formaldehyde and benzene from the air (Wolverton, 1989). Mishra (2021) reported that all plants purify air as they can take carbon dioxide and other volatile organic compounds through stomata of leaves and degrade them to form their food by photosynthesis. Every plant has different speed of purification level.
- **Improve mental freshness**: Participants who spent even 5 to 10 minutes in a room with a few houseplants felt happier and more satisfied than those in a room without plants (Shoemaker *et al.*, 1992).
- **Reduce fatigue:** Exposure to indoor plants helped people recover from mental fatigue.
- **Lower stress and anxiety:** A study found that caring for indoor plants reduced psychological and physiological stress anxiety in study participants. Interacting with plants helped suppress sympathetic nervous system activity

and diastolic blood pressure. Participants reported feelings of comfort when working with plants versus working on a computer (Burchett *et al.*, 2009).

- **Improve our sense of well-being:** Exposure to indoor plants improved feelings of well-being in participants compared to those who weren't exposed to plants (Tarran *et. al*, 2007). Feelings of well-being can help improve our physical health too.
- **Improve environmental wellness:** Environment has a major impact on overall wellness. All areas of wellness are interconnected. Living in an environment that is healthy and peaceful can be beneficial to overall wellness. Our home is our sanctuary and a place for us to recharge. As we spend a long time of our lives in indoors so, house plants can be an easy way to bring nature into our home.

Moreover, plants can:

- Reduce temperature of roof surface by up to 40 °F below conventional roofs
- Reduce city-wide ambient air temperatures by up to 5 °F.
- Improve office performance and focus.
- Support mental health.

The researchers found that a small green wall, containing a mix of indoor plants, were highly effective at removing harmful, cancer-causing pollutants, with 97 per cent of the most toxic compounds removed from the surrounding air in just eight hours (Burchett *et al.*, 2005).

Across a series of tests conducted before and after plants were present in offices, people showed:

- 30% reduction in confusion.
- 37% reduction in tension/anxiety.
- 4.5% increase in vigour.
- 38% reduction in fatigue.
- 58% reduction in depression/dejection.
- 44% reduction in anger/hostility (Source: Burchett *et al.*, 2009).

Considering profitability and eco-friendly nature of Floriculture sector, so many initiatives are being taken by the Government and Private Organizations

- As a public research Institute, Bangladesh Agricultural Research Institute (BARI) is the main research wing and conducting research on various aspects of flowers and ornamentals.
- BARI is also disseminating proven technologies and technical know how to the farmers, extension officers and workers and NGO personnel primarily through lecturing, training and field trials.

- BARI also provides the research facilities for M.Sc. and Ph.D students of different universities in the area of Floricultural research.
- Establishment of a Floriculture Research Centre is under process.
- Every year a Tree fair is organized at Dhaka spanning over a month emphasizing floricultural products and tree fairs are also organized at every divisional and districts level country wide.
- The Government is offering a 10 per cent discount on holding tax for property owners who grow rooftop gardens in city corporation and municipal areas across the country.
- Some Agricultural universities also have research facilities for providing support to the M.Sc and Ph.D students.
- Department of Agricultural Extension (DAE) is the key authority to disseminate the research findings/ technology among the flower growers.
- Private organizations and NGOs encourage the flower growers by providing propagating materials (bulbs/corms/seedlings), training *etc.*

## Challenges in Floriculture in Bangladesh

Flower industry in Bangladesh is suffering from various obstacles those are given bellow

- Lack of quality and required amount of propagating materials.
- Lack of good number of varieties and diversified flowers.
- The maximum flower growing area (>95%) is being practiced under open field conditions.
- Unavailability of various production and postharvest management inputs like UV poly film, agro shade net, packaging materials, liquid fertilizers, pesticides, flower cap *etc.*
- Lack of infrastructure for protective cultivation, propagation, conservation and post-harvest management facilities.
- Inadequate and under developed communication system and absence of cool chain facilities from production areas to market place.
- Lack of organized flower market having preservation facilities.
- Unavailability of large amount of industrial loan.
- Unavailability of soft loan.
- Importation of Artificial flower.
- No linkage yet developed with flower based Institution of any other developing and developed countries.
- Absence of Good Agricultural Practices in every step of production to marketing to ensure the quality of the flowers.

## Strategies

- Floriculture should be recognized as a priority area for development in the agriculture policies of the country.
- Commercial floriculture should be emerged as hi-tech activities are taking place under controlled climatic conditions inside green house.
- Quality planting materials should be ensured to produce quality flowers and flowers must be grown and packed according to the international standards.
- Good Agricultural Practices should be followed in every step of production to marketing to ensure the quality of the flowers.
- Effective cool chain facilities from field to market should be developed.
- Financial institutions like commercial bank should come forward to provide sufficient soft loan to flower growers with reduced interest rate.
- Specialized terminal flower market having preservation facilities in major city must be developed.
- Development of Marketing Information System (MkIS) for floricultural produces and make it easily accessible.
- The Government should contemplate duty exemption on the import of materials like UV poly film, production accessories like liquid fertilizer, pesticides flower cap *etc.* and packaging materials for promoting the floriculture industry.

## Conclusion

Floriculture has the potential of more expansion in Bangladesh in area and production which needs market need intensive effort for its marketing on the part of government and private entrepreneurs. Developing an integrated approach for floriculture including input needs, technology and guidance, resource management, infrastructure development, marketing facilities, financial assistance, export promotion, entrepreneur–friendly policies would lead to a balanced growth of the floriculture industry. It is the right time for research organization, universities, private organization and flower grower's association to come forward and work closely with each other for betterment of farmers, users as well as countries economy.

### References

Brown, S. K., M. R. Sim, M. J. Abramson, C.N. Gray. 1994. Concentrations of volatile organic compounds in indoor air-A review. *Indoor Air* 4(2):123-34.

Burchett, M. 2005. Improving indoor environmental quality through the use of indoor potted plants. Final Report to Horticulture Australia Ltd, Sydney.

Burchett, M., J. Tarran and F. Torpy. 2007. Potted-plants improve indoor environmental quality published by Faculty of Science, University of Technology Sydney. NSW, Australia

Burchett, M., F. Torpy and J. Brennan 2009. Towards Improving Indoor Air Quality With Potted-Plants A Multifactorial Investigation. Final Report to Horticulture Australia Ltd. (Section 2).

Campo, J.B. 2001. Export potentialities of ornamental plants from Bangladesh. In: Development of Chinese palm and other ornamental plants. Hortex foundation, Sher-e-Bangla Nagar, Dhaka

Carrer, P., D. Alcini, D. Cavallo. 1999. Home and workplace complaints and symptoms in office workers and correlation with indoor air pollution, Proceedings the 8th International Conference on Indoor Air Quality and Climate, Edinburgh, Scotland, Vol. 1. 129-134.

Dadlani, N.K. 2003. Global Positioning of Bangladesh Floriculture. A paper presented in International Floriculture Conference, 6 November 2003, BARC, Farmgate, Dhaka.

Dilta, B.S., Sharma, B.P., Baweja, H.S., and Kashyap, B. 2011. Flower drying techniques - A review. Int. J. Farm Sci. 1(2): 1-16.

Export Promotion Bureau. 2019. HS Code Wise Export Report of 2017-2019, Dhaka, Bangladesh.

Kallol, A.S. 2013. Custom duty hike may boost flower export. 8th June, Dhaka Tribune.

Khan, F.N. 2013. Production dynamics of Bangladesh Floriculture. A paper presented in the National Seminar on Floriculture Development in Bangladesh on 18th May, 2013 at BARC, Farmgate, Dhaka organized by SCDP, DAE.

Khan, M. R. A. 2013. Flower market development in Bangladesh. A paper presented in the National Seminar on Floriculture Development in Bangladesh on 18th May, 2013 at BARC, Farmgate, Dhaka organized by SCDP, DAE.

Mishra, S. 2021. Air Purifying Plants. DOI: 10.13140/RG.2.2.36137.93284

Price, M. 2023. Best indoor plants for low light, clean air and easy care. CNET HOME TIPS.

Shoemaker, C.A., K. Randall, P.D. Relf, and E.S. Geller. 1992. Relationships between plants, behavior, and attitudes in an office environment. Hort Technology 2: 205-206.

Tarran, J., F. Torpy and M. Burchett. 2007. Use of living pot-plants to cleanse indoor air – research review. Proceedings of Sixth International Conference on Indoor Air Quality, Ventilation & Energy Conservation in Buildings – Sustainable Built Environment, Sendai, Japan, Volume III, 249-256.

Wittstock, C. and M.B. Quinto 2008. Poisoned blossoms, withering hopes: The floriculture industry in Asia. South Asian Journal of Agriculture, 12(9), pp. 27-48.

Wolverton, B. C., A. Johnson and K. Bounds. 1989. Interior Landscape Plants for Indoor Air Pollution Abatement, Final Report, NASA Stennis Space Centre MS, USA.

Wood., R.A, M.D. Burchett, R. Alquezar, I. Ralph, Orwell, J. Tarran and F. Torpy. 2006. The potted-plant microcosm substantially reduces indoor air VOC pollution: i. office field-study. Water, Air, and Soil Pollution, 175: 163–180. DOI: 10.1007/s11270-006-9124-z

World Health Organisation (WHO). 2000. The Right to Healthy Indoor Air – Report on a WHO Meeting, Bilthoven, NL, European Health Targets 10. 13.

# CHAPTER 10

# FLORICULTURAL PRODUCTION THROUGH TECHNOLOGICAL INTERVENTION

**M. K. Singh, Devarai Lava Kumar and Namita**

*Division of Floriculture and Landscaping, ICAR- IARI, New Delhi-110012*
*Corresponding author: singh_markandey@yahoo.co.in*

---

The progress of horticultural technologies and methods is essential for addressing worldwide food security, employment and income generation demands while upholding environmental sustainability. Many high-tech technologies like protected cultivation, advanced irrigation systems, biopesticides, biofertilizers, precision agricultural techniques, breeding of resilient cultivars and vertical farming encompasses to enhance the horticultural production. Intensive greenhouse cultivation, integrating climate control, supplemental lighting, hydroponics and automation, facilitates year-round flowering, fruiting and vegetable yields which is about 20 times greater compared to open-field cultivation. Precision farming maximizes yields by utilizing data analytics, sensors and AI to optimize inputs and cultivation decisions for each sub-field zone. Organic biopesticides and biofertilizers, sourced from microbes, plants and minerals, enhance soil health and deter pests without leaving chemical residues. Advanced irrigation systems such as drip lines coupled with onscreen scheduling effectively reduces water usage. By integrating horticultural technologies, we can boost global food production without compromising the environment. This chapter explores recent innovations aimed at improving crop yields and resource efficiency and other necessities to achieve in horticulture sector.

Horticulture is an important sector and plays an important role in necessitating the food, employment and aesthetic demands of the individuals. With the rise in the nation's population, there is huge demand in boosting up the horticulture production.Yet, the horticultural sector faces significant obstacles in boosting yields, including the impacts of climate change, deteriorating soil health, water scarcity and diminishing arable land(Pretty and Bharucha, 2014).Urgent advancement in horticultural technologies and cultivation methods are imperative to tackle these compounded constraints.

## Advancements in breeding and genetics

Progression in plant breeding and genetic modification are ushering in a new era of horticultural crop varieties, characterized by heightened yield potential, enhanced nutritional quality and increased resilience to critical production challenges. Cutting-edge molecular tools facilitate the swift integration of diverse desirable traits into leading cultivars through marker-assisted breeding (MAB), genetic engineering, and gene editing techniques(Hasan *et al.*, 2021).This acceleration in development is leading to the creation of varieties specifically tailored to thrive in various localized conditions.

**High- yielding hybrids:** F1 hybrid enhance heterosis vigor by meticulously controlling the hybridization of genetically distinct inbred parental lines (Singh *et al.*, 2020).These typically exhibit significantly enhanced yield, quicker crop establishment, uniformity and resilience to stress compared to open-pollinated varieties. As an example, the displacement of open-pollinated cultivars with hybrid seeds has led to a doubling of tomato yields in diverse condition (Begna, 2022). However, reliability and production costs hinder hybrid adoption in developing regions. Public programs with subsidized pricing aid smallholder farmer's transition to hybrid crops (Shi *et al.*, 2023).

**Resistant varieties:** Plant pathogens and insect pests pose significant constraints on the health, productivity and stability of horticultural crop systems. Integrating natural plant resistance genes through breeding offers an ecological alternative for sustainable disease and pest control (Sharma *et al.*, 2018). For example, researchers transferred three powdery mildew-resistance genes from wild strawberries to commercial strawberry varieties, enhancing protection against prevalent fungal pathogens.

**Biofortification:** Biofortification improves the nutritional quality of food crops *via* agronomic practices, conventional breeding, or genetic engineering (Varshney and Dubey, 2009). Familiar successes include the production of vitamin A-enriched sweet potatoes, beans, and cassava to combat malnutrition in Africa, as well as iron-biofortified rice to tackle anemia in Asia (Ribaut and Hoington, 1998).

**Genetic engineering:** By 2016, genetically engineered (GE) crops covered over 457 million acres worldwide, showcasing valued traits such as insect, disease, or herbicide resistance (Collard *et al.*, 2008). Research initiatives are utilizing these tools to enhance horticultural yields, postharvest quality and shelf life. Many genetic engineered crops developed like improved perishability in tomatoes, blue-coloured flowers in chrysanthemums *etc.*

## Advances in cultural practices

**Vertical farming:** Vertical farming entails the use of stacked layers or vertical surfaces to cultivate crops indoors, often in urban environment, without relying on sunlight (Al-Kodmany, 2018). Rather than conventional horizontal farming, crops are cultivated vertically along walls or stacked in tall towers to optimize space utilization. Vertical farming has the capacity to achieve greater yields per square foot than traditional agriculture.

**Protected cultivation:** Protected agriculture involves cultivating crops beneath coverings that alter the growing conditions. Common protected cultivation structures include greenhouses, high or low tunnels, cold frames and shade houses (Holloway*et al.*, 2020).These methods provide different levels of control over the environment and shield crops from both non-living and living stresses. Greenhouse coverings serve to conserve warmth, offer shade and insulation and shield plants from rain and wind along with more sophisticated management of light, humidity and ventilation (Molin and Martin 2018).Protected structures allow for earlier planting, delayed harvesting, accelerated maturation, and typically yield much higher outputs compared to crops grown in open fields (Sun *et al.*, 2013).Sophisticated climate control, irrigation and automation facilitate year-round production with minimal human intervention, while grow lights such as LEDs supply the energy needed for photosynthesis (Javanmard, 2019).

**Precision farming:** Precision agriculture requires sophisticated farming management principles that revolve around observing, measuring and responding to variations in crops and fields. Leveraging emerging technologies, precision agriculture strives to enhance efficiency, productivity, sustainability, and profitability by optimizing field-level management practices (Gebbers, 2010).

**Hydroponics:** It refers to growing of the plants without the use of the soil, instead nutrient solutions are dissolved in water (Savvas, 2018), operated well in indoors like protected structures and aids the year-round production (Sharma *et al.*, 2018). Protected cultivation broadens agricultural possibilities for numerous regions and seasons that would otherwise struggle to sustain viable crops (Sun *et al.*, 2013).

**Aeroponics**: It is another soilless culture technique in which nutrient solution is misted on the suspended plant roots in air. This enables 30-50% of the faster growth rate (Austin and Austin, 2016).

**Aquaponics:** A type of hydroponic system that combines with aquafarming in a symbiotic recirculation system. Integrated systems like these can be operate efficiently and easily in small greenhouse setups as well asin large commercial scales (Bailey *et al.*, 2019).

**Reduced tillage systems:** Tillage refers to the agricultural practice of preparing and aerating the soil by turning or loosening it in preparation for planting crops. Excessive tillage can harm soil structure, deplete soil organic matter and lead to erosion problems over time (Palm *et al.*, 2013). Reduced tillage systems alleviate these issues by minimizing soil disturbance. Approaches encompass shallow non-inversion tillage, ridge tillage, strip tillage and no-till/zero-till methods (Farooq and Siddique, 2015).

**Remote sensing:** In agriculture, remote sensing employs aerial images and satellite data to identify crop variability, enhancing decision-making processes. Variations in soil properties, plant health, moisture levels and other factors can be detected through the use of multispectral, hyperspectral and thermal sensors, aiding in improved decision-making in agriculture (Mulla, 2013). Processing these remote sensing inputs with geographic information systems (GIS) software allows for accurate field mapping, revealing spatial patterns. Remote sensing also aids in scouting for weeds, diseases and insects by identifying outbreak hotspots, enabling farmers to precisely target control measures.

**Variable rate technologies:** While remote sensing, maps the variability, variable rate technologies (VRT) aim to manage fields differently based on their individual requirements. VRT entails the custom application of inputs such as seed, fertilizers, pesticides, or water at precise locations, determined by measured requirements (Stafford, 2000). VRT is closely integrated with data management systems containing field prescription maps, real-time sensing, and yield monitors, enabling swift adjustments by smart machines (McCarthy *et al.*, 2010).Combing these Internet of Things sensors, artificial intelligence and cloud computing become integrated, they promise increasingly responsive and optimized variability management, especially with the accelerating global adoption of VRT (Bongiovanni *et al.*, 2004).

**Automation and robotics:** Robotic technology holds significant potential for addressing the growing challenges of labor shortages and rising input costs in farming (Bechar and Vigneault, 2016). Autonomous tractors equipped with precision guidance systems are already capable of conducting various tasks such as plowing, planting, spraying, harvesting and more with exceptional accuracy, operating independently of human operators (Blackmore, 2007). Meanwhile, "agribot" pollinators and pest controllers mimic crucial ecosystem services (Duckett *et al.*, 2018). Automated fruit pickers delicately harvest sensitive produce, overcoming constraints posed by labor availability. When combined with artificial intelligence, agricultural robots could manage numerous complexes in-field tasks that currently overwhelm farmers.

**Biological controls:** Biological control utilizes natural enemies to suppress pests and diseases. This ecological approach bypasses health and environmental concerns associated with synthetic chemicals, while also uniquely permitting the self-replication of agents at pest hotspots (Dara, 2019). Key tools in biological control include biopesticides, conservation tactics that boost native enemies and classical biological control, which involves importing exotic predators to target invasive pests. Notable instances include the utilization of *Bacillus thuringiensis* to combat caterpillars and the infection of hundreds of pest species by the *Beauveria bassiana* fungus. Plant-derived biopesticides, in contrast, utilize extracted compounds or oils possessing antimicrobial, insecticidal, or antifeedant properties, such as the pressing of neem oil and the purification of azadirachtin from neem trees (Shields *et al.*, 2019). Classical biological control involves introducing specialist natural enemies from a pest's native range to achieve self-replicating, long-term control, thus circumventing problems associated with synthetic chemical resistance. Worldwide, this approach has enabled the extended, area-wide suppression of over 200 problematic invasive insect species and weeds in agriculture (Amundson *et al.*, 2015).

**Integrated pest and disease management:** Integrated Pest and Disease Management (IPDM) blends biological monitoring and control strategies with targeted chemical interventions, only exceeding economic thresholds when necessary(Blanco-Canqui *et al.*, 2015). Fundamental IPDM practices encompass scouting or sampling for informed decision-making, employing resistant cultivars, adjusting sowing density or timing to deter pests, intercropping with species that inhibit pests, and implementing conservation biological control. Studies on IPM adoption across various cropping systems indicate typical yield gains of 10-15% over conventional uniform pesticide application, coupled with reductions in chemical usage ranging from 35-65%.

**Organic amendments:** Compost comprises organic wastes such as crop residues, manures, leaves, or food scraps that have undergone biological decomposition, resulting in stabilized humus-like material (Finney *et al.*, 2016). Mature compost gradually releases nutrients, enhances soil structure, and improves water retention and microbiological activity. Thoughtfully integrating composts and organic soil amendments lays a crucial groundwork for nourishing soil microbial communities and revitalizing degraded farmlands (Mirsky *et al.*, 2011). Microbial inoculants of bacteria, fungi, protozoa drive more critically for the biogeochemical process like the nutrient transformations.

**Advanced irrigation techniques:** Conventional irrigation methods such as flood, furrow and sprinkler applications result in substantial water losses due to evaporation, runoff, wind drift and drainage beyond root zones (Solanki *et al.*, 2021). Drip irrigation systems deliver water directly and gradually to the root areas of crops through networks of valves, pipes, tubing, and emitters (Hart*et al.*, 2019). By precisely delivering moisture where it is needed, drip irrigation can reduce water usage by over 50% compared to conventional methods, all while enhancing yields and quality through improved plant health and fruit development (Frac *et al.*, 2018).

Additional efficiencies in drip irrigation stem includes, applying precise water quantities that match crop needs. Reduced weed growth and minimized disease transmission due to targeted water application without widespread wetting of soils and leaves. Enhanced fertilizer efficiency by directly applying dissolved nutrients in irrigation water to the roots. Capability for partial field watering and adjustable application rates across uneven topographies or soil conditions. Potential for automated irrigation scheduling using soil moisture sensors and interconnected controllers.

**Deficit Irrigation Scheduling:** Instead of maximizing yields per unit of land, deficit or regulated irrigation applies calculated, limited water below the full crop evapotranspiration requirements (Jagermeyr *et al.,* 2016). Decreasing irrigation during specific developmental stages may lead to slight biomass reductions, but it enhances water productivity per unit yield. For instance, temporarily restricting moisture during orchards' cell expansion phases limits fruitlet drops, concentrating sugars and nutrients in fewer mature fruits (Bar-Yosef, 1999). Efficient deficit irrigation scheduling involves integrating knowledge of crop growth stages, yield determinants, stress tolerance factors and soil conditions to determine optimal timings and thresholds for water limitations within adequate total seasonal irrigation. Responsive deficit applications are facilitated by precision technologies like soil moisture sensors, ET controllers and automated valves (Fereres and Soriano, 2007).

**Storage techniques:** Modified atmosphere (MA) and controlled atmosphere (CA) storage techniques regulate the environments of produce to slow down ripening and senescence processes, thus extending shelf-life after harvest (Indurthi *et al.,* 2024).Elevated carbon dioxide and/or reduced oxygen levels modify crops normal atmospheres, temporarily stalling their metabolism. Respiration converts sugars into $CO_2$ and water while consuming $O_2$; thus, suppressing respiration helps preserve quality.

The MA packaging systems passively alter in-pack conditions as products respire. Elevated $CO_2$ and reduced $O_2$ levels accumulate while condensation is absorbed (Jalota *et al.,* 2018). CA rooms actively monitor and control gases, temperature and humidity levels throughout entire cold storage facilities. MA systems are more economical for smaller traders, whereas CA allows for precise control of conditions for larger volumes. Optimal CA setpoints strike a balance between maximizing storage duration and avoiding anaerobic fermentation or chilling injury. Maintaining oxygen levels at 10-15% and carbon dioxide levels at 3-10% for durations ranging from 2 to 8 months preserves many fruits and vegetables with minimal disorders (Hamdy *et al.,* 2003).

1-MCP is a gaseous inhibitor of the plant hormone ethylene, which naturally triggers ripening in climacteric fruits such as apples, bananas, and tomatoes. 1-MCP irreversibly occupies ripening enzyme receptor sites, preventing ethylene from binding and initiating signal cascades (Perry *et al.,* 2009). Exposure to just 0.3-3 ppm of 1-MCP can extend shelf life from days to weeks, depending on the crop, cultivar, timing, temperature and other variables (Alsalhi, 2024).

Edible surface coatings function as barriers to moisture, gases, and aromas, while also serving as carriers for antimicrobials, nutraceuticals and texturizing compounds (Yildiz, 2017). Typical food-grade ingredients encompass polysaccharides such as starches, celluloses, alginates, and chitosan; proteins like those from milk, soy, collagen, and keratin; and lipids including waxes, acacia gum, and fatty acids. These ingredients form protective matrix structures that regulate the respiratory gases of produce, moisture loss, oxidation and microbial entry.

**Waste to energy techniques:** Various organic waste streams generated from horticulture operations can be transformed into energy through processes such as anaerobic digestion, gasification, pyrolysis and incineration (Salam *et al.*, 2021). Anaerobic digesters employ bacteria to decompose organic materials such as crop residues, manures and food waste. Anaerobic digesters harness bacteria to decompose organic materials such as crop residues, manures, and food waste. This process yields biogas composed of methane and carbon dioxide, which can be harnessed for electricity and heat production or refined into renewable natural gas. Thermal methods like gasification and pyrolysis involve heating biomass at elevated temperatures in a low-oxygen environment to generate syngas or bio-oil. Energy-from-waste systems decrease the amount of organic waste needing disposal by producing renewable energy and nutrients. This helps to minimize the reliance on additional fossil fuels and fertilizers.

**Advances in Horticulture horizon:** Promising horticultural advancements on the horizon encompass CRISPR gene editing customized to specific local conditions, integrated agritech ecosystems integrating automation and AI and predictive analytics harnessing big data. CRISPR facilitates swift and accurate modification of plant DNA, resulting in the development of uniquely tailored varieties that address specialized regional requirements (Torre *et al.*, 2019). Automated sensor networks, intelligent cultivation systems, autonomous robots and AI algorithms can adjust inputs and growing conditions with precision to enhance production and optimize resource efficiency (Hu *et al.*, 2021). These intelligent farm systems have the capability to regulate lighting, humidity, nutrient levels and harvest timing while simultaneously monitoring the health of plants.

**Support through government programs:** Government initiatives can foster sustainability both within India and globally by supporting research, offering incentives and implementing industry regulations. India has established ambitious renewable energy objectives, aiming for 175GW by 2022 and 450GW by 2030. This drive has led to initiatives such as the National Solar Mission, which seeks to deploy 100GW of solar capacity. This mission is supported by research, subsidies, and collaborations between the public and private sectors. India has also initiated the Green Grids Initiative in collaboration with the UK, with the objective of modernizing its grid infrastructure to facilitate the seamless integration of a greater proportion of renewable energy sources. Moreover, the government's Faster Adoption and Manufacturing of Hybrid and Electric Vehicles (FAME) scheme provides purchase incentives for electric vehicles, contributing to the reduction of emissions from transportation (Indurthi *et al.*, 2024).

On a global scale, governments invest in renewable energy research and development, offer incentives to promote energy efficiency in buildings, provide subsidies for electric vehicles, and enforce emissions standards to mitigate environmental impact. The US Department of Energy has allocated over $40 billion towards renewable energy research and development since 1978. Meanwhile, China maintains its position as the global leader in cleantech investment, with an annual expenditure exceeding $83billion. Numerous local governments across the globe provide streamlined permitting processes, rebates or tax incentives to encourage the construction of LEED/BREEAM-certified green buildings. Stricter fuel economy standards are accelerating the global electric vehicle market - EV sales are projected to hit 145 million by 2030.

Analysis indicates that US federal renewable energy incentives yielded $7 billion in net economic benefits between 2016 and 2020, alongside subsidizing $56 billion in industry investment. Globally, collaborations between public and private entities focused on sustainability facilitate advancements in areas such as conservation, infrastructure modernization, renewable energy, carbon removal technologies, grid management, and sustainable fuels.

## Conclusion

Sustaining global produce needs while preserving environmental integrity requires ongoing multi-disciplinary innovation in horticulture. The complete potential of emerging innovations relies on inclusive, holistic implementation that considers local contexts. Achieving climate-smart, regenerative food production necessitates involving farmers, laborers, and underserved communities to guarantee the fair adoption of new practices. Technology alone cannot revolutionize agriculture; systemic social and policy constraints must also be addressed. Yet, through deliberate leadership and concerted, conscientious collaboration spanning research, business, government, and civil society, 21st-century horticulture stands ready to provide nutritious, accessible food for all while rejuvenating agricultural landscapes globally.

## References:

Al-Kodmany, K., 2018. The vertical farm: A review of developments and implications for the vertical city. Buildings, 8(2), p.24.

Alsalhi, A., 2024. Applications of selected polysaccharides and proteins in dentistry: A review. International Journal of Biological Macromolecules, p.129215.

Amundson, R., Berhe, A.A., Hopmans, J.W., Olson, C., Sztein, A.E. and Sparks, D.L., 2015. Soil and human security in the 21st century. Science, 348(6235), p.1261071.

Austin, M. E., Austin S. A. 2016. Aeroponics: A production system and research tool. ISHS Acta Horticulturae 1134: International Symposium on Soilless Culture and Hydroponics.

Bailey, D. S., Ferrarezi, R. S, Testezlaf R, Yoo, K. H. 2018. Commercial aquaponics: Industry growth and constraints. Aquacultural Engineering. 87:10–17.

Bar-Yosef, B., 1999. Advances in fertigation. Advances in agronomy, 65, pp.1-77.

Bechar, A. and Vigneault, C., 2016. Agricultural robots for field operations: Concepts and components. Biosystems Engineering, 149, pp.94-111.

Begna, T., 2022. Speed breeding to accelerate crop improvement. Int J Agric Sc Food Technol, 8(2), pp.178-186.

Blackmore, B.S., 2007. A systems view of agricultural robots. In Precision agriculture'07 (pp. 21-31). Wageningen Academic.

Blanco-Canqui, H., Shaver, T.M., Lindquist, J.L., Shapiro, C.A., Elmore, R.W., Francis, C.A. and Hergert, G.W., 2015. Cover crops and ecosystem services: Insights from studies in temperate soils. Agronomy journal, 107(6), pp.2449-2474.

Bongiovanni, R. and Lowenberg-DeBoer, J., 2004. Precision agriculture and sustainability. Precision agriculture, 5, pp.359-387.

Collard, B.C. and Mackill, D.J., 2008. Marker-assisted selection: an approach for precision plant breeding in the twenty-first century. Philosophical Transactions of the Royal Society B: Biological Sciences, 363(1491), pp.557-572.

Dara, S.K., 2019. The new integrated pest management paradigm for the modern age. Journal of Integrated Pest Management, 10(1), p.12.

De La Torre 2021. Available:https://www.frontiersin.org/article s/10.3389/fsufs.2021.700157/full

Duckett, T., Pearson, S., Blackmore, S., Grieve, B., Chen, W.H., Cielniak, G., Cleaversmith, J., Dai, J., Davis, S., Fox, C. and From, P., 2018. Agricultural robotics: the future of robotic agriculture. arXiv preprint arXiv:1806.06762.

Farooq, M. and Siddique, K.H., 2015. Conservation agriculture: concepts, brief history, and impacts on agricultural systems (pp. 3-17). Springer International Publishing.

Fereres, E. and Soriano, M.A., 2007. Deficit irrigation for reducing agricultural water use. Journal of experimental botany, 58(2), pp.147-159.

Finney, D.M., White, C.M. and Kaye, J.P., 2016. Biomass production and carbon/nitrogen ratio influence ecosystem services from cover crop mixtures. Agronomy journal, 108(1), pp.39-52.

Fr c M, Jezierska-Tys S, Yaacoby T. The effect of agricultural techniques on soil microbiome. In Environmental chemistry and recent pollution control approaches. IntechOpen. 2018;33-56.

Gebbers, R. and Adamchuk, V.I., 2010. Precision agriculture and food security. Science, 327(5967), pp.828-831.

Hamdy, A., Ragab, R. and Scarascia-Mugnozza, E., 2003. Coping with water scarcity: water saving and increasing water productivity. Irrigation and Drainage: The Journal of the International Commission on Irrigation and Drainage, 52(1), pp.3-20.

Hart, M.M., Powell, J.R., Gulden, R.H., Dunfield, K.E., Pauls, K.P., Swanton, C.J., Klironomos, J.N., Antunes, P.M., Koch, A.M. and Trevors, J.T., 2009. Separating the effect of crop from herbicide on soil microbial communities in glyphosate-resistant corn. Pedobiologia, 52(4), pp.253-262.

Hasan, N., Choudhary, S., Naaz, N., Sharma, N. and Laskar, R.A., 2021. Recent advancements in molecular marker-assisted selection and applications in plant breeding programmes. Journal of Genetic Engineering and Biotechnology, 19(1), p.128.

Holloway, L., Starry, O., McClung, R. and Rosenstiel, T., 2020. The influence of microclimates created by photovoltaic panels and irrigation on green roof ecosystem properties. J Living Archit, 7(2), pp.62-80.

Hu, X., Zhang, Y., Zhang, J., Islam, S., She, M., Zhao, Y., Tang, G., Jiang, Y., Rong, J. and Ma, W., 2021. Consensus genetic linkage map construction based on one common parental line for QTL mapping in wheat. Agronomy, 11(2), p.227.

Indurthi, S., Sarma, I. and Vinod, D.V., 2024. Horticultural Innovations Elevating Crop Yields and Agricultural Sustainability for a Flourishing Future. Plant Cell Biotechnology and Molecular Biology, 25(1-2), pp.22-44.

Jagermeyr, J., Gerten, D., Schaphoff, S., Heinke, J., Lucht, W. and Rockström, J., 2016. Integrated crop water management might sustainably halve the global food gap. Environmental Research Letters, 11(2), p.025002.

Jalota SK, Vashisht BB, Sharma S, Kaur S. 2018. Water balance approach for assessing water use efficiency and water productivity of wheat-based production systems. Agricultural Water Management. 202:81-88.

Javanmard A. 2019. Automation and Control in Vertical Farming: A Review and Perspective. IFAC-PapersOnLine. 52(30):23–28.

Kaye, J.P. and Quemada, M., 2017. Using cover crops to mitigate and adapt to climate change. A review. Agronomy for sustainable development, 37, pp.1-17.

McCarthy, A.C., Hancock, N.H. and Raine, S.R., 2010. VARIwise: A general-purpose adaptive control simulation framework for spatially and temporally varied irrigation at sub-field scale. Computers and Electronics in Agriculture, 70(1), pp.117-128.

Mirsky, S.B., Curran, W.S., Mortenseny, D.M., Ryany, M.R. and Shumway, D.L., 2011. Timing of cover-crop management effects on weed suppression in no-till planted soybean using a roller-crimper. Weed Science, 59(3), pp.380-389.

Molin E, Martin M. 2018. Reviewing the use of advanced control systems for the optimization of environmental parameters for protected cultivation in the mediterranean region. Agronomy; 8(7):106.

Mulla, D.J., 2013. Twenty-five years of remote sensing in precision agriculture: Key advances and remaining knowledge gaps. Biosystems engineering, 114(4), pp.358-371.

Netafim USA. Netafim Drip Irrigation Handbook. Fresno, California; 2011.

Palm, C., Blanco-Canqui, H., DeClerck, F., Gatere, L. and Grace, P., 2014. Conservation agriculture and ecosystem services: An overview. Agriculture, Ecosystems & Environment, 187, pp.87-105.

Perry, C., Steduto, P., Allen, R.G. and Burt, C.M., 2009. Increasing productivity in irrigated agriculture: Agronomic constraints and hydrological realities. Agricultural water management, 96(11), pp.1517-1524.

Pretty, J. and Bharucha, Z.P., 2014. Sustainable intensification in agricultural systems. Annals of botany, 114(8), pp.1571-1596.

Ribaut, J.M. and Hoisington, D., 1998. Marker-assisted selection: new tools and strategies. Trends in Plant Science, 3(6), pp.236-239.

Salam, S Pandey, T Van Der Zaag, SB Salam. 2021. A review of the impact of anaerobic digestion on agriculture and soil quality. Journal of Cleaner Production. 280: 124255.

Savvas, D. and Gruda, N., 2018. Application of soilless culture technologies in the modern greenhouse industry—A review. Eur. J. Hortic. Sci, 83(5), pp.280-293.

Sharma, N., Acharya, S., Kumar, K., Singh, N. and Chaurasia, O.P., 2018. Hydroponics as an advanced technique for vegetable production: An overview. Journal of Soil and Water Conservation, 17(4), pp.364-371.

Shi, T., Gao, Y., Xu, A., Wang, R., Lyu, M., Sun, Y., Chen, L., Liu, Y., Luo, R., Wang, H. and Liu, J., 2023. A fast breeding strategy creates fragrance-and anthocyanin-enriched rice lines by marker-free gene-editing and hybridization. Molecular Breeding, 43(4), p.23.

Shields, M.W., Johnson, A.C., Pandey, S., Cullen, R., González-Chang, M., Wratten, S.D. and Gurr, G.M., 2019. History, current situation and challenges for conservation biological control. Biological control, 131, pp.25-35.

Singh, R.K., Prasad, A., Muthamilarasan, M., Parida, S.K. and Prasad, M., 2020. Breeding and biotechnological interventions for trait improvement: status and prospects. Planta, 252, pp.1-18.

Solanki MK, Singh RK, Srivastava S, Kumar S, Kashyap PL, Srivastava AK, Arora DK. 2021. Plant growth-promoting rhizobacteria augment agricultural sustainability by ameliorating nutrients uptake, stimulating antioxidant machinery, and suppressing pathogens. Frontiers in Microbiology; 11:3276.

Stafford, J.V., 2000. Implementing precision agriculture in the 21st century. Journal of agricultural engineering research, 76(3), pp.267-275.

Sun Y, Wang C, Engel BA. 2013. Simulation of land use-soil interactive effects on biomass production in Northern Indiana. Agriculture, Ecosystems & Environment. 165:90– 98.

Varshney, R.K. and Dubey, A., 2009. Novel genomic tools and modern genetic and breeding approaches for crop improvement. Journal of Plant Biochemistry and Biotechnology, 18, pp.127-138.

Yildiz, F., 2017. Initial preparation, handling, and distribution of minimally processed refrigerated fruits and vegetables. Minimally processed refrigerated fruits and vegetables, pp.53-92.

CHAPTER 11

# ICT IN POST-HARVEST MANAGEMENT OF HORTICULTURAL PRODUCE

**M Feza Ahmad* and Deep Lata****

*Directorate of Seed and Farms, Department of Food Science and Post Harvest Technology**,*
*Bihar Agricultural University, Sabour, Bhagalpur**
*Corresponding author: faze@rediffmail.com*

The application of Information and communication technologies (ICTs) in horticulture using networks, mobiles, digital devices, services, and their applications may enhance processing, post-harvest management, and exchange of data, information or knowledge with farmers. The application of ICTs in horticulture has revolutionized. ICT includes several converging technologies, including traditional telecommunications, television and video, radio, CD-ROMs, cell phones and smart devices, and other modern technologies such as computers and internet, sensors, geographic information systems (GIS), satellites, etc. The prime objective of ICT is to transfer information from one point to another and help farmers to get right value for their produce and helps consumers to get a premium product on optimum price.

## Impact of ICT on Horticulture

The need of modern farming system is to acknowledge the role of ICT as a decision support system for farmers. It helps and assists farmers to update their information with all recent data, and provides inclusive of data about weather condition, horticultural crops, new technologies and other advanced technologies to enhance production of quality produce.

Around 60% population has access to internet, and mobile internet is now the most widely-used channel for assessing information worldwide. Its large-scale use has facilitated better communication, ensuring better delivery of services and information to farmers.

The infusion of advanced technologies has allowed the global horticulture sector to rise and transform the cultivation practices, harvesting, and distribution channel. The use of technology in Indian horticulture, or e-horticulture, has enhanced the production and socio-economy condition. It has particularly revolutionized smallholder horticulture in several agrarian economies and also helped to overcome many challenges associated with the traditional horticulture (Fig. 1).

## Application of ICTs

**Preventing banana field using electronic fencing**: These fencing are made of galvanized wire along with the perimeter of field of 2-3 acres having a banana plantation. A high voltage is used to charge these fencing in such a way that it is not fatal but provides a mild shock to animals or even a human being. The alerts are sent to farmer's mobile phone if there is any breaching.

**Use of wild animal's voice/sound generator for areca nut and banana plantation**: In Goa and other states of India especially forest-prone area, where monkeys spoil the areca nut and banana fruits. They can easily cross the electronic fencing but are scared to sound of tiger or elephants. However, these monkeys are so clever that once they get acquainted with cycles of sounds, they are not scared. To overcome this challenge, some of ICTs in India are creating different sounds at different time cycles so as to confuse the monkeys. The sounds are generated by an amplifier powered by solar and controlled by Remote control.

**Weather forecast alert**: In India, weather change in December is a common problem and harms mango flowering which has just started blossoming. Sometimes sudden windy weather in March destroy mango flowers and leads to huge yield losses. Similarly, rains in December cause a huge loss in budding grapes. The ICT provides a weather forecast which helps farmers to proactively work on coverings, emergency remedies, temporary curtains etc. to avoid the loss.

**Use of moisture sensors in drip irrigation system connected to cloud (IOT)**: The adequate quantity of water helps in better yield of horticultural produce. Use of sensor controlled accurate water dissemination through drip and sprinkler irrigation system help grapes growers in Nasik region of Maharashtra. The amount of water can be measured per day in winter, summer as well as sensed by moisture sensors. There is a huge saving of water, soil minerals due to controlled irrigation. Wireless monitoring and controlling of irrigation has been achieved with the help of ICTs.

**Use of ICT in greenhouse**: The ICT has been used for weather control in greenhouse net sheds production of roses and strawberries in western Maharashtra. ICT intelligence-based equipment help in monitoring and control of temperature, irrigation, humidity, light radiations etc.

**Use of sequential timers for irrigation**: As per our need if 10 plots need to be irrigated with less capacity of the water pump, every plot is irrigated one by one with solenoid valve control and water sensors. After that all the data are transferred to the cloud for analytics.

**Prevention of electrical motor pump burning using alert devices**: Water pump burning can cause huge yield losses during peak period. There is a vast network of pumps and electrical cables in agriculture farms, burning of motors tend to not only problematic for irrigation but also cause expenses for repairing, logistics and time loss. In India, almost 80% water pumps are equipped with ICT protection devices to prevent this problem and sends alerts on mobile phones.

**Facilitates communication with overseas market**: The application of ICT helps farmers to establish a better communication with domestic and overseas market efficiently. They can take a wise decision about driving the products to logistics network.

**Plants and soil health-care**: Though, it is still at very ground stage, ICT for health care of plants and soil mineral contents are being on the way to farmers. The ICT plays an important role in diagnosis of minerals present in plants and soil. In India, sugarcane growing areas, overwatering cause loss of fertility in dilution of soil minerals. Different electronic sensors along with chemical processes are most useful in deciding the fertilizers input to the soil.

**Postharvest Management and storage**: The ICT based refrigeration and storage temperature control and monitoring during the logistics as well as static storage help in better management of fruits and vegetables. There are so many localized ICT innovations used in agriculture that few of them seem to be really scalable across the globe.

**Sorting and grading of fruits**: Manual sorting and grading of fruits as per colour, size, quality and smell is not accurate and need highly experienced and skilled workers. Sensors based photo sensing machines as well as fully automated size sorters help sorting out of fruits in different boxes along with an output in the form of digital data sent to the computer terminal.

## Distribution channel (geo-traceability)

The ICT based distribution channel provide certification of products and visibility by tracking consumers in product supply chain. It offers customized farm business plans, product tracing to promote sustainable and bio diverse practices, survey design and collection tools, SMS systems for small-holder farmers with simple phones, and data infrastructure. It also helps in developing the data collection, management, and supply tracing tool for a producer, e.g. coffee suppliers in Uganda.

**Cold chain assistancein supply chain**: The ICT enable smallholders to access sustainable cooling facilities through the servitization business model, and to benefit from available data and market information to optimize decisions on produce and farm management. This application will provide real-time instructions to guide farmers on how to control storage of products to minimise food loss and energy use, and on when to sell their produce to maximise market value.

There is need of information dissemination among farmers throughout the supply chain (Meitei and Devi, 2009; Narula and Nainwal, 2010) (Table 1). On time, accurate and representative market information is a powerful tool which empowers farmers in a liberalised marketing system (Tollens, 2006).

## Marketing

The ICTs can be used to determine current market prices and gain access to markets, to bargain and conduct transactions, and to send and receive money. It helps in up-to-date market information on commodity and input prices, as well as demand trends and boosts the farmers negotiating positions and informs decisions about when and where to buy and sell, what to produce, and the quantity and quality of future production. It can end the middlemen interference and help in fetching better farm-gate prices (Table 2).

**Farmers' satisfaction**: The ICT tools provides better pricing transparency information to farmers. It helps farmers to improve marketing logistics and reduce transportation costs and increases profitability by facilitating the farmers to achieve leverage economies of scale. It also helps farmers to control their product sales by finding new sources of demand, improving product quality to meet market conditions and to cut out intermediaries by selling directly to large-scale buyers, traders and processors.

**Consumer satisfaction**: The ICT based marketing and distribution channel provides better quality produce. It also helps consumers in picking right produce and aware them about adulteration. Consumers can buy produce directly from farm and also get premium quality on optimum price. It also facilitates door to door services and make purchase convenient.

## Differences in quality of decision on marketing practices

The quality perceptions of the farmers on various aspects of agricultural marketing were recorded on a five point Likert scale (Tonnyet al., 2019).From results it can be seen farmers both ICT user and non-user were not collecting information on all the marketing practices (Table-1). They were collecting only that information which were more needed to them. However, the usage of information on selling, sales price and input price practices were high among both ICT users and non-users. So, it is evident from the findings that the users of ICT based information were getting better quality information and hence, making significantly better decisions on all aspects of agricultural marketing. Jabir, (2011) found the same result where the livestock farmers who were ICT users were getting better quality information and hence, made better decisions on all aspects of livestock farming.

## Conclusion

Low-cost ICT tools such as mobile phones are having a lot of promise for horticultural extension. The ICT can play a great role in private sector agribusiness, market information and market intelligence. It can help an average Indian farmer to get relevant information regarding agro-inputs, crop production technologies, agro processing, market support, agro-finance and management of farm agri-business. Horticultural extension and horticultural development by formulating national and state level e-Agriculture policy, human resource development, strengthening ICT infrastructure etc.

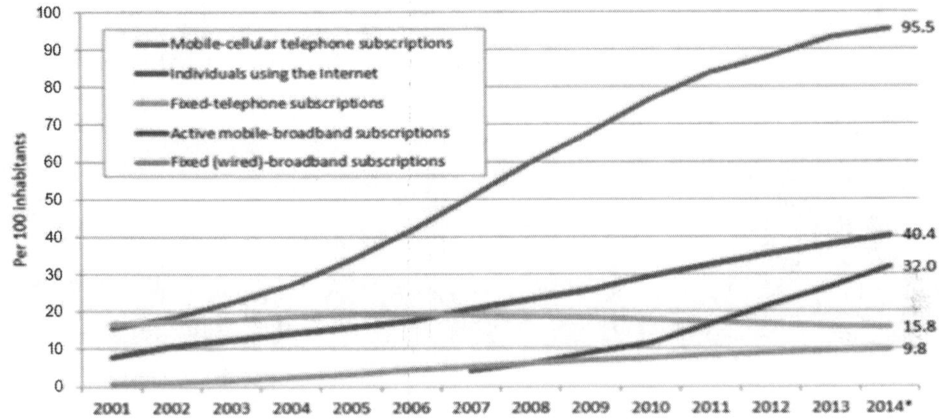


**Fig. 1. Global ICT developments, 2001–2014.**
**Source: Patel and Shukla, 2014.**

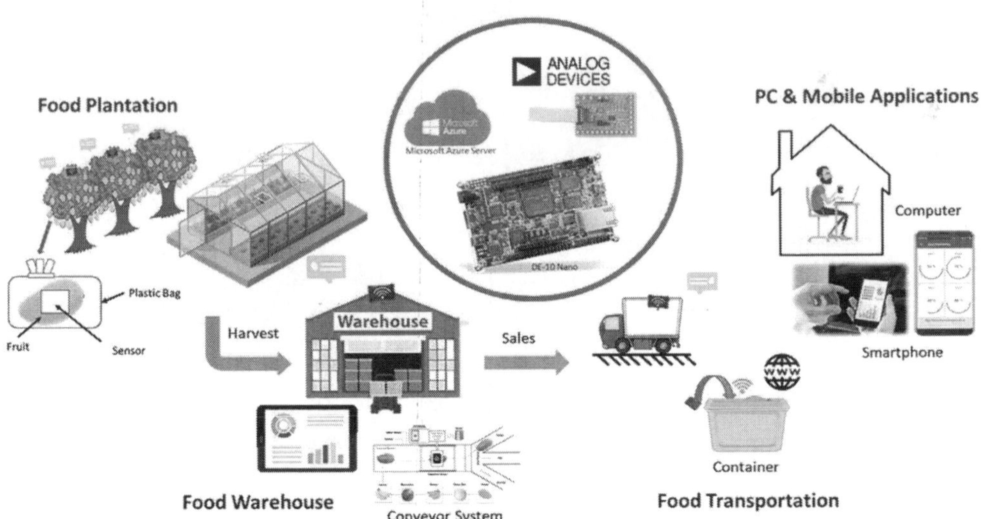


**Fig. 2. Role of ICT in postharvest handling operations**

**Fig. 3**

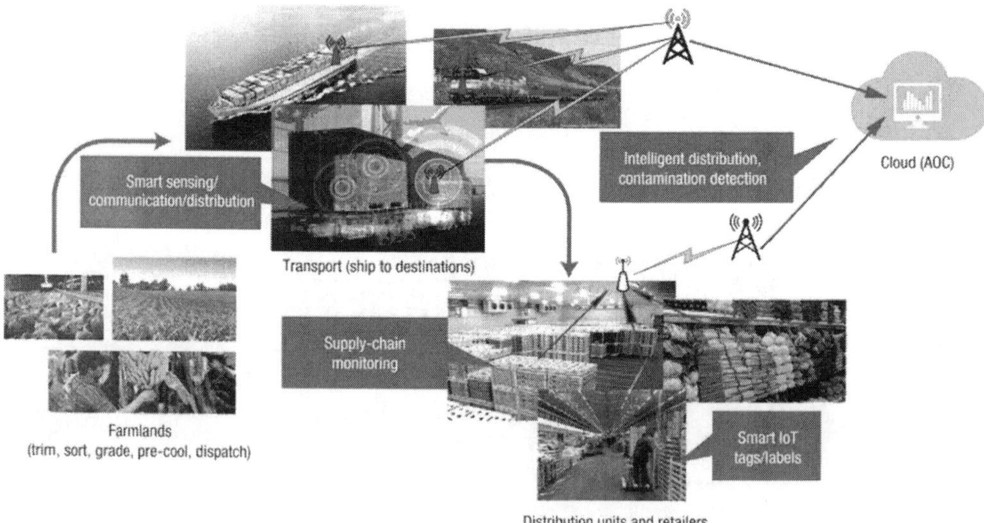


**Fig. 4. Role of ICT in postharvest distribution channel**

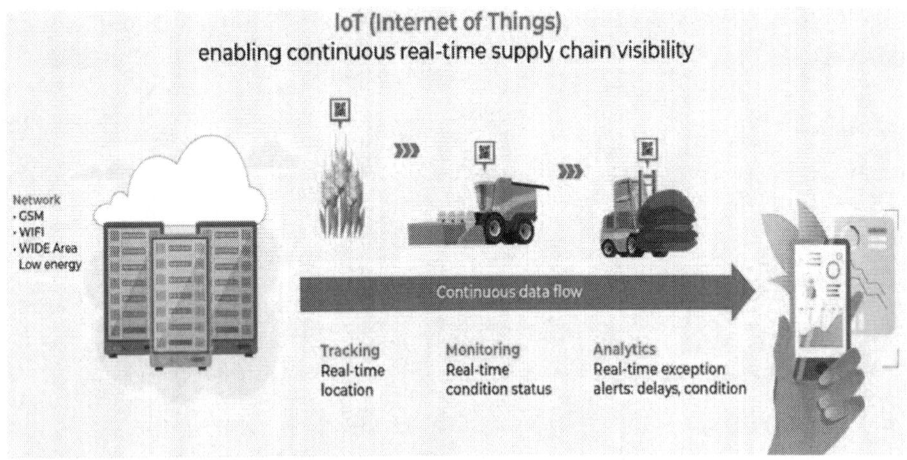


Fig. 5. Role of ICT in supply chain

Table 1: Information requirement of different players in supply chain

| Stage | Farmer-Producer | Trader | Retailer |
|---|---|---|---|
| Production | Inputs (prices & availability), soil testing, production technologies, good agricultural practices | Acreage under crops, production estimates | — |
| Post-harvest management | On-farm/ scientific storage, grading and sorting, packing, transportation | Storage facilities, grades and standards, transportation, packaging | Packaging |
| Marketing | Markets, price and arrivals, alternative buyers | Markets, price and arrivals, market and legal issues, international trade | Prices, arrivals, intra-mark-up prices, markets, transport |
| Others | Weather, credit | Weather, credit | Credit, festive-driven demand |

*Source:* Narula and Nainwal (2010) and field discussions.

Table 2. Sources of information for Indian farmers and their reliability.

| Source | Sources of information for farmers | | Reliability of information source (per cent farmers) | | |
|---|---|---|---|---|---|
| | Number | Per cent | Highly reliable | Moderately reliable | Unreliable/ undecided |
| Fellow farmers | 77 | 73.3 | 43.8 | 21.9 | 34.3 |
| Relatives | 64 | 61.0 | 27.6 | 26.7 | 45.7 |
| Extension service providers | 57 | 54.3 | 38.1 | 17.1 | 44.8 |
| Private extension service providers | 44 | 41.9 | 28.6 | 16.2 | 55.2 |
| Traders | 54 | 51.4 | 53.3 | 19.0 | 27.6 |
| Television | 47 | 44.8 | 51.4 | 25.7 | 22.9 |
| Radio | 33 | 31.4 | 42.9 | 22.9 | 34.3 |
| News papers-magazines | 66 | 62.9 | 40.0 | 16.2 | 43.8 |
| Cellular phone | 70 | 66.7 | 51.4 | 14.3 | 34.3 |
| Internet | 17 | 16.2 | 31.4 | 10.5 | 58.1 |

*Source:* Field Survey 2010-11

**Table 3 Differences in quality of decision between ICT user and non- ICT user**

| Marketing Practices | ICT User | | | Non-ICT user | | | F-Value |
|---|---|---|---|---|---|---|---|
| | N | Mean | Standard Deviation | N | Mean | Standard Deviation | |
| **Market Planning** | 35 | 1.74 | .701 | 28 | 4.11 | .832 | 53.81* |
| **Selling** | 45 | 1.82 | .535 | 35 | 3.80 | .933 | 54.44* |
| **Transportation** | 40 | 1.82 | .446 | 29 | 3.34 | .614 | 107.671* |
| **Storage** | 38 | 1.97 | .592 | 26 | 3.77 | .815 | 17.890* |
| **Sales price** | 45 | 1.22 | .420 | 35 | 2.73 | .822 | 66.74* |
| **Market Demand** | 44 | 1.41 | .542 | 29 | 3.72 | .528 | 90.90* |
| **Buyers Availability** | 42 | 1.57 | .630 | 28 | 3.68 | .670 | 46.09* |
| **Market Availability** | 43 | 1.70 | .708 | 27 | 3.59 | .694 | 44.95* |
| **Input Price** | 45 | 1.51 | .549 | 35 | 2.86 | .810 | 32.793* |

*Significant at the 0.01 level
Source: Field Survey 2017
Note: Responses in five points Likert scale indicated as very good=1, good=2, neutral=3, poor=4 and very poor=5

## References

Behera, B. S., Panda, B., Behera, R. A., Nayak, N., Behera, A. C., and Jena, S. (2015). Information communication technology promoting retail marketing in agriculture sector in India as a study. Procedia Computer Science, 48: 652-59.

Fare, Y., Dufumier, M., Loloum, M., Miss, F., Pouye, A., Khastalani, A., and Fall, A. (2017). Analysis and diagnosis of the agrarian system in the niayes region, northwest senegal (west africa). Agriculture 7(7): 59.

Gummagolmath, K. C., and Sharma, P. (2013). User centric ICT model for supply chain of horticultural crops in india. Agricultural Economics Research Review 26: (347-2016-17078), 91-100.

Jairath, M. S., &Yadav, H. (2012). Role of ICT in decision making in agricultural marketing–a case of Arid India. Indian Journal of Agricultural Economics 67: (902-2016-67841).

Meitei L. S. and Devi, Purnima (2009) Farmers informationneeds in rural Manipur: An assessment, Annals ofLibrary and Information Studies 56: 35-40.

Narula, S. A. and Nainwal, N. (2010) ICTs and agricultural supply chains: Opportunities and strategies for successful implementation, Information Technology inDeveloping Countries 20(1): 24-28.

Patel, D.J. and Shukla, K.K. 2014. Challenges and opportunities for ICT initiatives in agricultural marketing in India. Oriental Journal of Computer Science and Technology 7(3): pp. 377-81.

Okello, J. J., Kirui, O., Gitonga, Z. M., Njiraini, G. W., &Nzuma, J. M. (2014). Determinants of awareness and use ICT-based market information services in developing-country agriculture: The case of smallholder farmers in Kenya. Quarterly Journal of International Agriculture.

Rengarajan, V., VijayAnand, V., Sivakumar, M., Sankararaman, G., Natarajan, R., Velavan, N. V. D. M., &Kathirvelbabu, C. (2018). Role of ICT Based Services in Agricultural Marketing in Kanchipuram District–A Study Using One-Way ANOVA. International Journal of Pure and Applied Mathematics 119(15): 1025-34.

Tonny, N.B.W., Palash, M.S. and Moniruzzaman, M., 2019. Use of ICT in decision making of agricultural marketing: Factors determining of farmers' involvement: Use of ICT in agricultural marketing. Journal of the Bangladesh Agricultural University 17(2): 226-231.

CHAPTER 12

# POSTHARVEST TREATMENTS OF FRESH FRUITS & VEGETABLES: LATEST TRENDS

**Ram Asrey**

*Division of Food Science and Post-harvest Technology,*
*ICAR-Indian Agricultural Research Institute, New Delhi 110 012, India*
*Corresponding author: ramu_211@yahoo.com*

---

The fruits and vegetables are rich source of vitamin, minerals and phytopigments, but short shelf life limits their retailing period and availability to the consumers. This needs safe and sustainable molecules for treating fruits and vegetables after harvesting. The postharvest treatment of fruits and vegetables has evolved immensely in recent years, driven by the need to enhance quality, extend shelf life, and reduce losses in the supply chain along with the safety of consumers. Green elicitors are being commercialized and involved in postharvest supply chain management, including physical treatments (UV-C, ozone, electrolyzed water, ultrasonication), plant hormones (salicylic acid, brassinosteroids, jasmonates, melatonin), novel signalling compounds (Nitric oxide, polyamines, GABA, L-arginine), organic acids (ascorbic acid, oxalic acid) and edible coatings (chitosan, guar gum, gum arabica, starch, essential oil). The aim of green approaches is to maintain nutritional value, addressing food safety concerns, and minimizing the effect of environmental stresses. The topic also gives insight on the role of physical, biological, organic acids and edible coatings in retaining health benefitting compounds and preserving the freshness of fruits and vegetables. Emphatically, the content focused on various facets of postharvest treatments, their impact on retention of secondary metabolites and synergy with storage life and overall quality of the produce.

The postharvest phase of fruits and vegetables plays a pivotal role in determining their shelf life, quality, and nutritional value. With the global demand for fresh produce on the rise, there is an increasing need to enhance postharvest technologies and treatments to ensure that these essential commodities reach consumers in optimal condition. Recent years have witnessed significant advancements in postharvest treatments, ranging from innovative preservation techniques to the development of novel coatings. These breakthroughs have not only addressed the challenges posed by postharvest losses and food waste but have also paved the way for a more sustainable and efficient supply chain in the horticulture sector. The importance of postharvest treatments cannot be overstated, as they directly influence the overall economic and nutritional aspects of the fresh produce industry. Fruits and vegetables are highly perishable commodities that are susceptible to various physiological, biochemical, and microbial changes after being harvested. Factors such as moisture loss, enzymatic activity, pathogen invasion, and ethylene production can accelerate deterioration and compromise the quality of these products. Therefore, effective postharvest interventions are critical in preserving the freshness, appearance, and nutritional content of fruits and vegetables, ultimately contributing to food security and minimizing losses. In recent years, researchers and scientists have been diligently exploring innovative methods to enhance the postharvest life of fresh produce and improve their marketability. One of the prominent areas of advancement is the development of physical treatments (UV-C, ozone, electrolysed water, ultrasonication), plant hormones (salicylic acid, brassinosteroids, jasmonates, melatonin), novel signalling compounds (Nitric oxide, polyamines, GABA, L-arginine), organic acids (Ascorbic acid, oxalic acid) and edible coatings (chitosan, guar gum, gum arabica, starch, essential oil). The adoption of physical treatments underscores the industry's commitment to food safety and environmental consciousness. These methods not only extend shelf life but also address concerns related to microbial contamination and pesticide residues, ensuring that consumers can enjoy fruits and vegetables with greater confidence. Plant hormones and novel signalling compounds have emerged as powerful tools to fine-tune postharvest physiology and enhance the market appeal of produce. By harnessing the regulatory roles of these substances, postharvest researchers are unlocking innovative ways to delay senescence, improve color retention, and enhance antioxidant defenses. The integration of organic acids and their roles in preserving postharvest quality, particularly through combating enzymatic browning and oxidative stress, showcases the potential of natural compounds to replace or complement traditional chemical treatments. Overall, these techniques involve altering the metabolism they produce to regulate respiration rates, inhibit microbial growth, and delay ripening processes, leading to improved storage durations and quality retention of fruits and vegetables.

## Melatonin

Melatonin (MT) (N-acetyl-5-methoxytryptamine) is a versatile signalling chemical found in a wide range of plant species (Ze et al., 2021). It is an indoleamine compound with low molecular weight which was initially identified in *Hypericum monogynum*. Phytomelatonin is biosynthesized from tryptophan in four sequential enzymatic processes involving five primary enzymes in the catalytic process. MT is a safe and nontoxic molecule that has been investigated extensively in postharvest preservation for its function in extending the shelf life and quality of fruits and vegetables, as well as promoting crop output and maintaining food safety in an environmentally responsible manner. MT also acts as an ethylene antagonist. MT does not directly inhibit ethylene biosynthesis, but it may inhibit ethylene biosynthesis by increasing NO content. Pear fruit treated with MT (0.1 mM) resulted in reduce ethylene biosynthesis, maintain firmness, and downregulate the relative expression of PcACS1, PcACS2, PcACO1 and PcACO2 gene.

## Brassinosteroids

Brassinosteroids (BRs) represent a novel group of plant hormones, characterized as natural polyhydroxylated steroidal compounds. The initial discovery of these hormones originated from *Brassica napus* pollen, leading to the naming of the first isolated compound as "brassinolide" (Ali 2017). In 1982, a similar growth substance, resembling brassinolide, was subsequently identified in insect galls of chestnut and was termed 'castasterone'. Collectively, these two compounds are collectively referred to as brassinosteroids.

BRs like 24-epibrassinolide, brassinolide, and 28-homobrassinolide are frequently employed in various concentrations to modulate both pre- and post-harvest aspects. The exogenous application of BRs exerts multiple effects, including the inhibition of cell wall degradation and softening processes driven by enzymes like pectin esterase and polygalacturonase. BRs also play roles in regulating sugar and energy metabolism , curbing lipid peroxidation, maintaining robust membrane integrity , overseeing ripening and color metabolism , restraining peroxidase and polyphenol oxidase (PPO) coenzymes, ultimately leading to a delay in enzymatic brown, and reducing the production of reactive oxygen species (O-, H2O2) , significantly impacting on a variety of physiological responses throughout the postharvest lifespan.

Brassinosteroids (BRs) have emerged as effective agents in mitigating chilling injury in fruits by increasing membrane integrity through the reduction of phospholipase D (PLD) and lipoxygenase (LOX) activities, while simultaneously elevating the unsaturated fatty acid/saturated fatty acid (unSFA/SFA) ratio (Ahamad et al., 2024). Furthermore, BRs contribute to disease prevention by fortifying defense mechanisms and inducing systemic acquired resistance. Their impact on reducing fruit decay is attributed to enhanced disease resistance and delayed senescence, rather than direct fungal toxicity. In the context of fruit ripening, BRs facilitate the softening process by augmenting the activity of softening enzymes like pectin methylesterase (PME) and polygalacturonase (PG), along with modulating water-soluble pectin and protopectin content.

## γ-Aminobutyric acid (GABA)

γ-Aminobutyric acid (GABA) is a non-proteinaceous free amino acid which is prevalent across diverse organisms, spanning prokaryotes to eukaryotes . In plants, GABA serves as an endogenous component, participating in functions like pH regulation, nitrogen storage, and plant development, while affording protection against various abiotic stresses such as drought, oxidative stress, salinity, and cold conditions. The application of exogenous GABA has shown promise in increasing disease resistance and preservation of postharvest fruits and vegetables. This includes maximised activity of enzymes like POD, SOD, and CAT, leading to delayed pear browning . Moreover, exogenously applied GABA has demonstrated efficacy in inducing resistance against *A. alternata* induced black spot in tomato fruit (Yang et al., 2017).

## L-arginine

L-arginine is an amino acid with versatile functions and is considered essential due to its role in synthesizing important signalling molecules such as NO and PAs. This unique property enables L-arginine to enhance stress resistance in plants (Khan et al., 2023). The accumulation of these signaling molecules, facilitated by L-arginine, proves beneficial in mitigating chilling damage and extending the shelf life of horticultural products. In the context of postharvest applications, immersion of harvested produce in L-arginine solutions has shown promising results. This treatment has been found to effectively prevent browning and increase the storage duration of lettuce and oranges. Additionally, L-arginine has demonstrated its ability to impede disease development in tomatoes, delay senescence in broccoli, and counteract chilling injury in pomegranates. These findings highlight the potential of L-arginine as a valuable tool for enhancing the postharvest quality and longevity of various horticultural products. In mushrooms and persimmon fruit, L-arginine treatment has been shown to reduce chilling damage, resulting in decreased accumulation of MDA and reduced electrolyte leakage. This indicates that pre-storage treatments with L-arginine could offer a safe and commercially viable strategy, given its GRAS status, to address postharvest decay, mitigate chilling stress, and enhance the nutraceutical qualities of horticultural items, ultimately extending their shelf life.

## Cold plasma

Plasma is often described as the fourth state of matter, characterized by ionized gas containing free electrons, ions, reactive atoms, neutral molecules, and photons in a metastable state with near-zero net electrical charge. Low-temperature plasma can be categorized into thermal and nonthermal (cold) plasma based on thermodynamic equilibrium. Various electrical discharges such as corona discharge, glow discharge, radiofrequency (RF) discharge, dielectric barrier discharge (DBD), pulsed corona discharge, microwave discharge, and plasma jet are used to generate nonthermal plasma or cold plasma. Cold plasma has demonstrated effectiveness in pathogen inactivation on the surface of fresh produce. While cold plasma generates free radicals, these reactive species can lead to rapid oxidation, potentially causing lipid deterioration and the breakdown of natural antioxidants, which may negatively impact the taste and aroma of food (Sonawane et al., 2020). It is important to note that the reactive species present in cold plasma can interact with various components of fruits, leading to potential effects on both physicochemical and sensory attributes of different fresh produce.

Cold plasma has demonstrated effective inactivation of aerobic bacteria and molds on blueberries, leading to a reduction in decay rate during storage. Furthermore, cold plasma treatment has been shown to increase antioxidant content and enhance antioxidant enzyme activity in blueberries, contributing to the maintenance of lower levels of ROS in the fruit.

## Ultrasonication

Its potential has been recognized in food industry since the 1970s and recently, accepted as an effective treatment to destroy harmful microorganisms in fruits and vegetables owing to its cavitation effect. The formed cavity grows and collapse generating localized mechanical (generation of turbulence, compression, rarefaction, and shearing effect) and chemical energy (production of free radicals) and physicals effects include extreme temperature and pressure in and around the produce. Raising the intensity of US in a probe system has shown increasing antimicrobial efficacy, yet excessive intensities were linked to detrimental impacts on fruit quality during storage. Many previous works have suggested that US could enhance defense mechanisms, antioxidant activities, secondary metabolites biosynthesis, control postharvest quality and induce pathogenic resistance in fruits and vegetables (Aday et al., 2013).

US-driven cavitation plays a crucial role in breaking down contaminant molecules, aided by the collapse of cavitation bubbles, resulting in the breakdown of both contaminant and water molecules. This process generates reactive OH radicals with potent oxidation capabilities against organic pollutants.

## UV radiation

Ultraviolet (UV) treatment is widely used to control the decay of the fresh produce and minimally processed commodities. UV-radiation (100-400 nm) has been classified as UV-A (long waves, 400–315 nm), UV-B (medium waves, 315–280 nm) and UV-C (short waves, 280–100 nm). UV-C at 254 nm being more energetic and possessing highest germicidal activity is used for surface disinfection and has low penetration into solids (Vinod et al., 2023). Even the application of low or sub-lethal doses of UV-C light induces disease resistance in host (this phenomenon also called "hormesis"). Mechanisms include, pyrimidine dimers formation (DNA damage), disruption of cytoplasmic membrane integrity and cellular enzyme activity, increase in the level of antioxidants ( -tocopherol, β-carotene and AsA), phytoalexins (scoparone and scopoletin) and stimulates biosynthesis of phenols and defense related enzymes such as phenylalanine ammonia-lyase.

## Electrolyzed water

Electrolyzed water (EW) is produced through the electrolysis of neutral salt solutions in an electrolytic cell, resulting in either acidic or alkaline solutions. This process involves a cathode and an anode within the electrolytic cell, often separated by a diaphragm. When using NaCl as the electrolyte, electrolyzed water exhibits strong disinfection and cleaning properties, making it a commonly used form of electrolyzed water. EW's ability to hinder the activity of enzymes that degrade cell walls may be linked to its capacity to reduce ethylene production. Ethylene is a significant regulator in the ripening and senescence processes of postharvest fruits, influencing various metabolic pathways associated with maturation and aging. Consequently, the activity of enzymes that break down cell walls is directly influenced by ethylene levels (Hussien et al., 2017).

EW has gained significant traction in the field of horticultural technology owing to its eco-friendliness, cost-effectiveness, and user-friendly application. Initially utilized as a cleansing and sanitizing agent, EW has demonstrated remarkable potential in ensuring food safety, with the capability to swiftly eliminate various microorganisms such as viruses, bacteria, and fungi. Recent research highlights the multifaceted benefits of EW treatment beyond mere microbial control in postharvest fruits and vegetables. Notably, it has been shown to enhance the overall quality of stored fruits and vegetables by retarding senescence, managing diseases and mitigating chilling injury. Additionally, studies have indicated that EW treatment can effectively eliminate pesticide residues from fresh fruits and vegetables.

## Ozonation

Ozone (O3) is naturally formed from oxygen as a result of lightening or UV light interaction, which undergoes quick auto-decomposition into molecular oxygen leaving no chemical residues. In 1997, ozone was designated as a GRAS disinfectant and was licensed as an antimicrobial agent in the treatment, storage, processing, and transportation of meats and horticulture goods by the US FDA in 2001 (Vinod et al. 2023). Ozonation is done either in liquid or gaseous form which inactivates enzymes and microorganisms, oxidative destruction of hazardous pollutants, eradicate mycotoxins, enhance antioxidant capacity and delays fruit softening, thereby extending shelf life of the produce. Ozone being strong oxidant acts upon cell surface and oxidizes numerous cellular components, such as sulfhydryl groups of amino acids, unsaturated lipids and respiratory enzymes in the cell membrane, peptidoglycans in the cell envelope, enzymes and nucleic acids in cytoplasm, as well as protein and peptidoglycans in spore mantles and virus capsids. Ozone treatment has been shown to effectively suppress ethylene production in fruits. Ethylene biosynthesis enzymes, such as ACC synthase and ACC oxidase, are downregulated by ozone treatment, resulting in reduced ethylene production. This is mostly due to the ozone's strong oxidative capacity.

## Salicylic acid

Salicylic acid (SA) (2-hydroxybenzoic acid, C7H6O3), a ubiquitous phenylpropanoid compound, holds a pivotal role as an endogenous signaling molecule with robust metabolic and physiological influence on horticultural crop growth and development (Prajapati et al., 2024). SA biosynthesis in plants occurs through discrete enzymatic pathways: either from chorismate via isochorismate (chloroplast-localized) or from phenylalanine (cytoplasm-localized). SA has recently gained traction in postharvest technologies for various horticultural crops, earning the status of generally recognized as safe (GRAS). SA and its derivative, acetylsalicylic acid, have demonstrated their ability to curb ethylene biosynthesis in crops like pear, apple, and others, suggesting SA's role as an ethylene antagonist. Inhibition of ethylene biosynthesis and delay in fruit ripening is attributed to suppression of both amino cyclopropane-1-carboxylic acid synthase (ACS) and oxidase (ACO) production and activity during early stage of fruit ripening (Loay and Taher 2018). Additionally, SA has been linked to diminished lipid oxidation, membrane senescence, reduced chilling injury, and browning reactions, enhanced firmness, anthocyanin synthesis, and antioxidant capacity. Moreover, SA application pre- or post-harvest has exhibited efficacy against fungal decay by invoking defense responses and enhancing antioxidant enzymes.

## Nitric oxide

Nitric oxide (NO) is a gaseous free radical serving as a pivotal plant intrinsic signal regulator. Its high reactivity leads to the formation of other oxides (NO2, N2O3, and N2O4) when exposed to atmospheric oxygen. Application of a low concentration of NO gas in post-harvest treatments for fruits and vegetables is associated with the inhibition of ethylene biosynthesis, effectively delaying fruit ripening and senescence (Mukherjee 2019). Notably, NO's direct inhibition of ethylene is not its primary mechanism; it's suggested that an ACO-NO complex formation occurs, which then interacts with ACC, yielding a stable ACC-ACO-NO complex that directly curbs ethylene biosynthesis (Manjunatha et al. 2010). Furthermore, NO can directly react with ACO through S-nitrosylation, leading to reduced ACO activity and subsequently lowering overall ethylene production.

Prior investigations have asserted that NO has demonstrated utility in decreasing respiration rate, inhibiting ethylene production, reducing enzyme activity, mitigating disease occurrence, and altering skin coloration in fruits (Palma et al., 2019). NO offers a promising avenue for enhancing texture retention during storage. It facilitates improved preservation of cellular constituents, including pigments, titratable acidity, and antioxidants, by mitigating cellular membrane disintegration and reducing electrolyte leakage (Palma et al., 2019). Furthermore, NO displays high efficacy in inducing resistance against systemic infections and safeguarding against microbial pathogens like *Penicillium italicum, Rhizopus nigricans, Aspergillus niger*, and *Monilinia fructicola*. Consequently, treated fruits experience extended post-harvest shelf life. By preserving cellular compartmentation, NO prevents direct enzyme-substrate interaction, thereby curbing enzyme activity, such as polyphenol oxidase (PPO). This novel application of NO underscores its potential for eco-friendly enhancement of fruit shelf life, marking a progressive step in post-harvest research.

## Polyamines

Polyamines (PAs) represent a group of small aliphatic amines with low molecular weight, widely present in all plants. They play a pivotal role in various biological processes, encompassing plant growth, development, and stress reaction. Within the realm of plants, these molecules are intricately involved in growth, development, and responses to abiotic stressors. Among the prevalent PAs are Putrescine (PUT), Spermidine (SPD), and Spermine (SPM), ubiquitously present in plant cells at varying concentrations, from micromolar to millimolar. Their metabolism is closely governed by enzymatic regulations, heavily influenced by environmental factors, particularly stress conditions (Patel et al., 2019).

PAs have emerged as a novel class of plant growth regulators, contributing significantly to a wide array of physiological and biological processes within fruits and vegetables. Notably, free-form PAs have demonstrated anti-senescent properties, as evidenced by studies involving both internal and external applications. In postharvest fruit management, the impact of PAs is substantial, encompassing the inhibition of ethylene biosynthesis, enhancement of fruit firmness, reduction in respiration rate, mitigation of chilling injury, attenuation of color alterations, and mitigation of mechanical damage. Both pre- and post-harvest treatments with PAs have been demonstrated to suppress ethylene biosynthesis and associated responses across various systems. This regulation contributes to enhanced fruit firmness through the inhibition of cell wall enzymes (Prajapati et al., 2024).

Numerous research studies have highlighted various positive effects of interventions, including down regulation of ethylene biosynthetic enzymes, enhanced stability of cell membranes, increased antioxidant capacity, reduced occurrence of chilling injury and physiological disorders, as well as delayed senescence processes. Furthermore, the exogenous polyamine treatment helped maintain the structural integrity and energy levels of pericarp mitochondria.

## Jasmonates

Jasmonates, a class of intrinsic plant growth regulators, offer distinctive and potentially valuable attributes that impact plant growth and development under environmental stresses. The compounds methyl jasmonates (MeJA) and jasmonic acid (JA), which includes its free acid form, are collectively referred to as jasmonates. Plants synthesize jasmonic acid through the octadecanoid pathway, involving the activation of membrane lipids by phospholipase to release linolenic acid. Methyl Jasmonate (MeJA), initially identified in *Jasminium grandiflorum* flower extracts for its pleasant scent, has emerged with significant effects on postharvest fruit and vegetable management. US FDA has approved this compound as GRAS substance. Thus, it's use may have full potential for enhancing the synthesis of bioactive compounds, inducing resistance against chilling injury and prolonging the shelf-life (Per et al., 2018).

## Ascorbic acid

Ascorbic acid (AsA) protects cell membrane from lipid peroxidation and in conjunction with calcium salts and organic acids limit the activity of PPO, prevents browning and membrane disintegration. AsA based derivatives have mostly been employed as anti-oxidants, anti-browning agents, and anti-bacterial agents (Sogvar et al., 2016). Many studies have demonstrated that using right amount of AsA to treat horticulture produce can extend their shelf life.

## Oxalic acid

A natural organic acid that plays a key role in systemic defense, environmental restraint and main regulator of calcium concentrations in plants. Application of oxalic acid is a safe and promising strategy for preserving fruit quality and extending storage life through antioxidant and anti-browning activities. It plays a significant role in systemic strength, programmed cell death, homeostasis in plants and anti-senescence in harvested fruits. Oxalic acid reduces fruit softening and exo-PG enzyme activity while increasing antioxidative enzyme activity (SOD, CAT, and POD). On the other hand, it also inhibits PPO activity, which minimizes non-enzymatic browning. Antimicrobial activity is linked to up-regulation of several stress- and defense-related proteins, enhanced defense-related enzyme activity, phenolic compound buildup, and a delay in fruit ripening (Zheng et al., 2019).

## Surface coatings

Edible coating, a thin and uniform layer applied through spraying or dipping in accordance with good manufacturing practices is both colorless and odorless (de Vasconcellos Santos Batista et al., 2020). Comprising hydrocolloids, polysaccharides (like chitosan, starch, cellulose, alginate, gums, and gelatin), proteins, lipids (essential oils and waxes), as well as nanoparticles and their composites, it serves as a versatile tool. Notably, edible coatings significantly contribute to preserving postharvest quality and improved the storage life of fresh fruits. Beyond this, they fulfil roles as antimicrobial agents, antioxidants, texture enhancers, dyes, flavor enhancers, nutritional supplements, spices, surfactants, emulsifiers, and plasticizers. Moreover, their strong adhesion to product surfaces, especially when combined with other materials, further expands their application. Edible coating commonly used in fruits and vegetables are listed in Table 1.

The establishment of an additional layer around the fruit effectively seals the stomata, resulting in decreased transpiration and respiration rates. This process generates a modified atmosphere that serves as a semi-permeable barrier, hindering the movement of gases (O2, CO2), moisture , and volatile compound. Consequently, this mechanism acts to impede both ripening and senescence. Moreover, this technique proves beneficial in mitigating adverse chemical reactions, reducing deterioration, preventing microbial contamination, and controlling a range of postharvest diseases.

**Table 1: Specific coating applications for various fruits and vegetables**

| Coating material | Composition | Fruit/ Vegetable | Effect of coating |
|---|---|---|---|
| Semperfresh | High concentrations of short-chain unsaturated fatty acid esters in sucrose esters, sodium compounds of CMC, and mixed mono- and diglycerides | Apple, tomato | Decreased color changes, preserved acid, lengthened shelf life, and preserved quality |
| | | Banana | Delayed chlorophyll loss and reduced ethylene synthesis |
| | | zucchini | Less water loss and internal CO2 |
| Prolong | Mixture of sodium CMC, mono and diglycerides, and sugar fatty acid esters | Mango | Slows down ripening, weight loss, chlorophyll loss |
| | | Pear | Retaining firmness, maintaining green epidermis, and maintaining titratable acidity |
| Tal prolong | Mixture of sucrose fatty acid esters, sodium CMC, and mono and diglycerides | Mango | Delayed ripening with extended shelf life |
| Nature-seal | Cellulose-based edible coating | Mango and tomato | Delayed ripening |
| | | Carrot, pome fruit | Retard discoloration and carotene loss, and is a barrier for oxygen diffusion |
| Chitosan | Chitosan and Tween 80 | Strawberry, cucumber, and bell pepper | Antimicrobial coating |
| | | Apple, pear, plum, andpeach | Act as gas barrier |
| | | Litchi | Reduced weight loss and browning, improve storability |
| | | Carrot | Reduced decay and improve Appearance |
| | | Tomato | Reduced respiration rate, ethylene production, and increase tithable acidity |
| | | Pear | Low ethylene production and delayed ripening are both suggested by a high titratable acidity content. |
| zein | Corn zein protein | Tomato | Extended shelf life, delayed color change, loss of firmness and weight |

| Coating material | Composition | Fruit/ Vegetable | Effect of coating |
|---|---|---|---|
| Nutri-save | NO carboxyl methyl and chitosan | Apple, pear, pomegranate | Increased shelf life |
| Brilloshine | Sucrose esters and wax (shine | Apple, avocado, melons and citrus | Protected, increased shine and extended shelf life |
| Nu-coatFlo, Ban-seel | Sucrose esters of fatty acids and sodium salt of CMC | Apple, banana, cucumber, guava, melon, pear, and plum | Extended shelf life |
| Casein | Calcium caseinate | Bell pepper | Gas barrier that effectively prevents interior O2 and CO2, inhibits color change, and slows decay |
| Fresh Seal | Polyvinyl alcohol, starch and surfactant | Fruits | Extended shelf life |

## Conclusion

Latest advances in postharvest treatments for fruits and vegetables comprises a diverse array of strategies, each contributing to prolonged shelf life, enhanced quality, and reduced postharvest losses. In the face of global shifts towards sustainability and heightened consumer expectations, the insights gained from recent advancements will continue to guide the future of postharvest research and practice. By embracing these innovations and fostering interdisciplinary collaborations, we have the opportunity to shape a more resilient and efficient fresh produce supply chain, ultimately benefiting producers, distributors, and consumers alike. As we move forward, it is imperative that researchers, policymakers, and industry stakeholders continue to invest in and support these advancements, ensuring that the journey from farm to table is marked by freshness, quality, and a commitment to a more sustainable future food.

## References


Ahamad, S., Asrey, R., Vinod, B. R., Meena, N. K., Menaka, M., Prajapati, U., & Saurabh, V. (2024). Maintaining postharvest quality and enhancing shelf-life of bell pepper (*Capsicum annuum* L.) using brassinosteroids: A novel approach. *South African Journal of Botany*, 169, 402-412. https://doi.org/10.1016/j.sajb.2024.04.049

Aday, M. S., Temizkan, R., Büyükcan, M. B., & Caner, C. (2013). An innovative technique for extending shelf life of strawberry: Ultrasound. *LWT-Food Science and Technology*, *52*(2), 93-101. https://doi.org/10.1016/j.lwt.2012.09.013

Ali B (2017). Practical applications of brassinosteroids in horticulture—Some field perspectives. *Scientia Horticulturae* 225:15–21. https://doi.org/10.1016/j.scienta.2017.06.051

de Vasconcellos Santos Batista, D., Reis, R. C., Almeida, J. M., Rezende, B., Bragança, C. A. D., & da Silva, F. (2020). Edible coatings in post-harvest papaya: impact on physical–chemical and sensory characteristics. *Journal of Food Science and Technology, 57*, 274-281. https://doi.org/10.1007/s13197-019-04057-1

Hussien, A., Ahmed, Y., Al-Essawy, A. H., & Youssef, K. (2018). Evaluation of different salt-amended electrolysed water to control postharvest moulds of citrus. Tropical Plant Pathology 43:10–20. https://doi.org/10.1007/s40858-017-0179-8

Khan, M. A., Azam, M., Ahmad, S., & Atiq, M. (2023). Improvement of physicochemicals, antioxidant system and softening enzymes by postharvest L-arginine application leads to maintain persimmon fruit quality under low temperature storage. *Journal of Food Measurement and Characterization* 17:2964–2977. https://doi.org/10.1007/s11694-023-01835-6

Loay, A. A., & Taher, M. A. (2018). Influence of edible coatings chitosan/PVP blending with salicylic acid on biochemical fruit skin browning incidence and shelf life of guava fruits cv. 'Banati.' *Scientia Horticulturae* 235:424–436. https://doi.org/10.1016/j.scienta.2018.03.008

Mukherjee S (2019). Recent advancements in the mechanism of nitric oxide signaling associated with hydrogen sulfide and melatonin crosstalk during ethylene-induced fruit ripening in plants. *Nitric Oxide* 82:25–34. https://doi.org/10.1016/j.niox.2018.11.003

Palma, J. M., Freschi, L., Rodríguez-Ruiz, M., González-Gordo, S., & Corpas, F. J. (2019). Nitric oxide in the physiology and quality of fleshy fruits. *Journal of Experimental Botany* 70:4405–4417. https://doi.org/10.1093/jxb/erz350

Patel, N., Gantait, S., & Panigrahi, J. (2019). Extension of postharvest shelf-life in green bell pepper (*Capsicum annuum* L.) using exogenous application of polyamines (spermidine and putrescine). *Food chemistry*, 275, 681-687. https://doi.org/10.1016/j.foodchem.2018.09.154

Per, T. S., Khan, M. I. R., Anjum, N. A., Masood, A., Hussain, S. J., & Khan, N. A. (2018). Jasmonates in plants under abiotic stresses: Crosstalk with other phytohormones matters. *Environmental and Experimental Botany* 145:104–120. https://doi.org/10.1016/j.envexpbot.2017.11.004

Prajapati, U., Asrey, R., Sinha, S. K., Joshi, A., Varghese, E., & Meena, N. K. (2024). Salicylic acid and putrescine treatments reduce yellowing, decay and retain postharvest quality of bitter gourd (*Momordica charantia* L.) during cold storage. *South African Journal of Botany, 164*, 111-120. https://doi.org/10.1016/j.sajb.2023.11.043

Sogvar, O. B., Saba, M. K., & Emamifar, A. (2016). Aloe vera and ascorbic acid coatings maintain postharvest quality and reduce microbial load of strawberry fruit. *Postharvest biology and Technology, 114*, 29-35. https://doi.org/10.1016/j.postharvbio.2015.11.019

Sonawane, S. K., & Patil, S. (2020) Non-thermal plasma: An advanced technology for food industry. *Food Science and Technology International* 26:727–740. https://doi.org/10.1177/1082013220929474

Vinod, B. R., Asrey, R., Sethi, S., Prakash, J., Meena, N. K., Menaka, M., & Shivaswamy, G. (2023). Recent advances in physical treatments of papaya fruit for postharvest quality retention: A review. *eFood* 4. https://doi.org/10.1002/efd2.79

Yang, J., Sun, C., Zhang, Y., Fu, D., Zheng, X., & Yu, T. (2017). Induced resistance in tomato fruit by γ-aminobutyric acid for the control of alternaria rot caused by *Alternaria alternata*. *Food Chemistry* 221:1014–1020. https://doi.org/10.1016/j.foodchem.2016.11.061

Ze, Y., Gao, H., Li, T., Yang, B., & Jiang, Y. (2021). Insights into the roles of melatonin in maintaining quality and extending shelf life of postharvest fruits. *Trends in Food Science & Technology* 109:569–578. https://doi.org/10.1016/j.tifs.2021.01.051

Zheng, J., Li, S., Xu, Y., & Zheng, X. (2019). Effect of oxalic acid on edible quality of bamboo shoots (*Phyllostachys prominens*) without sheaths during cold storage. *Lwt, 109*, 194-200. https://doi.org/10.1016/j.lwt.2019.04.014

# CHAPTER 13

# SENSING TECHNOLOGIES FOR SUSTAINABLE HORTICULTURAL PRODUCTION

**N.V. Gowtham Deekshithulu, T. Janakiram, L. Naram Naidu, P. Ashok, M. Ravindra Babu and T. Thomson**

*Dr Y.S.R. Horticultural University, Venkataramannagudem, West Godavari*
*Corresponding author: ushajanaki2023@gmail.com*

---

Indian horticulture sector contributes about 33% to the agriculture Gross Value Added (GVA), making very significant contribution to the Indian economy. Over the last decade, the world has experienced rapid advancements in horticultural practices triggered by the increasing demand for efficient alternatives to human and labour in crop production. Also, automation in horticultural systems is a necessity to optimize water usage, reduce water wastage, and to implement modern technology in horticulture systems is the need of the hour. The objective of this chapter is to develop sensor based automated system to reduce water requirement and to increase the productivity of the crop. The significant advancements in horticultural technology have also fostered the automation of weed and pest control practices, this consequently enhancing farm practices' efficacy. This application has been adopted at Dr. YSR Horticultural University in crop protection to prevent pest infestation, improve product quality and maximize yield using robotic sprayer. The initiation towards automation for optimal pH and EC levels in hydroponic system using smartphone was also developed. Horticultural innovations, hence, provide essential conditions for meeting sustainable development goals of India.

Though India's horticulture sector is growing, country's share in global trade remains insignificant. There is only 1% of the global trade in vegetables and fruits. The horticulture sector faces many production challenges that stop it from realising its full potential, such as small operational landholdings, lack of irrigation and poor soil management.

Although in India, most of water diverted to agriculture is used for growing staple food crops and only about 10% of agricultural water is used for horticultural crops. As there is huge potential of increasing overall water productivity on basin level by improving water productivity of horticultural crops (Paroda, 2013). As majority of horticultural crops are perennial in nature, have deep and extensive root system capable of extracting water from deeper layers and large canopy to harvest optimum natural resources. Thus, opportunity exists for improving both blue water productivity and green water productivity of horticultural crops through suitable interventions (Malhotra and Manish, 2016).

Drashti et al., (2023) reviewed that most of the farmers use large portions of farming land and it becomes very difficult to reach and track each corner of large lands. In this scenario the Smart Irrigation System using Latest IoT technology is helpful and leads to ease of farming. Pawan et al., (2016) reported that "Machine to machine (M2M)" communication is an emerging technology which allows devices, objects etc to communicate among each other and send data to Server or Cloud through the Core Network.

Camille (2020) reported that about 38% of earth's non frozen land is used for growing food and by 2050, 593 million hectares of land will need to be transformed into agricultural land to meet the projected calorie needs of the global population.

This survey of reasons finds alternatives to our current agricultural system to meet the growing demands for food is far from exhaustive. Hydroponic farming offers a solution to many of the detriments of our world's current agricultural problems.

Hydroponic systems can be automated using Internet of Things technology is very beneficial in this regard by 2033.

Another relevant aspect is that farmers depend more and more on the application of agrochemicals in crops to reduce possible losses and guarantee productivity. Globally more than half of the pesticides are utilized in Asia. India stands 12$^{th}$ in pesticide use globally and 3$^{rd}$ in Asia after China and Turkey. While consumption in China, Japan and America was around 13.07, 11.76 and 3.57 kg/ha of pesticides respectively (Pragati and Hitesh, 2021). It is clear that India applies less amount of pesticides in per hectare of crop land area, but uncontrolled and haphazard pesticide usage is responsible for the presence of high pesticide residues in both natural and physical environment.

Therefore, new technologies have been developed to make agricultural spraying more effective, reducing the amount of pesticide applied and its dosing to the crop.

## Technologies Developed

IoT connected devices are creating a world in which data is exchanged between physical objects such as sensors, on-device software and adjacent technologies with other systems and devices. There are many different IoT connection types to choose from and there is likely to be an optimum solution for all IoT device types that matches IoT device requirements with the application, the system, software and devices it needs to connect to and which takes account of coverage and availability at the deployment location.

## Blynk application

Blynk is an IoT platform for iOS or Android smartphones that is used to control Arduino, Raspberry Pi and NodeMCU via the Internet (Fig. 1).

## Hardware components

**ESP32 Wireless Module SIM**: The ESP (Espressif32) is used for wireless communication. Wireless Transceiver Module is a self-contained SOC (security operations centre) with an integrated TCP/IP (Transmission Control Protocol/ Internet Protocol) stack that can give any microcontroller access to your WiFi network (Fig. 2).

**HC (half-duplex wireless serial communication module)-12 433 SI4463 Wireless Serial Module**: The HC-12 wireless RF UART (Radio Frequency universal asynchronous receiver/transmitter) communication module is a new generation of multi-channel embedded wireless data transmission module within a range of 1000m. This module has an onboard STM8S003F3P6 MCU which is communicating with the SI4463 so that the user can communicate with the HC12 module using a simple TTL 2 wire serial interface (RX, TX, GND) (Fig. 3).

**ATmega328P**: This Arduino-compatible ATmega328P 5V Nano V3.0 Development Board 16MHz CH340 Type-C micro-controller is a small, complete, micro-controller board. CH340 is a USB bus conversion chip, it can realize USB to UART interface or USB to printer interface. In serial UART mode, CH340 provides common MODEM liaison signal, used to expand UART interface of computer or upgrade the common serial device to USB bus directly (Fig. 4).

**LM2596S DC**: LM2596 Power Supply is a step-down(buck) switching regulator, capable of driving a 3-A load with excellent line and load regulation (Fig. 5).

**DFRobot Gravity Analog Capacitive Soil Moisture Sensor**: DFRobot Gravity Analog Capacitive Soil Moisture Sensor measures soil moisture levels by capacitive sensing, rather than resistive sensing like other types of moisture sensors. It is made of a corrosion-resistant material giving it long service life.

The product includes an onboard voltage regulator which gives it an operating voltage range of 3.3 ~ 5.5V (Fig. 6).

**Espressif ESP32**: Espressif ESP32 WiFi Bluetooth Module are powerful and generic Wi-Fi/Bluetooth low energy microcontroller modules that target a wide variety of applications. It is designed and optimized for the best power performance, RF performance, robustness, versatility, features, and reliability, for different power profiles (Fig. 7).

**Waterproof Ultrasonic Obstacle Sensor**: It comes in two separate parts one being the transducer which is the sensing element and the other being the control board. The module is capable of providing information of the objects between the distance range of 250 mm to 4500 mm. This Ultrasonic Distance Sensor is an industrial-grade sensor to measure distance (Fig. 8).

**Grove TDS Sensor**: The Grove – TDS Sensor detects the Total Dissolved Solids (TDS) levels in the water which can be used to indicate inorganic and organic substances present in the water. Typically, the higher the TDS value, the more substances dissolved in water. Hence, higher levels of Total Dissolved Solids (TDS) can indicate that water has more contaminants that can pose health risks (Fig. 9).

**24V 6 channel RELAY**: This is a 24V 6-Channel Relay interface board and each one of the individual relays needs 15-20mA Driver Current. The board features an indication LED for Relay output status and the standard interface that can be controlled directly by many widely used microcontrollers like Arduino etc.

## Components of smart automatic robotic pesticide sprayer

Pesticide spraying is a tedious job in agriculture as it requires various protection equipment to protect the farmer. The massive use of pesticides represents not only a high cost to the farmers but also a risk to their health, the environment and even to the safety of the food consumed by the peoples. In this context, new technology has been developed to make agricultural spraying more effective, reducing the amount of pesticide applied and dosing its use according to the need of the crop.

Thus, a low-cost proto type in a modular solution to automate existent agricultural sprayer has been developed that will decrease pesticide use and human health damage, allowing farmers to be protected and labour intensity can be reduced. This method controls the amount of pesticide distributed across the field. This robot has been created to improve application of precision and to increase yield.

## Salient features

- Reducing direct exposure to pesticides and the human body and improve production efficiency.
- The robot can traverse a variety of terrains and soils.
- To control machine remotely up to 1km across the field.

## Specifications

- DC motor: 12v,1400W, 10 amp, 12kg torque, 120 rpm
- Sprayer motor pump: 12v, 3.5 amp, 6 lpm
- Battery: 12v, 18,000 Mah, 18hr
- Flysky FS-i6X 2.4GHz 6 channel transmitter with FS-iA10B 10 channel receiver.

## Automatic irrigation system

Farmers start to utilize various monitoring and controlled system in order to increase the yield with help of automation of an agricultural parameters like temperature, humidity and soil moisture are monitored and control the system which can help the farmers to improve the yield.

This study includes an embedded system for automatic control of irrigation. This project has wireless sensor network for real-time sensing of an irrigation system. This system provides uniform and required level of water for the agricultural farm and it avoids water wastage. When the moisture level in the soil reaches below threshold value then system automatically switch ON the motor. When the water level reaches normal level the motor automatically switch OFF. The sensed parameters and current status of the motor will be displayed on user's android application.

The DHT11 is a basic, digital temperature and humidity sensor. It uses a capacitive humidity sensor and a thermistor to measure the surrounding air, and spits out a digital signal on the data pin (no analog pins needed). It is simple to use, but requires careful timing to grab data. Humidity sensors are used for measuring moisture content in the atmosphere. Then current temperature, humidity values are send to the microcontroller, those values will display in the users android app.

Capacitive moisture sensor is a sensor which senses the moisture in the soil. It has both digital and analog outputs. Digital output is simple to use, but it is not as accurate as analog output based on moisture level.

## Capacitive moisture sensor

A capacitor consists of 3 pieces. A positive plate, a negative plate and the gap between the plates known as di-electric. A capacitive moisture sensor works by measuring capacitance changes caused by the changes in the dielectric. It does not measure soil moisture directly (as pure water does not conduct electricity well), instead it measures the ions that are dissolved in the moisture. Capacitive measuring basically measures the dielectric that is formed by the soil and the water is the most important factor that affects the dielectric.

When the moisture level drops below a certain threshold (Field capacity 70%), the system sends a wireless message to the receiver unit. Once the microcontroller receives the message, the power supply to the water pump is altered to activate the pump, causing water to flow through the system and water the soil in that particular region. When the moisture level rises above a certain level, the system sends a wireless message to the receiver unit that alters the power supply to the water pump causing water to stop flowing through the system and the pump to stop allowing water from the water source flow into the irrigation pipelines. The automated irrigation system can be applied to a single land section requiring irrigation or multiple land sections with differing irrigation requirements by using number of sensors.

Similarly, the pump can be operated by using temperature mode at a threshold value of 32 either in sensor or manual modes. The real time data of soil moisture, temperature and humidity values can be monitored through blink application. The graphical representation of real time and past three months soil moisture, temperature and humidity data will be displayed in the mobile application.

## Automatic Hydroponic System

### EC and pH sensor:

The EC and pH sensors were tested and calibrated by connecting it to the Arduino microcontroller. The process of calibration was done by following the instruction in the datasheet of the manufacturer. The EC sensor was calibrated using dry, low, and high EC calibration using solutions with different EC level. It was noticed that the sensors produce very accurate results. The pH sensor was calibrated using solutions with different pH level and results were very accurate. EC and pH are important factors to sustain healthy growth of the plants.

### Water level sensor and imer:

The non-contact capacitive water level sensor was continuously shows reading in mobile application. The submersible motor will on/off based on the timer schedule that can be operated hours/day/weekly basis.

### Smart automatic robotic pesticide sprayer:

A low-cost prototype smart robotic pesticide sprayer has been designed for capsicum crop under naturally ventilated polyhouse. The sprayer was designed to move in between the two bed rows of capsicum. In this connection, the chassis of the sprayer was designed as 70 cm length and 40 cm width. The solution allows sprayer and pump on/off control, vehicle speed and nozzle pressure control up to 1 km radius. The block diagram representation of the developed prototype robotic sprayer has been shown in figure 19.

Remote control of the robot is possible through wireless technologies and command for controlling the robot movement and spraying of pesticides is done by using a remote controller. Hence control of the robot is user-friendly and is not very complicated; thus, farmers can easily control this intelligent vehicle. This robot mainly emphasizes on pesticide spraying by farmers from a remote location without directly meeting it.

### Description of proposed system

- The robot is installed in the farm and is powered on by battery.
- DC motors are used for the robot motions that are governed electronically using relays, ULN 2003 ICU, and controller units.
- When the transmitter was on, it sends the signals to receiver, thus DC motor and pump are switched ON.

### Working of robot:

Before spraying the pesticides with this robot, the farmer must follow the steps:

- Fill the required pesticides into tank
- Connect receiver to the transmitter
- Send commands using remote control
- Move sprayer on field
- Change direction of sprayer

## Conclusion

Overall, sensors in horticulture are a powerful tool to gather accurate information about crops, fields, and equipment in order to improve smart farming practices. Use of sensors and precision farming in horticulture aims at increasing productivity, decreasing production costs and minimizing the environmental impact of farming. Using sensors in horticulture can empower farmers to make informed decisions about their farms, maximizing their food growth with minimum resource use. Farmers can determine exactly what is working in their fields and plan ahead for changes that can be predicted with sensors. These benefits are making farmers to turn into smart farming practices, creating a path for improving continuous connectivity for future of farming.

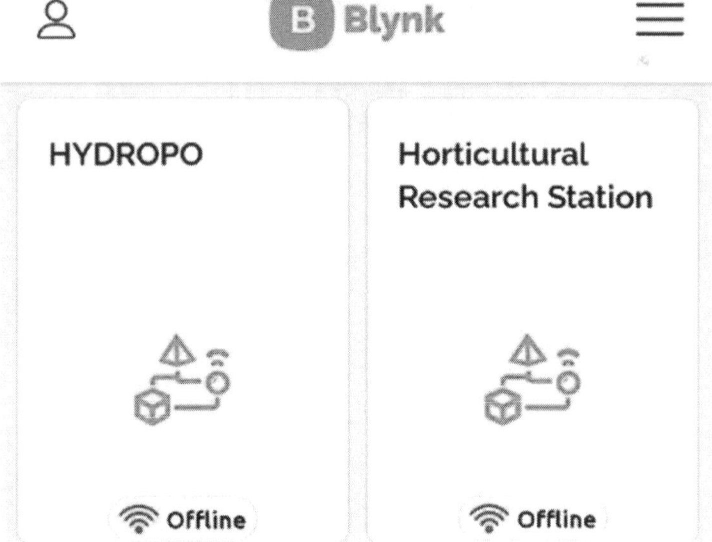


**Fig. 1. Blink application**

**Fig. 2. ESP32 wireless module SIM**

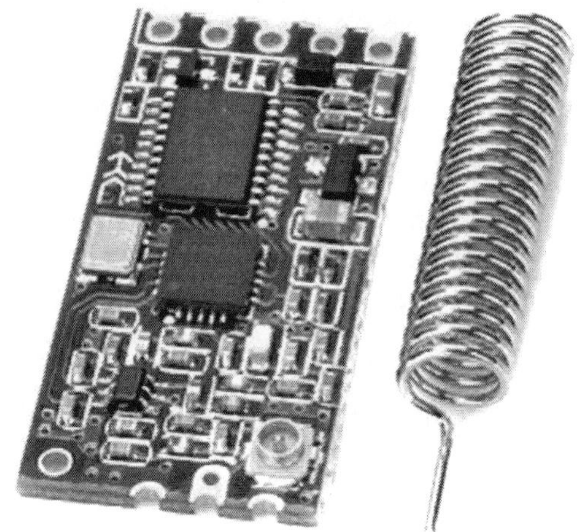

**Fig. 3. HC-12 433 SI4463 wireless serial module**

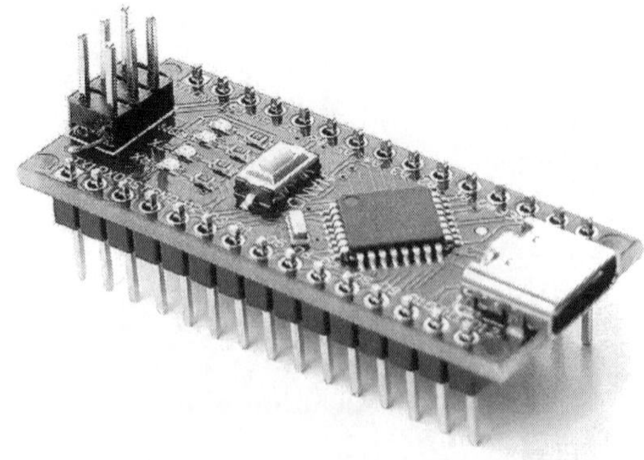

**Fig. 4. ATmega328P nano board**

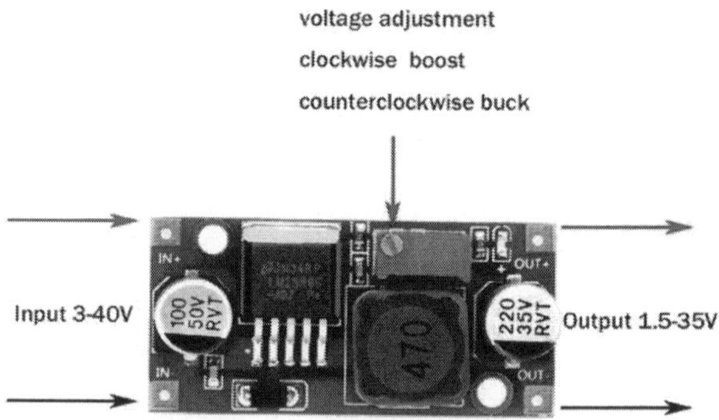


**Fig. 5. LM2596S DC power supply unit**

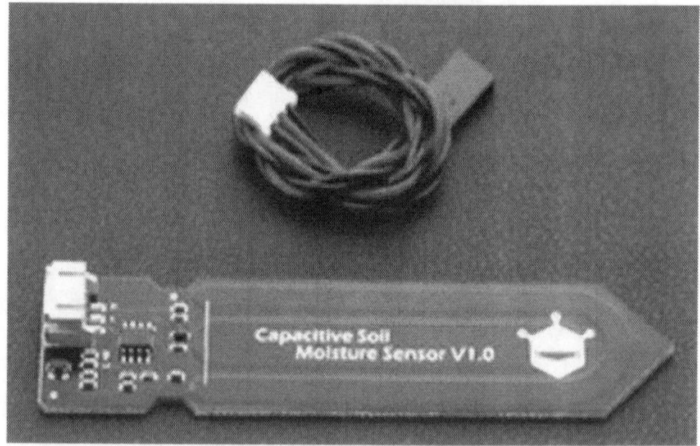


**Fig. 6. DFRobot gravity analog capacitive soil moisture sensor**

**Fig. 7. Espressif ESP32 wifi bluetooth module**

Fig. 8. Waterproof ultrasonic obstacle sensor

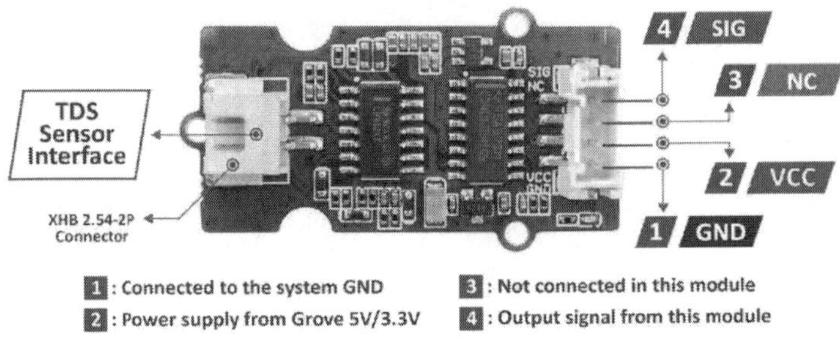


Fig. 9. Grove TDS sensor

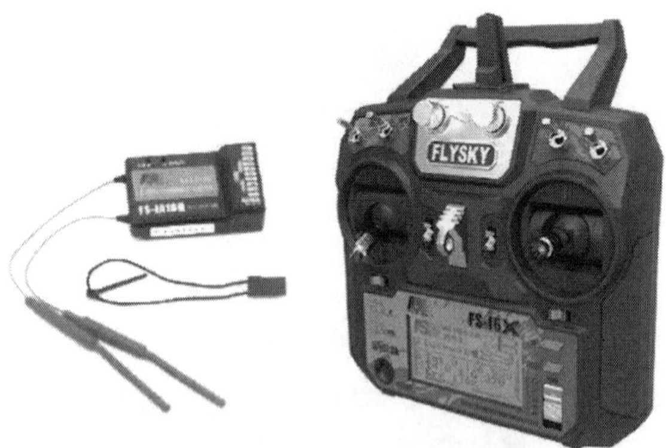


Fig. 10. Receiver and transmitter

**Fig. 11. Real time soil moisture, temperature and humidity data monitoring from mobile application**

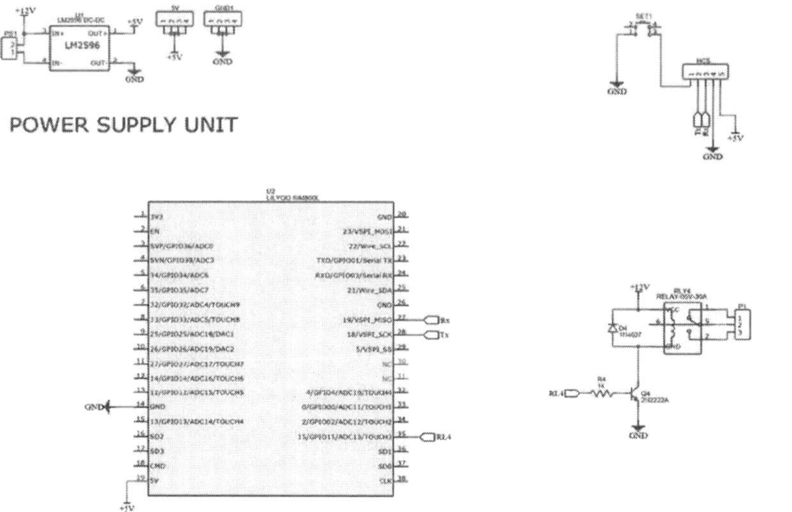


**Fig. 12. Schematic pin diagram of soil moisture receiver unit**

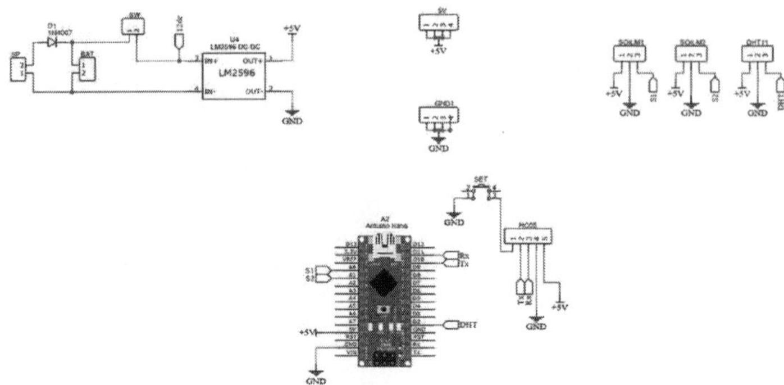


**Fig. 13. Schematic pin diagram of soil moisture transmitter unit**

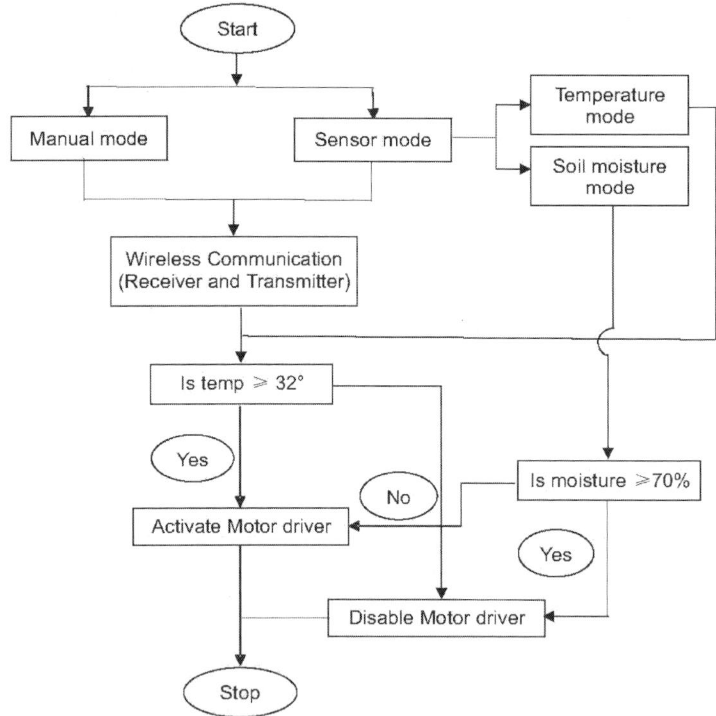


**Fig. 14. Flowchart of Automatic irrigation system using soil moisture sensor**

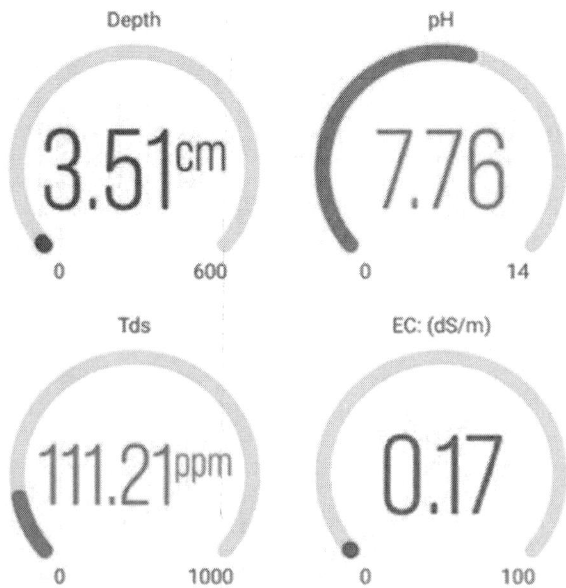


**Fig. 15. Real time pH, EC, TDS and water depth data monitoring from mobile application**

**Fig. 16. Different structural shapes of hydroponic system**

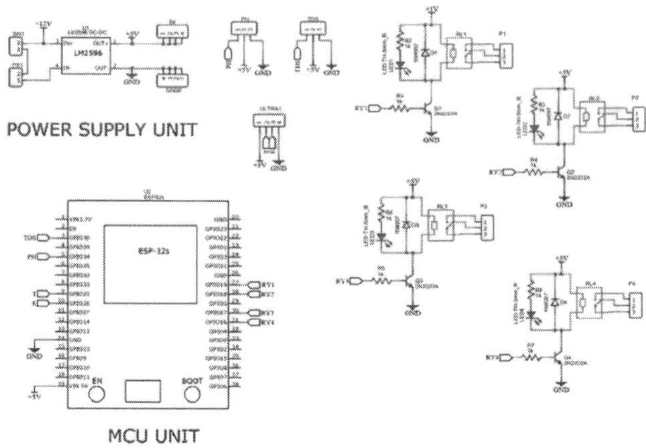


**Fig. 17. Schematic pin diagram of hydroponic system**

**Fig. 18. Working of smart robotic pesticide sprayer**

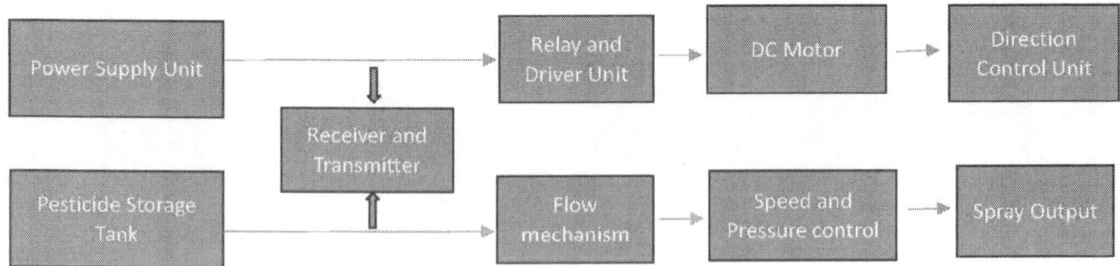


**Fig. 19. Block diagra3m of smart robotic pesticide sprayer**

## References

Ministry of Agriculture & Farmers Welfare. Second advance estimate for the year 2022-23 released for area and production of horticultural crops. 18th October, 2023. https://pib.gov.in.

Horticulture Overview. Department of agriculture & Farmers welfare. 18th October, 2023. https://agriwelfare.gov.in.

Paroda, R.S. Managing our water resource for increased efficiency. International Conference on Water Quality and Management for Climate Resilient Agriculture, May 28, 2013. Maharashtra, India.

Malhotra, S.K and Manish Das. 2016. Augmenting Water Use Efficiency in Horticulture. Souvenir, World Aqua Congress held from 28-30, December, 2016 at India Habitat Centre, New Delhi, pp 45-60.

Drashti, B., Bhargav, L., Yash, M., Mohmmadali, I.A and Manan, S. 2023. A comprehensive and systematic study in smart drip and sprinkler irrigation systems. Smart Agricultural Technology. 5:1-12.

Pawan, K.V., Rajesh, V., Arun, P., Ashish, A., Kshirasagar, N., Rajeev, T., Maazen, A., Tarek, K., Abdelkader, T and Abdulhakim, A. 2016. Machine-to-Machine (M2M) communications: A survey. Journal of Network and Computer Applications. 66: 83-105.

Camille, B. 2020. The future of farming: hydroponics. https://psci.princeton.edu.

Sharma, A. 2021. Status of hydroponics market in India. Express Pharma.

Pragati, N and Hitesh, S. 2021. Pesticides and Indian agriculture- A review. International journal of research-Granthaalayah. 9(5): 250-63.

Maria, T.P., Joan, A and Manel, G. 2009. Powering wireless sensor nodes: Primary batteries versus energy harvesting. I2MTC-International instrumentation and measurement technology conference, Singapore, 5-7 May, 2009, pp. 1-6.

Monalisha, P., Manoj, K., Singh, M., Singh, D.K., Susama, S., Arti, B and Rajeev, R. 2022. Automation of soil moisture sensor-based basin irrigation system. Smart Agricultural Technology. 2: 1-11.

Jacek, M., Marcin, K., Andrzej, W., Agnieszka, S., Arkadiusz, L., Aleksandra, W and Wajciech, S. 2021. Application of a dagger probe for soil dielectric permittivity measurement by TDR. Measurement. 178: 1-13.

Zhang, X., Yang, C and Wang, L. 2018. Research and application of a new soil moisture sensor. MATE web of conferences. 175: 1-3.

Indira Devi, P., Thomas, J and Rajesh, K.R. 2017. Pesticide consumption in India: A spatiotemporal analysis. Agricultural economics research review 30(1): 163-72.

# CHAPTER 14

# WASTELAND MANAGEMENT THROUGH SEMI-ARID HORTICULTURAL TECHNOLOGIES

**A.K. Singh, D. S. Mishra, L.P. Yadav, Gangadhara K., Anand Sahil, Anil and Jagdish Rane**

*Central Horticultural Experiment Station (ICAR-CIAH), Godhra, Gujarat*
*Corresponding author: aksbicar@gmail.com*

Nature is our friend, we have to conserve and preserve it at its best. We have to take all major steps to prevent the nature from getting destroyed. The soil is one of the precious and natural wealth, takes thousands of years to produce a rich soil which is essential for our environment. Usually, nature takes a lot of time for producing the things at its best. With our primary concern,there is a need to save the land from being wasted for the betterment of our future. The soil is the primary source of many primary things either directly or indirectly, need to conserve it to satisfy our needs for food, fuel, and other things. The total geographical area of the country is 329 m.ha where 167 m.ha belongs to wasteland with different kinds of degradation, water erosion, wind erosion, salinity, alkalinity, and flooding. Due to many misuses and deforestation, it has been degraded every year with 2.1 m. ha land. With an increasing population, there is a need of land which is suitable for cultivation. For that, we need to understand the concept of Wasteland Management and ways to prevent it. Out of these lands, approximately 50% land can be put into socially productive uses through strategic planning. Much of these lands are suitable for fruits, plantations and could be used to help increase supplies of fruits, which are not available in sufficient quantities to meet the current demand. Fruit orchard-based land-management strategies can contribute significantly to productive uses of wastelands and minimize the level of nutritional security. Suitable technologies (selection of fruit species, micro site improvement, planting method, in-situ moisture conservation, and propagation) have been developed on semi-arid fruit crops which can be used for wasteland management. Utilization of wasteland through fruit crops can contribute in some way towards maintaining a healthy society and land. An attempt has been made to compile the information available on rehabilitation of wastelands through semi-arid fruit plantations.

## Wasteland and its management

In simple terms, wasteland refers to land which is not in use owing to any obstruction is considered as wasteland. Wasteland is caused due to many factors. Areas that are severely affected by water-logging, ravine, gully erosion, ravine lands, salinity, and alkalinity, shifting and sand dunes, wind erosion, extreme moisture deficiency, overexploitation of natural resource, a dump of industrial and sewage wastes, soil erosion, deforestation are some of the causes of wasteland. Water, wind, man and other source are causing agents for wasteland. There are two types of wasteland (1) utilizable wasteland, (2) un- utilizable wasteland. Utilizable wastelands include marshy lands, saline soils, and undulating lands. Rocky land, glacier land and desert are part of un-utilizable wasteland. Managing these lands for proper use for different purposes is known as wasteland management. With the upcoming technology and advanced procedures, we can contribute towards utilizing the land to its full potential. It is estimated that the biomass production in the wasteland is below 20% due to which it is not considered suitable for use. Usually, when a land does not return anything or not serve any purpose for cultivation falls under this category. As such, fruit culture on wastelands and degraded forest lands has become need of hour due to increasing demand for fruits, and the realization that the semi-arid fruits can be grown successfully on degraded wastelands with the adoption of suitable technology. So far, there is relatively little experience in the rehabilitation process of such degraded soils through fruit plantation, and research is urgently needed on the standardization of suitable technologies for wasteland management. What is needed now is to strengthen on-going efforts at the national as well as regional (or ecological) level, providing not only nutritional security but also ensuring economic security. Plantation of fruit trees on wastelands requires a specialized approach to site-development, choice of species and level of management. Though, some fruit species have miraculous adaptability to adverse soil conditions and climate, even though establishing orchards on such sites requires special attention during initial phase of the plant. The stress faced by semi-arid fruit trees depends upon the nature and extent of the problem in soil, which itself is governed by the soil-reaction, nature and amount of soluble salts, physical barriers like hard pan, quality of underground water and depth of water table etc. (Madhu *et al.*, 2001). So, understanding the soil-problem is a pre-requisite for successful plantation in wastelands. Since the constraints faced by the trees differ depending upon the type of the soils, the planting techniques, method of orchard establishment and choice of fruit species also differ in different situations. The wasteland management has been summarized under following sub heads: 1. Selection of fruit species, 2. Micro site improvement, 3. Methods of in-situ moisture conservation, 4. Methods of planting and its intensity 5. Methods of orchard establishment and 6. Methods of propagation techniques.

## Selection of fruit crops

In order to obtain maximum production from wasteland, it is vital to choose the correct species and varieties. The choice of species depends on soil and agro-climatic conditions, intensity of management, degree of soil problems and social need. Based on local agro-climatic condition of the site, the fruit species which can tolerate adverse conditions, have higher rate of growth, can tolerate drought, improve soil-conditions and least care can be considered. The fruit crops, viz., ber (*Ziziphus mauritiana*), aonla (*Emblica officinalis*), bael (*Aegle marmelos*), pomegranate (*Punica granatum*), lasoda (*Cordia dichotoma*), khirni (*Monilkara hexendra*), karonda (Carissa congesta), jamun (*Syzygium cuminii*), chironji (*Buchanania lanzan*) tamarind (*Tamarindus ndica*), wood apple (*Feronia limonia*), custard apple (*Annonas quamosa*),fig (*Ficus carica*), phalsa (*Grewia subinaequalis*), mulberry (*Morus nigra*), acid lime (*Citrus aurantifolia*), manila tamarind (*Pithecellobium dulce*), date palm (*Phoenix ductylifera*), timru (*Diospyrus melenoxylon*), mahua (*Madhuca indica*),pilu (*Salvadora oleoides*), kair (*Capparis decidua*), wild noni (*Morinda tomentosa*) and palmyra palm (*Borassus flabellifer*) are suitable for plantation in wastelands. Varieties of different fruit crops suitable to grow in wasteland are given in Table 1.

**Table 1: Varieties of semi-arid fruit crops**

| Crop | Varieties |
|---|---|
| Bael | Goma Yashi, Thar Divya, Thar Neelkanth, Thar Srishti, Thar Prakriti, Thar Shivangi, Thar Gauri, Thar Bhavya, NB-5, NB-7,NB-8 NB-9,NB-10, NB-11, NB-16, NB-17, CISHB-1, CISHB-2, Pant Aparna, Pant Shivani, Pant Sujata and Pant Urvashi |
| Jamun | Goma Priyank, Thar Kranti, Konkan Bahadoli, Jamwant, Paras, Rajamun, Rajendra Jamun-1, Jamwant |
| Custard apple | Thar Amrit, Pink Mammoth, Balanagar, Mammoth, Red Sitaphal, Yellow Sitaphal, PhuleJanki, NMK Gold, and Sindhan |
| Mulberry | Thar Lohit, Thar Harit, Saharanpur Local-1, Saharanpur Local-2, S-13, S-34, S-146, S-7999, S-1635, Chak Majra |
| Karonda | Pant Manohar, Pant Sudarshn, Pant Suverna, Konkan Bold, Thar Kamal |
| Tamarind | Goma Prateek, Thar Rashmi, Prathisthan, PKM-1, T -263, Urigam, Ajanta, Yogeshwari, DTS 1 and DTS 2, Anant Rudhira |
| Lasoda | Thar Bold, Paras Gonda, Puskar Local, Maru Samridhi, Karan Lasoda |
| Khirni | Thar Rituraj |
| Phalsa | Thar Pragati |
| Chironj | Thar Priya |
| Mahua | Thar Madhu, NM-2, NM- 4, NM-7, NM- 9 |
| Manila tamarind | PKM (MT) 1 |
| Fig | Poona Fig, Dianna, Dinkar, Conadria, Excel, Chalisgaon |
| Wood apple | Thar Gaurav, Thar Prabha |

**Source: Verma** *et al.,* **2003**

## Micro-site improvement

Micro site improvement plays an important role in ameliorating the ill effects of harsh environments and enhancing the plant survival (Madhu*et al.*, 2001). The plants often give very poor survival and growth under adverse soil and climate conditions. Micro site improvement consists of soil profile modification. Raising trees on degraded soil requires modification in the root-environment by; a) breaking the hard pan by perforation so as to permit vertical growth of roots; b) Amending chemical nature of soil for optimum growth of roots leaching of salts and maximum retention of soil- moisture; and c) proper maintenance of soil-fertility through application of manures and biofertilzers. Since the site-preparation is the costliest and time consuming operation, considerable planning is required for optimizing this aspect. The technique to be adopted for rehabilitation of wastelands through fruit orchard plantation is mainly governed by site and soil conditions, fruit species to be planted and the purpose of the plantation. Based on the above concepts, development of fruit orchards on waste lands involves the following components of micro site improvement (site preparation, method of refilling pits/auger holes and composition of filling-mixture, time of planting) and methods of in-situ moisture conservation.

## In-situ moisture conservation

Micro-catchments are formed around the single plant or along rows of plants depending on the planting geometry and topography of land. These measures are adopted to shape the land surface to concentrate the rainwater around the base of the plant. For this mini–catchments or half-moon configurations are created around each plant. These mini-catchments around the plant can be created in many ways, triangular, rectangular, crescent, v-shaped, catch pits, etc., with an open end at upper side to concentrate the surface flow for higher infiltration into the root zone. Besides these, trench cum bund, staggered and contour trenches were found useful in improving the survival and growth of seedling plants. The beneficial effects of in-situ moisture conservation through rainwater harvesting and mulching the plant basin on plant growth have been observed in ber (Byun et al., 1989), bael (Singh et al., 2021), guava (Borthakur and Bhattacharya, 1992), aonla (Singh *et al.*, 2008 & 2010) and in pomegrananate (Chattopadhyay and Patra, 1992). The most effective methods for in-situ moisture conservation that can be employed on wastelands for plantation of fruit plants are: preparation of crescent contour trench across the slope, placement of submerged pitcher and pit depression.

## Planting on saline soils

Due to more evaporation from the exposed area and little leaching during monsoon causes excess accumulation of salts on the sides and on the top of the ridges. Further, as a result of more evaporation, moisture-content is generally lower in the ridges than in the original land-configuration. Thus, plant-roots face excess salts and low moisture conditions, making survival harder. However, this technique is suitable only near the sea coast, where rainfall is normally high and this helps in leaching of salts from the ridges, providing stress- free environment. The method of planting in saline soil includes: ridge-trench method, sub-surface planting in auger hole.

## Orchard establishment

The fruit plants often give poor establishment in fields on account of damaged roots while lifting from nursery and heavy mortality (40-60%) occurs after transplanting from nursery to field. Therefore, the method of orchard establishment is the most important factor in determining the survival and growth of fruit plants on wastelands condition (Singh *et al.*, 2003). The most promising methods of orchard estthe field in the pits which were filled with mixture of soil, sand, FYM and cakes, thereafter budding and grafting should be performed on the seedlings/ rootstocks. There should be gentle slope to facilitate proper irrigation and proper drainage to avoid the harmful effects of water stagnation during rainy season. Well-decomposed organic matter is mixed with soil and pits are filled before commencement of rain. Planting is done during rainy season when the soil in pits has already settled. While planting, one should be careful that the earth ball does not break and graft union remains well above the ground level. Soil should be pressed all around the stem properly to avoid the formation of air pockets. The plants should be irrigated immediately after planting. In the initial 2-3 years, it is advised to protect plants against low and high temperature injury and from hot desiccating winds by covering plants with some covers, leaving one side open. Shelter belt and wind breaks around the orchard should be planted to protect the plant from hot desiccating wind during summer. For this, 2-3 rows of fast growing drought hardy tree species should be planted in staggered manner. In arid and semi-arid region, *in-situ* patch budding and softwoodgrafting in the months of May- June are found more suitable for better semi-arid fruit orchard establishment.

## Organic mulches

Mulching is highly useful to conserve the soil moisture for long period, protecting the soil from erosion and also improving soil organic carbon status. Mulches not only conserve soil moisture but also impart manifold beneficial effect, like suppression of extreme fluctuation of soil temperature, reduced water loss through evaporation, resulting more stored soil moisture, maintenance of soil fertility, suppression of weed growth, improvement in growth and yield. Continuous use of organic mulches is helpful in improving the soil physico-chemical properties, microbial flora and soil aeration which ultimately resulted into better growth and yield of plant. Under rainfed condition, application of organic mulch (paddy straw) in tree basin area is very beneficial for successful fruit cultivation. It reduces the loss of moisture from the soil, enhances the rate of penetration of rainwater in the soil and controls the growth of weeds. Mulching with paddy straw, maize straw, grasses and rice husk, etc. reduces the weed population and conserves the moisture in the soil (Singh *et al.*, 2021). An organic mulching material improves soil properties like pH and electrical conductivity. Microbial and earthworm population in the basin soil increases with the use of organic mulch over long period (Singh *et al.*, 2020). Mulches should be applied in the tree basin (20 cm thick) after rainy season and undecomposed organic mulches should be incorporated and mixed with soil of tree basin in the forthcoming monsoon. Leaf litter under the canopy is not only effective to retain soil moisture during summer but also improves the soil properties.

## Green manuring and intercropping

Green manuring with leguminous crop in the basin of fruit tree is very beneficial for proper growth and development of plants grown in wasteland. At ICAR-CIAH, RS-CHES Godhra, Gujarat, sunhemp and green gram (mung) sown in the basin of bael tree after first shower which were incorporated in the basin in the first week of September, not only improves the physical and chemical properties of soil but also enhances the growth and development of tree. Intercropping in interspaces of plantation can be done by selecting compatible crops. Leguminous and short duration high yielding vegetable crops such as cluster bean, *bhindi*, pumpkin, cucumber, bottle gourd etc. can be taken easily as inter crop in orchard established in wasteland. Under arid and semi-arid conditions, intercropping is only advisable during *kharif* while in irrigated conditions vegetable crop can be grown during *Rabi* season also. Cover cropping with cowpea, moth bean has been found to increase water holding capacity of light soils as a result of increased organic carbon content in the soil.

## Canopy management

Canopy management of the fruit crops deals with the development and maintenance of their structure in relation to the size and shape for the maximum productivity and quality. Tree vigor, light, temperature and humidity play a vital role in production of quality fruits. The crux of canopy management lies in fact, as how best we manipulate the tree vigour and maximize the use of available sunlight and temperature to increase the productivity and minimize the adverse effect of weather. Pruning in fruit crops is done to improve and regulate tree size and shape to achieve desired architecture of the canopy and also to increase the foliage density by removing branches of the tree. For wasteland management through fruit crops, canopy management plays an important role to maintain the canopy at desired height which improves the root biomass than shoot thus resulting into better growth and development of plant in wasteland.

## Propagation

All the fruit plants suggested for waste land conditions have long been raised by seed. But, seed propagation, does not ensure 'true to type' plants in respect to growth, yield and fruit quality. Besides this, such plants take more time to bear fruit. However, Seedlings raised from seeds are used as rootstock. In order to overcome the disadvantages of seed propagation, multiplication of suitable varieties by vegetative propagation was found desirable (Gangwar et al., 1975; Amin, 1978). In fruit crops, the root system is very vigorous, therefore, it is disturbed during the process of planting of grafts, which ultimately affects growth and establishment adversely in the field conditions. Therefore, *in-situ* patch budding/softwood grafting has been tried in different semi-arid fruit crops at Central Horticultural Experiment Station, Godhra, Gujarat. The plants propagated through *in- situ* patch budding in the month of May-June (before onset of rain) recorded maximum success ranges between 90.14% - 98.82% in different fruit crops. In arid and semi-arid regions, *in- situ* budding/softwood grafting is the most successful method for establishing semi-arid fruit crop orchards in wasteland. This is done by sowing 2-3 seeds directly in the field or by planting seedlings. After one year, budding is done in the field. This method not only saves time but also ensures higher success. Generally, 2-3 rootstocks sprout developing from ground level, and ultimately only one of them is budded and allowed to grow which encourages the growth of scion shoots. Sprouts emerging from the rootstock portion should be removed from time to time for better growth and survival. Among the various vegetative propagation methods, budding and soft wood grafting have been found most economical and successful methods of propagation in fruit species reported and suitable for wasteland conditions. Propagation techniques for semi-arid fruits is given in Table 2.

**Table 2: Propagation techniques in semi-arid fruit crops**

| Fruit crop | Period of multiplication | Propagation methods |
|---|---|---|
| Aonla | May-June | Patch budding |
| Ber | Ring or patch budding | April-August |
| Bael | May-June | Soft wood grafting and patch budding |
| Chironji | July-August | Soft wood grafting |
| Jamun | April -May | Soft wood grafting, patch budding |
| Lasora | June-July | Seed, patch budding and micro-propagation |
| Tamarind | July-August | Soft wood grafting and patch budding |
| Mahua | March-April | Soft wood grafting |
| Custard apple | April- May | Soft wood grafting |
| Wood apple | May- June | Soft wood grafting, patch budding |
| Karonda | June -July | Seeds and cutting |
| Fig | October-November | Cutting |
| Wild noni | July-August | Air layering/seed |
| Manila tamarind | May-June | Seed and budding |
| Khirni | March-April | Softwood grafting |
| Timru | May-June | Seed and softwood grafting |
| Phalsa | July -August | Seed and cutting |

**Source: Singh** ***et al.*****, 2021a & b**

## Future Strategies

- Identification of semi-arid fruit crops suitable for different types of wasteland.
- Each district or state should be aware of technological causes supporting wasteland and various reclamation measures should be taught.
- There should be a programme started for the greening of wastelands which carries advanced methods for reclamation (reusing) of wastelands along with the structure of planting high-quality planting material of semi-arid fruit crops.
- Farmers should be encouraged to adopt different ways and techniques to improve wasteland by growing hardy fruit crops.
- Market linked production and shift in attitude from subsistence farming to commercial orcharding.
- Location-specific farming systems involving semi-arid fruit crops need to be developed and popularized to ensure its adoption by the farmers of the region.
- Creating awareness among the growers for marketing of produce through FPOs, or cooperative or growers' association so as to fetch better price of the produce for grower's prosperity.
- Emphasis should be given on post harvest technology to develop value added export oriented products. Small scale processing units should be established and promoted for its commercialization.
- Screening of genotypes for abiotic stresses especially drought resistant and moisture stress.
- Development of integrated crop management strategies for sustainable fruit production on wasteland.
- Fruit based cropping systems and cropping models should be developed to provide stability in income to the farmers.
- To create awareness among the people regarding arid fruits role in nutritional, health and environmental security for its commercialization in wasteland.
- It is need of the hour to develop herbal products useful against various ailments with scientific validation.

## References

Amin, R.S. (1978). Soft wood grafting – a new technique for hard wood plants. *Curr. Sci.*, 47: 468-469.

Borthakur, P.K. and Bhattacharya, R.K. (1992).Organic mulches in Guava orchards. Indian Hort., 37 (2): 43-44.

Byun, J.K., Do, J.H. and Chang, K.N. (1989).The effect of black polythene and straw mulch on growth of young jujube (Zizyphus jujube Miller) plants.*Korean Soc. Hort. Sci.*, 7(1) 138-139.

Chattopadhyay, P.K. and Patra, S.C. (1992).Effect of soil covers on growth flowering and yield of pomegranate (*Punicagranatum* L).*South Indian Hort.*, 40(6): 309-312.

Gangwar, R.P., Singh, D. and Chudawat, B.S. (1975).A note on the rejuvenation of aonla (*Emblicaofficinalis*) tree by budding.*Haryana J. Hort.Sci.*, 4 (3): 150-151.

Luchkov, P.G. Ponomarenko, G.A. and Kudaev, R. Kh. (1989). Mulching the soil in young orchard on slopes.*Sodovodstvo-ivinogradarstvo*, 4: 11-12.

Madhu, M., Sikka, A.K., Singh, D.D.V. and Tripathi, K.P.( 2001). Alternate land use system with improved micro site conditions for rehabilitation of wastelands. *South Indian Horticulture*, 49: 354-356.

Maninannan, S. and Desai, A.R. (2007).Effect of in-situ moisture conservation practices on runoff, soil loss and initial growth of cashew (*Anacardiumoccidentale*). *Indian J. Soil Cons.*, 35 (2): 147-150.

Negi, R.S., Baghel, B.S, Gupta, A.K. and Singh, Y.K. (2010).Standardization of method of orchard establishment and propagation in Aonla (*Emblicaofficinalis*Gaertn) for rehabilitation of degraded pasture/grazing lands. *Ann. Horti.*, 3(1): 39-46.

Singh A. K., Singh, R. S. and Lata, K. (2021a). Propagation techniques in underutilized fruits. In: Underutilized Fruits of India (Eds. Singh R. S., Singh, A. K., Maheshwari, S. K., Bhargava, R. and Saroj, P. L.), Brillion Publishing, New Delhi, pp.149-164.

Singh Sanjay, Singh, A K, Mishra D S and Saroj P L. (2021b).Quality planting materials for arid and semi-arid fruits. In: Current Horticulture: Improvement, Production, Plant Health management and Value-Addition, Vol. 1. Brillion Publishing, New Delhi, pp. 49-64.

Singh, A. K. Singh, Sanjay, Mishra, D. S. and Saroj, P. L.(2016). More crop with minimal water. *Indian Horticulture*, 61(6): 86-91.

Singh, A. K., Pandey, D. Singh, Sanjay, Singh, R. S., Misra, A. K. and Singh, R. K. (2019). The Bael in India. Published by ICAR-DKMA, New Delhi, pp.1-130, ISBN: 978-81-7164-186-4.

Singh, A. K., Singh Sanjay, Saroj, P. L., Mishra, D. S. VikasYadav and Raj Kumar (2020).Underutilized fruit crops of hot semi-arid region: Issues and Challenges-a review. *Current Horticulture*, 8(1):12-23.

Singh, A. K., Singh Sanjay, Saroj, P. L.and Singh, G. P. (2021).Improvement and production technology of bael (*Aeglemarmelose*) in India. *Current Horticulture*, 9(1):3-14.

Singh, A. K., Singh, Sanjay, AppaRao, V. V., Bagle, B. G. and More, T. A. (2010) Efficacy of organic mulches on soil properties, earthworm population, growth, and yield of aonla cv. NA-7 in semi-arid ecosystem.*Indian Journal of Hort.*, 67 (special issue):124-28.

Singh, A.K, Singh, Sanjay, AppaRao, V. V. and Meshram, D. T. (2008).Effect of mulching on soil properties, growth and yield of aonla (*Emblicaofficinalis*Gaertn) in semi arid ecosystem.*Indian Journal of Agricultural Sciences*, 78 (3) 193-7.

Singh, B.K., Sharma, S., Niwas, R. and Kumar, S. (2003). Effect of methods and time of budding on bud sprouting in aonla (*Emblicaofficinalis*Gaetn.) cv. Chakaiya. *Haryana J. Hort. Sci.*,32 (1/2): 27-28.

Singh, G.K. and Parmar, A.S. (1998).Studies on the effects of methods and dates of budding in aonla (EmblicaofficinalisGaertn).*Annals of Arid Zone*, 37(2): 199-201.

Singh, Sanjay and Singh, A. K. (2006).Standardization of method and time of propagation in jamun (*Syzygiumcuminii*) under semi-arid environment of western India.*Indian Journal of Agricultural Sciences*, 76 (4):242-5.

Singh, Sanjay and Singh, A. K. (2007). Standardization of method and time of vegetative propagation in Tamarind (*Tamarindusindica* L) under semi- arid environment of western India.*Indian Journal of Horticulture*, 64 (1): 45-49.

Singh, Sanjay, Singh, A. K., Mishra D. S., Singh, G. P. and Sharma, B. D. (2022). Advances in research in jamun: a review. *Current Horticulture*, 10 (1):8-13.

Srivastava, A.L., Singh, H.K. and Singh, I.S. (2002).Studies on propagation of aonla (Emblicaofficinalis). Haryana J. Hort. Sci., 31(3/4): 156- 158.

Verma, A. K., Haldhar, S. M., Singh, A. K. and Samadia, D. K. and Sharma, B. D.(2023). Inventory of CIAH Technologies for Arid and Semi-Arid Region. M/s Hitech Graphics, New Delhi, pp. 1-124. ISBN-978-81-956916-1-6.

CHAPTER 15

# INTRODUCTORY TELEMETRY SOIL AND CLIMATE MONITORING

**Akash Sharma[1] and Steven Falivene[2]**

*1 Advanced Centre for Horticulture Research, SKUAST-J, Udheywalla, Jammu, India*
*2 NSW-Department of Primary Industries Institute, Dareton- Australia*
*Corresponding author: - akashskuastj@gmail.com*

The application of new technologies such as the Internet of Things IoTs offers the opportunity to improve current agricultural development, facilitate daily tasks, and turn farms into efficient and sustainable production systems. The use of telemetry systems in SMART agriculture is an innovative approach which consists in the implementation of an information system able to provide data on irrigation parameters throughout a year, also taking into consideration other meteorological parameters. The need for a telemetry system for irrigation is emphasized by the market's interest in having access to fully automated monitoring and automation solutions for energy efficient and cost-effective agricultural crops. This chapter aims to present a telemetry system for monitoring crops with an improved architecture from the point of view of very low energy consumption, low management costs, scalability, forecasting functions, and diagnosis. IoT devices are needed in the agriculture sector to monitor plant growth.

Environmental changes caused by human beings could potentially cause conditions in which the development of new crops is not possible. Likewise, the growing urbanization decreases labor in areas typically involved in food production, increases costs and reduces the productive capacity of the sector. In view of this, smart farming is a new farm management concept that employs techniques and technologies at various levels and scales of agricultural production, enabling to overcome the challenges in food production demands and reduction in the workforce. For example smart farming may use different types of sensors to collect data (e.g., temperature, humidity, soil moisture, light, drip irrigation pressure, Sap flow, Trunk dendrometer, fruit size etc.), communication networks to send and receive data which is then managed and analyzed by management information systems and data analysis solutions. The use of the data provided by smart farming helps boosting productivity and minimizing waste by allowing necessary actions to be carried out at the right time, quantity and place. Moreover, recent technological developments in areas relevant to IoT facilitate an easier adoption and use of smart farming with IoT. Such technological developments include, for example, network communications, reduction of hardware size, optimization of power consumption and devices cheapening. Considering this context, this chapter aim; (i) to characterize the scientific knowledge about SF based on the main factors of development by country and over time and (ii) to describe current SF prospects in this field.

Advances in modern sensors, their physics and implementation strategies have been discussed vividly by Arridha et al., 2017. This can be used as a good manual for studying and choosing an appropriate sensor for a specific application. Different types of IoT, their designs, implementations, challenges, privacy issues are discussed by Koo et al., 2016, Hassan et al., 2020; Ji, C et al. 2015; Tawalbeh et al., 2020; Brewster et al., 2017; Hadzovic et al., 2021. Putting these characteristics altogether becomes very important to understand the most suitable type of IoT that could be viable for an application meeting the usage requirements.

The telemetry soil and climate monitoring showed the technology impact to increase the productivity and production in the country (Kashyap et al., 2021). The use of SF tools is possible due to the use of sensors in horticulture. A sensor is an electrotechnical device that measures physical quantities from the environment and converts these measurements into a signal that can be read by an instrument. Among the measurements read by sensors are the following: temperature, humidity, light, pressure, noise levels, presence or absence of certain types of objects, mechanical stress levels, speed, direction, and object size. The IoT allows objects to be controlled remotely via an existing network infrastructure, creating opportunities for more direct integration between the physical world and computer-based systems. The use of IoT depends on the internet infrastructure, and this presents several shortcomings, especially when dealing with a large number of network devices and the integration with other systems. SF tools introduce a new level of technology into agriculture, including robotics, mapping and geomatics technologies, decision making and statistical processes. The most promising SF technologies incorporate advances in sensors, data analysis, telemetry, and positioning technologies, but the development and dissemination of these technologies may require time and investment.

## Review of work Done

Many methodology such as Design of IoTs-based Monitoring System for Intelligence Indoor Micro-Climate Horticulture Farming in Indonesia, IoT based Horticulture Monitoring System, Smart Agriculture, IoT based smart sensors agriculture, IoT Based Smart Agriculture Monitoring System, an efficient water optimization for horticulture crops using IoT, sensor based automated irrigation system, design and implementation of automatic plant watering system has been proposed by many authors and the detailed explanation about these methods is as follows.

Emil Robert Kaburuana in 2019 proposed the micro-climate is very closely related to habitat and important for organisms on a micro scale. Light intensity, soil moisture, soil temperature, air temperature, soil acidity, wind direction, and $CO_2$ level are micro-climate factors that are very influential in aquaculture productivity (Balaji and Sudha, 2016).

Rabka et al., 2021 proposed the Remote monitoring and control of these farms is the obvious next step in trying to tackle the issues at hand. This is where the IoT comes into the fold. IoT is a network of Internet-connected devices that collect user-accessible data using sensors and processors and transfer it over wired or wireless networks. Controlled-Environment Agriculture will help make farmers' daily work more manageable by reducing the amount of time needed for physical monitoring (Subalakshmi and AnuAmal, 2016).

Suma et al., 2017 propose the monitoring the environmental factors is not the complete solution to increase the yield of crops. There are a number of other factors that decrease productivity to a greater extent. Hence automation must be implemented in agriculture to overcome these problems. In order to provide solutions to all such problems, it is necessary to develop an integrated system which will take care of all factors affecting productivity in every stage (Vandana et al., 2013) .

**Overview of Telemetry systems**: Telemetry systems play a pivotal role in modern horticulture, particularly in fruit crops, by providing real-time data on soil and climate conditions. These systems leverage advanced technologies to enhance precision farming practices.

Telemetry monitoring systems in orchards involve the use of tracking modules, servers, and networks of sensors for data acquisition and remote monitoring. These systems offer real-time information on environmental conditions, optimizing agricultural practices. Agriculture vehicle-mounted telemetry systems, such as automatic labor time determination and process large data amounts. Harvest Systems utilize solar-powered loggers with IoT sensors for remote monitoring in orchards. Telemetry proves useful in crop protection applications, as seen in a technical-economic analysis of telemetry systems in agriculture. Overall, these systems enhance orchard management through data-driven insights and automation.

Geocledian platform is used to monitor the crop fields on large scale with efficient results. Climate change impacts fruit crops significantly. Telemetry systems help monitor these changes, offering valuable insights into adapting cultivation methods to ensure optimal yield and quality.

**Automated irrigation protocols:** Implementation of telemetry-based automated irrigation protocols, such as those estimating water uptakes from soil, enhances water management efficiency in fruit crop cultivation. To assemble loggers for sensor installation in fruit orchards, adapt the assembly based on orchard-specific needs and the type of fruit cultivation as briefly mentioned below:

## Details of FHB logger Components list:-

### FHB Loggers:

- FHB logger for sensors and devices likely stands for "Fieldbus HART Logger."
- HART (Highway Addressable Remote Transducer) is a communication protocol widely used in industrial automation.
- A Fieldbus HART Logger would be a device capable of monitoring and recording data from sensors and devices connected through the HART communication protocol.
- These loggers capture real-time information like temperature, pressure, and other environmental parameters, storing it for analysis.

## FHB logger component purchasing list

### Light intensity lux sensor

- BH1750 BH1750FVI Chip Light Intensity Light Module

### Humidity sensor

- BME280 5V 3.3V Digital Sensor Temperature Humidity Barometric Pressure Sensor Module
- NOTE You must choose BME280, not BMP 280 and choose 3.3v version

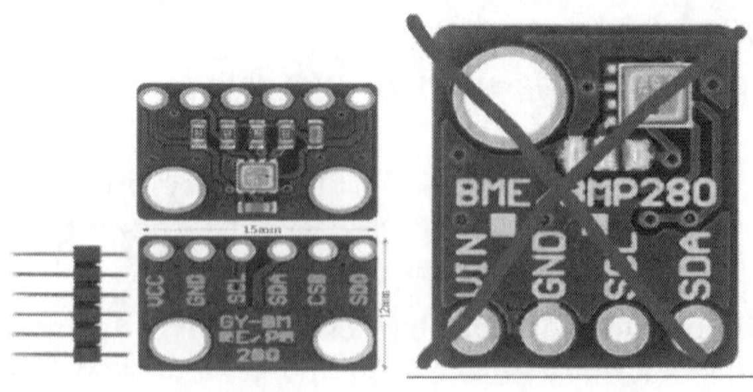

## Temperature sensor

- DS18B20 18B20 Stainless steel package Waterproof Probe Temperature Sensor

## Connectors

- Aviation Plug SP13 IP68 Waterproof Connector Male plug & Female Socket 1/2/3..5../9p Back Nut panel Mounted Wire Cable Connector
- 3X 2 pin Back nut model (shown below)
- 2X 4 pin Back nut model

## Connectors

- Waterproof Docking Connector Aviation Plug Plugs Sp13 2pin 3pin 4pin 5pin 6pin 7Pin IP68
- 6X 3 pin docking model

## Screw bar

- 2X 12 Position 3/6/10A Barrier Screw Terminal Block Wire Dual Connector Strip

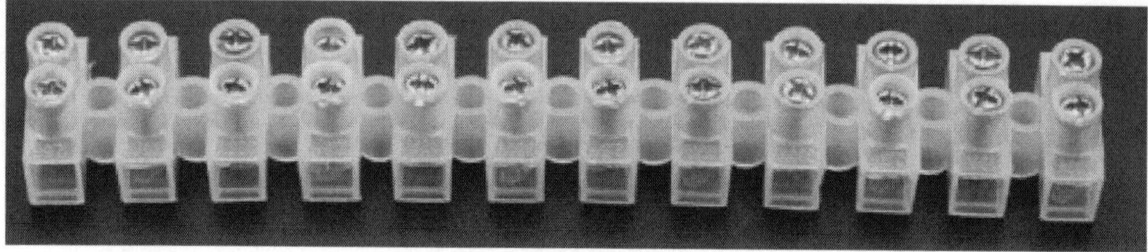

## Core Security/Alarm Cable (used for humidity and lux sensor)

- SEC7204 7/0 20(0.22mm2) There are four wires in the cable. Each wire has 7 strands of wire 0.22m2
- Tis wire can be used for all sensors (5v solar panel, 12v solar panel, Wind, light lux and humidity). 20m is used for on system. 2 core wires can be purchased to be used on the wind sensor and solar panel using 3.5 m for each.
- Can be purchased by the meter from electronics store. 300m roll can be purchased from electrical components suppliers (e.g. Rexel or L&H).
- There is a 14 strand/wire version of security cable and the extra thickness of the cable is not required.

## Cable glands

- Waterproof Cable Gland 10pcs Cable entry IP68 PG7 for 3-6.5mm PG9 PG11 PG13.5 PG16 PG19/21/ White Black Nylon Plastic Connector - 11X PG7 - 1X PG9

## Weather proof box for data logger

- Medium ABS IP66 Enclosure - Clear Cover
- Transparent cover - Medium 175(L) x 125(W) x 75(D)mm Cat HB-6412

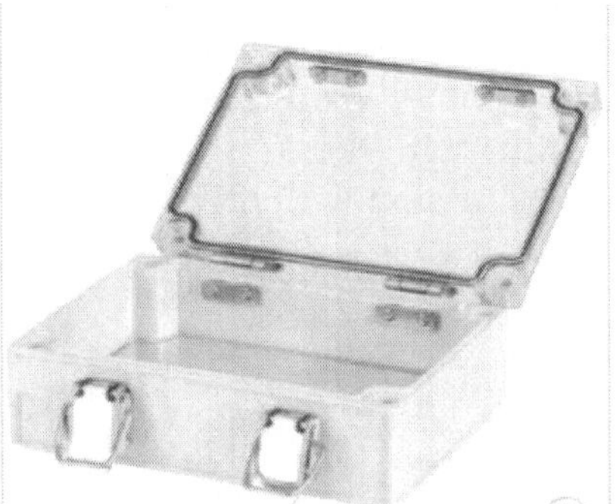

## Weather proof box for ground connections if using a sap flow sensor

- 200x120x75mm (or larger)

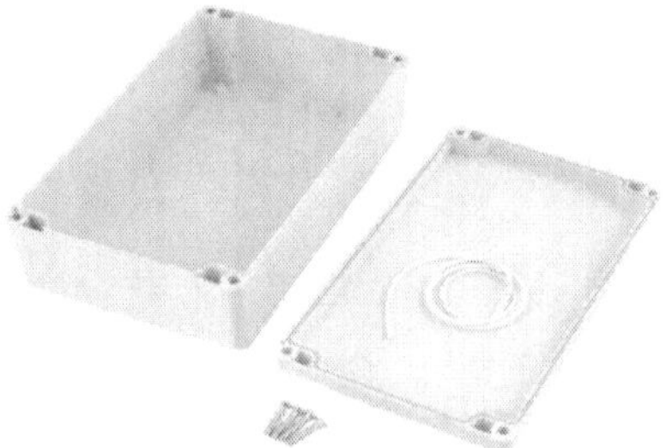

## 14 or 16 core or more wire

- Must not be thicker than 9 mm in diameter
- 10/12/14/16/20/24 core PVC sheathed pure copper core conductor cable signal wire 26/24/22AWG control wire flexible wire - 2m

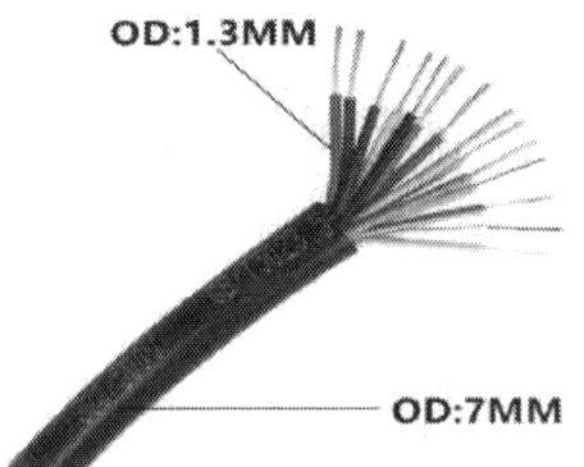

## Lora antenna for logger

- Antenna LoRa ...........MHz Hotspot Antennas for LoRawan .........MHz Antena RP SMA Gateway Aerial Long Range Commnucaiton
- **Important:** You need 3m of antenna cable, if the cable terminates in a SMA or RP-SMA connector. Which every version you buy you need to buy the corresponding connection type for the Lora connection on the logger box. I have had good results with the STARF 55 cm models, the 40cm are also good for short range situation with little obstruction (e.g. 600 m)

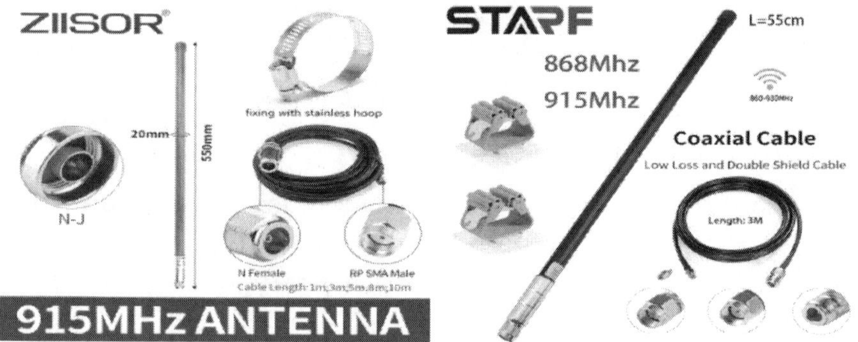


- Most often supply an adaptor to change between a SMA and RP-SMA connection which makes it easier to match with the rest of components

- Make sure the antenna terminates in the following head (N-male) and that the antenna cable has the corresponding fitting (N-female)

## Explanation SMA vs RP-SMA connections for antennas

- Antenna cables can terminate in a SMA or RP-SMA head. First buy the antenna and then buy the cable to match the antenna cable.

| | SMA | RPSMA |
|---|---|---|
| *Male* | | |
| *Female* | | |

- However there are adapters to convert between the two in case of incompatibility, I always carry each of these two versions in case of incompatibility.
- RF Adapter SMA Male plug to RP-SMA Female jack For Raido Antenna RF Coaxial Adapter connector Converter

SMA MALE TO RP SMA FEMALE    RP SMA MALE TO SMA FEMALE

**Pigtail cable for Lora communication – SMA connection**

- RG316 SMA Male Plug to SMA Female Jack Bulkhead Connector RF Coaxial Jumper Pigtail Cable 15 cm . This is used in the logger box to make a antenna connection between the external part of the box to the Lora module. There are different thicknesses of cable, The RG316 cable thickness is a good compromise between least signal loss (thicker the better) and flexibility to bend. The next grade of wire down is RG174.

RG316

## Antenna connection for Lora module on logger

- SMA female Jack right angle solder PCB mount RF connector
- The EZSCB Ant microcontroller does not have Lora communication, it only has WiFi. You need to build a Lora module and this is one part. Please note it is a SMA connector, not RP-SMA

## Lora module PCB plate

- ESP8266 Adapter Plate
- This is the base plate for the Lora communication module. It has three resistors on the bottom, these can be easily removed with a soldering iron.

## Lora module

- RFM95 RFM95W wireless transceiver module HopeRF
- Make sure you buy the 915 Mhz for Australia, or the corresponding Lora frequency for your country

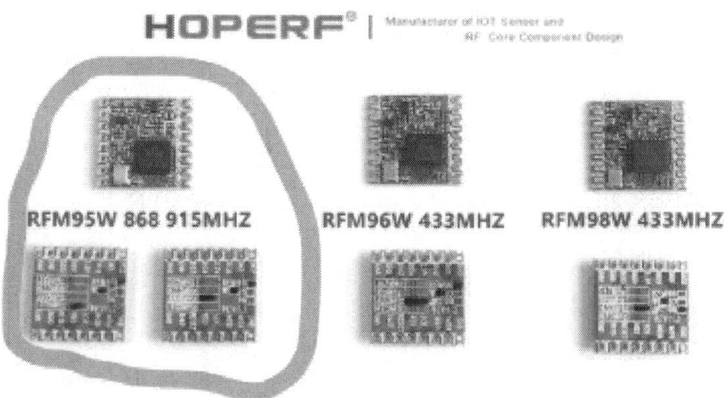


**WiFi antenna – If using WiFi communication between logger and WiFi base station**

- waterproof outdoor WiFi antenna screw mount 2.4GHz antenna
- This is the best model. Please note that need 3 m of cable and that the needs to be the same connection as the box pigtail connector. If the antenna ends in a Male RP-SMA, you need to buy a female RP-SMA connector. IF it is the standard SMA (no RP) then you buy a female SMA pigtail connector. Below is a picture showing the differences between the two versions of a SMA connector (the difference is in the pion in the middle)

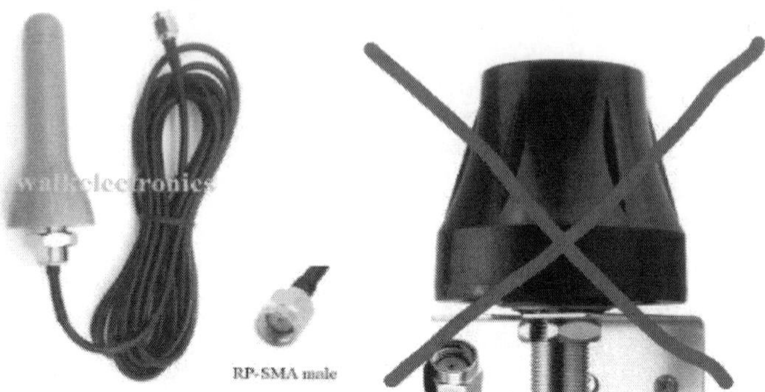


- Do not purchase the mushroom head version

**Wifi outdoor booster**

- This unit uses the wifi projected from your house or office (it needs to be close to the office/home WiFi e.g. within 10 m) and projects to WiFI over the orchard. IT needs to be 2.4 GHZ. Note that Wifi communication is only reliable in line of site and the units only project in about a 60 degree angle. If you need further distance or need to pass through obstructions such a wind break trees then the lora communication system is best suited. There are different strength of booster units. The price difference between them is only about $20, so I purchase the highest strength 3-5 km range models that will cater for all situations. However, I have used the shorter range moseld and they have rejected 400m. The most reliable distance achieved has been 700m and I assume I could get 1 km. However in most cases after one Km it is obstructions that cause problems rather than not enough signal strength. Also obstructions can give dropouts were reads are intermittent. Intermittent reception is not acceptable to ensure reliable data. Sometimes you might do a test and it works well for the first few readings and then it drop is out intermittent. When the signal strength recorded by the logger is above -80 db this often results in intermittent connections. -75 to -80db is fair whilst -75 to -60 is very good.

- 5KM Comfast Outdoor CPE Wifi Repeater 5GHz 300Mbps Wireless Wifi Router Extender Bridge Nano station 2*14dbi Antenna Wi fi

Make sure booster has a repeater mode where they can pick up a wii signal and boost it

**Sap Flow Sensor**

- **Prepare the tree**: Ensure the selected tree is suitable for installation, typically on the trunk. Clean the area where the sensor will be attached.
- **Sensor attachment** : Affix the sap flow sensor to the tree trunk, usually on the xylem tissue. Check for proper alignment and secure attachment.
- **Calibration and testing** : Calibrate the sensor according to the manufacturer's instructions. Test the sensor to ensure accurate sap flow measurements.

### Trunk Dendrometer Sensor

- **Orientation** : Mount the sensor arm perpendicular to the trunk or branch, ensuring that the sensor head makes linear contact with the trunk or branch
- **Contact surface** : Confirm that the sensor head is in direct contact with the trunk for precise measurements of micro-variations in trunk radius
- **Sensor type** : Consider using a highly precise incremental LVDT-based dendrometer, such as the DE-1T, known for its accuracy in monitoring micron-range variations in trunk radius.
- **Continuous monitoring** : Dendrometers continuously measure small fluctuations, including shrinkage and swelling, providing insights into the tree's water stress before visible cues.
- **Integration with sensors** : Consider combining dendrometer sensors with other tools, such as dynagage sensors for sap flow measurement, for a comprehensive understanding of tree health.

## Software details

### Introduction to Arduino

Arduino is an open-source electronic platform based on easy-to-use hardware and software. Arduino boards are able to read inputs - light on a sensor, a finger on a button, or a Twitter message - and turn it into an output - activating a motor, turning on an LED, publishing something online. The open-source Arduino Software (IDE) makes it easy to write code and upload it to the board. It runs on Windows, Mac OS and Linux. The environment is written in C and based on Processing and other open-source software. This software can be used with any Arduino board. The Arduino IDE uses a programming language based on C++. It is essentially a framework built on top of C++, allowing users to write code in C or C++. The IDE provides a reference guide organized into functions, variables, constants, and structure keywords for programming with Arduino. While the default language of the Arduino IDE is English, users have the option to change it in the preferences

## USB Powered Arduino

The USB port of the Arduino Uno can be connected to an USB device or port that can provide a stable 5V output like a computer or power bank or USB charger, etc. A USB 2.0 port can provide current up to 500mA and it is more than enough for an Arduino UNO. Load software onto esp32

## Embedded C:

Embedded C is a set of language extensions of C programming language. It includes a number of features not available in normal C, such as fixed point arithmetic, named address spaces, and basics I/O hardware addressing. Embedded C uses most of the syntax and semantics of standard C.

### Is the Arduino a microcontroller?

As Arduino is a microcontroller board, it contains the CPU, RAM, and ROM on the chip. The additional devices on the Arduino Board are to connect the power supply and to connect input/output devices with the board. Arduino software is an IDE (integrated development environment) through which developers write and upload the code to the microcontroller. The Arduino board enables developers to feed the programme in the simplified version of the C++ language, making it easier for them to learn and code.

## What is a Microcontroller?

A microcontroller is a compact integrated circuit designed to govern a specific operation in an embedded system. A typical microcontroller includes a processor, memory and input/output (I/O) peripherals on a single chip.

Sometimes referred to as an embedded controller or microcontroller unit (MCU), microcontrollers are found in vehicles, robots, office machines, medical devices, mobile radio transceivers, vending machines and home appliances, among other devices. They are essentially simple miniature personal computers (PCs) designed to control small features of a larger component, without a complex front-end operating system (OS).

## How do microcontrollers work?

A microcontroller is embedded inside of a system to control a singular function in a device. It does this by interpreting data it receives from its I/O peripherals using its central processor. The temporary information that the microcontroller receives is stored in its data memory, where the processor accesses it and uses instructions stored in its program memory to decipher and apply the incoming data.

It then uses its I/O peripherals to communicate and enact the appropriate action. Microcontrollers are used in a wide array of systems and devices. Devices often utilize multiple microcontrollers that work together within the device to handle their respective tasks.

## Proposed model

This horticulture system using IoT system is powered by ESP32 Micro-controller, it consists of temperature sensor, Moisture sensor, LDR sensor, pH Sensor. When the IoT based Horticulture monitoring system starts it checks the light intensity, water quality, humidity, temperature and moisture level.

If the temperature, humidity goes above the level, and light intensity values, and the soil moisture content are upload it to cloud. This all is displayed on the LCD display module. This all is also seen in IoT where it shows information of humidity, temperature, moisture, light intensity and water quality. An electronic sensors board is implemented to monitor air, water, and soil condition at the horticulture site. The results are collected into a monitoring database. This database is also be integrated with the IoT, which consists of weather data at the horticulture area. This study limitation is only on the design and development of monitoring system. Sensors sense the level of water if it goes down, it automatically starts to pump the water.

The data visible in the LCD will be sent to server. The output of Thing speak shows the temperature values vs date, humidity values vs Date, light intensity vs Date, motor on/off vs date, pH vs date. If the soil moisture is wet then motor automatically switches off and if the soil moisture is dry then motor automatically switches on.

## Instructions for installation of FHB logger alongwith sensors in orchard:

*Antenna:*-Wire glands for the solar power wire, wind anemometer, 13 core ground sensor and thermometer with 3d printed Stevenson screen. An external antenna is not required if the internal antenna will be used (e.g. within 200 m of unobstructed line of sight from a Wi-Fi booster antenna. A 158mm x 90mm x 60mm box could be used, however it will be very tight and difficult to fit components easily into the box. When drilling a large hole (greater than 8 mm ) It is best to first drill a small hole (e.g. 4 mm) and use it as a guide to drill the large hole. When drilling the large hole use a slow speed.

- Put the weatherproof sealing cord into the lip edge of the box. Make the join at the bottom
- Antenna: Drill a 6.5 mm hole in the top of the box 50 mm from the right and 45 mm from the rear of the box. Screw in the antenna wire. Some silicon sealant can be placed at the base of the antenna thread so as it is pushed into the box it forms a seal

- Thermometer gland; Drill a 12.5 mm hole in the bottom of the box 35 mm from the right and 20 mm from the rear of the box. This is the hole for the thermometer PG9 gland
- Wind gland: Drill a 12.5 mm hole in the bottom of the box 35 mm from the right and 12 mm from the front of the box. This is the hole for the wind wire PG9 gland
- Solar power gland: Drill a 12.5 mm hole in the bottom of the box 65 mm from the right and 12 mm and 20 mm from the rear of the box. This is the hole for the solar power wire PG9 gland
- Ground sensors gland: Drill a 15 mm hole in the bottom of the box 65 mm from the right and 12 mm and 15 mm from the top of the box. This is the hole for the 13 core wire PG11 gland for all of the ground sensors.
- Reset button: Drill an 11.5 mm hole in the right hand side of the box 70 mm from the bottom and 15 mm from the rear of the box. This is the hole for the reset. The reset button can be placed on the bottom of the box. If in the future you want to install more wires into the box the reset button can be moved to the side.
- Box attachment holes: Drill two 4 mm holes along the centre line of the rear of box about 35 mm from the top and bottom. These two holes will be screw holes to secure the box to the PVC pole
- Attach PCB to the rear of the box
- -Drill 2 mm holes into the top of four Self Adhesive Cable Wire Zip Tie Mounts
- Screw the Self Adhesive Cable Wire Zip Tie Mounts to the board with a 3.5 to 4 mm spacer in between the board and the mount cut from 3-4 mm inside diameter tube (e.g. 4 mm irrigation riser tube or 3 mm PVC tube). 3 mm tube is better than the 4 mm tube because it does not fall off the screw.
- Pull off the sticky tabs and stick the PCB to the top of the inside of the box
- Change the wires on the battery connector wire. Push a round headed pin inside the extruding head on the side of the clip all the way through and pull the wire out, switch the wires so they match the battery polarity
- The logger antenna, solar panel and wind sensor needs to be position at the top of the tree. Most trees are no larger than about 3.5m . A 3 m length of 60 mm class 9 or 12 PVC (50 mm pvc has an external diameter of about 60 mm, class 12 is thicker walled than 9 and stronger but more expensive, class 9 has been adequate to date, class 12 might be needed in consistently high wind regions) ,will be attached to a start post 0.5 m above the ground to provide a total height of 3.5 m height. Use a longer length of PVC for taller trees. A 3 m PVC length is convenient because PVC is mostly supplied in 6 m length and using a 3 m piece provides no wastage. The logger will be attached to the PVC. The PVC will be split in two with a joiner so it can be lowered for a mechanical topping operation.

## Conclusion

In conclusion, telemetry soil and climate monitoring systems are indispensable tools for increasing the productivity of fruit crop, offering data-driven insights for sustainable and adaptive agricultural practices. This presented a systematic review of the state-of-the-art of IoT adoption in smart agriculture and identified the main components and applicability of IoT solutions. From the identified applications of IoT for smart farming it was observed that the most common application is the monitoring of fruit crops. Another important future research direction could be the analysis of the edge and fog computing usage in smart agriculture as a way to deal with challenges associated with traditional centralized cloud solutions such as high communication latencies, lack of support for real-time reaction to detected events, large bandwidths, etc.

## References

Archana and Priya. 2016. Design and Implementation of Automatic Plant Watering System" presented at International Journal of Advanced Engineering and Global technology vol-04, Issue-01.

Arridha, R., Sukaridhoto, S., Pramadihanto, D., & Funabiki, N. 2017. Classification extension based on IoT-big data analytic for smart environment monitoring and analytic in real-time system. International Journal of Space-Based and Situated Computing, 7(2), 82-93

Brewster, C.; Roussaki, I.; Kalatzis, N.; Doolin, K.; Ellis, K. 2017. IoT in agriculture: Designing a Europe-wide large-scale pilot. *IEEE Commun. Mag. 55,* 26–33, doi:10.1109/ mcom. 2017.1600528.

Fisher, D.K.; Woodruff, L.K.; Anapalli, S.S.; Pinnamaneni, S.R. 2018. Open-source wireless cloud-connected agricultural sensor network. *J. Sens. Actuator Netw. 7,* 47, doi:10.3390/ jsan7040047.

G.Nisha and J.Megala. 2014. Wireless Sensor Network Based Automated Irrigation and Crop Field", Sixth International Conference on Advanced Computing ICOAC.

Gondchawar, N., Kawitkar, R.S. 2016. IoT based smart agriculture. Int. J. Adv. Res. Comput. Commun. Eng. 5(6) (2016). ISSN (Online) 2278-1021 ISSN (Print) 2319 5940.

Hadzovic, S.; Mrdovic, S. 2021. Radonjic, M. Identification of IoT actors. *Sensors, 21,* 2093, doi:10.3390/s21062093.

Hassan, R.; Qamar, F.; Hasan, M.K.; Aman, A.H.M.; Ahmed, A.S. 2020. Internet of Things and its applications: A comprehensive survey. *Symmetry. 12,* 1674, doi:10.3390/sym12101674.

Ji, C.; Lu, H.; Ji, C.; Yan, J. 2015. An IoT and mobile cloud-based architecture for smart planting. In Proceedings of the 2015 3rd International Conference on

Machinery, Materials and Information Technology Applications, Qingdao, China, 28–29 November 2015; pp. 1001–1005.

Joaquin Gutierrez, Juan Francisco Villa-Medina, and Alejandra Nieto-Garibay, Miguel Angel Porta-Gandara. 2013. Automated Irrigation System Using a Wireless Sensor Network and GPRS Module", IEEE Transaction on Instrumentation and Measurement.

Karan Kansara, Vishal Zaveri, Shreyans Shah, Sandip Delwadkar and Kaushal Jani. 2015. Sensor based Automated Irrigation System with IOT", International Journal of Computer Science and Information Technology Vol. 6, Issue 6.

Kashyap, B.; Kumar, R. 2021. Sensing methodologies in agriculture for soil moisture and nutrient monitoring. *IEEE Access, 9*, 14095–14121, doi:10.1109/access.2021.3052478.

Kavi anand G, Nivas V M, Kiruthika R and Lalitha S. 2016. Automated drip Irrigation system", IEEE International Conference on Technological Innovations in ICT for Agriculture and Rural Development.

Koo, D.D.; Lee, J.J.; Sebastiani, A.; Kim, J. 2016. An Internet-of-Things (IoT) system development and implementation for bathroom safety enhancement. *Procedia Eng. 145*, 396–403, doi:10.1016/j.proeng.2016.04.004.

R.Subalakshmi and Anu Amal. 2016. GSM Based Automated Irrigation using Sensors" presented at Special Issue published in International Journal of Trend in Research and Development (IJTRD), March.

Rajalakshmi, P., Mahalakshmi, S.D. 2016. IOT based crop-field monitoring and irrigation automation. In: 10th International Conference on Intelligent systems and control (ISCO).

Sonali.D.Gainwar and Dinesh.V. Rojatkar. 2015. Soil Parameters Monitoring with Automatic Irrigation System" presented at International Journal of Science, Engineering and Technology Research (IJSETR), vol-04, Issue 11, Nov.

Tawalbeh, L.; Muheidat, F.; Tawalbeh, M.; Quwaider, M. 2020. IoT privacy and security: Challenges and solutions. *Appl. Sci., 10*, 4102, doi:10.3390/app10124102.

V.R.Balaji and M.Sudha. 2016. "Solar Powered Auto Irrigation System"presented at International Journal of Emerging Technology in Computer Science and Electronics (IJETCSE), vol-20 Issue-2, Feb.

Vandana Dubey, Nilesh Dubey and Shailesh singh Chouchan. 2013. Wireless Sensor Network based Remote Irrigation Control System and Automation using DTMF Code", IEEE Transcation on Communication Systems and Network Technologies, July.

# CHAPTER 16

# STATUS AND PROSPECTS OF CULTIVATION OF GARCINIA SPECIES IN INDIA

**Prakash Chandra Tripathi*, R. Senthi Kumar**, G. Karunakaran**, C. Kanupriya** and Anuradha Sane****

**Horticultural Science Division, ICAR, KAB-II, Pusa, New Delhi*
***ICAR-IIHR, Bangalore*
*Corresponding author: prakaashtripathii2000@yahoo.com*

India is the centre of origin for many tropical fruit tree species. Some of them are still unexploited although these provide a significant source of food and nutrition for many ethnic communities. The Garcinia is one of the important genus of fruiting plants. The genus *Garcinia* is the largest genus within the family Clusiaceae (formerly Guttiferae). It comprises of nearly 250 species world over. The trees of this genus are generally small or medium sized and evergreen, These are distributed in tropical regions, with high species diversity in South East Asia particularly Malaysia. Some of these species are originated in India, Micronesian islands, tropical Africa and the Neotropics (Rogers and Sweeney 2007, Stevens, 2007, Jones, 1980, Sharma *et al.*, 2013, Nimanthika and Kaththriarchi, 2010). The genus name *Garcinia* honours the Dutch army doctor and naturalist Laurentius Garcin (1683-1752), who described the fruiting specimen of mangosteen collected from Moluccas, the Maluku islands, Indonesia (Garcin, 1733). The family Guttiferae was created by Jussieu (1789) based on the presence of the exudates secreted from cut stems and leaves. Anderson described that 30 species in found in British India (Anderson, 1874). In India, the genus *Garcinia* is represented by 43 species and 5 varieties, of which 37 species and 4 varieties occur in wild, whereas the rest were introduced into cultivation (Anderson, 1874, Maheshwari, 1964, Singh, 1993, Srivastava, 1994, Mohanan *et al.*, 1997, Sabu *et al.*, 2013, Sarma *et al.*, 2016). Among the 37 indigenous *Garcinia* taxa, 16 species and 4 varieties are endemic to the country. In India, *Garcinia* species are distributed mainly in three phyto-geographical zones; North East India, the Western Ghats and Andaman and Nicobar Islands. North East India hosts 17 species, of which 2 species and 1 variety are endemic to the region. The Western Ghats hosts 9 species and 2 varieties, of which 7 species and 2 varieties are endemic and the Andaman and Nicobar Islands hosts 15 taxa, of which 6 species and 1 variety are endemic(Table 1). Among the nine species indigenous in the Western Ghats, *G. gummi-gutta* is an economically important and widely cultivated fruit crop in Southern Western Ghats, while *G. indica* is cultivated widely in Central Western Ghats region for their fruits. Besides, 6 introduced species (*G. cowa* Roxb. ex. DC., *G. hombroniana* Pierre, *G. xanthochymus* Hook.f. ex T. Anderson, *G. cymosa* (K. Schum.) I. M. Turner and P. F. Stevens, *G. intermedia* (Pittier) Hammel, *G. mangostana* L.) are also reported to be cultivated in the Western Ghats region either as a fruit plants or as ornamental plants. *Garcinia mangostana* L., source of the edible fruit mangosteen, is native to South East Asia and now cultivated throughout the Western Ghats for their delicious fruits. *G. hombroniana,* an allied species is getting popular in the Western Ghats region as source of edible fruits. The introduced tree *G. xanthochymus* is also getting popular as a fruit crop and avenue tree. The trees of genus *Garcinia* provide spice, fruit, medicine, cooking butter, colour and polishing agents for metallic surfaces such as gold and silver. The fruits of the genus contain p-hydroxycitric acid which has inhibitory effect on lipogenesis and helps in controlling obesity in human beings. *Garcinia* possesses anticancer and antioxidant properties or can be used as bio preservatives. *Garcinia* has assumed great significance in the biological research arena today providing not only traditionally known products but new ones also. The fruits of *Garcinia* species are used for edible purpose or medicinal purpose. Some of the species commercially grown or utilised commercially in India are *Kokum (Garcinia indica),* mangosteen (*Garcinia mangostana* L), Malabar tamarind (*Garcinia gummigutta* ) and Yellow mangosteen (*Garcinia xanthochymus*(Hook).f.).

**Table 1. Different Garcinia species available in India**

| Species | Common name | Distribution |
|---|---|---|
| *Garcinia andamanica* King | - | Andaman Islands |
| *G. anomala* Planc. | - | Khasi Hills |
| *G. atroviridis* Griff. | - | North Eastern districts of Assam |
| *G. cornea* L | - | Bengal |
| G. cowa (G. kydia Roxb. | - | Eastern parts of India, Assam, Bihar, Bengal, Orissa, Andaman and Meghalaya |
| *G. dulcis* (Roxb.) Kurz | - | Introduced into India from Malaysia |
| *G. echinocarpa* Thw | - | Tirunelveli forests |
| *G. cambogia* (Gaetn.) Desr/*G. gummi-gutta* (L). Robsmall | Malabar tamarind | Western Ghats, Maharashtra, Goa, Karnataka, Kerala, Shola forests of Nilgiris |
| G. . *hanburyii* Hook | - | South India |

| Species | Common name | Distribution |
|---|---|---|
| *G. hombroniana* Pierre | - | Nicobar Island |
| *G. imbertti* Bourd | - | South India |
| *G. indica* Choicy | Kokum | South India, Konkan region |
| *G. lanceaefolia* Roxb. | - | Assam, Khasi Hills |
| *G. livingstonei* T. Anders | - | Introduced to India form East Africa |
| *G. malabarica* Talbot | - | South India |
| *G. mangostana* L. | Mangosteen | Introduced to South India |
| *G. microstigma* Kurz | - | Andaman Islands |
| *G. morella* Desr. | - | Assam, Khasi Hills, Western Ghats |
| *G. paniculata* Roxb. | - | Foot hills of Himalayas, Assam, Khasi Hills |
| *G. pedunculata* Roxb. | - | Assam, Manipur |
| *G. speciosa* Wall | - | Andaman islands |
| *G. spitcata* HooK | - | Western Ghats from Konkan Southwards |
| *G. stipulate* T. Anders. | - | Eastern Himalaya |
| *G. succifolia* Kurz | - | South India |
| *G. travancorica* Beddome | - | Western Ghats |
| *G. wightii* T. Anders | - | South India |
| *G. xanthochymus* Hook. Syn. *G. tinctoria* / *G. pictorius* Roxb | Yellow mangosteen | Eastern Himalayas (Assam and Meghalaya) Western Ghats, and Andaman islands |

***Source:* Krishnamoorthy *et al.* 2007**

Kokum found growing naturally in evergreen and semi-evergreen forests of the western parts of Maharashtra, Karnataka and Goa. Kokum is a small slender evergreen tree reaching to a height of 10-15 metre with spreading branches. It flowers in November -February and fruits ripe in April to June. Generally trees are either male or female. The ripe fruits are bright red, spherical or globose. The rind is thick and red is coloured. Pulp is white acidic and encases 5 to 8 seeds. The rind contains anti-obesity compounds, anthocyanin etc. It is bright red and has a sweetish acid taste when dried and used as a souring agent in traditional dishes in Konkan , Goa and Karwar region and also used for the preparation of syrup, juice, dried flakes, powder etc. White pulp is also edible. The rind has medicinal properties, for the treatment of obesity, piles, dysentery, tumors, and heart complaints. Seed contains high amount of oil and seed oil or butter is extracted. The oil is used as an ingredient in cosmetic products.

Mangosteen (*Garcinia mangostana* L) is native of South Eastern Asia introduced in India several decade ago. It is regarded as one of the best-flavoured fruits in the world and sometimes referred to as the " Queen of Fruits" due to its nutritive value and the presence of some compounds that offer various mangosteen health benefits. Malabar tamarind widely found in southern parts of Western Ghats of Karnataka to the *shola* forests of Nilgiri hills up to an altitude of 2000 meters. Mangosteen tree are medium to large in size with spreading branches is a small slender evergreen tree reaching to a height of 10-15 metre with spreading branches. It flowers in February to March and fruits ripe in July -August. Generally, trees are either male or female. The fruit of dark purple colour at maturity. It is one of the tropical fruits where taste is much like other known fruits and is easily accepted by the consumers in Western markets.

Malabar tamarind is a small slender evergreen tree reaching to a height of 10-15 metre with spreading branches. It flowers in February to March and fruits ripe in July -August. Generally trees are either male or female. The fruit of Malabar tamarind resembles small pumpkin and is green to pale yellow in colour. The dried rind is used in the preparation of non vegetarian dishes. The fruits are used for making dried flakes, vinegar etc. The fruit has lot of medicinal properties. The dried flakes and powder is used in several anti- obesity products. This species is traditionally used for polishing silver and gold .

Yellow mangosteen widely distributed throughout south and East and North Eastern India. Yellow mangosteen tree have rounded attractive crown that grows up to 8 m in height. The leaves are large and leathery and are oblong to lance shaped. It produce round, beaked, bright yellow-orange fruit of 5 cm to 8.9 cm diameter. The fruit pulp is yellow, edible, and sweet sour to sour with 1-5 seeds at the centre. The fruits are used for making squash, jam, dried flakes etc. The fruit has lot of medicinal properties

## Composition and Uses

Among the *Garcinia* species, mangosteen is used as fresh fruits while Kokum is used for making squash and syrups. The dried rind of Kokum and Malabar tamarind and yellow mangosteen is used as souring agents in various preparations since long in entire Western Ghats region. The rind of these fruits contains hydroxycitric acid (HCA), garcinol and several other compounds which are known for their antioxidant, anti-inflammatory and anti-carcinogenic activity. The fruits are used in traditional food preparation and have a distinct taste. The dried rind is used as a condiment in Kerala and Karnataka in place of tamarind or lime. The fruits have anti-scorbutic, astringent, demulcent and antiseptic properties. Some health benefits of fruits are lowering of blood pressure, anti-inflammatory properties, anti-microbial, promote skin regeneration, strengthening of bones, treatment of flu and coughs,. alleviates allergies ,treatment of arthritis, reduce stress, fight depression, lower cholesterol, regulate blood sugar,. increase metabolism, boost energy levels. Fruit rind is a rich source of alpha-HCA (hydroxy citric acid) which prevents fat accumulation in body cells. The HCA content in the fruits of Malabar tamarind is reported to have three times higher than kokum. This natural HCA is a potent metabolic regulator of obesity and it also lowers blood lipids such as cholesterol and triglycerides by triggering the fatty acid oxidation in the liver via thermogenesis. It mobilizes body's fat stores and dissolves fat in the liver and also throughout the body paving way for weight management. These fruits have also been historically used to treat gastric ulcers. Garcinol present in the fruits is known to lower acidity in the stomach and protects the gastric mucosa. The anti-obesity compound, i.e. Hydroxycitric acid (HCA) present in Kokum and Malabar tamarind has gained much attention in recent years for use in weight loss and several anti obesity products are being prepared. The Kernel fat is edible, used as a component in ointments, soaps, confectionaries, cosmetics and also for culinary purposes. The gum is used to make good varnish. In Ayurveda, the ancient Indian system of medicine, it is mentioned as antidiabetic, anthelmintic, cardiotonic, and anti-obesity properties and used in treating piles, dysentery, tumours, pain, and heart ailments (Swami *et al*, 2014). The dried rind of *G. indica* is traditionally used as a culinary spice to impart a sour taste to food and is also widely used in preparing soft drinks. The sour taste is mainly due to hydroxycitric acid (HCA), a potential anti-obesity agent and widely used in anti-obesity medicines (Jayaprakasha & Sakariah, 2002; Nainegali *et al*, 2019; Nayak *et al*, 2010). The purple colour of the rind is mainly due to anthocyanins, which are responsible for the red, blue, and purple colours of many fruits, vegetables, flowers, and even cereal grains. Anthocyanin is considered as potential replacement for synthetic colours because as its brightness and water solubility make it easy to incorporate it into edible products Apart from anthocyanin (Nayak, Srinivas, & Rastogi, 2010; Nayak *et al*, 2010), *G. indica* also contains phenolic compounds (Bommayya *et al*, 2011) such as phenolic acids and flavonoids and can serve as a natural source for these compounds as well. These bioactive phenolic compounds have a wide range of biological activity and have shown antioxidant, anti-microbial, antityrosinase, anticholinesterase, anti-inflammatory, anti-cancer, antihyperglycemic or antidiabetic, and antinociceptive activity and useful home remedies.

**Table 2. Chemical composition of fruits of different *Garcinia* species**

| Nutrient | Kokum rind (moisture free basis) | Mangosteen (100 g fresh wt.) | Malabar Tamarind Rind (100g edible portion) | Yellow mangosteen (100g edible portion) |
|---|---|---|---|---|
| Moisture (%) | - | 79.7 | 86.91 | 80.96 |
| Energy (calories) | - | 34.0 | - | 96.48 kj |
| Protein (%) | 1.92 | 0.6 | 0.61 | 5.01 |
| Crude Fat(%) | 10.0 (hexane ext extract) | 1.0 | 1.4 | 1.65 |
| Carbohydrate (%) | 36.4 | 5.6 | 8.6 | 11.44% |

| Nutrient | Kokum rind (moisture free basis) | Mangosteen (100 g fresh wt.) | Malabar Tamarind Rind (100g edible portion) | Yellow mangosteen (100g edible portion) |
|---|---|---|---|---|
| Ash (%) | 2.57 | 0.1 | - | 1.65 |
| Crude fibre (%) | 14.28 | 5.1 | 3.1 | 2.73 |
| Pectin | 5.71% | - | - | - |
| Anthocyanin | 2.40 | - | - | 1.62 mg |
| Ascorbic acid | 0.06% | 4.2 mg | - | 11.62 mg |
| Starch | 1% | - | - | - |
| Caroteniods | - | - | - | 33.3 ug |
| Hydrocyanic acid | 22.8 % | 15.2 % | - | - |
| Tannins | 2.85% | - | - | - |
| TSS | - | 14.20Brix | - | 12.3 0Brix |
| Calcium (mg) | - | 7.0 | - | 1.39%* |
| Magnesium (mg) | - | 13.0 | - | - |
| Phosphorous (mg) | - | 13.0 | 169.7 | - |
| Sodium (mg) | - | 7.0 | 2.1 | - |
| Potassium (mg) | - | 45.0 | - | 0.5 %* |
| Iron (g) | - | 1.0 | - | - |
| Vitamin B1(g) | - | 0.03 | - | 56 ppm* |
| Manganese | - | - | - | 2.9 ppm* |
| Zinc | - | - | - | 9.2 ppm* |
| Copper | - | - | - | 8.4 ppm* |
| Acidity | - | - | - | 4.88 % |
| Vitamin B2 (g) | - | 0.03 | - | - |
| Niacin (g) | - | 0.3 | - | - |
| Reducing sugars | - | - | 2.67 | - |
| Non Reducing sugars | - | - | 5.62 | - |
| Tannins | - | - | 1.7 | - |

***on dry weight basis**

The primary active components of the mangosteen fruit are called xanthones which have several benefits including anti- inflammatory properties, anti-allergic and anti-convulsant. In the Philippines people used to make a decoction of mangosteen leaves, bark and rind which was beneficial in bringing down body temperature and also to treat dysentery, diarrhoea and urinary disorders (www.engineeredililifestyle.org/mangosteen.html). The rind is reported to contain 7-15 % tannin and is used to tan leather and to dye fabric black. It is also reported to use as an ingredient in soap, shampoo and conditioner. The fruit is mainly consumed fresh and can be canned, frozen or processed in to jam, sweetened preserved, syrup and candy.

The fruits of yellow mangosteen are used for edible purpose and contain lot of nutrients and have peculiar taste. The fruits are rich in carbohydrates, proteins, calcium, manganese and zinc. The fruits contain garcinol, xanthones have lot of nutritional as well as pharmaceutical values. The fruits of yellow mangosteen are highly acidic and used for preparing pickle, chutney and flavour curries. The fruit rind is dried and used as a substitute for tamarind. In North Eastern region of India, fruits are used for culinary purposes as well as folk medicine. Ripe fruits are eaten raw or cooked with other vegetables by the people of Dhubri district of Assam. The fruit is also used as fixative with alum in the dyeing of silk. The leaves are used for thatching of roof and said to have anti termite properties. The latex is used as a lubricant during weaving. The bark of the tree and latex of unripe fruits are used to make yellow dye. Wood is hard and good for making house furniture and traditional rice mill 'Dheki' in Assam. The plant shape is very attractive and can be used as avenue trees. This species contain high levels of secondary compounds called xanthones that are reported to possess antibacterial and antimalarial properties. The fruit contain several phytochemicals such as xanthones, flavonoides, saponins, tannins, alkaloids, lipids, benzophenones and biflavonoids. Methanol extract of fruit contains benzophenones, guttiferone H and gambogenone. These compounds play major role for prevention of colon cancer and breast cancer. An extract of the bark was found to stimulate the growth of neurons or nerve tissues in culture studies. It is rich source of oxygenated and prenylated xanthones. The leaves have anti-inflammatory activity. The decoction from leaves and roots is used in the treatment of earaches.

## Taxonomy

*Garcinia* species were characterized by their monopodial branching form, where secondary shoots or branches arise behind the growing point but remain subsidiary to the main stem, which continues to grow indefinitely (Tootil, 1984). Hence *Garcinia* species usually exhibited horizontal spreading branching pattern. However, *G. gummi-gutta* var. *gummi-gutta* and *G. morella* showed pendulous drooping branchlets whereas *G. indica* showed crown shaped canopy ending with horizontal branchlets. Bark is usually grey to brown, inner bark is yellow or occasionally white. The stem and twigs produce yellow, white or cream exudates, known as 'Gamboge'. Gamboge is used as a pigment in paint and varnishes. The colour of the exudates varies from yellow to white and is a characteristic identification feature for *Garcinia* species. Gamboge of *G. morella* is widely used in the preparation of golden coloured water colours and spirit varnishes for metals and also for dyeing silk fabrics. A golden yellow coloured ink was prepared from the gamboge of *G. morella* for writing on black paper (Anonymous, 1950). Leaves of *Garcinia* species are opposite, usually thick and characterized by the presence of a foveola (an excavation with an extension resembling a ligule) at the base of the petiole. While majority of *Garcinia* species are dioecious, a few species or races are reported as hermaphrodite (Dunthorn, 2004). *Garcinia* species generally display an unusual evolutionary plasticity and there are many unresolved phylogenic issues surrounding the genus. Male and female flowers are seen on different trees (dioecious). The basic inflorescence type of *Garcinia* is a simple cyme or few flowered (2 to 16) clusters in fascicles. Flowers of *Garcinia* are generally sessile. Flowers are solitary or in fascicles, terminal or axillary and variously coloured. Sepals and petals 4-5, stamens usually numerous, very variable in arrangement and structure, sometime with pistillode; ovary 1-12 loculed with a single apical ovule per locule, ovule 1 in each locule; stigma conspicuous and variously lobed, usually peltate. Inflorescence of male flowers are observed both in terminal and axillary positions. The sepals are usually orbicular and green or yellowish in colour. The petals, however, have brighter colour, from yellow (*G. gummi-gutta, G. indica*) to white (*G. talbotii, G. wightii*), cream (*G. morella*), pink or red (*G. pushpangadaniana*), pale greenish (*G. travancorica*) and green (*G. rubro-echinata*). The stamens are always united in a bundle at the centre of the flower. The female flowers are fewer compared to male flowers. The female flowers have shorter, stouter pedicels and peduncles comparatively smaller than the male flowers. In general, the ovary in *Garcinia* is superior and very few species have constant locule numbers. Most of the *Garcinia* species have 4 or 5 locules. Generally, ovary is globose to ovoid. The stigma is usually sessile. Fruits are fleshy to woody berry; seated on the usually persistent calyx. Seed 1-12, often flattened and enclosed in pulp. Regarding fruit size, *G. wightii* has the smallest (10-15 gm), while the largest is that of *G. pushpangadaniana,* weighing upto 750 gm. Most of the fruits are globose in shape Species like *G. imberti, G. wightii, G. travancorica, G. morella* and *G. indica* have pulpy aril while the aril of *G. pushpangadaniana* is crispy and that of *G. rubro-echinata* is fibrous. The seed shape was oblong in most of the *Garcinia* species, except plano-convex for *G. pushpangadania*, ovoid-reniform for *G. morella* and *G. gummi-gutta.* The fruit colour is a characteristic distinguishing feature which varies from yellowish green in *G. travancorica, G. imberti* and *G. talbotii,* brownish yellow in *G. pushpangadaniana,* yellow in *G. gummi-gutta,* red in *G. wightii* and *G. morella* and purple in *G. indica.*

**Table 3: Collection and sale of *Garcinia gummigutta* rind in Karnataka***

| District | Collection (tones) | Value (Rs) |
|---|---|---|
| Uttara Kannada | 48 | 8,64,000 |
| Dakshin Kannada | 113 | 20,034,00 |
| Hassan | 136 | 24,48,000 |
| Kodagu | 972 | 1,64,15,000 |
| Total | 1206 | 21,762,000 |

***Report of Forest Dept ., Karnataka (2014), based on the market collectors information**

## Climate and soil

Garcinia is originated in humid tropical region. The plants grow well in humid tropical and subtropical climatic condition. These flourishes very well up to an elevation of about 800 m from msl and high fall ranges from 150 to 250 cm. Mangosteen is more specific to high humid tropical regions while other species can also be grown in subtropical region with lower humidity. The optimum temperature is 25-350°C and relative humidity (RH) over 80% with areas of low elevation *i.e* less than 500 metes above sea level. Mangosteen prefers deep, fertile, well-drained and slightly acidic clay loam soil rich in organic matter. It is not adapted to limestone soils, sandy alluvial soil or sandy soil low in humus. The plants grow well in soils with a high pH of 5.5-6.8.Other *Garcinia* species can be grown lateritic and alluvial soils.

Most of these species require good rainfall and humidity for growth and fruiting. Thus these are well adapted to heavy and well distributed rainfall of 1,200 mm or more without prolonged dry periods. However, the plant requires a short dry spell of 15-30 days for flower induction. These species grow well in dry or occasionally water logged or flooded soils, but good drainage required for optimum growth.

## Genetic resources

The variability with respect to morphological features such as leaf parameters, flowering, fruit set, fruit shape, colour, physicochemical composition and yield has been reported in *Garcinia* ssp. Some efforts have been made to collect these variability and conserve in field gene bank. Two types of mangosteen are available, the large leaves and fruits of variable size and small leaves and small fruits. In general variability is low in mangosteen due to less parthinocarpy. Nine Mangosteen collections are under evaluation at CHES (IIHR), Chettalli. Some of exotic varieties have been brought by the nurserymen's from South East Asian countries which are cultivated in some areas of Kerala and Karnataka. Kokum is cross pollinated crop and lot of variability is observed in the natural population. The kokum collections have been made at Dr. B. S. Sawant Konkan Krishi Vidyapeeth, Dapoli, Central Horticultural Experiment Station (IIHR), Chettalli and many other research centres to identify the elite lines of kokum. Two improved kokum varieties namely Konkan Amruta and Konkan Hatis have been released by Dr. B.S. Konkan Krishi Vidyapeeth, Dapoli. Few promising lines of Kokum have been identified by CCARI, Goa and CHES, Chettalli. The germplasm of mangosteen is being evaluated at CHES-Chetalli. The results showed wide variations for yield and fruit characteristics. The yield varied from 2.5 kg to 24 kg per plant. Various phenological stages in mangosteen are also being studied. Lot of variability in *Garcinia gummigutta* was observed in terms of vegetative, floral, fruiting and biochemical character of the fruits by NBPGR, RS, Thrissur , IISR, Calicut, CHES, Chettalli and IIHR, Bengaluru and some collections were made. The growth and yield characteristics of 103 accessions of yellow mangosteen were recorded at CHES, Chettalli.The plant height ranged from 1.70 m to 3.95m... The yield ranged from 1.0 to 15.6 kg/tree and fruit weight ranged from 97 to 175 g. Among the accessions GX-IV-6, GX-V-1 and GX-VI-8 showed better performance than other accessions. At IIHR, morphological and fruit characteristics of 13 collections of yellow mangosteen were recorded. The yield was 6 to 123 fruit/ tree while fruit weight ranged from 82 g to 172 g. Some collections of Garcinia species have been made at ICAR- ICAITI, Port Blair, RPRC, Bhubaneshwar etc (Table 4 ).

**Table 4. Germplasm of Garcinia spp at different institution**

| Institution | No of collections | | | |
|---|---|---|---|---|
| | Mangosteen | Kokum | Malabar tamarind | Yellow mangosteen |
| CHES(IIHR), Chettalli | 9 | 45 | 53 | 55 |
| IIHR, Bangalore | 1 | 31 | 8 | 11 |
| BSKKV , Dapoli | - | 300 | - | - |
| CCARI, Goa | - | 268 | - | - |
| College of Forestry , Sirsi | - | - | - | - |
| IISRRS, Appangla | - | - | 50 | - |
| NBPGRRS, Thrissur | - | - | 124 | - |
| IISR, Calicut | - | - | 20 | - |

## Varieties and elite lines

### Kokum

Kokum is indigenous tree of India and a lot of variability is observed in the natural population with respect to morphological features such as leaf parameters, flowering, fruit set, fruit shape, colour, physicochemical composition and yield. The kokum collections have been made at Dr.B. S. Sawant Konkan Krishi Vidyapeeth, Dapoli, Central Horticultural Experiment Station (IIHR), Chettalli and many other research centres to identify the elite lines of kokum. Two improved kokum varieties namely Konkan Amruta and Konkan Hatis have been released by Dr. B.S. Konkan Krishi Vidyapeeth, Dapoli. Few promising lines of Kokum have been identified by CCARI, Goa and CHES, Chettalli .

### Konkan Amruta

This variety was released from Dr. B.S. Konkan Krishi Vidyapeeth, Dapoli, Maharashtra in 1997. This is the first ever released variety of Kokam. It flowers in the month of November under Konkan conditions and fruits ripens in the month of March-April. The fruits are medium in size (35 g weight) with 6-7 seeds/fruit. Fruits are apple shaped. The rind weight is around 17.55 g. The total soluble solid are 9.0°Brix and acidity is 5.12 percent.

### Konkan Hatis

This variety released from Dr. B.S. Konkan Krishi Vidyapeeth, Dapoli, Maharashtra. It flowers in the month of December under Konkan conditions and fruits ripens in the month of April-May. The fruits are big size with 90-95 g weight with 5-6 seeds/fruit. Fruits are apple shaped. The rind weight is around 48-50g. The total soluble solid are 9.2°Brix and acidity is 5.10 percent.

## Kokum CHESGI-V-4 [IC-0612477]

It is seedling selection by CHES (IIHR), Chettalli. The trees are medium spreading. The fruit weight is 45 to 50g. The rind recovery is 48 percent. The fruit rind has TSS-14.5°Brix, Vitamin C is 36 mg /100g pulp. Average yield is 1000-1250 fruits / tree/year.

## Kokum CHES GI-VIII-5 [IC-0612478]

It is seedling selection by CHES (IIHR), Chettalli. The trees are medium and spreading. The fruit weight is 50 to 60g. The rind recovery is 59.5 percent. The fruit rind has TSS-14.1°Brix, Vitamin C is 64 mg /100g pulp. Average yield is 1000-1200 fruits / tree/year.

## IIHR- GI-7-9

It is seedling selection by IIHR, Bengaluru. The trees are big and have upright growth. The fruit weight is 30-35g. The rind recovery is 56.5 percent with pulp recovery of 36.3 percent. The fruit rind has TSS-16.5°Brix with pulp TSS of 150Brix, Vitamin C is 36 mg /100g pulp. Average yield is 1000-1500 fruits / tree/year.

## Mangosteen

The National Seed Industry Council (NSIC) of Philippines has registered two varieties of mangosteen as UPLB Sweet of Los Baos, Laguna and Roxas Purple of Kailuan, Laguna. The germplasm of mangosteen is being evaluated at CHES-Chetalli. Sixteen accessions resulted in the year 2021. The results showed wide variations for fruit yield characteristics. The yield varied from 2.5 kg to 24 kg per plant. Various phenological stages in mangosteen are also being studied. A line (CHESMS-1) was found performing better than others.

## Malabar tamarind

No variety has been released in India. However few promising lines have been identified at NBPGR RS , Thrissur; IISR, Calicut and CHES, Chettalli.

**IC244110** is promising for fresh fruit weight (200.0g), fresh rind weight (125.5g), fresh rind thickness (13.6mm), fresh seed weight (2.4g) and fresh and dry kernel weight (0.6g). It is an accession with the highest length of protrusion (nipple) in fruit.

**IC244100-2** (INGR No. 04061) is early bearing (6 years after transplanting), high yield in number of fruits (1496), high fresh fruit yield (104.2 kg), single fresh fruit weight (82.6 g), high number of seeds per fruit (6), high seed weight (1.6 g), high dry rind thickness (3.5 mm) and high yielding during the last 9 years.

**IC244111-1** (INGR No. 04062) is early bearing (7 years after transplanting), high yield in number of fruits (1470), high fresh fruit yield (99.3 kg), single fresh fruit weight (84.4 g), high number of seeds per fruit (6) and high seed weight (2.4 g), high dry rind thickness (2.5 mm) and high yielding during the last 9 years.

### CHES-GG-V/1 (IC-06124709)

This is a promising seedling selection with higher fruit weight. The tree has upright growth, medium sized. The fruits ripe in the months of July. The fruit weight ranges between 75 to 85g. Fruits are dull yellow coloured, 6-7 seeds/fruit, rind yellow, flesh white, 8 segments, juicy and acidic. The rind recovery is 62 per cent. The fruit contains 13.50 Brix total soluble solids and 62 mg Vitamin - C /100g pulp. The average yield is 400-700 fruits per tree .

### IIHR-GG-2-09

This collection is high yielding and regular bearer. The tree has upright growth, medium sized. The fruit ripe in July-August under Bengaluru conditions. The fruit weight ranges between 75 to 80g. Fruits are dull yellow coloured, 6-7 seeds/fruit, rind yellow, flesh white, 8 segments, juicy and acidic. The rind recovery is 56 per cent. The fruit contains 130 Brix total soluble solids. The average yield is 200-250 fruits per tree.

### Yellow mangosteen

There is no released variety of yellow mangosteen in India. However, few promising lines have been identified at NBPGR, RS, Thrissur, CHES, Chettalli and IIHR, Bangalore.

### CHES-GX-II-5

Tree up right growth, Fruit weight -175 g, yield -15.64 kg, fruit colour –glossy, smooth, bright yellow, fruit round with prominent break.

### IIHR –GX-VI-403

This collection is high yielding and regular bearer. The tree has upright growth, medium sized. The fruit ripe in September to March under Bengaluru conditions. The fruit weight ranges between 150 to 200 g. Fruits are glossy, smooth, bright yellow, fruit round with prominent break, 1-2 seeds/fruit, rind bright yellow, flesh dark yellow, juicy and acidic. The rind recovery is 52.1 per cent with pulp weight of 42.1 percent. The fruit Pulp has 15°Brix (total soluble solids) while rind TSS is 13°Brix. The average yield is 200-250 fruits per tree.

## Propagation

Most of the Garcinia species are multiped through seeds. Recently the propagation of some of the species have been started by vegetative methods such as grafting and root suckers.

## Seed propagation

Seed propagation is the easiest method to multiply plants. Most of the Garcinia seed germinate easily but there are short duration in some species. The fresh seed have good germination but the germination percent decreases rapidly during room temperature storage. The seeds should be extracted from fully ripe fruits. Seeds of more than 1 gram are to be selected for good germination. Newly extracted seeds must be sown right away to obtain high percent germination. Sowing is usually done in a seedbed. A typical seedbed may be of wood or cement and the sowing medium is a mixture of sand and soil in 3:1 ratio by volume. The sowing medium needs to be moisture-retentive and well drained. Seeds are planted 5-10 mm in depth and spaced 2-3 cm apart and are covered with fine sand. Seeds germinates 30 days after sowing. Seedlings are pricked to the polyethylene bags using same mixture as germination media while the cotyledon are still attached to the seedlings. Water the seedlings regularly after pricking. Seedlings are ready for field planting in 9 to 18 months. Some of the species of Garcinia produce parthinocarpic seed which can be used for multiplying true to type plants. In mangosteen and yellow mangosteen, the seedling of 24 to 36 months are used for planting which increases survival rate and minimize maintenance expense in the field as young seedlings require utmost care. The seedling propagation is not recommended for commercial planting of kokum and Malabar tamarind as 50 percent of seedlings turns to male plant. But seedling plants may be used for grafting or budding purpose. The seedling trees may be used as pollinizers in commercial orchards of elite lines/varieties.

Various vegetative methods of propagation such as air layering, approach grafting, veneer grafting, softwood grafting and root cutting were attempted. The studies conducted under Karnataka conditions revealed that among the cleft, whip, tongue and epicotyl grafting, the survival was lowest in epicotyls grafting and maximum in cleft grafting. For grating scion wood should be taken from high yielding trees. Scion wood should be taken from 0.5 -0.6cm thick greenish brown colour terminal shoots. Healthy and vigorous kokum seedlings of 10-15cm height, 4-5 months are selected for rootstock purpose. October to November and March to August was found to be the best season for softwood grafting in kokum in Konkan region of Maharashtra. Cleft grafting is widely used methods of asexual propagation. Vegetative propagation through cutting was found successful in yellow mangosteen in Assam. The grafted plants became ready for planting in 6-7 months after grafting.

## Planting

The land clearing can be done by removal of bushes and large plants together with their roots. Ploughing followed by harrowing/rotavating is sufficient to attain the desired tilth. High organic matter in the soil is recommended. Organic matter may be incorporated into the planting site 1-2 months prior to planting. Planting in done at 6 to 8-meter distance both ways. The pits of 0.6 to 1 $m^3$ size are dug before rainy season in the month of May. These pits are filled with a mixture of top soil, 10 kg farmyard manure and 1 kg single super phosphate. Field planting should be done anytime of the year if water source is available. However, the best time for planting at the onset of the rainy season. Provide shade to the newly transplanted seedlings. Shade is maintained 1-2 years and this is gradually reduced to allow full sun exposure. Slow growth makes the seedlings vulnerable and they can be quickly overtake by weeds, hence weeding is necessary. Irrigate and mulch the seedlings with coconut husk or grasses to conserve moisture during dry period. Garcinia plants grow slowly in initial years. Therefore, these require special care and maintenance in initial years of planting.

In Konkan region and coastal region of Karnataka, kokum is planted as a mixed crop in coconut and areca nut plantations. In many parts of Karnataka, kokum plants are grown with border of plantations. In those case, the distance of planting may differ. Intercropping with short duration crop such as banana to provide shade to mangosteen seedlings. Mangosteen trees pass through a juvenile phase which can last anything up to 12-20 years, although good husbandry result in trees fruiting at 5-7 years after planting.

## Manure and fertilizer

Most of the Garcinia trees except mangosteen are still found in forest and periphery of other plantations. Thus inorganic fertilizers are generally not used in these crops. Some farmers of coastal Karnataka and Konkan region used organic manures such as farmyard manures or leaf compost for higher yield in kokum . In Konkan region of Maharashtra application of 2 kg farmyard manure, 50 g nitrogen, 25 g phosphorus and 25 g potassium is recommended for one year kokum plants. This dose is increased in same proportion every year and up to ten years and 20 kg farmyard manure, 500g nitrogen, 250g phosphorus and 250 g potassium is recommended for ten year old plants. Fertilizer are applied in the month of August September after harvesting of fruits. Fertilizers may be given in 30 cm deep and 30 -40 cm wide circular trench around plant and covered with soil.

Application of a nitrogenous fertilizer produces faster vegetative growth of the mangosteen plants. Fertilizer can be applied in a ring around the base at the edge of the canopy. Application of foliar fertilizer is needed and may be applied singly or mixed with pesticide materials to save on labour cost. The fertilizer doses use din Philippines are given in Table 5.

**Table 5. Recommended doses of fertilizer for Mangosteen (ATI-FTC, Tupi, South Cotabato)**

| Age of trees (year) | Kg. Tree/Year of Commercial Fertilizer | | |
|---|---|---|---|
| | 45-0-0 | 14-14-14 | 0-0-60 |
| 1 | 0.22 | 0.35 | - |
| 2 | 0.44 | 0.70 | - |
| 3 | 0.56 | 1.40 | - |
| 4 | - | 2.80 | - |

| Age of trees (year) | Kg. Tree/Year of Commercial Fertilizer | | |
|---|---|---|---|
| | 45-0-0 | 14-14-14 | 0-0-60 |
| 5 | - | 3.60 | - |
| 6 | - | 4.20 | 0.30 |
| 7 | - | 5.70 | 0.30 |
| 8 | - | 7.10 | 0.30 |
| 9 above | - | 8.50 | 0.50 |

Most of the Garcinia plants are still found in forest and periphery of other plantations. Thus inorganic fertilizers are generally not used. Use of organic manures such as farmyard manures or leaf compost may give higher yield. Apply 10 kg farm yard manure or compost per plant during the first year. Gradually increase the quantity so that a well-grown tree of 15 years and above receives 50 kg of organic manure per year. There is no recommendation for inorganic fertilizer so far.

## Irrigation

Irrigation should be given immediately after planting to ensure better establishment of plants. The irrigation is essential for young plants for first three years during dry months. In Coorg region, irrigation is generally required during January to March for better fruit quality. In other areas, regular supply of water is after October- November to ensure higher yield and quality of fruits. Irrigation system is recommended to supply water during dry months. The mangosteen root system lack root hairs and the plant requires a constant and abundant supply of water in the soil. A continues dry period of at least one month followed by heavy rain can imposed plants to induce flowering but mangosteen can flower twice a year or sporadically and erratically. Regular watering is then required during fruit growth and development.

## Training and pruning

Garcinia plants are evergreen. The plant has the tendency to grow slender and tall. This creates problem in harvesting at later stages of tree. It is better to restrict the size of tree by training in the initial year and pinching the main branch in later years. So that the plants remain shorter and the fruiting zone remain within the reach for easy harvesting. The drooping, interlocking and disease and dried branches must be removed every year for better yield and quality of fruits. It is observed in grafted plants that only one branch grows in certain direction. This growth should be prevented by regular pinching for production of branches in all the directions. The suckers from rootstock below graft union should be removed regularly. For bearing trees, no pruning is needed except to remove broken or dead branches. The plants which are about 8 years or more, small inner branches are pruned from old, unproductive plants to stimulate bearing. The seedling plant has the tendency to grow upright so that the apical branches may be trimmed in initial years to make open centre system. The drooping and interlocking branches must be removed every year for better yield and quality of fruits. The suckers should be removed regularly. Yellow mangosteen plants have tendency to grow compact and dense .

## Flowering, Pollination and Fruiting

The seedlings start flowering in 7 to 8 years of planting whereas grafted plants starts flowering after 3-4 years. Generally kokum plant flowers during December to February depending on the climatic conditions. In coastal areas, it flowers in November -December while it flowers in January - February in the hilly regions. Flowers are borne singly or as fascicular cymes on leaf axils and are tetramerous. The period from flower bud appearance to initiation of flowering is about 30 days. Pollination is through wind. The fruits mature in the month of April-June. The immature fruits are green and turn to dull red at maturity. The mangosteen flowers in the month of February to March in south Indian conditions. The time of flowering may vary as per the altitude. Late flowering is observed in higher altitude as compared to warm coastal region. The Mangosteen flowers rise from the upper part of tree crown down to the lower branch tip of the tree. The fruits mature in the month of June to August. The immature fruits are green and turn to dark brown at maturity. Mangosteen fruit is formed without fertilization or parthenocarpy. Fruit character has various phenotypes. In Malabar tamarind, flowering takes places in the month of March- April and fruits ripe by July– August. The pollination is supposed to be done by wind but parthinocarpic fruit set may also be there in few cases. There is some fruit drop if there is moisture stress. The fruit drop of mature fruits is very common if not harvested timely. Fruit take 110 to 125 days to mature. In Yellow mangosteen, flowering takes places in the month of May and continue up to August -September. It is reported that the fruits are parthinocarpic but pollination may require for stimulating the fruit set but it is not proved.

## Harvesting and Yield

The fruits are harvested after about 120-150 days of fruit set depending on the species and the climatic conditions. Kokum fruits are ready for harvesting from the month of April to May. The mangosteen fruits mature in the month of June to August while Malabar tamarind fruits ripe by July–August. The yellow mangosteen fruits ripe from September to March. All fruits on a tree are not ready for harvesting at the time and hence periodical plucking is done. The number of pickings varies from tree to tree. Generally, 6 - 8 picking are required in high yielding plants. Fully ripe are plucked by hand. Generally, a 15-year-old seedling plant produces 30 -50 kg fruits per plant.

## Post-harvest handling, Value-addition, Storage and Marketing

The fruits may have various degrees of injury as a result of harvesting and in field handling before the fruits reach the packing house and processing. Such injury should be minimized in subsequent handling, including long distance transportation, to avoid further damage and loss in quality. Mechanical damage in mangosteen is reduced by careful handling avoiding any compression. Market experts emphasize that growers must understand how to harvest, pack, store and transport so that the consumers will receive a product of high quality. Therefore, growers, shippers, carriers and receivers should be familiar with quality standards and strive to reduce postharvest losses. Mangosteen fruits are transported to the packing house to prepare the fruits before being transported to local market. The mangosteen fruits are sorted according to quality and size before packing. The grades are small ( 16-18 fruits/kg), medium ( 12-14 fruits/kg), large ( 8-10 fruits/kg) . The selected fruits are placed in wooden crates or packed in corrugated cartons. Mangosteen fruits can be stored in cool temperature of 8-100 C for 8 weeks.

Kokum and Malabar tamarind fruits are hard and can be stored for 6-7 days at room temperature. The fruits are very sour and not used for table purpose. The fruit is made into, dried flakes, powder, vinegar etc. The dried flakes are used as a substitute for tamarind in curries. The ripe fruits are cut into slices, dried in the sun and sold in market as a sour relish in curries as well as for dressing fish. The dried flakes have lot of demand in *Malnad* and Kerala and sold at 300-800 per kg. The seed contains high amount of oil and use to extract Garcinia butter. The butter has lot of demand for cosmetic purposes. The mature fruits of yellow mangosteen are hard and can be stored for 6-7 days at room temperature. The fruit is made into preserve, jam, chutney, curries, squash, dried flakes etc. A kind of cool drink is prepared from fruits. In Malaysia, fruit is used as a substitute for tamarind in curries. .

## Processing and Value additional

The shelf life of Garcinia fruits is 4 -5 days under ambient temperature. It can be extended up to 28 days if stored at 13°C and 86 per cent relative humidity. The products such as kokum syrup, salted kokum syrup and dried rind are traditionally prepared from rind of fruit. Seed butter is extracted from seeds in Konkan region of Maharashtra and coastal region of Karnataka. Dried rind is prepared by sun drying .

### Jam

Jam can be prepared from mangosteen and yellow mangosteen but the production of these fruits in India is very low as the demand is very less.

### Kokum Syrup

Kokum fruits have certain medicinal properties. Juice extracted from this fruit is sweet and sour and thus liked by many. A glass of cold kokum syrup is refreshing and it also improves the digestive system. The product is popular in the state since long and now it is sold in nearby states as well. Since it is a natural fruit extract, it is preferred by many people. In traditional method kokum rind is separated by removing fruit pulp and seeds. Pulp and seeds are not used for squash production. Equal quantity of sugar mixed with kokum rind in a wide mouth vessel. This mixture of sugar and kokum kept open for sun rays up to eight to ten days. In this process juice comes out from kokum rind and already sugar is mixed with juice. If balance sugar accumulation is found at the bottom of the vessel no additional sugar is to be added. Filtering of juice and rind is done with the helps of cotton cloth. This clean Kokum Syrup is to be filled in clean glass bottle. For long term preservation these filled bottles again kept under sun light for another ten days without capping. The bottles are capped at end of the process.

### Kokum Agal (Kokum Juice Concentrate)

Kokum fruits are selected randomly and washed with clean water. The fruits are crushed seed and pulp from the fruit is removed manually. Salt is added in the hollow space of the rind and the mixture of kokum rind and salt is kept in available plastic drum or container and is covered with cloth. Juice from the container is collected and filled in glass bottles of 750 ml to 1000 ml size. The colour and quality of *agal* varies from location to location as the proportion of slat used ranges from 10 to 20 per cent and it depends on the quality of fruits and process. Kokum *Sorbet* is prepared by using the kokum syrup. The water is added into kokum syrup at 1:5 proportions and some salt and cumin powder is added to make the kokum sarbat. Kokum fruits is ready to use for preparation of *Solkadh*i when mixed with coconut milk in Goa and Konkan region. In this process the grinded green chilli, cumin paste, salt and jaggery are used for the preparation of kokum *solkadh*i.

### Dried flakes and Dried Salted Rind

The Dried flakes are prepared by sun drying of rind of kokum, Malabar tamarind and yellow mangosteen. The fruits are washed in running water and ait dried. Fruits are manually cut into halves/ segments and the fleshy portion containing the seed is removed. The outer skin is sun dried 3-5 days. The **dried salted rind** (*amsu*l) is made from the peel of the fruits 'Kokum'. Amsul is a fine ingredient in veg-dishes and curries for the sour taste also it is useful for skin boils and irritation, if rubbed against the irritated parts. In traditional process, fresh kokum fruit washed properly and cut into two halves to separate the seed, pulp and the rind. The seed and the pulp are mixed with around 10 percent salt. The salt solution leached out from this mixture is used for dipping of the separated rind. The rind then placed for sun drying during daytime. The next day the dried sample is again dipped in the salt solution, which was leached on the second day from the salt seed mixture. Then the rind is again placed for the drying. The process of dipping and drying is repeated for 4 to 5 times to get the *amsul.*

### Garcinia Butter

Generally seeds of kokum and Malabar tamarind are the waste product of processing. The seeds contain oil and oil content ranged from 22.12 to 48.30 percent . The seed oil can be used for edible purposes and still used for preparations of special dishes such as *holige* and *panchakajaye* (local names of the dishes) in Karnataka and Maharashtra (Table 6). The seeds contain oleic acid rich edible fat. Total of five fatty acid methyl esters are present in the oil. Among the five fatty acid esters present in the Malabar tamarind seed oil, Methyl 16-methyl heptadecanoate (54.57 %) was found to be the predominant compound. Traditionally seed *butter* is obtained in a very laborious *process*. The seeds are dried in the sun, chopped, and boiled in water and the churning and separation of butter. The butter can be used in various cosmetic products similar to kokum butter. Kokum and Malabar tamarind butter has emollient property, it is used as a natural moisturizer to keep skin supple and silky smooth. It is effective for treatment for severely dry skin, ulceration and fissures of lips, hands, feet, etc. Kokum butter is also nutritive and astringent. Uses include candle, soap, ointments and other pharmaceutical purposes.

### Malabar tamarind Vinegar

Malabar tamarind Vinegar (*Kachampuli* vinegar in *Kodagu*) is routinely used in non-vegetarian preparations especially pork for emulsification of fat, to add flavour and to improve taste in Kodagu. For generations in Kodagu, this is used at the rate of half teaspoon for 1 kg of meat. Reddish colour of pork after cooking with vinegar will turn to brownish black. Individuals consuming 3-4 kgs of pork every week, even then are slim. It is also used in making fish, chicken and mutton curries. For preparation of vinegar, fruits are allowed to ferment in bamboo basket for 10-15 days after harvesting. Yellowish green juice coming out of the fruits (*panakke* in Kodagu) is collected in mud pot. This mud pot is gently boiled for 4-5 hours, after repeated agitation, to get the water evaporated and to get a sticky and black residue like honey, which is the concentrated form of vinegar. Depending upon the dryness of the vinegar, it can be stored in bottles for 5-6 years without spoiling.

**Table 6. Processed products from *Garcinia* species**

| Fruit | Products |
|---|---|
| Mangosteen | Jam, Squash, Jelly, Canned fruits |
| Kokum | Squash, Dried rind, Rind powder, Seed butter, Gamboge |
| Malabar Tamarind | Dried rind, Rind powder, Seed butter, Gamboge |
| Yellow mangosteen | Jam, Chutney, Dried rind, Gamboge |

Malabar tamarind and kokum are known for anti-obesity properties and other health benefits. The powder and tablets are prepared by many pharmaceutical companies and sold in the market. There is a lot of demand of processed products of *Garcinia* in Kerala and coastal parts of Karnataka.

## Plant Protection

Mangosteen does not appear to suffer from serious attacks of major pests and diseases.

### Insect Pests and Management

#### Thrips

They prefer to feed on the tender, young plant tissue such as flower buds, opening blooms and unopened leaves. Spray Dimethoate two to three times at weekly intervals.

**Tussock Caterpillar (*Eupterote favia*)**

Larval stage eat on the young leaves. Spray Malathion or Fenvalerate

**Mites**

Attack the fruit surface, deface the fruits with small bites and make it unattractive for market. Spray Profenefos or any acaricide.

**Mealybugs**

Attack on young flushes and fruits. Spray Carbaryl or Dimethoate.

### Diseases and Management

**1. Sooty mould**

Attack on young leaves. Improving aeration and sunlight penetration by pruning overlapping branches will reduce infection. Spray with common fungicides.

**2. Anthracnose**

It affects the leaves. Spray with common fungicides to control the disease.

**3. Bacterial Leaf Sheath**

It affects the leaves .Spray with common fungicides to control the disease.

### Physiological Disorder and Management

A major physiological disorder called gummosis is found in mangosteen. This is evidenced by the oozing of latex as yellow spots on the fruit surfaces or skin. Physical damage to the latex vessel can be caused by sucking insects, strong winds and rough harvesting or handling. Fruits exposed to strong sun.

## Future prospects

Garcinia fruits has tremendous potential as crop of next generation in humid tropical and subtropical regions. It has lot of potential for pharmaceutical and processing industries. Some woks have been done on identification of superior lines but very limited research works has been done on the production technologies. The cultivation of most of the Garcinia species is almost negligible and almost all quantity is collected from forests or home gardens. The deforestation and low regeneration have resulted in depletion of the high yielding trees. Thus, production is decreasing while demand is increasing day by day. Thus, it is high time that the commercial cultivation of these species may be started as a part of mixed cropping system. The forest departments should also include these species in their afforestation programmes in their native locations.

## References

Anderson T. 1874. Guttiferae. *In*: Hooker JD. (ed.) *Flora of British India*. 1. L. Reeve and

Anonymous. 1950. *Wealth of India Raw Materials*, 4. CSIR, New Delhi. 99-108.

Anonymous. 2014. Plant Discoveries, New genera, species and new records. Botanical Assam, North East India. *Phytotaxa,* 252 (1), 73-76.

Bommayya, H., Kusum, R., & Ramachandran, H. D. (2011). Antioxidant capacities of fruit extracts of *Garcinia indica* with different assays and maturity stages. *Vegetos, 24* (2), 86–90.

Dunthorn M. 2004. Cryptic dioecy in *Mammea* (Clusiaceae). *Plant Syst. Evol.,* 249, 191-196.

Engler A. 1925. Guttiferae. *In*: Engler A and Prantl K (eds.), Die natürlichen Pflanzenfamilien. 21, Engelmann W, Leipzig, 154-237.

Jayaprakasha, G. K., & Sakariah, K. K. 1998. Determination of organic acids in *Garcinia cambogia* (Desr.) by HPLC. *Journal of Chromatography A, 806,* 337–339.

Jayaprakasha, G. K., & Sakariah, K. K. 2002. Determination of organic acids in leaves and rinds of *Garcinia indica* by LC. *Journal of Pharmaceutical and Biomedical Analysis, 28,* 379-384.

JNTBGRI, Diversity of *Garcinia* species in the Western Ghats: Phytochemical Perspective, JNTBGRI, Thiruvananthapuram.

Maheshwari JK. 1964. Taxonomic studies on Indian Guttiferae III. The genus *Garcinia* L.*Bulletin of the Botanical Survey of India,* (2-4), 107-135.

Nainegali, B. S., Prasanna, R., & Belur, D. (2019). Simultaneous extraction of four different bioactive compounds from *Garcinia indica* and their enrichment using aqueous two-phase systems. *Food and Bioproducts Processing, 114,* 185–195. https:// doi.org/10.1016/j.fbp.2019.01.002

Nayak, C. A., Rastogi, N. K., & Raghavarao, K. S. M. S. (2010). Bioactive constituents present in *Garcinia indica* choisy and its potential food applications: A review. *International Journal of Food Properties, 13*(3), 441–453.

Nayar, T.S., Beegam, A.R. and Sibi, M. 2014. Flowering plants of the Western Ghats, India.

Nimanthika, W.J.and Kaththriarachchi, H.S.. 2010. Systematics of genus *Garcinia* L. (Clusiaceae) of *Garcinia* L. and its reproductive ecology. *Folia Malaysiana,* 14 (1), 63-76.

Roberts, E. 1984. Vegetable Materia Medica of India and Ceylon. Bishen Singh Mahendra Pal Singh, Dehra Dun, India.

Sabu, T., Mohanan, N., Krishnaraj, M.V., Shareef, S.M., Shameer, P.S. and Roy, P.E. 2013. *Garcinia pushpangadaniana,* (Clusiaceae) a new species from southern Western Ghats, India.

Semwal, R. B., Semwal, D. K., Vermaak, I., & Viljoen, A. (2015). A comprehensive scientific overview of Garcinia cambogia. *Fitoterapia, 102,* 134–148.

Senthil Kumar, R. ,Tripathi, P. C., Sankar, V. and Karunakaran ,G., 2014. Kokum – A potential fruits of Western Ghats. Technical Bulletin 6-/2014, ICAR-IIHR, Central Horticultural Experiment Station, Chettalli, Kodagu, Karnataka10.

Senthil kumar,R., Tripathi, P.C. , Sankar, V. , Karunakaran, G. , Sakthivel,T., Kishore Kumar,M., Venkatravappa,V. ,Priti Sonavane, Ravisankar,H. and Chithraichelvan, R.2015. Fifteen Year of Research on Garcinia at CHES Chettalli. In souvenir and Abstract of workshop on Garcinia held at CHES Chettalli on August 20. P.7-9.

Singh P, Roy TK, Kanupriya C, Tripathi PC, Kumar P, Shivashankara KS. (2021). Evaluation of bioactive constituents of Garcinia indica (kokum) as a potential source of hydroxycitric acid, anthocyanin, and phenolic compounds. LWT-Food Science and Technology. 156: 112999.

Swami, S. B., Thakor, N. J., & Patil, S. C. (2014). Kokum (*Garcinia indica*) and its many functional components as related to the human health: A review. *Journal of Food Research and Technology, 2*(4), 130–142.

Tripathi P.C., Shetti ,D .L. and Rupa ,T.R. 2019. Nutrient content of some underutilized fruits . Progressive Horticulture. 50(2):115-22.

Tripathi, P C, Yogeesha,H S and Shetti, D L 2022. Standardization of propagation methods in minor wild fruit crops Current Horticulture 10 (1): 32-36.

Tripathi, P. C . 2021. Kokum- an underutilized fruit with medicinal importance. Agro India 26(6):21 , 10-13.

Tripathi, P. C. 2021. Medicinal and theraptic properties of minor fruits - A Review. International Journal of Minor Fruits, Medicinal and Aromatic Plants. 7 (2) : 01- 28.

Tripathi, P. C. 2021. *Garcinia*: A genus of nutraceutical fruits. Indian Hortic. 66(6): II &3-8.

Tripathi, P. C . 2021. Yellow mangosteen-a future fruit with medicinal value, Agro India 26(6):20-22.

Tripathi, P.C. , Karunakaran,G., Sankar,V. and Senthil kumar, R..2015. Garcinia in India –An Introduction, In souvenir and Abstract of workshop on Garcinia held at CHES Chettalli on August 20. P.1-6.

Tripathi, P.C. 2021. Malabar Tamarind : A potential fruit, condiment and medicine. Spice India 34(7): 4-10.

Tripathi, P.C. Anuradha sane, Kanupriya, Preetee singh and B.M. Muralidhara. 2022. Mangosteen : A fruit with splendid taste . Agro India December, pp.10-13.

Vasanth Kumar, T., Tripathi , P.C. , Lokesh A. N. and Rohini, M. R. .2015. Medicinal properties of Garcinia - a Review . In souvenir and Abstract of workshop on Garcinia held at CHES Chettalli on August 20. P.70-79.

CHAPTER 17

# RECENT ADVANCES AND INNOVATIONS IN POTATO SEED PRODUCTION

**Subhash S[1*], Vishwanath Bidarmali[1], Tanuja Buckseth[2] and Rajesh Kumar Singh[1*]**

*[1] ICAR-Central Potato Research Institute, Regional Station, Modipuram, Uttar Pradesh*
*[2] ICAR-Central Potato Research Institute, Shimla, Himachal Pradesh*
** Correspondence: subhashsram@gmail.com; rjan_1971@yahoo.co.in*

Potato seed production has evolved significantly with the integration of modern technologies and sustainable practices. This chapter explores recent advancements such as micropropagation, synthetic seeds, aeroponics, hydroponics, and Apical Rooted Cuttings (ARC), emphasizing their impact on improving seed tuber quality and multiplication efficiency, because traditional methods faces limitations including low multiplication rates and disease accumulation. New techniques, such as aeroponics and ARC, offer higher productivity and disease resistance, supporting year-round and large-scale production. Furthermore, innovations in seed storage, handling, and transportation, including controlled atmosphere storage and biocontrol agents, are enhancing seed longevity and quality. This chapter also covers case studies from India and the Netherlands which provide practical insights into successful implementations of these technologies and concludes by addressing current challenges, including high initial costs and regulatory hurdles, and also discuss future directions in research and development to optimize and scale these technologies.

Potato (*Solanum tuberosum* L.) is a critical staple crop globally, valued for its high nutritional content and adaptability to diverse climatic conditions (Chakrabarti, 2021). Quality seed is the cornerstone of successful potato production, directly influencing yield, disease resistance, and overall crop health. Seed costs account for about 30-50% of the total production cost in potato farming (Singh et al., 2021).

Poor-quality seed can lead to reduced yields and increased susceptibility to diseases, affecting overall productivity and profitability (Buckseth et al., 2022a).

## Global and National Significance of Potato

Globally, potato is the fourth most important food crop after rice, wheat, and maize (Jansky and Spooner, 2018). Its versatility and high yield per unit area make it critical for food security, especially in densely populated regions (NAAS, 2021). India is the second-largest producer of potatoes, contributes about 12.5% of the world's production, cultivating the crop across various agro-climatic zones from the Indo-Gangetic plains to high-altitude regions (Singh et al., 2021; Sood et al., 2022). The potato's ability to produce more edible energy, protein, and dry matter per unit area compared to cereals underscores its importance in India's agricultural landscape. The FAO has identified potato as "food for the future" due to its potential to ensure global food and nutritional security (Chakrabarti, 2021; Sood et al., 2022).

## Traditional Methods of Seed Production and Their Limitations

Traditionally, potato seed production relies on vegetative propagation using tubers. This involves selecting healthy tubers, often indexed against viruses, and multiplying them over several generations under controlled conditions (Singh et al., 2021). This method has limitations, including a low multiplication rate and the risk of viral disease accumulation with each field exposure (Raigond et al., 2019; Chakrabarti, 2021). The conventional seed production system in India includes different stages-Breeder seed, Foundation seed, and Certified seed-to ensure a pipeline of quality seed material. However, challenges such as limited land availability, slow multiplication rates, and disease accumulation necessitate innovations in seed production to meet the growing demand for quality seed (NAAS, 2021).

## Relevance of Recent Advances and Innovations in Potato Seed Production

Recent advances in potato seed production are critical in addressing the increasing global demand for food driven by population growth and shifting dietary patterns. Traditional methods of potato seed production are reliable but have limitations like slow multiplication rates and the progressive accumulation of diseases over successive generations (Chakrabarti, 2021). Innovations such as aeroponics and hydroponics have transformed the production of disease-free seed tubers, offering significantly higher multiplication rates and better-quality control (Singh et al., 2021). These technologies ensure faster and more efficient seed production, contributing to higher yields and improved crop health.

Additionally, the adoption of sustainable practices like Integrated Pest Management (IPM) and precision irrigation has become increasingly important in the context of climate change. These practices not only enhance resource efficiency but also mitigate the adverse environmental impacts of agriculture, ensuring the long-term viability of potato farming (NAAS, 2021).

# Basics of Potato Seed Production

## Seed Types and Quality Standards

### Classification of Potato Seeds

Potato seed production follows a multi-tiered system to ensure high-quality seeds at each stage of multiplication. The classification into nuclear seed, breeder seed, foundation seed, and certified seed is crucial for maintaining the quality and integrity of seed potatoes in commercial farming (Fig. 1).

**Nuclear Seed:** This is the initial stage of seed production. Nuclear seeds are produced through advanced techniques like tissue culture and micropropagation to ensure that the planting material is free from diseases, especially viral pathogens. The use of meristem culture in producing nuclear seed ensures that the seed stock is genetically pure and free from pathogens (Buckseth et al., 2016; Naik and Karihaloo, 2007).

**Breeder Seed:** Once nuclear seed is obtained, it is multiplied under controlled conditions to produce breeder seed, maintaining genetic purity and health (Singh et al., 2021). Breeder seed serves as the primary source for further multiplication into foundation and certified seeds.

**Foundation Seed:** The breeder seed is further multiplied to produce foundation seed, which is divided into two classes: Foundation Seed I and Foundation Seed II. Foundation seeds are produced under strict certification standards and are used to produce certified seeds. The production of foundation seed ensures a higher multiplication rate while maintaining seed quality (Singh et al., 2021).

**Certified Seed:** Certified seed is the final product in the seed production chain and is distributed to farmers for commercial potato production. Certified seeds must meet stringent quality standards, including freedom from diseases and pests, and must maintain high genetic purity (Chakrabarti et al., 2017). These seeds are typically produced by government agencies or licensed private companies under the guidelines provided by the Indian Minimum Seed Certification Standards (IMSCS) (Singh et al., 2021).

### Quality Standards and Certification Procedures

Quality standards and certification procedures are essential to ensuring that potato seeds meet the necessary criteria for health, vigor, and genetic purity. The certification process involves several steps, from field inspections to laboratory testing, to guarantee that the seed meets the stringent requirements set by regulatory bodies.

- **Field Inspections:** Field inspections are conducted at various stages of seed production to monitor the health of the crop. Inspectors check for the presence of diseases, pests, and off-types that could compromise the quality of the seed. Inspections are carried out during the growing season, and any fields that do not meet the required standards are rejected (Kumar, 2019; Buckseth et al., 2022).

- **Laboratory Testing:** Samples of seed potatoes are tested in laboratories for diseases, particularly viral infections that can severely impact yield. Techniques such as ELISA (Enzyme-Linked Immunosorbent Assay) are used to detect the presence of viruses like PVX, PVY, and PLRV, which are common in potato crops (Singh et al., 2021).
- **Certification Standards:** The Indian Minimum Seed Certification Standards (IMSCS) outline the specific requirements for seed certification. These standards cover aspects such as seed purity, germination rates, disease tolerance levels, and the maximum permissible limits for seed-borne pathogens. Only seeds that meet these standards are certified and labelled for commercial distribution (Chakrabarti et al., 2017).
- **Labelling and Distribution:** Certified seeds are labelled with a certification tag that provides information on the variety, class of seed, and the certifying agency. This ensures that farmers receive high-quality seed that has been rigorously tested and certified (Singh et al., 2021).

## Traditional Seed Production Techniques

### Conventional Methods of Seed Tuber Production

Traditional potato seed production predominantly relies on vegetative propagation through tubers, known as clonal multiplication. This method involves selecting healthy, disease-free tubers to propagate subsequent generations of seed. Clonal multiplication has been widely practiced in countries such as India, the Netherlands, South Africa, and Kenya (Singh et al., 2019a, b). The system depends on visual inspection of seed fields to remove diseased plants a process termed as rouging which is critical for maintaining crop health. However, this approach has notable limitations, especially in regions with high disease pressure, where pathogens can accumulate over successive generations, resulting in a decline in seed quality (Naik and Buckseth, 2018a).

The traditional system includes multiple field multiplications of seed tubers, often over several years. For example, the production of Breeder Seed from Nucleus Seed involves four generations of multiplication, followed by additional multiplications to produce Foundation Seed and Certified Seed. This multi-generational approach is designed to produce sufficient certified seed but is time-consuming and requires stringent quality control at each stage (NAAS, 2021).

In India, this traditional method has been practiced since the 1930s, initially in hilly regions where cooler climates and lower aphid infestations slowed seed degeneration. The method was later adapted for the plains with the introduction of the Seed Plot Technique in 1965, enhancing the process to facilitate seed production in areas with higher disease pressure (Pushkarnath, 1959). This technique involves the strategic timing of planting and harvesting seed crops to avoid peak aphid populations, thereby minimizing the risk of virus transmission (Pushkarnath and Nirula, 1970; Buckseth et al., 2016).

**Limitations of Traditional Techniques**

While traditional methods of seed tuber production have been foundational in the development of the global potato industry, they are accompanied by significant limitations. The primary challenges include:

- **Low Multiplication Rate**: The clonal multiplication method has a slow multiplication rate, approximately 1:6, meaning it takes several years to produce sufficient quantities of seed to meet commercial demand (Singh et al., 2019a). This low rate is particularly problematic in regions where the demand for seed is high, and the availability of disease-free propagules is limited.
- **Disease Accumulation**: A critical issue with traditional methods is the progressive accumulation of viral diseases with each field exposure. As seed tubers are propagated over multiple generations, the risk of disease buildup increases, leading to a gradual decline in seed quality (Venkataslam et al., 2017). Visual inspection and rouging is necessary, but not entirely effective; because infected plants may remain undetected and exacerbating the issue further.
- **Slow Process of Generating Healthy Seed Stock**: Producing 100% healthy seed stock from infected material is a labor-intensive and time-consuming process. It involves meticulous selection and indexing of tubers to ensure only disease-free materials are used for further propagation (Naik and Karihaloo, 2007). This process can span several years, especially when starting with material previously exposed to field conditions.
- **Requirement for Extensive Land**: The traditional system demands large tracts of land for the multiple field multiplications necessary to produce sufficient seed quantities. This requirement poses a significant constraint, particularly in areas where land availability is limited or where land costs are high (Buckseth et al., 2020).

These limitations highlight the need for more advanced and efficient seed production techniques to overcome the challenges associated with traditional methods. Innovations such as micropropagation, synthetic seeds, aeroponics, and apical root cuttings provide promising alternatives by offering faster multiplication rates, reducing field exposure, and producing disease-free planting material in controlled environments (Fig. 2).

# Recent Advances in Potato Seed Production

## Innovations in Seed Tuber Production

### Micropropagation Techniques

Micropropagation, a tissue culture-based technique, has become central to the production of high-quality, disease-free potato seed tubers. This method involves the *in vitro* clonal propagation of plants using small tissue sections called explants, which are cultured in sterile media (Naik and Buckseth, 2018b). The process often starts with meristem culture, taking advantage of the fact that viruses and other pathogens are less likely to infect the rapidly growing tips of shoots. By culturing these meristematic tissues under controlled conditions, virus-free plants can be produced and then multiplied to generate large quantities of disease-free seed stock (Fig. 3a) (Naik and Buckseth, 2018b; Venkataslam et al., 2017).

Meristem culture is typically combined with thermotherapy or chemotherapy to increase the likelihood of virus elimination. The meristem, a small growing point approximately 0.1-0.3 mm in size, is excised and placed in a nutrient medium. This method has proven particularly effective in producing virus-free clones, which are rigorously tested using techniques such as ELISA (Enzyme-Linked Immunosorbent Assay) and PCR (Polymerase Chain Reaction) before being utilized for large-scale micropropagation (Fig. 3b) (Sarkar et al., 2011).

The micropropagation process can be conducted using either semi-solid or liquid culture media. The Murashige and Skoog (MS) medium, supplemented with specific growth regulators like D-calcium pantothenate and sucrose, is commonly employed for large-scale *in vitro* potato propagation. This approach allows for the rapid multiplication of plants, with a single culture tube potentially producing over 59,000 microplants within 7-8 months under controlled conditions (Naik and Buckseth, 2018b; Buckseth et al., 2016).

Micro-tuber production is another technique used in plant biotechnology and tissue culture to rapidly produce small tubers (micro-tubers) in a controlled environment. These micro-tubers are generated from *in vitro* cultured potato plants and serve as high-quality seed material for field planting (Sharma et al., 2021).

### Advantages of Micropropagation in Seed Production

Micropropagation offers several significant advantages in seed production, particularly for potato cultivation:

- **Rapid Multiplication of Seed Stock:** Traditional potato seed production methods are often slow and insufficient to meet the increasing demand for high-quality seed. In contrast, micropropagation enables rapid multiplication of seed stock, crucial for meeting the high demand for certified seed potatoes. The multiplication rate in micropropagation is significantly higher than that of conventional methods (Naik and Sarkar, 2001).

- **Production of Disease-Free Planting Material:** A critical benefit of micropropagation is its ability to produce disease-free planting material. Potato crop is highly susceptible to viral diseases, which can accumulate over successive generations using traditional methods. Micropropagation ensures that the initial seed stock is pathogen-free, significantly reducing the risk of disease transmission and improving overall crop health (Naik and Buckseth, 2018b).
- **Year-Round Production:** Unlike traditional methods, which are often constrained by seasonal limitations, micropropagation can be conducted year-round in controlled environments. This flexibility ensures a continuous supply of seed potatoes, which is essential for maintaining stable potato production, especially in regions where the crop is a staple food (Naik et al., 2001).
- **High Genetic Uniformity:** Micropropagation ensures high genetic uniformity among propagated plants, as it involves the clonal reproduction of a single genotype. Such uniformity is crucial for maintaining the quality and consistency of seed potatoes (Buckseth et al., 2016).
- **Scalability:** Micropropagation techniques are highly scalable, enabling the production of large quantities of seed potatoes within a relatively short timeframe. This scalability is particularly beneficial in regions with high seed demand, where traditional methods may be inadequate to meet farmers' needs (Venkataslam et al., 2017).

**Synthetic or Artificial potato seed:**

Synthetic seeds, also known as artificial or encapsulated seeds, are artificially created seeds that encapsulate somatic embryos, shoot buds, microtubers, or other tissues capable of developing into a complete plant. These synthetic seeds are an effective method for propagating and conserving plant species, particularly in crops like potatoes where traditional seed production is challenging (Ghanbarali et al., 2016). In potato cultivation, which typically relies on vegetative propagation through tubers, synthetic seeds are particularly valuable due to the challenges associated with traditional methods, such as disease transmission and storage difficulties. Sarkar and Naik (1998) demonstrated for the first time that *in vitro* derived potato nodal segments encapsulated in whole bead alginate capsules could be used for potato propagule production. Later, synthetic seeds of potato were produced using encapsulated shoot tips in hollow alginate beads. Nyende et al. (2003) also used potato shoot tips encapsulated in hollow alginate beads, recording germination rates of 100% in non-sterile soil in the greenhouse and a 100% survival rate after 270 days of storage.

This technique offers significant opportunities for producing large quantities of disease-free potato propagules from *in vitro* plantlets under protected conditions, using limited time and space. Moreover, the method is cost-effective; however, further research is required for its application at the field level.

## Application of Synthetic seeds in Potato Seed Production

- **Production of Disease-Free Seeds:** Synthetic seeds provide a means to produce disease-free planting material. By using somatic embryos or microtubers derived from tissue culture techniques, synthetic seeds ensure the resulting plants are free from pathogens. Naik and Sarkar (1997) demonstrated the successful production of synthetic seeds using encapsulated somatic embryos of potatoes, which could be stored for several months and still produce disease-free plants upon germination.
- **Uniformity and Clonal Propagation:** Synthetic seeds enable the clonal propagation of potato plants, ensuring that all plants produced from these seeds are genetically identical to the parent plant. This uniformity is essential for maintaining specific traits such as disease resistance, yield, and tuber quality. Raemakers et al. (1995) successfully used synthetic seeds of potatoes to produce clonal plants with high uniformity, crucial for commercial production.
- **Storage and Transport:** A significant advantage of synthetic seeds is their ability to be stored for extended periods under controlled conditions. This storage capability reduces the need for immediate planting, allowing for better planning and logistics in potato production. Furthermore, synthetic seeds can be easily transported without the risk of spreading soil-borne diseases, making them an attractive option for international trade and distribution. Ghosh and Sen (1994) found that synthetic seeds of potato remained viable after several months of storage, retaining their ability to germinate and produce healthy plants.

Synthetic seeds represent a promising technology in potato seed production, offering the potential for disease-free, uniform, and easily transportable planting material. However, further research and development are necessary to overcome current challenges and make this technology more accessible and cost-effective.

## Aeroponics and Hydroponics Systems

### Introduction to Soilless Cultivation Methods

Soilless cultivation methods, particularly aeroponics and hydroponics, represent a significant innovation in the production of seed tubers for potato cultivation. These methods are characterized by the cultivation of plants without soil, using nutrient solutions to deliver essential minerals directly to the plant roots. This approach is particularly advantageous for producing disease-free, high-quality seed tubers.

Aeroponics involves suspending plant roots in the air within a closed or semi-closed environment and misting them with a nutrient-rich solution. The absence of soil reduces the risk of soil-borne diseases and allows for greater control over the growing environment, leading to healthier and more uniform plant growth (Buckseth et al., 2016; Tiwari et al., 2019). The roots are exposed to air for optimal oxygenation, while the misting system provides the necessary water and nutrients. This method has been increasingly adopted to produce potato mini-tubers due to its efficiency in generating large numbers of uniform and disease-free tubers (Fig. 4) (Naik and Buckseth, 2018a).

Hydroponics, in contrast, involves growing plants in a nutrient solution without soil, often using techniques such as the Nutrient Film Technique (NFT). In this system, the roots of the plants are either fully submerged or intermittently exposed to the nutrient solution. Hydroponics allows for precise control over the nutrient supply, leading to more consistent and faster plant growth compared to traditional soil-based cultivation (Boersig and Wagner, 1988; Singh et al., 2019a).

## Benefits and Challenges of Aeroponics and Hydroponics in Seed Production

### Benefits:

- **High Multiplication Rate:** Aeroponics and hydroponics offer a high multiplication rate of seed tubers. In aeroponics, a single plant can produce an average of 40-60 mini-tubers, significantly higher than the 8-10 mini-tubers typically produced under conventional net house conditions (Buckseth et al., 2016; Naik and Buckseth, 2018a). This high-rate aids in rapidly scaling up seed production, crucial for meeting the growing demand for quality seed potatoes.
- **Disease-Free Production:** Both systems produce disease-free planting material. The soilless environment minimizes the risk of soil-borne diseases, and the controlled conditions reduce pathogen exposure, leading to healthier and more vigorous seed tubers (Buckseth et al., 2020).
- **Year-Round Production:** These methods allow for year-round seed tuber production, regardless of the season. This is particularly advantageous in regions with limited growing seasons or unfavorable environmental conditions for traditional farming practices (Singh et al., 2019). The ability to produce seed tubers throughout the year stabilizes supply and reduces the risk of shortages.
- **Efficient Use of Resources:** Aeroponics and hydroponics are highly efficient in their use of water and nutrients. The closed-loop systems ensure nutrient recycling and minimize water loss, making these methods more sustainable compared to traditional soil-based farming, especially in water-scarce regions (Tiwari et al., 2019).

### Challenges:

- **High Initial Investment:** A significant challenge associated with these systems is the high initial investment required. The cost of equipment, including nutrient solution reservoirs, misting systems, and growth chambers, can be prohibitive, particularly for small-scale farmers (Buckseth et al., 2016).

- **Technical Expertise:** Operating and maintaining aeroponics and hydroponics systems require a high level of technical expertise. Proper management of the nutrient solution, pH levels, temperature, and system functionality are critical for success. This technical requirement may be a barrier for farmers unfamiliar with these systems (Naik and Buckseth, 2018a).
- **Risk of System Failures:** These systems heavily rely on technology, making them vulnerable to failures such as power outages or pump malfunctions, which can lead to rapid crop deterioration. Implementing robust backup systems and regular maintenance adds to the overall cost (Singh et al., 2019b).
- **Optimization Needs:** Although aeroponics and hydroponics have demonstrated benefits, optimization is still necessary, especially in potato seed production. Factors such as nutrient solution composition, misting frequency and duration in aeroponics, and environmental conditions for different potato varieties need fine-tuning to maximize yields (Tiwari et al., 2019).

## Apical Rooted Cuttings (ARC)

### Introduction to Apical Rooted Cuttings (ARC)

Apical Rooted Cuttings (ARC) is an innovative and increasingly popular method in potato seed production, representing a significant advancement in agricultural biotechnology. ARC utilizes apical cuttings-growing tips of potato plants-carefully excised from mother plants and rooted under controlled conditions. This method enables the rapid multiplication of potato plants and the production of disease-free seed tubers, enhancing both the quality and quantity of seed potato production.

The ARC technique is particularly effective in producing high-quality planting material because it capitalizes on the genetic stability and vigor of the mother plants. Once rooted, these cuttings grow into full plants that can either produce mini-tubers or be directly used for further field multiplication (Fig. 5). Due to its efficiency, cost-effectiveness, and reduced risk of disease transmission compared to traditional methods, ARC is gaining popularity (Bryan et al., 1981; Buckseth et al., 2019).

### Benefits and Challenges of Apical Rooted Cuttings in Seed Production

**Benefits:**

- **Rapid Multiplication:** ARC enables the rapid multiplication of potato plants, with each mother plant providing multiple apical cuttings. This significantly increases the number of plants that can be produced within a short period, crucial for meeting the growing demand for seed potatoes, particularly in regions where traditional methods are insufficient (Buckseth et al., 2022).
- **Disease-Free Seed Production:** As apical cuttings are taken from the growing tips of plants, they are less likely to harbor pathogens, especially

viruses that frequently impact seed potato production. This makes ARC an excellent method for producing disease-free planting material, critical for maintaining healthy crops and achieving high yields (Buckseth et al., 2019).

- **Cost-Effectiveness:** Compared to more complex tissue culture methods, ARC is relatively cost-effective, requiring less sophisticated infrastructure. It is accessible for small and medium-scale farmers, making it an appealing option in regions with limited resources but high demand for quality seed potatoes (Singh et al., 2021).
- **Scalability:** ARC is highly scalable, making it suitable for both small-scale and commercial seed production. It can be integrated into existing seed production systems, complementing other methods such as micropropagation and mini-tuber production. This scalability is crucial for expanding seed production to meet market demands (Tiwari et al., 2019).

**Challenges:**

- **Technical Expertise:** While ARC is less complex than other advanced methods, it still requires a certain level of technical expertise to ensure success. Proper handling of cuttings, rooting techniques, and environmental control are essential for achieving high rooting success rates and maintaining plant health (Buckseth et al., 2020).
- **Environmental Control:** The success of ARC is highly dependent on maintaining optimal environmental conditions during the rooting and growth phases. Factors such as humidity, temperature, and light must be carefully regulated to prevent stress on the cuttings and ensure successful rooting. Any deviation from these conditions can result in poor rooting success and reduced plant vigor (Naik and Buckseth, 2018a).
- **Market Penetration:** Despite its advantages, the adoption of ARC may be slow in certain regions due to a lack of awareness or initial skepticism among farmers. Effective extension services and farmer education programs are needed to promote the benefits of ARC and ensure its widespread adoption (Buckseth et al., 2019).

## Mini-tuber Production

### Techniques for Mini-tuber Production

Mini-tuber production is a crucial step in the seed potato production process, serving as an intermediary between *in vitro* propagation and large-scale field multiplication. Mini-tubers are small seed potato tubers produced under controlled conditions, typically within greenhouses or net-houses (Pandey and Singh, 2014). They are derived from micro-tubers or micro-plants propagated *in vitro* to ensure that the planting material is free from pathogens (Naik and Sarkar, 2001).

The production of mini tubers can be achieved through several techniques:

- **Soil-Based Production:** In soil-based systems, virus-free micro-plants or micro-tubers are transplanted into nursery beds under aphid-proof net-houses. This technique, widely practiced in India, allows for the production of approximately 2,000 mini-tubers per square meter. The mini-tubers are grown in soil under controlled environmental conditions to minimize disease pressure and are usually multiplied over three successive generations (Generation 1, 2, and 3) before being supplied to farmers as seed (Naik and Buckseth, 2018a; Singh et al., 2019a).
- **Aeroponic Systems:** This system has been successfully adapted for mini-tuber production. In aeroponics, mini-tubers can be harvested repeatedly from the same plants over several cycles, significantly increasing yield. A single plant can produce 40-60 mini-tubers in an aeroponic system, compared to 8-10 mini-tubers in traditional soil-based systems. Aeroponics is particularly advantageous as it supports high-density planting and reduces the risk of soil-borne diseases (Buckseth et al., 2016; Venkataslam et al., 2017).
- **Hydroponic Systems:** Hydroponics is especially suitable for mini-tuber production in environments where soil quality is poor or where there is a high risk of soil-borne diseases. This method uses an inert medium such as perlite or vermiculite, with plant roots being regularly irrigated with a nutrient solution. The controlled environment in hydroponics allows for optimal growth conditions, resulting in high-quality mini-tubers (Singh et al., 2019a; Rajendran et al., 2024).
- **Apical Rooted Cuttings (ARC):** It involves taking apical cuttings from mother plants and rooting them in controlled environments, such as pro-trays with cocopeat. The rooted cuttings are then transplanted to produce mini-tubers. This method is cost-effective and easy to implement, particularly in resource-limited settings. The ARC technique allows for rapid multiplication of planting material, making it especially beneficial in areas where traditional seed production methods are less feasible due to high disease pressure or limited resources (Bryan et al., 1981; Buckseth et al., 2019; NAAS, 2021).

## Scaling up for Commercial Seed Production

Scaling up the production of mini tubers for commercial seed production involves several key considerations.

- **Infrastructure and Investment:** Successful commercial production of mini-tubers requires significant investment in infrastructure, including greenhouses, aeroponic or hydroponic systems, and storage facilities. Although initial costs are high, the ability to produce large quantities of disease-free seed potatoes justifies the investment. The scalability of these systems also allows for continuous production, enabling seed producers to meet market demand efficiently (Naik and Buckseth, 2018a).

- **Optimization of Growth Conditions:** For large-scale production, it is essential to optimize growth conditions specific to different potato varieties. This includes fine-tuning nutrient solution composition in hydroponic and aeroponic systems, controlling environmental parameters such as temperature, humidity, and light, and ensuring proper timing of nutrient delivery. Tailoring these conditions to the specific needs of each potato variety can significantly enhance the yield and quality of mini-tubers (Tiwari *et al.*, 2019).
- **Disease Management:** Scaling up production requires stringent disease management practices to prevent pathogen introduction and spread. Using pathogen-tested nuclear stock is crucial to start the production cycle. Additionally, regular monitoring and testing for viruses and other diseases using techniques such as ELISA and PCR are essential to maintain crop health throughout the production process (Singh *et al.*, 2019b).
- **Market Integration and Supply Chain Management:** Integrating mini-tubers into existing markets and supply chains is vital for the commercial viability of production. This includes establishing reliable distribution networks, educating farmers about the benefits of using high-quality seed potatoes, and ensuring that mini-tubers are stored and transported under optimal conditions to maintain quality until they reach end users (Buckseth *et al.*, 2016; Buckseth *et al.*, 2022).
- **Adoption of Apical Rooted Cuttings:** The ARC method is particularly promising for scaling up mini-tuber production due to its low cost and ease of adoption. The use of rooted cuttings allows for rapid multiplication and can be seamlessly integrated into existing seed production systems. As the demand for quality seed potatoes continues to grow, ARC offers a scalable solution that can meet this demand while maintaining high-quality standards (Buckseth *et al.*, 2019; Buckseth *et al.*, 2022).

## Disease-Free Seed Production

### Techniques for Producing Virus-Free and Disease-Free Seed Tubers

- Producing disease-free seed tubers is essential in potato cultivation due to the crop's susceptibility to viral, bacterial, and fungal diseases, especially with vegetative propagation methods. Modern techniques have been developed to produce virus-free and disease-free seed tubers, ensuring higher yields and better-quality crops.
- **Meristem Culture:** Meristem culture is one of the most effective techniques for producing virus-free seed potatoes. It involves excising a small piece of meristem tissue, which is less likely to be infected by viruses, and growing it in a sterile culture medium. This technique is often combined with thermotherapy (heat treatment) or chemotherapy (application of antiviral chemicals) to eliminate viruses from the plant tissue. The resulting plants,

called mericlones, are rigorously tested using methods like ELISA (Enzyme-Linked Immunosorbent Assay) and PCR (Polymerase Chain Reaction) to ensure they are virus-free before being used in seed production (Sarkar *et al.*, 2011; NAAS, 2021).

- **Micropropagation:** Micropropagation involves the *in vitro* multiplication of disease-free plants from a small amount of plant tissue. This method enables the rapid production of large numbers of uniform, disease-free plants, which can be further multiplied in the field. In potato seed production, micropropagation serves as a reliable source of pathogen-free planting material. The plants generated through micropropagation are frequently used to produce micro-tubers or mini-tubers under controlled conditions (Buckseth et al., 2016; Singh, 2019a).
- **Seed Plot Technique:** The Seed Plot Technique is a field-based method used to produce healthy seed potatoes. It involves planting the seed crop during periods of low aphid activity to minimize the risk of virus transmission. The seed plots are carefully monitored, and any diseased plants are removed (rogued) to prevent virus spread. This technique also includes practices such as applying systemic insecticides to control aphid populations and using certified seed to reduce the introduction of diseases into the seed plot (Pushkarnath, 1959; Sharma et al., 2021).
- **Virus Indexing and Tuber Testing:** Virus indexing ensures that seed tubers are free from viruses before further multiplication. This process involves testing a sample of tubers using diagnostic tools like ELISA and PCR to detect viral presence. Tubers that test positive for viruses are discarded, while virus-free tubers are used for further multiplication. This method helps maintain the health of the seed stock over successive generations, reducing viral disease spread (Singh et al., 2019b; NAAS, 2021).

### Use of Diagnostics and Clean Seed Programs

- **Advanced Diagnostic Techniques:** The development of advanced diagnostic techniques has significantly improved the ability to detect pathogens in seed tubers. ELISA and PCR are the most commonly used techniques for virus detection in seed potatoes. ELISA is a serological test that detects specific proteins associated with viruses, while PCR amplifies viral DNA or RNA to identify pathogen presence. These techniques are highly sensitive and can detect infections at low levels, making them essential tools in clean seed programs (Boonham et al., 2014; Kumar et al., 2019).
- **Clean Seed Programs:** Clean seed programs are organized efforts to produce and distribute disease-free seed potatoes. These programs typically involve the use of certified seed, rigorous pathogen testing, and the implementation of best practices in seed production. They also include education and training for farmers on maintaining seed stock health and preventing disease

introduction. Clean seed programs are crucial for maintaining high yields and quality in potato production (Singh et al., 2019b; Kumar, 2019).

## Innovations in Seed Storage and Handling

### Advances in Seed Storage Techniques

Effective storage of seed potatoes is crucial for maintaining their quality and viability until the next planting season. Innovations in seed storage techniques have significantly contributed to reducing losses and improving the quality of stored seed potatoes. Two key advancements in this area are controlled atmosphere storage and the use of biocontrol agents to prevent spoilage (Kumar, 2019; Singh *et al.*, 2020).

### Controlled Atmosphere Storage

Controlled atmosphere (CA) storage involves regulating oxygen ($O_2$), carbon dioxide ($CO_2$), and nitrogen ($N_2$) levels in the storage environment, along with maintaining optimal temperature and humidity conditions. This method effectively extends the shelf life of seed potatoes by slowing down their metabolic processes, reducing sprouting, and minimizing the risk of disease development (Yahia et al., 2019). In CA storage, the concentration of $O_2$ is typically reduced to below 5%, while $CO_2$ levels are maintained between 3% and 5%. This environment slows respiration rates in the tubers, thereby reducing decay and maintaining seed potato quality. Additionally, lower oxygen levels inhibit the growth of aerobic microorganisms, reducing spoilage and disease during storage (Yahia *et al.*, 2019).

CA storage has proven highly effective in maintaining the quality of seed potatoes for extended periods, particularly in regions where long-term storage is necessary due to climatic conditions or market demands. The method also provides better control over the sprouting process, which is essential for maintaining seed viability for the next planting season. The ability to store seed potatoes for longer periods without significant quality degradation is a major advantage, making CA storage increasingly popular among commercial seed producers (Yahia et al., 2019).

### Use of Biocontrol Agents to Prevent Spoilage

Another significant advancement in seed storage is the use of biocontrol agents to prevent spoilage and disease during storage. Biocontrol agents are natural organisms or compounds that inhibit pathogen growth responsible for post-harvest diseases in potatoes. These agents offer a sustainable and environmentally friendly alternative to chemical treatments, which may leave harmful residues and affect seed potato safety (Ajay and Sunaina, 2005).

One of the most promising biocontrol agents for seed potato storage is *Pseudomonas fluorescens,* a bacterium known for its antagonistic properties against a range of fungal and bacterial pathogens. When applied to seed potatoes before storage, it colonizes the tuber surface, creating a protective barrier that prevents the establishment and growth of harmful pathogens. This bacterium has been particularly effective in controlling diseases like soft rot and dry rot, which are major causes of storage losses in seed potatoes (Stegli ska et al., 2022).

In addition to bacteria, certain fungal species such as *Trichoderma* spp. are also used as biocontrol agents. These fungi outcompete pathogenic fungi for nutrients and space on the potato surface, thereby preventing spoilage. The use of *Trichoderma* spp. has shown effectiveness in reducing the incidence of common storage diseases and improving the overall quality of stored seed potatoes (Axel et al., 2012).

Integrating biocontrol agents into seed potato storage protocols enhances the quality and longevity of stored tubers and aligns with the growing demand for sustainable agricultural practices. By reducing reliance on chemical treatments, biocontrol agents contribute to safer and more environmentally friendly seed production systems.

## Handling and Transportation

### Innovations in Packaging Materials

The packaging of seed potatoes plays a critical role in preserving their quality during storage and transportation. Innovations in packaging materials focus on enhancing protection against mechanical damage, moisture loss, and contamination while ensuring adequate ventilation.

- **Breathable Bags and Perforated Packaging:** Traditional jute bags have largely been replaced by breathable bags made from synthetic materials, offering better protection against moisture loss and improved ventilation. These bags are often perforated to facilitate air circulation, helping to prevent humidity buildup that can lead to rot and sprouting during storage and transportation. Perforated packaging solutions have become industry standards due to their effectiveness in maintaining tuber quality over extended periods (Almenar, 2020).
- **Anti-Microbial and Anti-Fungal Packaging:** To further protect seed tubers from spoilage during transportation, new packaging materials are being developed with integrated anti-microbial and anti-fungal agents. These materials help reduce the risk of fungal infections, such as *Fusarium*, which can cause dry rot during storage. Incorporating these protective agents into packaging materials represents a significant advancement in preserving seed potato quality during handling and transportation (NAAS, 2021).

### Cold Chain Management for Seed Tubers

Maintaining an effective cold chain is crucial for ensuring seed potato quality and viability during transportation, especially over long distances or in regions with high ambient temperatures. Cold chain management involves using refrigerated storage and transportation solutions to keep seed tubers at optimal temperatures throughout the supply chain.

- **Refrigerated Transportation:** Seed potatoes are often transported in refrigerated trucks that maintain temperatures between 2°C and 4°C, an ideal range for minimizing metabolic activity, reducing sprouting, and preventing decay. Properly managed refrigerated transportation is essential to prevent temperature fluctuations that could cause cold injury or sprouting, ensuring that seed tubers remain in optimal condition until they reach their destination (Mehta et al., 2014; Kumar, 2019).

- **Monitoring and Control Systems:** Advanced monitoring systems are now commonly used in cold chain management to track temperature and humidity levels within storage and transportation units. These systems provide real-time data, allowing for immediate adjustments if conditions deviate from the optimal range. This level of control mitigates risks associated with temperature fluctuations, ensuring that seed tubers are maintained at the desired conditions throughout transportation (Mehta et al., 2014; Kumar, 2019).

### Ensuring Seed Quality During Transportation

The transportation of seed potatoes involves several critical practices to maintain quality and prevent damage:

- **Proper Loading Techniques:** During loading, care must be taken to avoid compressing or damaging the seed potatoes. Workers should refrain from walking on the bags, and bags should be stacked in a manner that ensures adequate ventilation. Proper loading techniques help reduce the risk of mechanical damage and ensure that seed tubers are evenly distributed within the transport vehicle.
- **Temperature Control During Transit:** Maintaining a consistent temperature during transit is essential to prevent stress that could lead to sprouting or decay. In addition to refrigerated transportation, insulated containers can help maintain a stable environment, especially in regions with fluctuating external temperatures. Transporting seed potatoes during cooler times of day is also recommended to minimize temperature variations.
- **Ventilation and Moisture Management:** Adequate ventilation is necessary to prevent humidity buildup within the transport vehicle, which can lead to rot and other moisture-related issues. Using ventilated packaging and ensuring proper air circulation within the transport vehicle are key strategies for managing moisture levels. Additionally, keeping the seed potatoes dry before loading and avoiding exposure to rain or condensation during transit are critical practices for maintaining seed quality.

## Case Studies and Success Stories

### Case Study 1: Successful Implementation of Aeroponics in India

Aeroponics, a soilless cultivation method, has revolutionized seed potato production in India, addressing challenges associated with traditional seed production, such as low multiplication rates and increased pathogen exposure. This method involves growing plants in an air or mist environment, with roots suspended in a growth chamber and periodically misted with a nutrient-rich solution.

The ICAR-Central Potato Research Institute (CPRI) in Shimla spearheaded the introduction of aeroponics in India. Recognizing the limitations of soil-based propagation, CPRI launched projects to optimize aeroponic systems for potato seed production. The primary objective was to enhance the production of high-quality, disease-free mini-tubers, essential for certified seed potato production. This initiative included setting up aeroponic chambers within insect-proof net houses and conducting experiments to determine optimal nutrient formulations and environmental conditions for various potato varieties (Buckseth et al., 2016).

### Outcomes and Lessons Learned

The implementation of aeroponics in India has delivered significant positive outcomes. One of the most remarkable achievements was the dramatic increase in the multiplication rate of seed potatoes. Traditional soil-based methods typically yield 8-10 mini-tubers per plant, whereas aeroponic systems have demonstrated the potential to produce 40-60 mini-tubers per plant. This substantial increase in productivity has improved the availability of high-quality seed potatoes, particularly in regions where disease pressure is high.

Moreover, aeroponics has proven to be highly efficient in terms of resource use. The controlled environment within the aeroponic chambers allows for precise management of water and nutrients, reducing waste compared to conventional methods. The closed-loop system minimizes contamination risks from soil-borne pathogens, resulting in healthier and more robust plants (Buckseth et al., 2022).

Despite the advantages, the adoption of aeroponics in India has not been without challenges. The initial setup costs for aeroponic systems are relatively high, and the technology requires specialized knowledge for effective operation. Training for farmers and seed producers is crucial, particularly for managing nutrient solutions and ensuring the proper functioning of the misting systems. Nevertheless, the long-term benefits of aeroponics, including higher yields and improved seed quality, make it a promising technology for the future of potato seed production in India (Buckseth et al., 2022; NAAS, 2021).

## Case Study 2: Disease-Free Seed Production in the Netherlands

The Netherlands has long been recognized as a global leader in producing high-quality, disease-free seed potatoes. The country's success is attributed to a combination of rigorous seed certification standards, advanced diagnostic techniques, and innovative cultivation practices that ensure the production of pathogen-free seed tubers (Haverkort and Zaag, 1989).

- **Seed Certification and Quality Control:** The Dutch seed potato production system is founded on a stringent certification process encompassing multiple stages of field inspections, laboratory testing, and post-harvest evaluations. The process begins with the selection of high-quality parent material, which undergoes intensive monitoring throughout the growing season. Field inspections focus on identifying and removing (rouging) plants showing disease symptoms, while laboratory tests like ELISA (Enzyme-Linked Immunosorbent Assay) and PCR (Polymerase Chain Reaction) detect viral infections that may not be visually apparent (Louwaars, 2023; NAAS, 2021).

- **Clonal Multiplication and Tissue Culture:** To maintain seed stock purity and health, the Netherlands employs clonal multiplication techniques using meristem culture to produce virus-free plantlets. These plantlets are then multiplied through tissue culture to produce mini-tubers, forming the foundation for further field multiplication. This method significantly reduces the risk of viral infections, a major concern in traditional seed production systems (Haverkort and Zaag, 1989; Goffart et al., 2022).
- **Use of Innovative Technologies:** The Netherlands has also led in integrating innovative technologies into seed potato production. Precision agriculture tools, such as GPS-guided equipment and drones, allow for accurate field management and monitoring, reducing disease incidence and improving crop health. Additionally, advanced soil fumigation techniques and strategic crop rotations minimize soil-borne pathogens, enhancing the quality of seed tubers (Louwaars, 2023).

The success of disease-free seed production in the Netherlands has profoundly impacted both the domestic and global potato industries. The rigorous standards and advanced methodologies have enabled the Netherlands to become one of the largest exporters of seed potatoes globally, with Dutch seed potatoes being highly sought after for their quality and reliability.

- **Domestic Benefits:** Within the Netherlands, the production of disease-free seed potatoes has led to increased yields, reduced reliance on chemical inputs, and higher profitability for farmers. Access to high-quality seed has also contributed to the sustainability of potato production, ensuring that Dutch farmers consistently meet domestic and international market demands (Vinterhalter et al., 2008).
- **Global Influence:** The Netherlands' approach to disease-free seed production has served as a model for other countries, influencing global seed potato standards and practices. Dutch seed potatoes are exported to over 80 countries, and the methodologies developed in the Netherlands have been adapted and implemented worldwide, contributing to improved potato yields and disease management on a global scale (Louwaars, 2023).

The success of the Netherlands in producing disease-free seed potatoes through stringent certification, advanced diagnostics, and innovative farming practices demonstrates the importance of maintaining high standards and adopting cutting-edge technologies for sustainable and profitable seed production.

## Challenges and Future Directions

### Current Challenges in Potato Seed Production

#### Challenges in Scaling Up New Technologies

The adoption of innovative technologies in potato seed production, such as aeroponics, micropropagation, and apical rooted cuttings, presents significant advantages in terms of yield and quality. However, scaling up these technologies to meet large-scale seed production demands comes with several challenges:

- **High Initial Costs:** Technologies like aeroponics and advanced tissue culture systems involve substantial initial setup costs, requiring specialized equipment, controlled environments, and trained personnel. For small and medium-scale farmers, these capital investments are often prohibitive, and even larger commercial operations may face financial barriers due to the long return on investment period (Chakrabarti, 2021).
- **Technical Expertise:** Advanced seed production technologies demand a high level of technical expertise. Training is essential for farmers and operators in maintaining sterile conditions, managing nutrients, and monitoring for diseases. The lack of readily available training programs and skilled labor is a significant barrier to scaling these technologies across different regions (Singh *et al.*, 2021).
- **Infrastructure and Maintenance:** Scaling up requires substantial infrastructure investments, such as controlled environment facilities, reliable power supply, and advanced irrigation systems. Maintaining these facilities, especially in rural or remote areas, can be challenging due to inconsistent electricity supply, water availability, and limited access to spare parts for repairs. The reliability of infrastructure is critical for the success of scaling these technologies (Chakrabarti, 2021).

#### Regulatory and Policy Hurdles

- **Regulatory Compliance:** Potato seed production is subject to stringent regulations to ensure seed quality and health. These regulations include certification standards, disease-free status verification, and adherence to specific growing conditions. For new technologies like aeroponics or micropropagation, the lack of clear regulatory guidelines can slow adoption, as integrating these methods into existing certification frameworks can be complex and time-consuming (Buckseth *et al.*, 2022).
- **Policy Support:** While the importance of innovative agricultural technologies is increasingly recognized, the policy support necessary for adoption is often inadequate. Support mechanisms such as financial incentives, research funding, and extension services are lacking. Without robust policy frameworks, small and marginal farmers may struggle to adopt these technologies, missing out on productivity and disease resistance benefits (NAAS, 2021).

- **Market Access and Seed Distribution:** Efficient distribution of high-quality seed potatoes is essential for reaching farmers. Current networks may not accommodate the specific needs of seeds produced through advanced technologies, such as the storage requirements for tissue-cultured plantlets or mini-tubers. Additionally, farmers may be hesitant to adopt seed produced through non-traditional methods due to familiarity with conventional sources and perceived risks (Singh *et al.*, 2021).

## Future Research and Development

### Potential Areas for Future Research

As potato seed production continues to evolve, there are several promising areas for further research and development, focusing on genetic improvements, disease management, and optimizing seed production systems:

- **Genetic Improvements:** Future research is likely to prioritize developing new potato varieties with enhanced resistance to diseases, pests, and environmental stresses. Gene-editing technologies like CRISPR hold potential for targeted modifications to improve disease resistance and yield. Breeding efforts may also focus on varieties adapted to changing climates, such as those tolerant to heat and drought (Tiwari *et al.*, 2022).
- **Disease Management:** While significant progress has been made in producing disease-free seed tubers, challenges remain in managing viral, bacterial, and fungal diseases. Future research could explore more effective integration of biocontrol agents, such as beneficial microbes, into seed production systems. Advances in diagnostic tools for early pathogen detection would also improve the ability to maintain healthy seed stock (Buckseth et al., 2016; Singh et al., 2019b).
- **Optimization of Seed Production Systems:** As demand for high-quality seed potatoes grows, optimizing seed production systems is crucial. Research into soilless cultivation methods, such as aeroponics and hydroponics, is expected to continue, focusing on enhancing efficiency and cost reduction. Additionally, exploring true potato seed (TPS) technology, which produces seed potatoes from botanical seeds, could provide a sustainable alternative to traditional tuber-based systems, especially in regions with limited access to quality tubers (Buckseth *et al.*, 2022; Kumar *et al.*, 2019).

### Emerging Trends and Technologies

The future of potato seed production will be influenced by emerging trends and technologies aimed at enhancing efficiency, sustainability, and productivity:

- **Digital Agriculture and Precision Farming:** The integration of digital technologies, such as IoT (Internet of Things), remote sensing, and AI (Artificial Intelligence), into potato farming is expected to revolutionize seed production. These technologies enable precision farming, providing data-driven insights for monitoring crop growth, soil health, and environmental conditions. For example, drones equipped with multispectral cameras offer real-time data on crop health, allowing farmers to optimize inputs like water, fertilizers, and pesticides (Ahmad *et al.*, 2023).

- **Sustainable Practices and Circular Economy:** Sustainability will remain a key focus in potato seed production. Circular economy principles, such as recycling agricultural waste into biofertilizers or energy, and reducing the carbon footprint through efficient resource use, are gaining traction. The use of renewable energy sources, like solar and wind power, to minimize fossil fuel dependence in seed production and storage is also becoming increasingly important (Ahmad *et al.*, 2023).
- **Advances in Storage and Transportation:** Developing new storage technologies, such as controlled atmosphere storage and biocontrol agents to prevent spoilage, will continue to evolve. Research into improving cold chain logistics for seed potatoes, particularly in hot climates, is crucial for maintaining seed quality during transportation (Buckseth *et al.*, 2022).

## Conclusion

The transformation of potato seed production through innovative technologies and sustainable practices has significantly enhanced the efficiency, quality, and scalability of seed production systems. Techniques like micropropagation, aeroponics, and synthetic seeds offer high multiplication rates and disease-free planting materials, addressing the challenges posed by traditional methods. The successful application of these technologies, as seen in case studies from India and the Netherlands, emphasizes their potential to meet the growing global demand for quality potato seed.

Despite these advancements, challenges remain, particularly in scaling up these technologies due to high costs, technical expertise requirements, and regulatory barriers. To overcome these hurdles, continued investment in research, policy support, and farmer education is necessary. Future research should focus on optimizing these systems, integrating digital agriculture tools, and developing new potato varieties with improved disease resistance and climate adaptability.

By advancing seed storage methods and employing sustainable practices, such as IPM and precision irrigation, the potato industry can further reduce its environmental impact. The integration of renewable energy and circular economy principles presents an opportunity to enhance the sustainability and profitability of potato seed production.

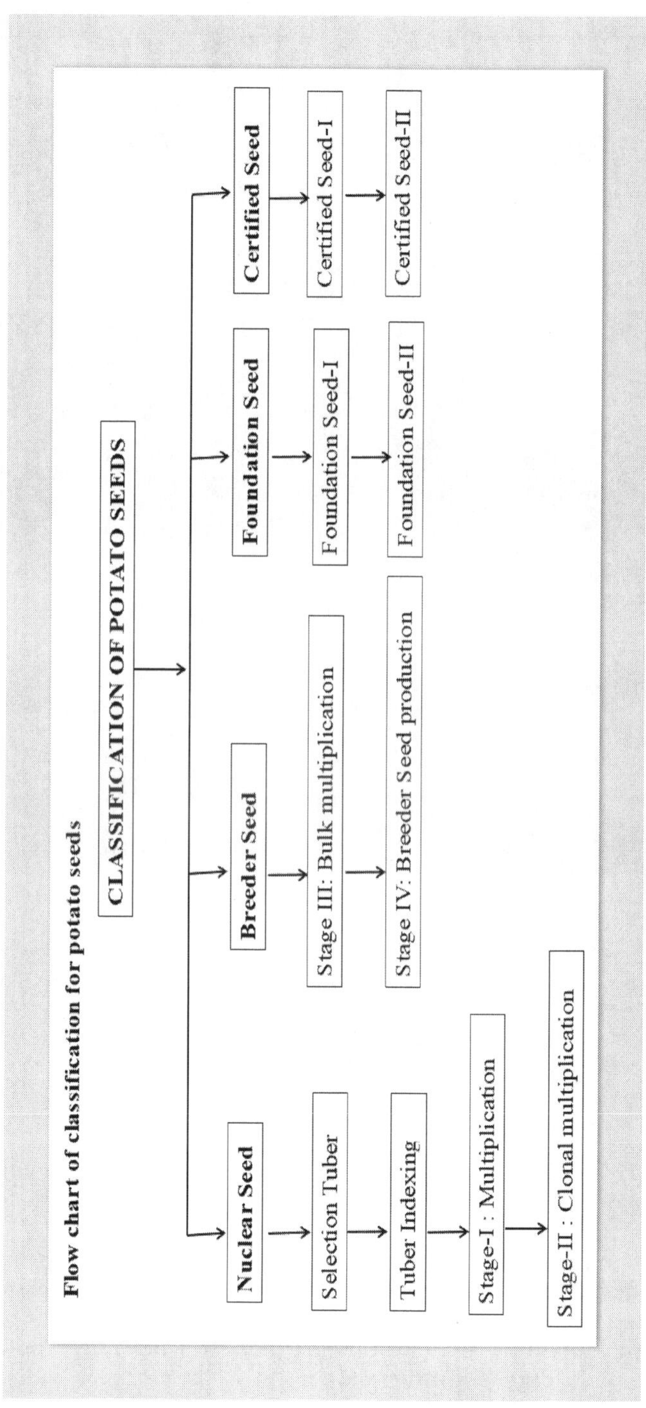


**Fig. 1. Classification of Potato Seed Stages in Seed Production Systems**

**Fig. 2: Innovations in Potato Seed Production (a) Disease-free microplants. (b) Mini-tubers directly produced from microplants in net houses. (c) Mini-tubers developed in aeroponic chambers. (d) Mini-tubers produced through aeroponics. (e) Apical rooted cuttings ready for transplanting. (f) Mini-tubers produced from apical rooted cuttings. (g) Mini-tuber crop in the field.**

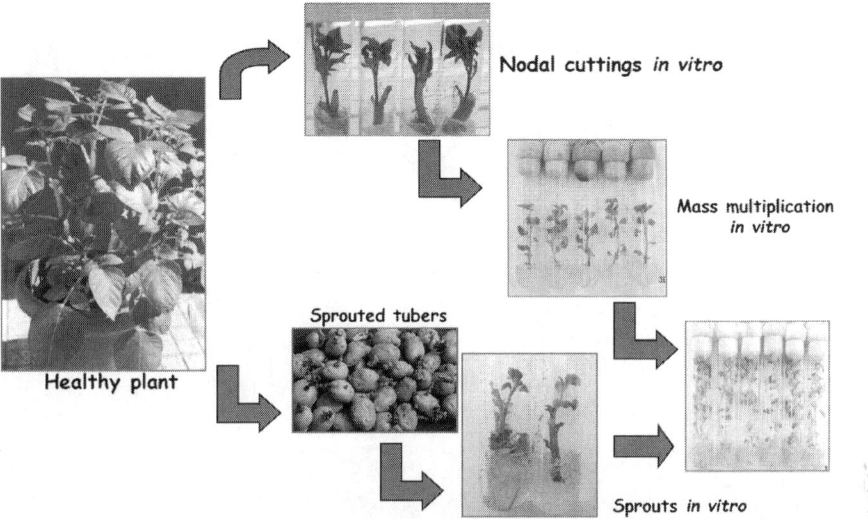


**(a) Virus elimination and meristem culture techniques for producing virus-free starting material.**

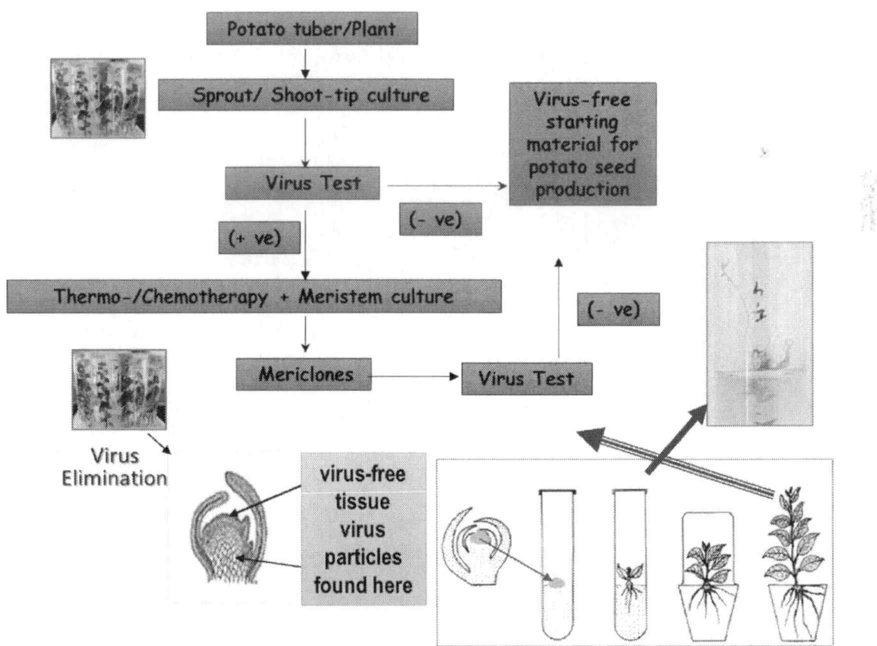


**(b) Stepwise process for mass multiplication of virus-free plants through in vitro nodal cuttings and sprouted tubers.**

**Figure 3: Micropropagation in Potato Production**

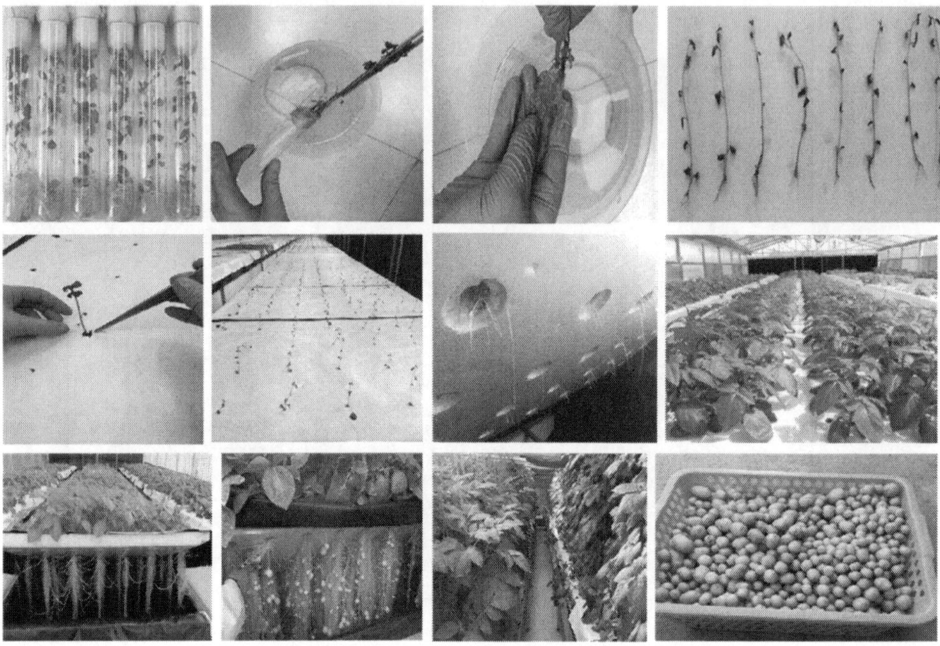

**Fig. 4: Stepwise Aeroponic-Based Seed Production System - illustrates the stepwise aeroponic process for producing mini-tubers.**

**Fig. 5: Apical Rooted Cuttings (ARC) Method in Potato Seed Production - depicts the ARC technique for potato seed production, showing the stages from rooting apical cuttings to the development of mini-tubers ready for transplanting in the field.**

## References

Ahmad, S., Mishra, S., & Sharma, V. (2023). Green computing for sustainable future technologies and its applications. In *Contemporary Studies of Risks in Emerging Technology, Part A* (pp. 241-256). Emerald Publishing Limited.

Ajay, S., & Sunaina, V. (2005). Direct inhibition of *Phytophthora infestans*, the causal organism of late blight of potato by *Bacillus* antagonists. *Potato Journal, 32*(3–4), 179–180.

Almenar, E. (2020). Innovations in packaging technologies for produce. In *Controlled and Modified Atmospheres for Fresh and Fresh-Cut Produce* (pp. 211-264). Academic Press.

Axel, C., Zannini, E., Coffey, A., Guo, J., Waters, D. M., & Arendt, E. K. (2012). Ecofriendly control of potato late blight causative agent and the potential role of lactic acid bacteria: A review. *Applied Microbiology and Biotechnology, 96*, 37-48.

Boersig, M. R., & Wagner, S. A. (1988). Hydroponic system for production of seed tubers. *American Potato Journal, 65*, 470–471.

Boonham, N., Kreuze, J., Winter, S., van der Vlugt, R., Bergervoet, J., Tomlinson, J., & Mumford, R. (2014). Methods in virus diagnostics: From ELISA to next generation sequencing. *Virus Research, 186*, 20-31.

Bryan, J., Jackson, M., & Meléndez, N. (1981). Rapid multiplication techniques for potatoes. Lima: International Potato Center (CIP).

Buckseth, T., Singh, R. K., Challam, C., Tiwari, J. K., & Chakrabarti, S. K. (2019). Apical root cutting: A novel technique for the production of quality seed potato. *Indian Horticulture, 65*(6), 63–64.

Buckseth, T., Sharma, A. K., Pandey, K. K., Singh, B. P., & Muthuraj, R. (2016). Methods of pre-basic seed potato production with special reference to aeroponics—a review. *Scientia Horticulturae, 204*, 79–87.

Buckseth, T., Sharma, A. K., Pandey, K. K., Singh, B. P., & Muthuraj, R. (2022). Advances in innovative seed potato production systems in India. *Frontiers in Agronomy, 4*, 956667.

Buckseth, T., Singh, R. K., Sharma, S., & Chakrabarti, S. K. (2020). Training-cum-laboratory manual on hi-tech seed potato production. ICAR-Central Potato Research Institute, Shimla, Himachal Pradesh, India.

Buckseth, T., Singh, R. K., Tiwari, J. K., Sharma, A. K., Singh, S., & Chakrabarti, S. K. (2020). A novel sustainable aeroponic system for healthy seed production in India—an update. *Indian Journal of Agricultural Sciences, 90*(2), 243–248.

Chakrabarti, S. K. (2021). *Advancements in Potato Production Technology*. ICAR-Central Potato Research Institute.

Chakrabarti, S. K., Singh, R. K., Buckseth, T., Sharma, A. K., & Kumar, M. (2017). Seed potato production in India. In *Current Horticulture: Improvement, Production, Plant Health Management, and Value-Addition* (pp. 78-85). ICAR Publications.

Ghanbarali, S., Abdollahi, M. R., Zolnorian, H., Moosavi, S. S., & Seguí-Simarro, J. M. (2016). Optimization of the conditions for production of synthetic seeds by encapsulation of axillary buds derived from minituber sprouts in potato (*Solanum tuberosum*). *Plant Cell, Tissue and Organ Culture (PCTOC, 126*, 449-458.

Ghosh, P. D., & Sen, S. (1994). Synthetic seeds for micropropagation of potato. *Journal of Plant Physiology, 144*(4), 527-531.

Goffart, J. P., Haverkort, A., Storey, M., Haase, M., Martin, M., Lebrun, P., Ryckmans, S., Florins, D., & Demeulemeester, K. (2022). Potato production in northwestern Europe (Germany, France, the Netherlands, United Kingdom, Belgium): Characteristics, issues, challenges, and opportunities. *Potato Research, 65*, 503-547.

Haverkort, A. J., & van der Zaag, D. E. (1989). Innovative techniques in seed potato production in the Netherlands (No. 124). CABO.

Jansky, S. H., & Spooner, D. M. (2018). The evolution of potato breeding. *Plant Breeding Reviews, 41*, 169-214.

Kumar, R., Tiwari, R. K., Jeevalatha, A., Kaundal, P., Sharma, S., & Chakrabarti, S. K. (2019). Potato viruses and their diagnostic techniques: An overview. *Journal of Pharmacognosy and Phytochemistry, 8*(6), 1932-1944.

Louwaars, N. P. (2023). Potato in the Netherlands–a remarkable story. In *Impact of Hybrid Potato* (pp. 45-58). Wageningen Academic.

Mehta, A., Singh, B., Ezekiel, R., & Minhas, J. S. (2014). Processing quality comparisons in potatoes stored under refrigerated and non-refrigerated conditions. *Indian Journal of Plant Physiology, 19*, 149-155.

National Academy of Agricultural Sciences (NAAS). (2021). *Innovations in Potato Seed Production, Strategy Paper No. 14*. National Academy of Agricultural Sciences, New Delhi.

Naik, P. S., & Buckseth, T. (2018a). Innovations in seed potato production: Challenges and future prospects. In *Current Horticulture: Improvement, Production, Plant Health Management, and Value-Addition* (pp. 78-85). ICAR Publications.

Naik, P. S., & Karihaloo, J. L. (2007). *Micropropagation for Production of Quality Potato Seed in Asia-Pacific*. Asia-Pacific Consortium on Agricultural Biotechnology (APCoAB), New Delhi, India.

Naik, P. S., & Sarkar, D. (1997). Synthetic seed production from encapsulated somatic embryos of potato. *Plant Cell Reports, 16*(6), 367-371.

Naik, P. S., & Buckseth, T. (2018b). Micropropagation and its role in potato seed production. *Journal of Agricultural Biotechnology, 21*(4), 345-352.

Naik, P. S., & Sarkar, D. (2001). Microtubers: Innovative propagules for virus-free potato seed production. In S. Nagarajan & D. P. Singh (Eds.), *Role of Resistance in Intensive Agriculture* (pp. 12-28). Kalyani Publishers.

Nyende, A. B., Schittenhelm, S., Mix-Wagner, G., & Greef, J. M. (2003). Production, storability, and regeneration of shoot tips of potato (*Solanum tuberosum* L.) encapsulated in calcium alginate hollow beads. *In Vitro Cellular and Developmental Biology-Plant, 39*, 540-544.

Pandey, K. K., & Singh, B. P. (2014). Effect of bower system on minituber yield of different varieties in aeroponic system. In *Abstract in National Seminar on Emerging Problems of Potato* (p. 193–194). Indian Potato Association.

Pushkarnath. (1959). Producing healthy seed potatoes in the plains: A new approach. *Indian Potato Journal, 1*, 63-72.

Pushkarnath, & Nirula, K. K. (1970). Aphid-warning for production of seed potato in subtropical plains of India. *Indian Journal of Agricultural Sciences, 40*, 1061-1070.

Raemakers, C. J., Jacobsen, E., & Visser, R. G. (1995). Synthetic seeds for encapsulation and regeneration of potato plants. *Plant Cell Reports, 14*(4), 268-271.

Raigond, B., Verma, A., Roach, S., Kochhar, T., Shilpa, Arjunan, J., & Chakrabarti, S. K. (2019). One-step reverse transcription loop-mediated isothermal amplification: A simple, sensitive and rapid assay for detection of potato virus X in potato leaves and tubers. *Indian Phytopathology, 72*, 321-328.

Rajendran, S., Domalachenpa, T., Arora, H., Li, P., Sharma, A., & Rajauria, G. (2024). Hydroponics: Exploring innovative sustainable technologies and applications across crop production, with emphasis on potato mini-tuber cultivation. *Heliyon, 10*, e26823.

Sarkar, D., & Naik, P. S. (1998). Synseeds in potato: An investigation using nutrient-encapsulated in vitro nodal segments. *Scientia Horticulturae, 73*(2-3), 179-184.

Sarkar, D., Naik, P. S., & Buckseth, T. (2011). Virus elimination through meristem culture in potato. *Plant Cell Tissue and Organ Culture, 105*(1), 99-107.

Sharma, A. K., Singh, S. P., Kumar, D., Singh, V., Buckseth, T., & Singh, R. K. (2021). Seed plot technique. In *Potato in Subtropics: A Saga of Success* (pp. 34-49). International Books Periodical Supply Services.

Singh, B., Raigond, P., Jaiswal, A. K., & Kumar, D. (2020). Potato post-harvest management strategies. In Singh, A. K., Chakrabarti, S. K., Singh, B., Sharma, J., & Dua, V. K. (Eds.), *Potato Science and Technology in Sub-Tropics* (pp. 343–360). NIPA. ISBN: 978-93-89571-93-6.

Singh, R. K., Buckseth, T., Tiwari, J. K., Sharma, A. K., & Singh, V. (2019a). Seed potato production systems in India: A chronological outlook. *Indian Journal of Agricultural Sciences, 89*(4), 578–587.

Singh, R. K., Tiwari, J. K., & Buckseth, T. (2019b). Clean seed programs: Ensuring virus-free seed production. *Journal of Plant Pathology, 45*(2), 123-130.

Singh, R. K., Tiwari, J. K., Buckseth, T., & Kumar, M. (2021). Potato production: Developments and challenges. In Singh, B., Tomar, B. S., Ranjan, J. K., & Dutt, S. (Eds.), *Current Horticulture: Improvement, Production, Plant Health Management, and Value-Addition* (pp. 78-85). ICAR Publications.

Sood, S., Bhardwaj, V., Kumar, V., Das, R., Gupta, V. K., Mangal, V., & Singh, A. K. (2022). Genetic progress in 50 years of potato breeding in India: Where do we stand? *European Journal of Agronomy, 141*, 126620.

Stegli ska, A., Kołtuniak, A., Berłowska, J., Czy owska, A., Szulc, J., Cieciura-Włoch, W., Okrasa, M., Kr giel, D., & Gutarowska, B. (2022). *Metschnikowia pulcherrima* as a biocontrol agent against potato (*Solanum tuberosum*) pathogens. *Agronomy, 12*(10), 2546.

Tiwari, J. K., Buckseth, T., Challam, C., Zinta, R., Bhatia, N., Dalamu, D., & Kumar, M. (2022). CRISPR/Cas genome editing in potato: Current status and future perspectives. *Frontiers in Genetics, 13*, 827808.

Tiwari, J. K., Buckseth, T., Singh, R. K., & Chakrabarti, S. K. (2019). Aeroponics: A novel precision system to study nutrient biology in potato. *Indian Horticulture, 64*(5), 33-34.

Venkataslam, P., et al. (2017). Challenges in potato seed production. *Journal of Potato Research, 45*(2), 123-130.

Vinterhalter, D., Dragicevic, I., & Vinterhalter, B. (2008). Potato in vitro culture techniques and biotechnology. *Fruit, Vegetable and Cereal Science and Biotechnology, 2*(1), 16-45.

Yahia, E. M., Fadanelli, L., Mattè, P., & Brecht, J. K. (2019). Controlled atmosphere storage. In *Postharvest Technology of Perishable Horticultural Commodities* (pp. 439-479). Woodhead Publishing.

# CHAPTER 18

# MEDICINAL PLANTS: LINKING WITH CHALLENGES, OPPORTUNITIES AND SUSTAINABLE PRACTICES

**Manish Das[1*] and Saravanan Raju[2]**

*[1]ICAR-Directorate of Medicinal and Aromatic Plants Research (ICAR-DMAPR), Boriavi, Anand, Gujarat, India*

*[2] ICAR-Central Tuber Crops Research Institute, Thiruvananthapuram, Kerala, India*

*Corresponding author: *manishdas50@gmail.com*

---

A profound thought process prevails involving medicinal plants for its universal usage and value-based expansion. A comprehensive examination of the medicinal plants sector in India needs to be dealt with in terms of its existence, depletion from its natural existence, its conservation requirement, augmentation, cultivation and its medicinal and commercial application. It begins by exploring the historical significance of medicinal plants in the country and their vital role in traditional and modern healthcare systems. To begin with there is need to understand its economic importance in relation to industry application. The challenges faced by the sector, including biodiversity conservation issues, quality control and standardization challenges, regulatory obstacles, and market dynamics need to be understood and dealt with. On the positive side, it also uncovers opportunities, such as the growing global demand for natural products, expanding international trade possibilities, advancements in research and development, and promoting sustainable agricultural practices. This chapter provides a valuable insight into the complexities and prospects of India's medicinal plants with regard to industry.

Medicinal plants in India are universal and known since time immemorial spanning millennia, showcasing the deep connection between humanity and the plant world (Jamshidi-Kia *et al.*, 2017). Our understanding has evolved into a sophisticated cultivation and usage of traditional medicine system. Significantly it started with the emergence of Ayurveda, a 5,000-year-old healing system (Prakash, 2019). Often referred to as the "Science of Life," Ayurveda strongly emphasizes inner balance, viewing health as a harmonious equilibrium of three elemental and constitutional substances known as doshas. There is need to understand other components of Indian System of medicine like India has the unique distinction of having six recognized systems of medicine in this category. They are Ayurveda, Siddha, Unani and Yoga, Naturopathy and Homoeopathy (Ravishankar and Shukla, 2007; Sen *et al.*, 2011). Gradually this heritage has gained global recognition as it attracts everybody's attention world over for natural and traditional remedies, India being a natural custodian of MAPs.

## Synergy between traditional and modern Systems

Medicinal plants are linked with wisdom and modern science which is so vital in India's healthcare, (Rupani and Chavez, 2018). Ayurveda, for example, focuses on doshas (Vata, Pitta, Kapha), using medicinal plants to harmonize these elements for wellness. This becomes very important in the contemporary era and medicinal plants enjoin the mainstream scientific research and pharmaceutical development (Dubey *et al.*, 2004). They hold promise for drug discovery and herbal pharmaceuticals, merging ancient wisdom with modern healthcare. These plants contribute to nutraceuticals, herbal pharmaceuticals, and herbal cosmetics, addressing the evolving healthcare needs (Sharma *et al.*, 2008). They stand at the intersection of tradition and innovation, preserving health, preventing disease, and advancing therapeutics, showcasing their enduring importance in our evolving healthcare systems.

## Medicinal Plants growing globally

An estimate reveals that the global medicinal plants industry has rapidly expanded and was valued at approximately USD 165.66 billion in 2022 (Grand View Research, 2023). It is grown due to the increasing demand for natural and organic products, the rise in chronic diseases, resulting into greater acceptance of traditional medicine. A projection indicates that by 2030, the industry could reach an impressive USD 347.50 billion, with a remarkable CAGR of 11.16% (Fortune Business Insights, 2023).

There is a clearcut increased trends in industry demand that encompasses the development of new plant-based medicines linking traditional healing into modern healthcare and the incorporating of medicinal plants into food and beverages, and a strong emphasis on sustainable sourcing. Trade patterns reveal distinctions, as China primarily exports to neighbouring countries, while India engages in long-distance trade, notably with the US and Germany. Countries like Luxembourg, Iceland, and Tanzania, which have shown substantial export growth also.

## Aligning medicinal plants with sustainability

There is no doubt that anything done with medicinal plants are going to benefit India's medicinal plants industry and thereby providing livelihoods to many, including farmers, herbalists, and manufacturers. Notably India's exports, encompassing raw herbs, extracts, essential oils, and herbal formulations has been appreciated (Dubey *et al.*, 2004). India's commitment to sustainable agriculture is promoting the cultivation of medicinal plants through eco-friendly practices (Das *et al.*, 2016). India's medicinal plants industry stands sound for further growth and driving for economic development.

India's rich biodiversity and centuries-old tradition of using medicinal flora position the country become as a global leader in the herbal medicine market. This sector includes many products from herbal medicines and health supplements to cosmetics and nutraceuticals. With the global preference for natural and sustainable healthcare solutions on the rise, the economic value of medicinal plants has seen as an exponential growth. India's exports of Ayurvedic and herbal products have generally been on an upward trajectory, with exceptions such as the impact of the COVID-19 pandemic in 2020. In the fiscal year 2022-23, India exported these products worth US$628.25 million, marking a 2.6% increase from the previous year (Statista, 2023). Key export destinations include the United States, the United Arab Emirates, the United Kingdom, Germany, and Singapore. This growth can be attributed to increasing awareness of traditional medicine benefits, rising disposable incomes, and a growing demand for natural and organic products. Consequently, sufficient care with regard to financial support, trade events, and research and development initiatives, are expected to boost the ayurvedic and herbal sector. We need to capitalise on this aspect for accelerating the growth pattern to sustain Ayurveda heritage and abundant medicinal plant resources.

## Sustainability challenges in MAPs

The major challenge is biodiversity conservation: Following are the challenges under this sub-head:

- **Over-exploitation:** Excessive harvesting of medicinal plants driven by global demand for herbal remedies. Traditional medicines like Ayurveda and Unani have deeply rooted traditions in India, creating immense pressure on medicinal plant populations. Unregulated harvesting and extraction can lead to population depletion and even the extinction of rare species (Kala et al., 2006).
- **Habitat loss:** This disrupts the delicate ecological balance supporting these plant species, endangering the plants and the rich biodiversity dependent on them (Kala et al., 2006).
- **Unscientific harvesting practices:** Demand for medicinal plants now exceeds their capacity. Unscientific harvesting practices, like uprooting entire plants instead of selectively harvesting leaves or fruits, exacerbate the threat to rare

species. It amounts to disruption natural regeneration and complete loss of the resource (Kala et al., 2006).

- **Lack of regulation:** Stringent guidelines and enforcement make conserving endangered species easier. Medicinal plant industry often needs comprehensive regulation despite its economic value. It has resulted in unsustainable harvesting, adulteration of herbal products, and poor quality control (Sharma and Thokchom, 2014).
- **Biopiracy:** Biopiracy has become an important factor for consideration, where foreign entities exploit India's biodiversity for profit without fair compensation or benefit-sharing. India's stand of its natural resources needs strengthening for strong footage (Sen et al., 2011).

## Impact of Over-Harvesting and Unscientific Methods

Impact of over-harvesting and unscientific methods could be serious once it is not controlled or not taken into cognization. We need to consider factors such as (i) Habitat destruction, (ii) Depletion of natural population, (iii) Unsustainable extraction, (iv) Climate change and (v) Biodiversity loss. The consequences of these challenges extend to the broader issue of biodiversity loss. As medicinal plants face habitat degradation and over-harvesting, the delicate balance of ecosystems is further destabilized, impacting various species within these ecosystems (Das *et al.*, 2016).

## Quality Control and Standardization

There is need to assess the quality and effectiveness in herbal medicine which depend heavily on robust quality control and standardization procedures. The herbal medicine sector faces a pressing challenge due to the need for standardized plant identification and processing protocols. This gap undermines the uniformity and trustworthiness of herbal products, with significant implications for public health. Also, accurate identification of medicinal plants is a fundamental step in herbal medicine production. Unfortunately, the absence of universally accepted standards for plant identification leads to discrepancies and inconsistencies (Sahoo *et al.*, 2010). Adding complexity is the prevalent taxonomic confusion within medicinal plants. This confusion needs to be scientifically solved for authentication. Beyond identification, the processing of medicinal plants introduces another layer of non-standardization. Achieving consistency and control in herbal medicine characteristics becomes challenging without standardized procedures.

Further, the absence of standardized procedures also exposes vulnerabilities to adulteration and contamination. Unscrupulous practices, such as incorporating unrelated plant material or using pesticides and contaminants, pose significant risks to the safety and effectiveness of herbal products, endangering public health (Nafiu *et al.*, 2017).

## Adulteration and Contamination in Herbal Products

There are challenges related to adulteration and contamination (Posadzki *et al.*, 2013; Ernst, 2003; Hoban *et al.*, 2018). Adulteration can manifest in various forms, from unintentional incidents to deliberate actions for economic gain. On the other hand, contamination arises from factors such as pesticide residues, heavy metals, and microbial contaminants (Wojcikowski *et al.*, 2004; Zhang *et al.*, 2012). Consumer health risks include allergies, poisonings, and infections, making it an urgent public health concern. Beyond health implications, these issues erode consumer trust in herbal products and pose regulatory challenges to ensure adherence to safety and efficacy standards (Ernst, 2002).

Advancements made in analytical techniques have revealed the extent of adulteration and contamination in herbal products (Ichim, 2019). Also, DNA-based authentication of commercial herbal products has unveiled widespread adulteration on a global scale. Simultaneous determinations of various phytochemicals are essential for herbal products quality control as they ensure the comprehensive assessment of multiple bioactive compounds, enabling a more accurate evaluation of the product's efficacy and safety (Gajbhiye *et al.*, 2016; Makasana *et al.*, 2023). Consequently, the integration of DNA sequencing techniques has become instrumental in uncovering cases of adulteration that are not easily detectable through chemical analysis alone. This innovative approach provides a valuable tool for quality control and monitoring efforts in the herbal medicine industry, enabling a more comprehensive assessment of product authenticity and safety (Ichim, 2019).

### Regulatory Hurdles

To guarantee safety, quality, and effectiveness, particularly given the expanding herbal medicinal products market regulating herbal drugs is essential. A significant challenge in regulating herbal drugs requires stricter guidelines for assessing safety, efficacy, quality control, safety monitoring, and traditional medicine/ complementary and alternative medicine knowledge. Since establishing regulatory norms for herbal drugs is a complex and evolving process, it necessitates addressing categorization, quality control, safety monitoring, and knowledge dissemination challenges. Standardizing definitions and regulatory frameworks, alongside international harmonization, are pivotal steps in ensuring the safety and efficacy of herbal medicines.

## Compliance with International Regulations for herbal exports

India's role for complying with international regulations for exporting herbal products poses a complex challenge, particularly for significant producers. We need to harmonise regulation with countries like Europe and the United States (Verma, 2013). Each region has its distinct regulatory framework, creating hurdles for exporters striving to meet diverse compliance requirements. A global standardized regulations is the need of the hour for herbal medicinal products.

## Steering with opportunities in Medicinal Plants Sector

Recent research trends reveal a renewed interest in medicinal plants worldwide, particularly in countries like China and India, where traditional medicine rooted in these plants has a deep history (Salmerón-Manzano *et al.*, 2020). With a vast array of medicinal plants, India is well-positioned to capitalize on this trend (Shakya, 2016). As the world increasingly embraces natural remedies for health and well-being, India's abundance of medicinal plants becomes a significant advantage. The utilization of traditional remedies and the promotion of herbal medicines have gained momentum and global attention. Medicinal plants can be cultivated sustainably, contributing to biodiversity conservation and offering ecological benefits. The global trend toward herbal and natural remedies thus gives India a unique opportunity to leverage its rich heritage of medicinal plants. India can position itself as a leading supplier of high-quality herbal products while championing sustainable and eco-friendly practices.

## Nation's international trade and exports

India's medicinal plants hold great promise for the nation's herbal industry. India has a rich tradition of using these plants for therapeutic purposes, and its diverse biodiversity boasts numerous plant species with potential medicinal value. The global market demands consistent and high-quality products. Therefore, Indian herbal exporters must prioritize the followings: Good Agricultural and Collection Practices (GACP) and Good Manufacturing Practices (GMP). These practices help maintain the purity and effectiveness of herbal products, aligning with WHO guidelines on herbal medicine (Vasisht *et al.*, 2016). Another critical consideration involves documenting traditional knowledge and complying with intellectual property rights. Many medicinal plant species in India have been integral to local communities' traditions for generations. Therefore, it is vital to document and patent indigenous herbal remedies while respecting the rights of indigenous communities (Zahra *et al.*, 2020) to protect against biopiracy and preserve traditional knowledge.

## Importance of traditional knowledge in promoting exports

For promotion of exports of medicinal plants and herbal products, there is need to understand traditional knowledge (Keswani *et al.*, 2017). It represents the accumulated wisdom, practices, and expertise passed down through generations within indigenous communities. Moreover, traditional knowledge encompasses sustainable harvesting practices developed over centuries. Indigenous communities have devised methods to ensure that the harvesting and cultivation of medicinal plants do not deplete natural populations (Bhat *et al.*, 2013).

## Focussing on research and development

Focus is on innovation in herbal drug discovery and development in recent years, offering exciting opportunities for both the pharmaceutical industry and herbal medicine practitioners. A deepening understanding of plant-based natural product drug discovery and advancements in various scientific and technological fields drives these innovations. Additionally, the convergence of herbal and precision medicine enables personalized herbal treatments based on individual genetic and biomarker information, enhancing their effectiveness and safety (Govindaraghavan and Sucher, 2015).

The natural products, including those derived from medicinal plants, continue to be valuable sources of therapeutic agents for drug discovery. Biosynthesis of molecules are vital to the researchers screening of natural products and harnessing their potential for innovative drug development. To address the innovation gap in the pharmaceutical industry, reverse pharmacology and systems approaches are employed which integrate traditional and alternative medicines with advanced scientific techniques, facilitating a more comprehensive and efficient exploration of herbal remedies (Jordan *et al.*, 2010).

An interplay for collaborations involving traditional healers, biomedical health practitioners, and research institutions in medicinal plants play a crucial role in the holistic development of herbal medicine. Role of traditional healers become more important who use the indigenous knowledge, towards making significant contributions to plant collection and bridging the gap between research areas. Such collaborations and partnerships can promote sustainable practices without harming wild plant populations. Using this approach an understanding between scientific research and practical clinical applications could be realised and promote the evidence-based development of herbal medicine. Consequently, it helps in preserving cultural heritage while upholding intellectual property rights.

## Sustainable Agriculture Practices

Sustainable agriculture practices play a pivotal role in the cultivation of medicinal plants. The adoption of organic farming and agroforestry methods is rising, emphasizing eco-friendly and chemical-free approaches. These practices not only ensure the production of best-quality medicinal herbs but also contribute to the preservation of biodiversity and the prevention of environmental degradation. Organic certifications can ensure that medicinal plants are grown using environmentally sustainable methods, without the use of chemical pesticides or fertilizers and promoting the production of clean and natural herbal products. These certifications enhance the quality and sustainability of medicinal plant cultivation and meet the increasing global demand for ethical and environmentally responsible herbal products.

## Conservation Efforts

Conserving medicinal plants is paramount to preserve the valuable resources and their potential benefits for humanity. One crucial strategy is the establishment of botanical gardens and ex-situ conservation programs. India, recognized as a biodiversity treasure house that substantially contributed to medicinal plant conservation. The country's botanical gardens and seed conservation have played a pivotal role in conserving of India's medicinal plant species (Das and Trivedi, 2021). Conservation efforts extend beyond preservation alone and encompass sustainable utilization. Innovative approaches, including *in vitro* technology, facilitate the sustainable use of medicinal plants while minimizing the impact on wild populations.

Efforts such as participatory restoration involving common lands holds promise for biodiversity conservation and socio-economic development. Such initiatives often involve planting medicinal plants and valuable species (Rawat *et al.*, 2017). By integrating the expertise of botanical gardens and the knowledge of ethnopharmacology into reforestation and habitat restoration programs, these endeavours gain scientific credibility and enhance the selection of appropriate plant species. These integrated efforts are pivotal for addressing ecological challenges, preserving biodiversity, and promoting holistic sustainable development.

## Recommendations for strengthening MAP sector

There is need to invest more in research and development with focus is required on new plant species for conservation and cultivation. Rigorous quality control measures and certifications like GACP and GMP can enhance product quality and safety and need consideration. Sustainable sourcing practices, promoting biodiversity conservation, and ensuring fair compensation for local communities need consideration and implementation. Targeting on research through market driven approach to adapt to consumer preferences, and building strong ties with regulatory authorities is must. Digitalization platform has to be grabbed for international outreach, capacity building, understanding local communities, and to collaborate with other similar agencies. Understanding the trade and engaging with export promotion councils, diversifying product ranges and value-added items is essential. All these aspects need to be thoroughly understood for implementing approaches to MAP cultivation and trading for better and improved linkage. These strategies are going to strengthen India's position in the global herbal and medicinal export market.

## Conclusion

Medicinal and aromatic plants are most vital entity in the growth status of Indian economy and offers great potential while facing challenges like habitat degradation and biopiracy. A sustainable approach is vital to address this, involving sustainable cultivation, conservation, and ethical trade. This also calls for international standards, ensuring quality, and preserving traditional knowledge. Collaborations between academia, industry and traditional practitioners show promise through innovations in herbal drug discovery. It is altogether the responsibility involving all stakeholders: governments for creating a conducive environment for sustainable policies, investment in research and ethical sourcing, and tapping of potential for growth. Such diverse action with strong policies and collaboration is critical to the overall growth of MAP sector for sustainable conservation and cultivation towards prosperous industry.

## References

Bhat JA, Kumar M and Bussmann RW (2013). Ecological status and traditional knowledge of medicinal plants in Kedarnath Wildlife Sanctuary of Garhwal Himalaya, India. *J. Ethnobiol. Ethnomed.*, 9:1-18.

Das Manish, Jain Vanita and Malhotra S K (2016). Impact of climate change on medicinal and aromatic plants. *Indian Journal of Agricultural Sciences,* 86:1375-1382.

Das Manish and Trivedi A P (2021). Essentiality of seed standards for cultivation of medicinal plants - a review. *Current Horticulture*, 9:9-16.

Dubey NK, Kumar R, Tripathi P and Upadhyay RK (2004). Global promotion of herbal medicine: India's opportunity. *Curr. Sci.*, 86:37–41.

Ernst E (2003). Cardiovascular adverse effects of herbal medicines: a systematic review of the recent literature. *Can. J. Cardiol.*, 19:818–827.

Fortune Business Insights (2023). Medicinal plants market size, share, growth analysis, regional trends, and key players forecast., 2023-2030.

Gajbhiye NA, Makasana J, Dhanani T and Saravanan R (2016). Development and validation of LC—ESI—MS/MS method for simultaneous determination of four coumarin derivatives and an alkaloid from root and stem bark of *Aegle marmelos* Correa. *Acta Chromatogr., 28*:473-488.

Govindaraghavan S and Sucher NJ (2015). Quality assessment of medicinal herbs and their extracts: Criteria and prerequisites for consistent safety and efficacy of herbal medicines. *Epilepsy & Behavior., 52*: 363-371.

Grand View Research (2023). Medicinal plants market size, share & trends analysis report by product, by application, by distribution channel, by region and segment forecasts., 2023 - 2030.

Hoban CL, Musgrave IF, Coghlan ML, Power MW, Byard RW, Nash C and Bunce M (2018). Adulterants and contaminants in psychotropic herbal medicines detected with mass spectrometry and next-generation DNA sequencing. *Pharm. Med.*, 32:429-444.

Ichim MC (2019). The DNA-based authentication of commercial herbal products reveals their globally widespread adulteration. *Front. in Pharmacol.*, 10: 1227.

Jamshidi Kia F, Lorigooini Z and Amini-Khoei H (2017). Medicinal plants: Past history and future perspective. *J. Herbmed Pharmacol.*, 7:1–7.

Jordan SA, Cunningham DG and Marles RJ (2010). Assessment of herbal medicinal products: challenges, and opportunities to increase the knowledge base for safety assessment. *Toxicology and applied pharmacology.*, 243(2): 198-216.

Kala CP, Dhyani PP and Sajwan BS (2006). Developing the medicinal plants sector in northern India: Challenges and opportunities. *J. Ethnobiol. Ethnomed.*, 2: 1-15.

Keswani C, Bisen K, Singh SP and Singh HB (2017). Traditional knowledge and medicinal plants of India in the intellectual property landscape. *Med. Plants-Int. J. Phytomed. Related Ind.*, 9:1–11.

Makasana J, Gajbhiye N, Kumar Bishoyi A, Savaliya M, Raju S, Bansod S, and Dholakiya B (2023). Exploration of HPTLC Technology for Rapid Chemical Fingerprinting and Simultaneous Determination of Bioactive Constituents from *Clitoria ternatea* Linn. *Chem Select.*, 8(13): e202203217.

Nafiu MO, Hamid AA, Muritala HF and Adeyemi SB (2017). Preparation, Standardization, and quality control of medicinal plants in Africa. *Med. Spices Veg. Afr.*, 171-204.

Posadzki P, Watson L and Ernst E (2013). Contamination and adulteration of herbal medicinal products (HMPs): an overview of systematic reviews. *Eur. J. Clin. Pharmacol.*, 69:295-307.

Prakash B (2019). Use of purified mercury in ayurveda and its safety evaluation. *VCP Cancer Research Foundation.*, Dehradun, Uttrakhand, India, 1-12.

Ravishankar B and Shukla V J (2007). Indian Systems of Medicine: A Brief Profile. *Afr J of Tradit Compliment Altern Med.*, 4(3): 319–337.

Rawat LS, Maikhuri RK, Dhyani D, Bahuguna YM and Pharswan DS (2017). Ecological restoration of village common degraded land through participatory approach for biodiversity conservation and socio-economic development in Indian Himalayan Region. *Acta Ecol. Sinica.*, 37:240-252.

Rupani R and Chavez A (2018). Medicinal plants with traditional use: Ethnobotany in the Indian subcontinent., *Clin. Dermatol.*, 36:306-309.

Sahoo N, Manchikanti P and Dey S (2010). Herbal drugs: standards and regulation. *Fitoterapia.*, 81:462-471.

Salmerón-Manzano E and Garrido-Cardenas JA (2020). Worldwide research trends on medicinal plants. *Int. J. Environ. Res. Public Health.*, 17:33-76.

Sen S, Chakraborty R and De B (2011). Challenges and opportunities in the advancement of herbal medicine: India's position and role in a global context. *J. Herbal Med.*, 1:67-75.

Sharma RK, Dash SK and Sharma S (2008). Medicinal plants for livelihood security and biodiversity conservation. *Indian J. Tradit. Knowl.*, 7:327–334.

Sharma S and Thokchom R (2014). A review on endangered medicinal plants of India and their conservation. *J. Crop Weed.*, 10:205–218.

Statista (2023). India: Value of ayurvedic and herbal products exports 2023. Available at: https://www.statista.com/statistics/652418/export-value-of-ayurvedic-and-herbal-products-india.

Vasisht K, Sharma N and Karan M (2016). Current perspective in the international trade of medicinal plants material: an update. *Curr. Pharm. Des.*, 22:4288–4336.

Verma N (2013). Herbal medicines: regulation and practice in Europe, United States and India. *Int J Herb Med.*, 1:1–5.

Wojcikowski KE, Johnson DW and Gobe G (2004). Medicinal herbal extracts–renal friend or foe? Part one: The toxicities of medicinal herbs. *Nephrology.*, 9:313-318.

Zahra W, Rai SN, Birla H, Singh SS, Rathore AS, Dilnashin H and Singh SP (2020). Economic importance of medicinal plants in Asian countries. *Bioeconomy for sustainable development.*, 13:359-377.

Zhang J, Wider B, Shang H, Li X and Ernst E (2012). Quality of herbal medicines: challenges and solutions. Complement. *Ther. Med.*, 20:100-106.

# CHAPTER 19

# NATIONAL AND EXPORT MARKET: DRIVEN FRUIT FLY MANAGEMENT IN MANGO

**Abraham Verghese[1] and M.A Rashmi[2]**

[1]*Former Director, ICAR-National Bureau of Agricultural Insect Resources, Bangalore, 560024, and Former Head, Entomology, ICAR-Indian Institute of Horticultural Research, Bangalore-560089, Karnataka*

[2]*Rashvee-International Phytosanitary Research and Services Pvt. Ltd., Bengaluru*

*Corresponding author: abraham.avergis@gmail.com*

---

Many fruit flies (Tephritidae: Diptera) are global pests that cause high losses to fruits and vegetables. They are highly mobile, capable of flying short distances but can be transported to long distances through commodities (Allwood and Drew, 1997). Besides causing direct losses in yield and marketability, they are barriers to the international trade of fresh fruits and vegetables. Climate changes too have impacted their distributions (Rashmi et al., 2020). Mainly *Bactrocera dorsalis* on mango and other *Bactrocera* spp. that attack mango have been discussed. Fruit fly management assumes significant as India is the largest producer of mangoes in the world and has a large opportunity to tap global trade. Besides, a relatively long period of availability (March-August), diversity of varieties, economic liberalization and priority to the export of fresh products have opened up the possibility of boosting mango exports from India. Both fresh mango and processed mango products are exported.

Oriental fruit fly, *Bactrocera dorsalis* (Hendel) (Diptera: Tephritidae), is a major pest of mango, particularly in commercial varieties (Verghese et al., 2004). It is distributed throughout India and undergoes hibernation during winter in North India. It is active throughout the year in southern parts of India, especially when minimum temperature surges in summer (Verghese and Devi, 1998). It is widespread in different parts of the world and is reported on more than 150 hosts. The yield loss due to *B. dorsalis* in India ranges from 1 to 31% with a mean of 16%. This fly is especially serious in south India (Verghese et al., 2002,2020).

Oriental fruit fly breeds continuously in tropical conditions. Females are capable of producing 1,200-1,500 eggs in their lifetime (1 to <3 months in the field), and population growth may be very rapid. Females lay batches of 1-20 eggs in single young mango fruits in which maggots feed, rendering the host unfit for consumption due to rotting (White and Elson-Harris, 1992). Pre-mating period varied from 15-26 days (Rashmi et al., 2020).

European and Mediterranean Plant Protection Organization (EPPO) lists *Bactrocera dorsalis* as an A1 quarantine pest (OEPP/EPPO, 1983) within Plant Health Committee of the broad category "non-European Tephritidae"; it is also of quarantine significance to Asia and Pacific Plant Protection Commission (APPPC). Southern Cone (COSAVE), Caribbean Plant Protection Commission (CPPC), Inter-African Phytosanitary Council (IAPSC), Cartagena Agreement Commission (JUNAC) and International Regional Organization for Agricultural Health (OIRSA). *Bactrocera dorsalis* is indigenous to Asia, but like other *Bactrocera* spp. is known by experience to have the potential to establish adventive populations in various other tropical areas.

Besides *B. dorsalis, B. zonata* is quite important especially in the western and northern parts of India (Appendix I, some important fruit fly of global significance).

## Integrated Fruit Fly Management

Earlier the Indian Institute of Horticultural Research (IIHR), Bengaluru, had worked out cost-effective and residue-free control strategies for integrated pest management (IPM). This was vastly improved upon subsequently by International Phytosanitary and Research Services: (IPRS recognized by GoI and DPPQ) and briefly are as follows.

Orchard and equipment sanitation from flowering, Basin raking to till tree canopy drip, placement of fruit fly traps, developed with the know-how of IIHR @ 15-20 per hectare. Traps should be fastened well on lower branches, between 2-m height, at least 30 days prior to harvest or earlier. The IPRS placement geometry should be followed. Bag with cloth-lined covers four weeks prior to harvest to prevent infestation. Ovipositional deterrence sprays consisting of Phyto-volatile Soap + Azadirachtin is to be given at 30 days prior to harvest. Bait sprays on the trunk should be given if population is high (Verghese et al., 2020).

## Post-harvest disinfestation

In India, cold treatment experiments at 8°C exposure for a week, showed zero percent infestation in Banganapalli and Totapuri mango, while Alphonso showed 10 per cent infestation. This treatment has the potential to act as an ovicide. However, one should be vary of chilling injury.

Hot water immersion quarantine treatments are relatively easy to engineer. Its use as a quarantine system was intensified after the banning of ethylene di bromide. In Mexico, this quarantine system started in 1988 for the control of fruit fly (Anastrepha ludens and A. obliqua) for mango fruits exported to USA and

Japan, and consists of immersion of fruit in water at 46.1 °C (115 °F) for 65, 75 or 90 minutes, depending on the weight of the fruit. This system is now widely used in different mango growing regions in Mexico, Central and South America, and the West Indies and India.

After quarantine treatment, fruit must be introduced to an isolated, insect-proof area to prevent re-infestation. From packaging and transport until the fruit is received at the importing end, secure measures are implemented to prevent reinfestation. After heat treatment, fruit is usually cooled with water at ambient temperature and/or forced air cooling. In Mexico, fruit is first cooled in ambient water (at temperatures of < 21 °C), packed, and then cooled with forced-air cooling.

For exports of mango, treatments for the Oriental fruit fly, *Bactrocera dorsalis* should ensure 99.99 per cent control (By 'probit 9' standards). Heat treatment of fruits should ensure this standard (Verghese et al., 2011). It has been reported that the eggs showed total mortality at temperature- time regimes of 45°C -60 min and 48°C – 45 min and 60 min. The first instar maggot was completely affected at 48°C -60 min while the second instar maggot succumbed to 100 per cent mortality at 45°C and 48 °C for 45°C min and 60 min. Further the temperature regime 47°C ± 1°C for 75-90 min was validated in different varieties namely Alphonso, Banganapally, Neelam and Totapuri. This is the most cost-effective method to disinfest exportable fresh mangoes provided it is done within 24 hrs after harvested (Verghese et al., 2011).

Hot water plants for fruit fly are commercialized in India and are about 52 units of which there are 2 registered facilities in Karnataka. The approximate cost of a one tonne capacity thermal tank will vary depending on type of material used (zinc sheet, steel, fibre glass etc.,) and source of energy (electricity/ solar). To treat one kg of fruit, the approximate cost, including labour has been reported to be less than Rs.2.0.

Vapour heat, the oldest of the quarantine heat treatment, consists of heating the fruits by moving hot air saturated with water vapour over the fruit surface. When the temperature of the mango fruit is at or below the dew point, condensation of atmospheric moisture occurs on the surface of the fruit. In this way, fruits are heated by conductive energy transfer. The heat from the fruit surface is transferred towards the fruit center. Commercial facilities are operated in Philippines, Thailand, the United States and Australia. The VHT disinfestations protocol accepted for mango access to the high-value markets in Japan include:

- 46°C fruit core temperature held for 10 min for Philippine 'Carabao' mango.
- 46.5°C fruit core temperature held for 30 min for Taiwanese 'Irwin' and 'Haden' mangoes.
- 46.5°C fruit core temperature held for 10 min for Tailand 'Nang Klang Wun' mangoes.
- 47°C fruit core temperature held for 10 min for 'Nam Doc Mai', 'Pimsen Dang' and 'Rad' mangoes.

Vapour heat treatment requires that mango be treated with vapor heat until the surface temperature is 46 °C and seed temperature reaches 46.5 °C and held that way for 10 minutes. A vapour heat treatment has also been approved for the control of the Queensland fruit fly in Kensington mango exported from Australia to Japan. The treatment consists in raising the 46.5 °C core temperature held at 10 minutes. The USA also approved a quarantine vapour heat treatment for the control of Mexican fruit fly (*Anastrepha ludens*) and other Anastrepha species in Manila mango, and for mango from Taiwan infested with the oriental fruit fly. Vapour heat treatment has also been approved for cv. Alphonso, for export to Japan from India. Through efforts of ICAR and APEDA, Vapour Heat Treatment Plants have been set up at important mango belts to ensure export to Japan. The current cost of the equipment is Rs.2-4 crores, and the equipment is supplied by Japan.

Forced hot-air, also known as high-temperature forced air, is a modification of the vapour heat treatment developed by Armstrong et al., (1989) to kill Mediterranean fruit fly, melon fly and Oriental fruit fly eggs and larvae in papaya. This is not a treatment available in India, hence economics of this are not available. Irradiation has been amply shown to have potential as a quarantine treatment.

**Number of Registered Phytosanitary Treatment providers in India**

| Facility | Numbers |
|---|---|
| Hot Water Treatment (NSPM 15) | 52 |
| Irradiation (NSPM 21) | 04 |
| Vapor Heat Treatment (VHT) (NSPM 20) | 09 |
| Heat Treatment (NSPM 18) | 02 |

## Appendix

Some examples of selected global fruit flies of quarantine importance in different countries.

***Bactrocera dorsalis* (Hendel), Oriental fruit fly**: India, Sri Lanka, Nepal, Bhutan, southern China (Sichuan, Guizhou, Hunan, Fujian, Yunnan, Guangxi, Guangdong, Hong Kong, Hainan), Burma, Thailand, Laos, Vietnam, Cambodia, Taiwan.

***Bactrocera papaya* (Drew & Hancock)**: Papaya fruit fly: Southern Thailand, Malaysia, Singapore, Brunei, Indonesia (Kalimantan, Sulawesi, Java, Bali, Lombok, Sumbawa, Flores, Timor, Maluku, Irian Jaya).

***Bactrocera carambolae* (Drew & Hancock) Carambola fruit fly**: Southern Thailand, India (Andaman and Nicobar Is.), Singapore, Malaysia (Sabah), Brunei, Indonesia (Java, Lombok, Sumbawa).

***Zeugodacus cucurbitae* (Coquillett) Melon fly**: Probably originates from Asia; present in Afghanistan, Pakistan, India, Sri Lanka, Nepal, Bangladesh, Burma, southern China, Thailand, Laos, Cambodia, Vietnam, Malaysia, Singapore, Taiwan,

Philippines, Brunei, Indonesia (Sumatra, Java, Kalimantan, Sulawesi, Sumba, Timor, Irian Jaya).

***Bactrocera latifrons* (Hendel) Solanum fruit fly**: Pakistan, India, Sri Lanka, Burma, China (Fujian, Yunnan, Hong Kong, Hainan), Thailand, Laos, Vietnam, western Malaysia, Singapore, Taiwan, Brunei.

***Bactrocera minax* (Enderlein) Chinese citrus fruit fly**: East India (Sikkim, W. Bengal), Bhutan, China (Jiangsu, Sichuan, Hubei, Hunan, Guangxi, Guizhou, Yunnan).

***Bactrocera occipitalis* (Bezzi)**: Philippines, Brunei, Malaysia (Sabah).

***Bactrocera philippinensis* (Drew & Hancock)**: Philippines (Luzon, Panay, Negros, Cebu, Mindanao).

***Bactrocera tau* (Walker)**: India (Andaman & Nicobar Is., northeast India), Bhutan, Burma, China (north to Xizang, Sichuan, Hubei and Zhejiang), Taiwan, Thailand, Laos, Cambodia, Vietnam, Malaysia (western Malaysia and Sarawak), Singapore, Brunei, Indonesia Bactrocera umbrosa (Fabricius): Southern Thailand, Malaysia (western Malaysia and Sabah), Singapore, Philippines, Brunei, Indonesia (Sumatra, Java, Kalimantan, Sulawesi).

***Bactrocera zonata* (Saunders)**: Peach fruit fly. Pakistan, India, Sri Lanka, Bangladesh, Burma, Thailand, Laos, Vietnam.

Other economically important genera of Dacinae include:

**Dacus** – species particularly in Africa and in some parts of Asia.

**Ceratitis** – mainly in Europe and Africa, though the most damaging species (the Mediterranean fruit fly or Medfly, Ceratitis capitata Wiedemann), has spread to some other areas including America and Australia.

**Anastrepha** – a genus with a number of economically important species attacking various fruits in North and South America. Some other genera also have economic importance in temperate to cold areas of Europe, North America and North Asia.

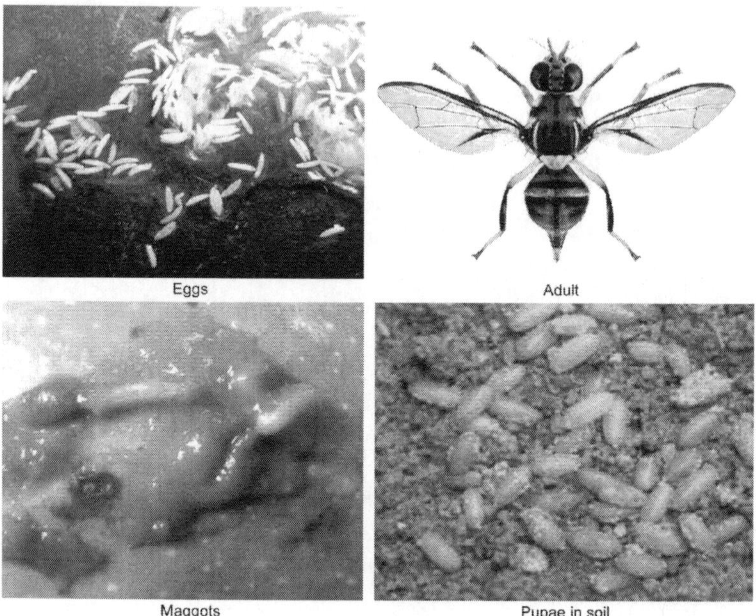


**Fig. 1. Different stages of B. dorsalis associated with mango**

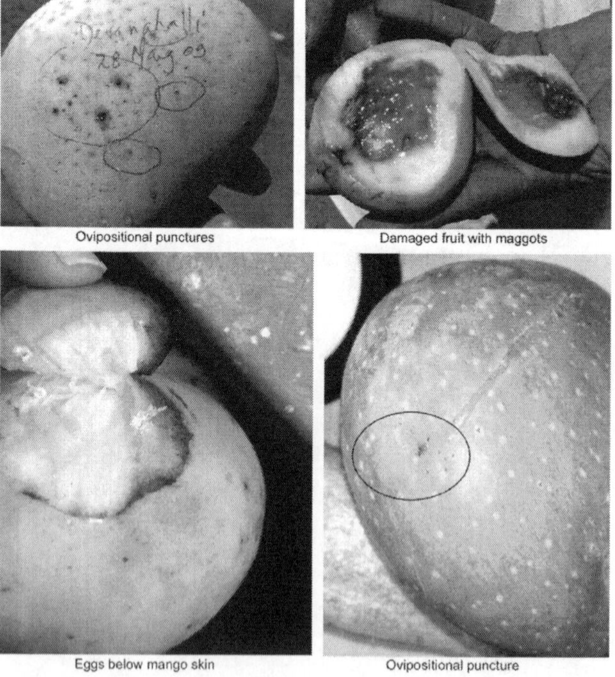


**Fig. 2. Damage symptoms of B. dorsalis in mango**

**Fig. 3. Hot Water Treatment (Source: DPPQS, 2023)**

**Fig. 4. Vapour Heat Treatment System**

## References

Allwood, A. J. and Drew, R. A. I. 1997. Fruit fly management in the Pacific. ACIAR Proceedings No 76, 267 pp.

Armstrong, J.W., J.D. Hansen, B.K.S. Hu, and Brown, S.A. 1989. High-temperature forced-air quarantine treatment for papayas infested with tephritid fruit flies (Dipterna: Tephritidae). J. Econ. Entomol. 82:1667-74.

CFFP. 1999. Carambola Fruit Fly Programme. (http://www.carambolafly.com)

Madura, H.S. and Verghese, A. 2004. A guide to identification of some common fruit flies (Bactrocera spp.) (Diptera: Tephritidae: Dacinae). Pest Management in Horticultural systems, 10: 1-10.

OEPP/EPPO (1983) Data sheets on quarantine organisms No. 41, Trypetidae (non-European). Bulletin OEPP/EPPO Bulletin 13 (1).

Rashmi, M.A., Abraham Verghese, P.D. Kamala Jayanthi, A.K. Chakravarthy, Arathi Kirubha and Subhash B. Kandakoor, 2020. Effect of climate change on biology of oriental fruit fly, Bactrocera dorsalis hendel (Diptera: Tephritidae). Journal of Entomology and Zoology Studies 8(3): 935-40.

Verghese, A. and Devi, S. 1998. Relation between traps catch of Bactrocera dorsalis Hendel and abiotic factors. Pp. 15-18. In: Proc. First Nat. Sym. Pest Management in horticultural crops: Environmental Implications and Thrusts, Reddy, P.P., Krishna kumar, N.K. and Verghese, A (Eds.). Association for advancement of pest management in horticultural ecosystem, Bangalore.

Verghese, A., Madhura, H. S., Jayanthi, P. D. K and Stonehouse, J. M. 2002. Fruit flies of economic significance in India, with special reference to Bactrocera dorsalis (Hendel). Proceedings of 6th International Fruit fly Symposium, held during 6–10 May 2002, Stellenbosch, South Africa, pp. 317–24.

Verghese, A., Nagaraju. D. K. and Sreedevi, K. 2011. Hot water as an effective postharvest disinfection for the oriental fruit fly, Bactrocera dorsalis (Hendel) on mango. Pest Management in Horticultural Ecosystems 17(2): 63-68.

Verghese, A Rakshitha Mouly, T. N. Shivananda, C. B. Soumya and M.A. Rashmi., 2020. A push-pull strategy for the management of the Oriental fruit fly, Bactrocera dorsalis (Hendel) in mango. Pest Management in Horticultural Ecosystems 26(2):269-71.

CHAPTER 20

# DISEASE MANAGEMENT IN ARID HORTICULTURAL CROPS

**S. K. Maheshwari, B. R. Choudhary, Ramkesh Meena, Ramesh Kumar and Anil**

*Central Institute for Arid Horticulture, Bikaner-334 006 (Rajasthan)*
*Corresponding author: maheshwariskciah@gmail.com*

---

The crops like aonla, ber, bael, date palm, pomegranate, mulberry, bottle gourd, longmelon, muskmelon, sponge gourd and watermelon are important arid including horticultural crops grown in arid regions. These are infected by several diseases alternaria leaf blight, cercospora leaf spot, powdery mildew, wilt and mosaic disease. These diseases pose serious threat in their cultivation, which inflict significant losses to these crops every year.

Disease management, includes cultural, biological, host resistance and chemical control strategies in a holistic way rather than using a single component strategy proved to be more effective. Success of disease management strategies needs appropriate policies in place that cover a wide range of themes such as plant protection, private sector investment, awareness and agricultural extension. Wide adoption of disease management practices in horticultural crops is a pre-requisite for achieving impact at field level. Adoption and support for using participatory approaches helps farmers in fruit disease management, reducing costs and improving production efficiency. To reduce yield losses caused by diseases, a complete knowledge of disease management strategies in fruit and vegetables in arid region are essential.

## Integrated Disease Management of Arid Fruit Crops

## Fruit Crops

### Aonla (*Emblica officinalis* Gaertn.)

Anthracnose cause by *Colletotrichum gloeosporioides* is the serious disease in aonla-growing areas. This disease appears on leaflets and fruits in August-September

(Mishra and Shivpuri, 1983). At beginning, symptoms are minute, circular, brown to grey spots with yellowish margin in leaflets. Leaves dried up at advance stages. In fruits, pin head like spots appear with dark brown to pink having yellow halo, while on fruits, depressed lesions develop which later turn dark. In severe cases, many spots coalesce with each other to form bigger lesions. The central areas of spots remain grayish raised with dot like fruiting bodies- the acervuli arranged in rings. Under severe condition, plenty spots result the fruits become shrivel and rot.

Discard the affected fruits and leaves at initial stage from the orchard. Proper pruning should be done in orchard to improve air circulation within the crown and plant trees on a wide spacing and keep the surrounding area clear of vegetation.

Spraying of Carbendazim (0.1%) or Difolatan (0.2%) can reduce the disease (Bhardwaj and Sharma, 1999).

### Ber (*Ziziphus mauritiana* Lamk.)

Many fungal diseases such as powdery mildew, black leaf spot, fruit rot and *Alternaria* leaf spot in ber have been identified in arid and semi- arid regions.

Powdery mildew is the most serious diseases of ber in semi-arid and subtropical regions. Maximum severity (35.2%) was observed in orchard of Hathnoda (Chomu, Jaipur) and it is a hot spot for powdery mildew disease.

Initial symptoms appear on young fruits at pea stage of development. White specks may appear on immature fruit and later may cover the entire fruit. Mildew produces symptoms as whitish powdery mass on all the aerial plant parts causing pre-mature drop of flower buds. The infected fruits become misshapen and corky. Fruits finally drop. In severe conditions, floral parts, whole fruits, tender branches and leaves would appear with powdery mass of fungal conidia. Whole tree would appear unhealthy with less fruit setting and malformed fruits. Such *ber* orchards are devastated completely. Infected fruits fail to develop, show discolouration and become mummified.

Remove and destruct infected leaves, twigs and collateral hosts from orchard. Ber orchards should be free from wild species of ber (*Ziziphus nummularia*). Summer ploughing in the orchard is also advised.

Maximum disease control has been found by combined application of *Psendomonas fluorescens* (1%) and Karathane @ 0.05% (Nallathambi *et al.*, 2003). Three sprays of karathane (0.1%) followed by difenoconazole (0.05%) at 15 days interval proved 51.0% and 40.3% disease control against powdery mildew respectively.

Spraying of *Trichoderma* spp. (CIAH- 240) @ 5% also reduced the disease.

### Black leaf spot

Gupta and Madan (1977) reported this disease from Haryana for the first time. Verma *et al.*, (1992) also reported this disease from Punjab. Disease incidence is also noted from Rajasthan and Uttar Pradesh.

Sooty tuft like circular to irregular black spots develop on lower surface of leaves. At advance stage, larger part of lower surface of leaves is also covered with spots, while upper leaf surface shows brownish discolouration. Leaves and twigs are dried rapidly under severe infection. *Isariopsis indica* var. *ziziphi* causes the disease.

- Removal and destruction of infected leaves and twigs from orchard.
- Keep the ber orchard free from all weeds.
- Pruning should be done properly in the orchard.
- Fungicidal sprays at 15 days interval (mancozeb or copper oxychloride @ 0.2%) (Verma and Cheema, 1988).
- Minimum disease intensity (10.60%) is recorded with 02 sprays of propiconazole 0.1%, followed by difenconazole 0.1% (11.70%) and carbendazim 0.1% (12.81%) at an interval of 15 days, starting from initiation of disease, i.e. first week of September.
- Use resistant varieties such as ZG-3, Seo, Bahadurgarhi and Safeda Rohtak.

## Date palm (*Phoenix dactylifera* L.)

### Alternaria leaf spot

It has been observed in severe form at Date Palm Research Centre, SKRAU, Bikaner, with disease incidence of 14.16-72.50%, depending upon cultivars and climatic conditions.

The spots are most common on lower leaves (pinnae) of plants, whereas on upper leaves spots are few and small size. The disease cause heavy losses in both quality and quantity of production (Pal *et al.*, 2006). *Alternaria alternata* causes the disease.

- Removal and destruction of infected leaves, twigs and weeds from orchard. Pruning should done. To disinfect all tools in pruning, *etc.* and cut surfaces.
- Use resistant varieties, Sriganganagar and Medini (Pal *et al.*, 2006).
- Fungicidal sprays at 15 days interval (mancozeb @ 0.2%) minimize the disease. Copper oxychloride (0.2%) is also useful for retarding the fungal growth.

## Graphiola leaf spot

It is the most widely spread disease occurs wherever the date palm is cultivated under humid conditions.

Symptoms of the disease appear as sub-epidermal spots on both sides of the pinnae (leaf flat) and on rachis with small black sori (fruiting bodies) developing in abundance on old fronds. Fruiting structures emerge as small-yellow/brown to black sori. These sori are abundant on three year-old leaves. The sori are 1-3 mm in diameter. *Graphiola phoenicis* causes the disease.

- Removal and destruction of infected leaves, twigs and weeds from orchard.
- Pruning should be done in orchard.
- To disinfect all tools in pruning, *etc.* and cut surfaces.
- Use disease free and quality planting material of date palm.
- Grow genetically tolerant date palm varieties (Barhee, Adbad, Rahman, Gizaz, Iteema, Khastawy and Tadala).
- Use variety, Hatemi (Pundir *et al.,* 2006). Variety Khadrawi is resistant under field conditions of Haryana.
- Carbendazim is most effective followed by thiophanatemethyl (Mehta *et al.,* 1989). Two sprays and soil drenching of copper oxychloride (0.2%) at fortnight interval in checking the infection and further disease spread.

## Pomegranate (*Punica granatum*)

### Bacterial leaf and fruit spot

It is a major constraints for productivity, and quality of fruits in Tamil Nadu, Maharashtra and Karnataka as well as in some parts of Gujarat and Rajasthan. It appeared as an epidemic form in Bangalore, Karnataka, causing 60-80% fruit yield losses. Dhandar *et al* (2004) have made in extensive survey in Solapur and Sangola districts of Maharashtra against this disease and other aspects.

Small, transluscent water soaked spots appear on leaves and became necrotic and finally may turn dark brown. Many spots may coalesce and cover big area which are surrounded by chlorotic halo. Leaves are distorted and defoliated. Water soaked spots on fruits may turn dark brown. The spots are oily in appearance. *Xanthomonas axonopodis* pv. *punicae* causes the disease.

- Removal of weed hosts and other infected plant parts.
- Pruning should be done in pomegranate orchard.

Spraying of 200 ppm streptocycline can be controlled followed by Pausamycin (0.05%) + Copper oxychloride (0.2%) with 3 sprays at fortnightly intervals (Suriachandraselvan *et al.,* 1993).

Jalore Seedless is resistant to this disease.

## Citrus (*Citrus* spp.)

### Citrus canker

Cankerous lesions appear on leaves, twigs and fruits as small, round, watery and translucent spots. They first develop on the lower surface of the leaf and then on both the surface. The surface of spots becomes white or greyish and finally ruptures in the centre giving a rough, corky and crater-like appearance. The rough lesions are surrounded by a yellowish-brown to green raised margin and watery yellow halo. Cankers on fruits are similar to those on leaves except that yellow halo is absent and a crater like depression in the centre is more prominent. *Xanthomonas axonopodis pv. citri* causes the disease.

Two pruning of infected plant parts along with 4 sprays of copper oxychloride (0.5 %) or bordeaux mixture (1%) have been reported effective and economical.

Application of *neem* cake solution (3-5%) on foliage has also been reported effective against this disease in nurseries.

## Karonda (*Carissa congesta*)

### Anthracnose disease

The small lesion is most economically important. The disease can be seen in the form of black grayish spots on leaves. *Colletotrichum inamdrii* causes the disease.

- Collecting and destroying affected parts of plants.
- Spray of copper fungicides or copper sulphate (0.1%) can effectively control.

## Vegetable Crops

### Bottle gourd (*Lagenaria siceraria*)

It is an important crop of Rajasthan and widely grown in open field conditions as well as in river beds to harvest early crop (Maheshwari and Haldhar, 2018).

Alternaria leaf blight has highest disease incidence (35.33%) and disease severity (24.0%).

The symptoms first appear on leaves as small, circular and light brown spots of different size which later enlarge in a concentric rings and margins appear. These spots mix together to form larger necrotic areas on leaves. Disease starts from margin of lower leaves which gradually increase and covere lamina. It is caused by a fungal pathogen (*Alternaria cucumerina*).

- Removal of infected crop debris, collateral hosts and weeds.
- Use of certified seed.
- Deep ploughing of the soil.
- Follow a 3-year crop rotation with non-cucurbitaceous crop.
- Combined treatment of carbendazim @ 0.1% (seed treatment) + mancozeb @ 0.25% (foliar spray) + *Pseudomonas fluorescens* @ 5% (foliar spray) + neem leaf extract @ 5% (foliar spray) is most effective with minimum disease incidence of 9.25%, minimum disease severity of 7.07% and maximum disease control (78.23%) against Alternaria leaf blight in bottle gourd (Maheshwari *et al.*, 2017).
- Three sprays of indofil M-45 (0.2%) at regular interval of 10 days were very effective against this disease (Maheshwari *et al.*, 2019).
- Use of moderately resistant varieties, as Pusa Naveen, Pusa Santushti, Pusa Sandesh, Pusa Samridhi, PSPL and Arka Bahar.

### Powdery mildew

The disease is destructive among cucurbits in most areas of world and can be a major production problem causing yield losses of 30%–50% (El-Naggar *et al.*, 2012). Maximum disease incidence (up to 58%) was observed in different bottle gourd genotypes at ICAR-CIAH, Bikaner (Maheshwari and Choudhary, 2018).

The symptoms appear as small floury patches on leaves first on the lower surface followed by upper surface. Fungal growth on these spots results in production of powdery growth. Powdery mass on the leaves decreases the photosynthetic rate (Queiroga *et al.*, 2008), causing reduction in plant growth, premature defoliation, and consequently reduction in yield. The infected plant parts remain stunted and distorted. Severely affected leaves lose their dark green colour and become pale yellow green to brown and are shriveled, drying and collapse. *Sphaerotheca fuliginea* causes the disease.

- Removal of infected crop debris.
- Deep ploughing may burry the pathogen where it may be killed by soil microflora of the pathogen present in below soil layers.
- Crop sequence with appropriate crops may be the most effective management practices in eliminating soil borne pathogens.
- Introduction of non-host crops in rotation helps in disease reduction.
- Use of moderately resistant varieties such as Pusa Naveen, Pusa Santushti, Arka Bahra and Pusa Sandesh against powdery mildew (Maheshwari *et al.*, 2012).
- Three sprays of hexaconazole (0.05%) were found effective (Gupta and Gupta, 2001).
- Three foliar sprays of non-target chemicals (mono-potassium phosphate, potassium nitrate) and neem kernel extract (1-5%) provide effective against powdery mildew.
- Three foliar sprays of neem seed kernel extract (5%) were effective against this disease (Maheshwari *et al.*, 2021)

## Muskmelon (*Cucumis melo* L.)

Yield losses up to 80% have been reported in the worst affected areas in cucurbits (Gupta *et al.*, 2001). The disease incidence in muskmelon was also observed from 5.0 to 60% at experimental field of ICAR-CIAH, Bikaner (Maheshwari and Choudhary, 2018).

The leaves are accompanied by yellowing and marginal necrosis. Older plants may first exhibit temporary wilting only during the heat of mid-day but will die within a few days. The infection results in lesion formation on the collar region and infected areas appear brown and water soaked. Wilting begins sudden after disease initiation. As a result of softening of the tissue, the plants shriveled and after that rapid plant mortality. The older plants wither and die during the growing season. *Fusarium acuminatum* causes the disease which is identified from IARI, New Delhi.

- All infected plant should be collected and burnt. Crop rotation with garlic, radish, onion has been found to retard it (Sen, 1986).
- Deep ploughing should be practised during summer.
- Growing crop with non-host crop.
- Certain soil amendments in soil which destroy the soil inoculums, can also be used to control disease.
- Durgapura Madhu and Punjab Sunehri are resistant. Pusa Madhuras' is moderately resistant against this disease.
- Drench the soil around roots with carbendazim @ 0.1%.
- Treat seeds with bavistin/ captan @ 2 g/ kg seed before sowing.
- Drench the soil with captan (0.2%) or bavistin (0.1%).
- Minimum wilt incidence (7.40%) in muskmelon was found in neem cake @ 5 q/ ha + biofertilizer followed by neem cake @ 5 q/ ha with incidence of 8.12%.
- b) *Trichoderma harzianum/T. viride* when applied as seed treatment (5 gm/ kg seed) followed by soil application along with FYM/ neem cake were effective against this disease.
- Wilt in cucurbits can be suppressed by *Pseudomonads* isolates.

## Watermelon (*Citrullus lanatus*)

Mosaic disease has become potential threat to the cultivation of watermelon/ *mateera* in western Rajasthan and other watermelon growing states of India. It is recorded from 14.28-35.0 % disease incidence in Bikaner and nearby areas of Rajasthan.

Infected plants showed characteristic symptoms of dry, yellow mosaic or mottling with chlorotic patches on foliage, development of vein clearing, reduction of leaf size and stunting of the plants. Systemic symptoms consist of downward leaf curling or rolling and mosaic. Disease incidence is found low in this crop due to unfavorable environmental conditions. This disease is caused by WMV/CMV.

- Remove infected plants as early as virus symptoms are observed to prevent/ minimize spread of virus by aphids.
- Destruction of weed hosts.
- Use of virus free seeds.
- Using reflective mulch that reduces the early season infection from aphids and gives an additional 2-4 weeks of a virus-free cucurbit field. Once the plants cover the plastic reflective mulch, it ceases to be an effective deterrent.

- 'Asahi Yamato' has tolerance.
- Spray at seedling stage with acephate (0.15%) gives good results. Among 11 treatments, imidacloprid (0.05%) was found the most efficient treatment against mosaic disease in ridge gourd with minimum per cent disease index (15.90%) and per cent disease reduction (61.69%) followed by neem leaf extract (Maheshwari *et al.*, 2020).

### Longmelon

**Gummy stem blight:** It appear as grayish-green, circular spots between the veins on leaves. Lesions develop first on the vines at the nodes and elongate into water-soaked streaks that become pale brown to gray with time. The fungus damages the tissues of the stem at all stages of development. The infection often begins at leaf margins (Koike 1997). Symptoms included stem necrosis with characteristic exudation of gummy material and stem tissue often cracks. Main stem lesions enlarge and slowly girdle the main stem. Infected vines die eventually, and entire plants die occasionally. The fungus, *Didymella bryoniae* is the causal organism of gummy stem blights in long melone.

- Remove and destroy infected fruits, vines and weed hosts at the end of the season.
- Deep ploughing.
- Crop rotation with non-host crops should be followed.
- Procure disease-free seeds from reliable sources.
- Apply fungicides like chlorothalonil and mancozeb (0.2%) in a preventative manner are effective and there are many effective systemic fungicides for its management.
- Treat the seeds with thiram (2 g/kg seed). Spraying of bavistin (0.1%) as soon as the disease is noticed. In case, the disease is not con-trolled, spraying of propiconazole (0.1%).
- Biofungicides (*B. subtilis*) alternated with chlorothalonil also minimized use of synthetic fungicides.

## Chilli

Chilli (*Capsicum annum* L.) is fourth major crop cultivated worldwide and is one of the most important constituent of the cuisines of tropical and subtropical countries.

Anthracnose disease causes yiels loss of 10–54% in chilli yield to both mature fruits in the field as well as during storage (Lakshmesha *et al.*, 2005).

The small lesion is the most important typical symtoms of the disease which lowers its market value thereby affecting the profitable yield of the crop. The disease is reported to affect almost all aerial parts of the plant. Chiefly, it causes fruit rot at both green and red stages primarily attacking ripe fruits, hence is also known by the name ripe fruit rot of chilli (Agrios, 2005). The disease is seed borne, soil borne, water borne and airborne and hence may lead to damage at the seedling stage or on the aerial parts of the plants. *Colletotrichum capsici* is the cause organism of this disease.

- Crop rotation and removal of any infected plant parts in the field.
- Proper distance between the plants should be maintained so as to reduce dense canopy.
- Transplanting raised from disease free seeds of the chilli variety. The transplants should be kept weed free and away from other solanaceous crops.
- Difenoconazole (1 ml/lit.) is effective to control.
- Use of resistant varieties (BS-35, BS-20, BS-28, Punjab Lal, Taiwan-2, and Pant C-1 are recommended (Garg *et al.*, 2014).
- The bioagents (*Psuedomonas fluorescens, Trichoderma* harzianum and *Bacillus subtilis)* are effective against *Colletotrichum* spp.
- Leaf curl disease is prevalent in all chilli growing areas including Bikaner, Jodhpur and Jaisalmer district of Rajasthan and nearby areas. This disease causes heavy yield losses.
- Leaves are yellowing, followed by slight curling. The interveinal areas puckered, shortening of the internodes and reduction in leaves size giving witch's broom appearance. Plants remain stunted and appear as bushy. Chilli leaf curl symptoms are caused by begomovirus.
- Removal of infected plant and weeds.
- Use seeds from healthy plants of previous season and growing nursery in protected structures (Kenyon *et al.*, 2014) is recommended.
- Grow maize crop as barrier crop and also trap crop for reducing the disease.
- Use of yellow traps.
- Imidacloprid (0.05%) and acephate (0.05-0.1%) was found effective against this viral disease (Ahmed and Ram, 2016).
- Neem seed kernel extract (5%) was also effective against this disease (Pandey *et al.*, 2010).
- Raw cow milk and *Trichoderma* spp. reduces this disease (Kumar, 2006).

## References

Agrios, G N. 2005. *Plant Pathology*. St. Louis, MO: Academic Press.

Ahmed A A and Ram M R. 2016. Medicinal plant extracts for the management of leaf curl disease of chilli (*C. annuum* Linn). *World Journal of Pharm Pharmaceutical Science* 5: 1916-24.

Bhardwaj, S. S. and Sharma, I. M. (1999). Diseases of minor fruits. In: *Diseases of Horticultural Crops* (Verma, L. R. and Sharma, R. C. (Eds)). Indus Publ. Comp., New Delhi, pp. 541-562.

Choudhary B R, Saroj P L, Haldhar S M, Maheshwari S K and Singh, D. 2018. Thar Sheetal fetches premium price. *Indian Horticulture* 63 (05): 58-59.

Dhandar, D. G., Nallathambi, P., Rawal, R. D. and Sawant, D. M. 2004. Bacterial leaf and fruit spot: a new threat to pomegranate orchards in Maharashtra. Proc. on National Seminar of Indian Society of Mycology and Plant Pathology held at University of Goa.

El-Naggar M, El-Deeh H, and Ragab S. 2012. Applied approach for controlling powdery mildew disease of cucumber under plastic houses. *Pakistan Journal of Agriculture* 28: 52–61.

Garg R, Loganathan M, Saha S and Roy B K. 2014. Chilli Anthracnose: a review of causal organism, resistance source and mapping of gene. *In*: Microbial Diversity and Biotechnology in Food Security, Eds Kharwar R. N., Upadhyay R., Dubey N., Raguwanshi R., editors. (Springer;), 589–610.

Gupta A K, Sharma R C and Sharma S. 2001. Fungal diseases of cucurbits. *In*: Diseases of vegetable crops, V. K. Gupta and Y. S. Paul (eds.). Kalyani Publishers, New Delhi, India, pp. 71-86.

Gupta, P. C. and Madan, R. L. (1977). Diseases of fruits from Haryana- a new leaf spot disease of ber. *Curr. Sci.* 46: 237-38.

Kenyon L, Kumar S, Tsai W S and Hughes J A. 2014. Virus diseases of peppers (Capsicum spp.) and their control. *Advance Virus Research* 90: 297–354

Khoury W El and Makkouk K. 2010. Integrated plant disease management in developing countries. *Journal of Plant Pathology* 92 (4): 35-42.

Koike S T. 1997. First report of gummy stem blight, caused by *Didymella bryoniae* on watermelon transplants in California. *Plant Disease* 81: 1331.

Kumar A. 2006. On-farm management of leaf curl disease in chilli under arid farming system. JFAE 4:180–182.

Lakshmesha K, Lakshmidevi K, Aradhya N and Mallikarjuna S. 2005. Changes in pectinase and cellulase activity of *Colletotrichum capsici* mutants and their effect on anthracnose disease on *Capsicum* fruit. *Archives of Phytopathology and Plant Protection* 38: 267–79.

Maheshwari S K and Choudhary B R. 2015. Field evaluation of bottle gourd genotypes for resistance against *Alternaria* blight in western Rajasthan. *Vegetos* 28: 103-5.

Maheshwari S K and Choudhary B R. 2018. Managing diseases in cucurbits- the eco-friendly way. *Indian Horticulture* 63(05): 114-17.

Maheshwari S K, Choudhary B R, Saroj P L and Sharma, B D. 2020. Field efficacy of botanicals and inorganic salts against mosaic disease on ridge gourd in western Rajasthan. *International Journal of Current Microbiology & Applied Sciences* 9(7): 1300-4.

Maheshwari S K, Choudhary B R, Sharma B D and Saroj P L. 2017. Management of *Alternaria* leaf blight of bottle gourd in western Rajasthan, India. *International Journal of Current Microbiology & Applied Sciences* 6 (6): 1272-77.

Maheshwari S K and Haldhar S M. 2018. Disease management in arid horticultural crops. CIAH/Tech./Pub. No. 68, ICAR-Central Institute for Arid Horticulture, Bikaner, Rajasthan, India, p. 42.

Maheshwari S K, Haldhar S M, Ramyashree, D G S and Choudhary B R. 2021. Integrated disease management in dryland cucurbitaceous crops. *In*: Pest Management in dryland horticultural crops. (*Eds*. Haldhar, S. M. and Maheshwari, S. K.). Published by Biotech Books, New Delhi, pp. 319-327. (ISBN: 978-81-7622-491-8)

Maheshwari S K, Ramyashree, D G S, Haldhar S M, Choudhary B R and Saroj P L (2019). Integrated disease management in arid cucurbitaceous crops. *In*: Integrated pests management of arid fruit and vegetable crops. (*Eds*. Haldhar, S. M. and Maheshwari, S.K.), Training Manual, ICAR-CIAH, Bikaner, pp. 92-98.

Maheshwari S K, Singh D, Choudhary B R and Sharma S K. (2012). Screening of bottle gourd germplasm for against powdery mildew under hot arid region of Rajasthan. *Crop Improvement* 39: 1073-74.

Mehta, N., Gupta, P. C., Thareja, R. K. and Dang, J. K. 1989. Varietal behaviour and efficacy of different fungicides for the control of date palm leaf spot caused by *Graphiola phoenicis. Tropical Pest Manage.* 35 (2): 117-19.

Misra, A.and Shivpuri,A. (1983). Anthracnose, a new disease of aonla. *Ind. Phytopath* 36: 406-7.

Nallathambi, P., Umamaheswari, C., Joshi, H. K. and Dhandar, D. G. (2003). Management of ber (*Ziziphus mauritiana Lamk.)* powdery mildew using Fluorescent Pseudomonads. *In:* Proc. of 6th International Workshop on Plant Growth Rhizobacteria held during 5-10 October, 2003 at IISR, Calicut, Kerala, pp. 184-87.

Pal, V., Rathore, G. S. and Godara, S. L. (2006). Screening of genotypes of date palm against *Alternaria* leaf spot. *Indian Journal of Arid Horticulture* 2: 62-63.

Pandey K S, Mathur A C and Srivastava M. 2010. Management of leaf curl disease of chilli (Capsicum annuum L.). *International Journal of Virology* 6: 246–50.

Pareek, O.P. (2001). *Ber*: International Centre for Underutilized crops, Southampton, U.K. pp. 248-66.

Pundir, J. P. S., Rathore, G. S., Naqvi, A. R. and Porwal, R. 2006. Date palm. In: *Advances in Arid Horticulture:* (Vol. II-Production technology of Arid and Semi-arid Fruits), International Book Distributing Co., Lukhnow.

Queiroga R, Puiatti M, Fontes P C R and Cecon P R. 2008. Produtividade e qualidade de frtos de meloeirovariandonumero de frutos de folhasporplanta. *Horticultura Brasileira* 26: 209–15.

Radhakrishnan P and Sen B. 1985. Efficacy of different methods of inoculation of *Fusarium oxysporum* and *F. solani* for inducing wilt in muskmelon. *Indian Phytopathology* 38: 70-73.

Sen, B. (1986). Cultural management of soil borne diseases. *In*: A. Varma and J. P. Verma(eds.). Vistas in Plant Pathology. Malhotra Publishing House, New Delhi, pp. 367-81.

Suriachandraslevan, M., Jayasekhar, M. and Aubu, S. 1993. Chemical control of bacterial leaf spot and fruit spot of pomegranate. *South Indian Hort.* 41: 228-29.

Verma, K. S. and Cheema, S. S. 1988. Chemical control trials against mouldy leaf spot of ber caused by *Isariopsis indica* var. *ziziphi*. *Plant Dis. Res.* 3(1): 32-36.

Verma, K. S., Cheema, S. S. and Kumar, S. 1992. Some physico-pathological studies on *Isariopsis indica* var. *ziziphi*. The incitant of mouly leaf spot of ber. *Pl. Dis. Res.* 7 (2): 210-14.

# CHAPTER 21

# BOTTOM-UP EFFECT ON PLANT-INSECT INTERACTION IN UNDERUTILIZED HORTICULTURAL CROPS FOR SUSTAINABLE PRODUCTION

**Shravan Manbhar Haldhar**

*Head, Department of Entomology, College of Agriculture, Iroisemba*
*Central Agricultural University, Imphal, Manipur-795004*
*Corresponding author: haldhar80@gmail.com*

Plants and insects have been coexisting for more than 350 million years. In this co-evolution, both have evolved strategies to avoid each other's defense systems. This evolutionary arms race between plants and insects has resulted in the development of an elegant defense system in plants that can recognize non-self-molecules or signals from damaged cells, much like animals. It activates the plant's immune response against herbivores. Differences in genotypes of plant characters may affect insect-plant herbivore interactions, and variations in genotype traits are responsible for modifying the bottom-up effects. Recent evidence shows that the simultaneous occurrence of abiotic and biotic stress can positively affect plant performance by reducing the susceptibility to biotic stress, a positive sign for pest management. Plant responses to these stresses are multifaceted and involve copious antibiosis, physiological, antixenotic, molecular, and cellular adaptations. Plants with antibiosis characteristics such as flavonoids, phenols, tannins, alkaloids, *etc.*, may cause reduced insect survival, prolonged development time, decreased size, and reduced new generation fitness. Quality and quantity of constitutive secondary metabolites production are species and cultivar-specific and can be expressed as the signature of a particular plant or species and leads to the phenomenon of host-plant resistance. Hence, such plant resistance mechanisms have been effectively and widely used for managing insect pests in fields of underutilized crops. Natural defences are mediated through plant characteristics that affect insect biology, such as mechanical protection on the surface of the plants (e.g., hairs, trichomes, thorns, spines, and thicker leaves) that either kill or retard the development of the insects. The phenomena of host plant resistance to insects can be exploited for the development of crop cultivars that readily produce an inducible response upon mild infestation. These cultivars can perform as one of the integrated pest management strategies for sustainable crop production.

The use of host plant resistance to insects can be an effective strategy for sustainable crop production. This resistance can be induced in crop cultivars that readily produce an inducible response upon mild infestation. Allelochemicals and antixenosis are two mechanisms that plants use to resist insect herbivores (Han *et al.*, 2016; Haldhar *et al.*, 2023a). Allelochemicals are secondary metabolites produced by plants that can be toxic or repellent to insects. Antixenosis is the ability of plants to reduce the attractiveness of themselves to insects. Plants and insects have been coexisting for more than 350 million years. In this co-evolution, both have evolved strategies to avoid each other's defense systems. This evolutionary arms race between plants and insects has resulted in the development of an elegant defense system in plants that can recognize non-self-molecules or signals from damaged cells, much like animals. It activates the plant's immune response against herbivores. Plant-arthropod interactions are of utmost importance for understanding the dynamics of ecological communities (Haldhar *et al.*, 2018a & d).

Plant defense strategies against insect herbivores may involve synthesizing many biologically active compounds (allelochemicals), which are phylogenetically conserved in specific plant families or genera. Plants frequently display genetic variation within and between populations for traits that influence the preference and non-preference of insects on their hosts, which are resistance traits (Table 1). It has been widely recognized that biological diversity plays a vital role in structuring community ecosystem processes (Gogi *et al.*, 2010; Johnson and Agrawal, 2005). The genotypic variation may influence the distribution and damage levels of herbivores on focal plants through processes referred to as associational resistance or susceptibility. Host plants play an essential role in determining insect populations with respect to concentrations and proportions of nutrients, which differ among species. Plants having antibiosis characteristics such as flavonoids, phenols, tannins, *etc.*, may cause reduced insect survival, prolonged development time, decreased size, and reduced fitness of new generation adults. Several plant characteristics affect the growth and development of arthropods. The plants defend against the attack by pests by synthesizing allelochemicals and displaying antixenosis (Tews *et al.*, 2004; Haldhar *et al.*, 2013a).

Host plants are essential in determining insect populations concerning concentrations and proportions of nutrients, which differ among species. Natural defenses are mediated by plant characteristics that affect the herbivore's biology, such as mechanical protection on the surface of the plants (hairs, trichomes, thorns, spines, and thicker leaves) that retard the development of the herbivores. Antixenosis refers to the potential plant characteristics/traits or morphological that impair or alter insect behaviour towards the host preference in such a way as to lessen the chances of insects using a host plant for oviposition food, damage or shelter (War *et al.*, 2012). Plant resistance mechanisms have been used to manage insect pests in horticultural crops. The utilization of native and introduced genetic material of arid horticulture crops for breeding varieties over a long period in the country resulted in the generation of many new genotypes/lines in the form of selections and, to some extent, through hybridization (Haldhar *et al.*, 2013a). Thereby, considerable generated material/variations may have existed in targeted species regarding the plant type, flowering, fruiting, morphological and physiological of fruits, reaction to biotic and abiotic factors and eco-adaptations. Since the detailed evaluation of identified genotypes is essential for further breeding programmes and incorporating desirable gene(s) through combination breeding, wild and relative species gene pool possesses unique traits that can be exploited by direct selection for use in combination breeding or improvement through advanced approaches. The need to promote the systematic utilization of a wild gene pool in strategic breeding work for developing genotypes having biotic and abiotic stress resistance or tolerance and their conservation as relative species. This chapter has a major focus on underutilized horticultural crops for the morphological and biochemical basis of plant-insect interaction.

## Screening of datepalm (*Phoenix dactylifera* L.) genotypes

Many arthropod species are known as pests of the date palm (*Phoenix dactylifera* L.) worldwide. Over 50 species of insects and mites are known to be pests of date palms, with approximately 25% of these species considered serious pests in Israel. The most dominant and economically important pests are scale insects, a mealybug, the lesser date moth, the greater date moth, red palm weevil, the rhinoceros beetle, and the termite. The lesser date moth is a pest that damages fruits in both fields and stores.

Screening of 32 date palm varieties/genotypes against fruit borers in a hot arid region showed that the percentage of fruit infestation of fruit borers varied significantly among the tested varieties/genotypes of date palm. The varieties/genotypes; Migra, Tayer, Javantri, and Medjool were found to be highly tolerant, while the varieties/genotypes like Zahidi, Suriya, Khasab, and Khuneizi had high fruit infestation and were declared as highly infestation varieties/genotypes to the fruit borers. Lower values of host plant susceptibility indices based on fruit infestation were recorded on resistance varieties/genotypes like Migra, Tayer, Javantri, and Medjool, which could be used as a source of resistance for developing date palm varieties/genotypes resistant to fruit borers (Haldhar *et al.*, 2021). The popular date palm cultivar, Khalas had the least antixenotic effect on female RPW adults, where a high degree of attraction to palm tissue volatiles was recorded, which was statistically similar to the cultivars; Reziz, Sheshi, and Hatmi. The cultivars Khasab, Shahal, and Gaar exhibited a high degree of non-preference (antixenosis). Further, Reziz registered the highest egg lay by red palm weevil and was statistically on par with the cultivars Khalas and Sheshi (Faleiro *et al.*, 2014). A study that tested the resistance of seven date palm cultivars to the red palm weevil (RPW) the results showed that there were no significant differences in the resistance of the cultivars tested. A study conducted in Qatar using microsatellite markers to assess the genetic diversity among 10 major date palm cultivars revealed two distinct groups (Elmeer *et al.*, 2011). Among the cultivars studied for resistance to RPW, Khalas, Sheshi, and Reziz were placed in one cluster of the six cultivars, while Khasab and Shahal were placed in another cluster of four cultivars. The RPW-susceptible and resistant cultivars may have evolved from two separate seeding date palm progenies with distinctly different genes for resistance to RPW (Johnson *et al.*, 2013).

The genotypes that were showed the maximum HPSI based on fruit infestation against fruit borer, (*B. amydraula)*: Zahidi (177.77%) and Suriya (168.55%). The genotypes showed minimum HPSI based on fruit infestation against fruit borer are Migra (31.63%) and Tayer (49.97%). The genotypes showed maximum HPSI based on fruit infestation against fruit borer were Zahidi (191.54%) and Khuneizi (184.89%). The genotypes showed minimum HPSI based on fruit infestation against fruit borer, (*A. sabella)* were Migra (40.16%) and Tayer (48.81%). Based on the percentage of fruit infestation, the highest HPSI was recorded in Zahidi categorized as highly susceptible to fruit borers and the lowest HPSI was found in Migra categorized as resistant to fruit borers (Haldhar *et al.*, 2021). According to Shahid *et al.*, (2014), the genotype Cool Sun-70 showed 27% HPSI followed by Cauliflower Desi 26%. The minimum HPSI was calculated at 9% for Pari F1 and Shumila F1. Cashmere and White Island genotypes were categorized as intermediate with 14% and 15% HPSI, respectively.

## Tingid bug resistance in Indian Cherry (*Cordia myxa* L.)

*Cordia myxa* L., commonly called Indian cherry (synonyms: clammy cherry, fragrant manjack, lasora, shell, cinnanakkeru, bahubara, chokri, lahsoda, lahsua, gonda, gondi), belongs to Boraginaceae family. AHCM-22, AHCM-25 and AHCM-34 were found to be resistant; AHCM-14, AHCM-30 and AHCM-31 were moderately resistant; AHCM-16 and AHCM-09 were moderately susceptible; AHCM-33 and AHCM-08 susceptible and AHCM-01 and AHCM-26 were highly susceptible germplasm of Indian cherry against tingid bug (Haldhar *et al.*, 2019). The allelochemical compounds of the leaf were significantly different among the tested germplasm accessions of Indian cherry. The flavonoid content was positively correlated to tannins, phenols, and total alkaloid content. The negative correlations were observed with the per cent bug infestation, bug density per leaf and free amino acid (Haldhar *et al.*, 2019). The tannin content of tested germplasm accession was the maximum in AHCM-22, followed by AHCM-25 and the minimum was found in AHCM-01. The maximum length of the leaf was found in AHCM-22 followed by AHCM-25 and minimum in AHCM-01, and the high length in resistant and the minimum length in susceptible germplasm. A positive correlation was observed between the length and the leaf's width, and negative correlations with the per cent bug infestation and bug density per leaf. The maximum width of the leaf was recorded in AHCM-14 and the minimum in AHCM-26. A negative correlation was observed between the width of the leaf and the per cent bug infestation and bug density per leaf. The resistant accession of Indian cherry (AHCM-22, AHCM-25 and AHCM-34) had the maximum leaf size, high rough and hairiness (Haldhar *et al.*, 2019). The first line of plant defense against insect pests is the erection of a physical barrier either through forming a waxy cuticle and developing spines, setae, and trichomes (Sharma *et al.*, 2009).

## Ber (*Ziziphus mauritiana* Lamark)

The ber, *Ziziphus mauritiana,* is native to the Province of Yunnan in southern China to Afghanistan, Malaysia and Queensland, Australia. It is native to South and Central Asia, found throughout the arid and semi-arid tracts. It is cultivated to some extent throughout its natural range on a commercial scale and has received much horticultural attention in India. *Z. mauritiana* is a gregarious spiny shrub or a small tree ends of branches curved or drooping. The crop suffered significant losses due to insect pests and diseases, and more than 130 insect pests were found to attach the crops in India. Balikai *et al.*, (2013) reported 22 insect and non-insect species; likewise, Kavitha and Savithri (2002) documented about 23 insect species on ber.

## Stone weevil (*Aubeus himalayanus*)

The stone weevil is an emerging threat to ber production in India, especially in Northern India. The ber varieties/ genotypes like Kali, Katha, Illaichi and Tikadi were found to be resistant; Akharota, Dandan, Gola, Goma Kirti, Sanaur-1, Seb, Umran and ZG-3 were moderately resistant; Badami, Banarasi Karaka, Gularvasi, Jogia, Kaithli, Mundia, Reshmi, Sanaur-3, Thar Bhubhraj, Thar Sevika and Thornless were susceptible and Banarsi Pebandi, Chhuhara, Sanaur-2, were Sanaur-4 and Sanaur-5 highly susceptible (Haldhar *et al.*, 2018c). The per cent fruit infestation was found highest in Sanaur-2 (49.44 % in retained fruit and 73.77 % in dropped fruits) and the lowest in Tikadi (8.80 % in retained fruits and 13.31 % in fallen fruits), followed by Katha (9.07 % in retained fruits and 13.71% in dropped fruits). Plant-arthropod interactions are thought to be of utmost importance for understanding the dynamics of ecological communities. Plant defense strategies against insect herbivores may involve synthesizing many biologically active compounds (allelochemicals), which are phylogenetically conserved in specific plant families or genera. Plants frequently display genetic variation within and between populations for traits that influence the preference of insects on their hosts, that is, resistance traits (Johnson & Agrawal 2005; Haldhar *et al.*, 2013c).

## Fruit fly (*C. vesuviana*)

The fruit fly, *Carpomyia vesuviana* Costa, is India's most destructive pest of ber. It is a monophagous pest that infests *Zizyphus* species only and contributes towards low-yield and poor-quality fruits (Haldhar *et al.*, 2013b; Haldhar *et al.*, 2016a; Dhileepan 2017). The fruit fly causes yield losses of up to 80% under severe infestation when no control measures as taken. The ber cultivars such as Tikadi, Katha and Illaichi were found to be resistant; BS-75-1, Safeda, Dandan, Gola, Goma Kirti, Jogia, Narma, Mundia, Reshmi, Seb, ZG-3, Umran and Akharota were found to be moderately resistant; Banarasi Karaka, Banarasi Pawandi, Chhuhara, Kaithli, Thar Sevika and Thar Bhubraj were susceptible, whereas Sanaur-3, Sanaur-4 and Sanaur-5 were highly susceptible to the fruit fly, *C. vesuviana* in both the studied seasons (Haldhar *et al.*, 2018a). Hosagoudar *et al.*, (1999) reported that fruit fly, *C. vesuviana* infestation was high in the cultivar Sanaur-2, followed by Umran and Sanaur-6 and the lowest infestation recorded in Illaichi. The *flavonoid* content (179.0 mg/100 g) was found to be maximum in Safeda, followed by Tikadi (176.5 mg/100 g) and minimum in Chhuhara (40.7 mg/100 g). The tannin content (511.6 mg/100 g) was found to be the highest in Safeda, followed by Tikadi (502.8 mg/100 g) and the lowest in Chhuhara (264.8 mg/100 g). Phenols content was highest in Safeda (239.0 mg/100 g), followed by Tikadi (232.0 mg/100 g) and lowest in Sanaur-4 (113.0 mg/100 g), with values significantly higher in resistant and lower in susceptible cultivars. The percentage of fruit infestation with flavonoids (-0.914), tannins (-0.914) and phenols (-0.947) had a significant negative correlation (Haldhar *et al.*, 2018a).

## Fruit-borer (*Meridarchis scyrodes*)

The ber genotypes under study indicated significantly low incidence of fruit borer, *M. scyrodes* in Safeda (13.27%) followed by Tikadi (14.01%). Significantly greater incidence of fruit borer was registered in Sanaur-5, Chhuhara and Sanaur-2 with a magnitude of 75.09, 72.90 and 71.78 %, respectively. Basis of fruit borer, *M. scyrodes* incidence; Safeda, Tikadi, Darakhi and Illaichi were considered as resistant; BS-75-1, Gola, Goma Kirti, Seb and Umran were found moderately resistant; Dandan, Mahawali, Jogia, Sukavani, Narma, Reshmi and ZG-3 were found to be susceptible whereas, Banarasi Karaka, Banarasi Pawandi, Chhuhara, Kaithali, Mundia, Sanaur-2 and Sanaur-5 were highly susceptible to fruit borer (Haldhar *et al.*, 2023). The quality and quantity of resources available for plants triggers on plant-insect herbivore interaction, *i.e.*, plants are affected by environmental factors, which in turn influence the performance of insect herbivores (Han *et al.*, 2016). Different studies have shown that genotypes of the same species could differ significantly in their resistance to insects (Vir *et al.*, 2015) and it is influenced by morphological and biochemical traits of plants (Visschers *et al.*, 2019).

## Fruit borers (*Meridarchis scyrodes*) and *Dudua aprobola* resistance in jamun (*Syzygium cumini*)

The jamun is native to India, Burma, Ceylon and the Andaman Islands and available throughout Indian plains as well as in Kumaon hills up to 1,600 m. The vinegar prepared from the juice of the slightly unripe fruit of jamun is stomachic, carminative, diuretic, cooling and digestive properties. The fruit borer, *Meridarchis scyrodes* (Lepidoptera: Carposinidae), is a serious pest of jamun. The moths lay eggs on fruits at the pea stage, and upon hatching, the newly emerged caterpillars bore into fruits, feed on the near pulp seed, and accumulate faecal. The borer causes up to 60% yield loss under severe infestation. The leaf roller, *Dudua aprobola* (Lepidoptera: Tortricidae), is reported to be a major pest of jamun in Rajasthan and Gujarat. It mainly damaged tender leaves and inflorescence, but has recently been damaged to fruits of jamun. Eggs were laid in axils of leaves or flowering stalks or on fruits. GJ-26, GJ-27, GJ-19, GJ-17 and GJ-15 were found to be resistant; GJ-3, GJ-6, GJ-7, GJ-9, GJ-13, GJ-16, GJ-18, GJ-20, GJ-23, GJ-24 and GJ-25 moderately resistant; GJ-4, GJ-8, GJ-10, GJ-12, GJ-14, GJ-21 and GJ-22 were susceptible and GJ-1 and GJ-GJ-11 highly susceptible genotypes. The per cent fruit infestation of fruit borer, *M. scyrodes,* was highest in genotype GJ-1 (64.70%) followed by GJ-11 (62.07%). The minimum per cent fruit infestation of fruit borer, *M. scyrodes,* was observed in GJ-27 (8.23%), followed by GJ-17 (10.03%). The per cent infestation of *D. aprobola* was highest in GJ-1 (76.30 %) and lowest in GJ-27 (10.80 %), followed by GJ-26 (11.13 %) (Haldhar *et al.*, 2016c). The allelochemical compounds of the fruit differed significantly among the tested jamun genotypes. Tannins, phenols, total alkaloids and flavonoids had significant negative correlations with per cent fruit infestation of fruit borer, *M. scyrodes* and *D. aprobola* (Haldhar *et al.*, 2016c). The antixenotic fruit traits were significantly different among the tested jamun genotypes. The fruit length, fruit width, pulp thickness and pulp: stone ratio were found to be minimum in the GJ-27 genotype, but maximum fruit length in GJ-4, fruit width in GJ-8, pulp thickness in GJ-11 and pulp: stone ratio in GJ-11 and GJ-24 genotypes, respectively. The fruit length, width, pulp thickness and pulp: stone ratio showed significant positive correlations with the per cent fruit infestation of fruit borer, *M. scyrodes* and *D. aprobola* (Haldhar *et al.*, 2016c).

## Mite (*Tenuipalpus punicae*) resistance in pomegranate

Pomegranate (*Punica granatum* L.) is an important fruit crop in arid and semi-arid regions of India. It has the capability to withstand extreme environmental conditions of heat, drought and moisture deficit situations. Mite, *T. punicae* is most active during dry spells from April to October. Shiny white or brown patches can be seen on the undersurface of affected leaves, which may further curl and fall. Adults and nymphs nourish on the lower leaf surface by sucking sap. The leaf is reddish on severe infestation (Haldhar and Maheshwari 2021; Nidhi & Haldhar 2021). The pomegranate cultivars; HP Collec., Speen Sacarin, Gule Shah Rose Pink and AHPG-C3 cultivars were found to be resistant; Sur Sukker, Tujetis EC 4347, Bassin Seedless, Dholka, Uthkal, Agah, Achik Dana and Goma Khata cultivars were found to be moderately resistant; Jalore Seedless, Jodhpur Red, Saharanpur and Bhagwa were susceptible whereas Dorsata Malus, Kajaki Anar, Saih Sirin, P-23, IC-318712 and Phule Arakta were highly susceptible against *T. punicae*. The leaf infestation ranged from 5.17 to 56.33% in 2019 whereas it was 5.33 to 57.5% during 2020. Pooled data of leaf infestation in both years (5.25- 56.92%) was significantly lower in resistant and higher in susceptible cultivars. In pooled data, the per cent leaf infestation was the highest in Kajaki Anar (56.92 %) and the lowest in HP Collec. (5.25 %) followed by Gule Shah Rose Pink (7.17 %) and Speen Sacarin (7.33%) (Haldhar *et al.*, 2022). The more closely related study is probably on carob moth susceptibility, which was assessed on 19 pomegranate varieties of Iranian origin (Sobhani *et al.*, 2015). More recently, a demographic laboratory study on the same insect also revealed significant differences among the 11 Iranian varieties under investigation (Abedi *et al.*, 2019). The flavinoid content (4.48 mg cat.Equi./g) was found to be maximum in HP Collec. followed by Speen Sacarin (4.05 mg cat.Equi./g) and minimum in Bassin Seedless (2.47 mg cat.Equi./g). Phenols content was highest in HP Collec. (49.00 GAE/g) followed by Gule Shah Rose Pink (48.94 GAE/g) and lowest in Kajaki Anar (34.10 GAE/g). The tannin content (9.87 mg cat.Equi./g) was found to be the highest in HP Collec. followed by Speen Sacarin (8.89 mg cat.Equi./g) and the lowest in AHPG-C3 (3.45 mg cat.Equi./g). Total antioxidant activity on CUPRAC scale was highest in Gule Shah Rose Pink (571.45 mg AAE/g) followed by HP Collec. (522.14 mg AAE/g) and lowest in IC-318712 (289.71 mg AAE/g) with values significantly higher in resistant and lower in susceptible cultivars (Haldhar *et al.*, 2022).

## Fruit fly resistance in snap melon

Snapmelon (*Cucumis melo* L. var. *momordica* (Roxb.) belongs to the family *Cucurbitaceae*, which is a native of India and is used as a vegetable in a variety of ways. The melon fruit fly has been observed on 81 host plants, but snap melon is one of the most preferred hosts and has been a major limiting factor in obtaining good quality fruits and high yield. The genotypes IC-430190, DKS-AHS 2011/4, and DKS-AHS 2011/3 were resistant; IC-430160, IC-430162, IC-430175, IC-430179, IC-430180, IC-430185, IC-369788, and DKS-AHS 2011/2 were moderately resistant whereas IC-430155, IC-430164, IC-430169, IC-430171, IC-430172 and IC-430184 were found the susceptible genotypes to fruit fly infestation (Haldhar *et al.*, 2016b & 2018b). The free amino acid and the total soluble solid were the lowest in resistant and the highest in susceptible genotypes. In contrast, flavonoids, tannins, phenols, and total alkaloid contents were the highest resistant and lowest in susceptible genotypes of snap melon (Haldhar *et al.*, 2018b). Fruit length, flesh thickness and fruit diameter had significant positive correlations. In contrast, rind hardness at an immature stage, rind hardness at a mature stage, pericarp thickness and length of ovary pubescence had significant negative correlations with the per cent fruit infestation and larval density (Haldhar *et al.*, 2018b). Glandular trichomes secrete secondary metabolites, including flavonoids, terpenoids, and alkaloids, that could be poisonous, repellent, or trap insects and other organisms, thus forming a combination of structural and chemical defense (Sharma *et al.*, 2009).

## Fruit fly resistance in muskmelon

Musk melon (*Cucumis melo* L.) is one of the economically impotant horticultural crops worldwide and plays a vital role in international trade. Different forms of melon are known that are morphologically different. The genotypes AHMM/BR-1, RM-50 and AHMM/BR-8 were the most resistant; MHY-5, *D. Mad*hu and *P. Sarabati* were moderately resistant; AHMM/BR-13, P. Madhuras and Arka

Jeet were susceptible whereas Arka Rajhans and GMM-3 were highly susceptible varieties/ genotypes of muskmelon to fruit fly (Haldhar *et al.*, 2013a). Ismail *et al.*, (2010) reported that the cantaloupe flesh extract afforded the highest yield (89.6 ± 0.3%), whilst the lowest yield was obtained from the seed (13.7 ± 0.5%). The leaf extract showed the highest total phenolic content (26.4 ± 0.3 mg GAE/g extract) and entire flavonoid content (69.7 ± 3.37 µg RE/g extract). Total sugar, reducing sugar, and non-reducing sugar of different varieties/ genotypes of muskmelon ranged from 309 to 553.27, 62.07 to 124.27 and 246.93 to 429 (mg/g on a dry weight basis), respectively, with values significantly lower in resistant varieties/ genotypes and higher in susceptible varieties/ genotypes. The pH was significantly highest in Arka Rajhans (6.56) and lowest in RM-50 (5.67). Tannins, phenols, total alkaloid and flavonoid contents ranged from 0.02 to 0.12 mg/g, 15.27 to 39.13 mg/g, 0.24 to 1.25 % and 0.40 to 1.05 mg/g, respectively, with values significantly higher in resistant varieties/ genotypes and lower in susceptible varieties/ genotypes. Total sugar, reducing sugar, non-reducing sugar and pH of fruit had a significant positive correlation (P = 0.01), whereas tannins, phenols, alkaloids and flavonoid contents had significant negative correlations with the percentage fruit infestation and the larval density per fruit (Haldhar *et al.*, 2013a; Bhargava *et al.*, 2016).

## Fruit fly resistance in watermelon

Watermelon (*Citrullus lanatus*) is a popular dessert crop throughout the world's tropics and Mediterranean regions. The watermelon varieties/genotypes Asahi Yamato, Thar Manak, and AHW/BR-16 were resistant; AHW/BR-12, Arka Manik, Charleston Grey, AHW-65, AHW-19, Sugar Baby, and Durgapura Lal were moderately resistant; and AHW/BR-137, AHW/BR-9, IC 582909, BSM-1, and AHW/BR-60 were susceptible (Haldhar *et al.*, 2015b). The free amino acid content of fruit had a significant positive correlation, whereas flavonoid, tannin, total alkaloid, phenol, and ascorbic acid contents had a significant negative correlation with the percentage of fruit infestation and larval density per fruit. Backward stepwise regression analysis indicated that flavonoid and total alkaloid contents explained 88.4% of the total variation in fruit fly infestation. The maximum variation in fruit infestation was explained by flavonoid content (69.7%) followed by total alkaloid (18.7%), phenol (3.3%), ascorbic acid (1.4%), tannin (1.0%), and free amino acid contents (0.3%) (Haldhar *et al.*, 2015b). Total soluble solids and pH of fruit had a significant positive correlation. In contrast, tannin, phenol, alkaloid, and flavonoid contents had a significant negative correlation with the percentage of fruit infestation and larval density per fruit (Gogi *et al.*, 2010). The antixenotic mechanisms of fruit traits significantly differed among the tested watermelon varieties/ genotypes. Fruit length, fruit diameter and days to first fruit harvest had significant positive correlations. In contrast, rind hardness, rind thickness and length of ovary pubescence had significant negative correlations with the per cent fruit infestation and larval density. In these findings, biophysical fruit traits were also significantly different among genotypes (Gogi *et al.*, 2010). Pubescence consists of the layer of hairs (trichomes) extending from the epidermis of the above-ground plant parts, including stem, leaves, and even fruits. It occurs in several forms such as straight, spiral, stellate, hooked, and glandular (Hanley *et al.*, 2007).

## Fruit fly resistance in ridged gourd

Ridge gourd (*Luffa acutangula*) is a warm-season cucurbitaceous vegetable crop grown in different parts of India and in the tropical countries of Asia and Africa. The ridge gourd varieties/ genotypes; AHRG-57, Pusa Nasdar, and AHRG-29 were resistant; AHRG-35, Arka Sujata, AHRG-41, AHRG-36, S. Manjari, and S. Uphar were moderately resistant; AHRG-49, AHRG-33, AHRG-42, and AHRG-30 were susceptible whereas AHRG-47 and AHRG-31 were the highly susceptible varieties/ genotypes. Pooled data of larval density per fruit in both seasons (13.23- 28.5 larvae per fruit) was significantly lower in resistant and higher in susceptible varieties/ genotypes. The per cent fruit infestation was highest in AHRG-31 (79.72 %) and lowest in AHRG-57 (15.92 %), followed by AHRG-29 (17.67 %) (Haldhar *et al.*, 2013c; Haldhar *et al.*, 2015b). The phenotypic fruit trait mechanisms significantly differed among the tested ridge gourd varieties/ genotypes. Fruit length and fruit diameter had significant positive correlations, whereas rind hardness, rind thickness, fibre content and length of ovary pubescence had significant negative correlations with the per cent fruit infestation and larval density. Biophysical fruit traits were also found to be significantly different among genotypes (Gogi *et al.*, 2010). The allelochemical compounds of fruit significantly differed among the tested ridge gourd varieties/ genotypes. The free amino acid was lowest in resistant and highest in susceptible varieties/ genotypes. In contrast, flavonoid, tannins, phenols, and ascorbic acid contents were highest in resistant and lowest in susceptible varieties/ genotypes (Haldhar *et al.*, 2015a).

## Fruit fly resistance in kachri (*Cucumis melo* var. callosus)

*Cucumis* species is an important genus of cucurbitaceous vegetable crops and is widely grown for their fresh fruits at various stages. Kachri, a non-desertic form of *Cucumis melo* var. *callosus,* is an under-exploited drought-hardy cucurbit vegetable of the Indian Thar Desert. The kachri genotypes IC-350933 and IC-373479 were found to be highly resistant; IC-350953, IC-351005, IC-351088, IC-258131 and DKS 2011/01 were found to be resistant, whereas IC-351258, DKS 2011/02 and DKS 2011/03 were highly susceptible to melon fruit fly. The larval densities ranged from 4.87 to 15.50 larvae per fruit and were significantly lower in resistant genotypes than in the susceptible genotypes. The larval density was the highest in genotype DKS 2011/03 (15.5 larvae/ fruit), followed by IC-351258 (15.3 larvae/ fruit). The minimum larval density was found in IC-370479 (4.9 larvae/ fruit), followed by IC-350933 (5.2 larvae/ fruit). The per cent fruit infestation was the highest in IC-351258 (76.9 %) and the lowest in IC-350933 (7.8 %), followed by IC-370479 (8.5 %). The fruit infestation ranged from 7.8 to 76.9 %, significantly lower in resistant genotypes and higher in susceptible genotypes (Haldhar *et al.*, 2017a). The phenotypic (antixenotic) fruit trait mechanisms significantly differed among the tested kachri genotype. The length of ovary pubescence, rind hardness, rind thickness, fruit length, and fruit diameter had significant negative correlations with the per cent fruit infestation and larval density. In these findings, biophysical fruit traits were also significantly different among the genotypes (Gogi *et al.*, 2010; Haldhar *et al.*, 2015b). The allelochemical compounds of the fruit differed significantly among the tested kachri genotypes. The flavonoid, tannins, phenols, and total alkaloid contents were the highest in resistant and lower in the susceptible genotypes of kachri. The percentage of fruit infestation and the larval density per fruit with flavonoid (-0.955 & -0.938), tannins (-0.891 & -0.902), phenols (-0.903 & -0.896) and total alkaloid (-0.797 & -0.759) had a significant negative correlation. Backward stepwise regression analysis indicated that flavonoid and tannin contents explained 93.6% of the total variation in fruit fly infestation. The maximum variation in fruit infestation was explained by flavonoid content (91.2%), followed by tannins (2.4%), phenols (0.4%), and total alkaloid contents (0.1%) (Haldhar *et al.*, 2017a).

## Conclusion

The chapter concludes that HPR is a promising and sustainable strategy for managing insect pests and diseases in underutilized horticultural crops, and that more research and development efforts are needed to exploit the potential of HPR in these crops. The chapter also suggests some future directions and recommendations for HPR research and development, such as germplasm collection, characterization, and conservation, screening and evaluation of HPR, breeding and biotechnology for HPR improvement, molecular and biochemical studies of HPR mechanisms, integration of HPR with other pest management tactics, and extension and dissemination of HPR technologies.

**Table 1. Genotypes of underutilized horticultural crops resistant to insect pests in arid and semi-arid regions of India**

| Crop | Insect species | Resistant genotypes | References |
|---|---|---|---|
| Ber, *Z. mauritiana* | Stone weevil, *A. imalayanus* | Resistant: Kali, Katha, Illaichi and Tikadi<br>Moderately resistant: Akharota, Dandan, Gola, Goma Kirti, Sanaur-1, Seb, Umran and ZG-3 | Haldhar *et al.*, 2018c |
| | Fruit fly, *C. vesuviana* | Resistant: The ber cultivars such as Tikadi, Katha and Illaichi<br>Moderately resistant: BS-75-1, Safeda, Dandan, Gola, Goma Kirti, Jogia, Narma, Mundia, Reshmi, Seb, ZG-3, Umran and Akharota | Haldhar *et al.*, 2018a |
| | Fruit borer, *M. scyrodes* | Resistant: Safeda, Tikadi, Darakhi and Illaichi<br>Moderately resistant: BS-75-1, Gola, Goma Kirti, Seb and Umran | Haldhar *et al.*, 2013b |

| Crop | Insect species | Resistant genotypes | References |
|---|---|---|---|
| Indian Cherry, *C. myxa* | Tingid bug, *D. cheriani* | Resistant: AHCM-22, AHCM-25 and AHCM-34<br>Moderately resistant: AHCM-14, AHCM-30 and AHCM-31 | Haldhar *et al.*, 2019 |
| Jamun, *S. cumini* | Fruit borers, *M. scyrodes* and *D. aprobola* | Resistant: GJ-26, GJ-27, GJ-19, GJ-17 and GJ-15<br>Moderately resistant: GJ-3, GJ-6, GJ-7, GJ-9, GJ-13, GJ-16, GJ-18, GJ-20, GJ-23, GJ-24 and GJ-25 | Haldhar *et al.*, 2016c |
| Datepalm, *P. dactylifera* | Fruit borers, *B. amydraula* and *A. sabella* | Resistant: Migra, Tayer, Javantri and Medjool | Haldhar *et al.*, 2021 |
| Pomegranate, *P. granatum* | Mite, *T. punicae* | Resistant: HP Collec., Speen Sacarin, Gule Shah Rose Pink and AHPG-C3<br>Moderately resistant: Sur Sukker, Tujetis EC 4347, Bassin Seedless, Dholka, Uthkal, Agah, Achik Dana and Goma Khata | Haldhar *et al.*, 2022 |
| Snapmelon, *C. melo* L. var. *Momordica* | Fruit fly, *B. cucurbitae* | Resistant: IC-430190, DKS-AHS 2011/4, and DKS-AHS 2011/3<br>Moderately resistant: IC-430160, IC-430162, IC-430175, IC-430179, IC-430180, IC-430185, IC-369788, and DKS-AHS 2011/2 | Haldhar *et al.*, 2018b |
| Musk melon, *C. melo* | Fruit fly, *B. cucurbitae* | Resistant: AHMM/BR-1, RM-50 and AHMM/BR-8<br>Moderately resistant: MHY-5, D. Madhu and P. Sarabati | Haldhar *et al.*, 2013a |
| Watermelon, *C. lanatus* | Fruit fly, *B. cucurbitae* | Resistant: Asahi Yamato, Thar Manak, and AHW/BR-16<br>Moderately resistant: AHW/BR-12, Arka Manik, Charleston Grey, AHW-65, AHW-19, Sugar Baby, and Durgapura Lal | Haldhar *et al.*, 2015a |
| Ridge gourd, *L. acutangula* | Fruit fly, *B. cucurbitae* | Resistant: AHRG-57, Pusa Nasdar, and AHRG-29<br>Moderately resistant: AHRG-35, Arka Sujata, AHRG-41, AHRG-36, S. Manjari, and S. Uphar | Haldhar *et al.*, 2015b; Haldhar *et al.*, 2013c |
| Kachri, *C. melo* var. *callosus* | Fruit fly, *B. cucurbitae* | Highly resistant: IC-350933and IC-373479<br>Resistant: IC-350953, IC-351005, IC-351088, IC-258131 and DKS 2011/01 | Haldhar *et al.*, 2017a |

*Source:* **Haldhar** *et al.*, **2023a**

## References

Abedi Z, Golizadeh A, Soufbaf M, Hassanpour M, Jafari-Nodoushan A & Akhavan HR. 2019. Relationship between performance of carob moth, *Ectomyelois ceratoniae* Zeller (Lepidoptera: Pyralidae) and phytochemical metabolites in various pomegranate cultivars. *Front. Physiol.* 10: 1425.

Balikai RA, Kotikal YK & Prasanna PM. 2013. Global scenario of insect and non-insect pests of jujube and their management options. *Acta Horticul.* 993: 253-77.

Dhileepan K. 2017. Biological control of *Ziziphus mauritiana* (Rhamnaceae), feasibility, prospective agents and research gaps. *Annal Appl Biol* 170(3): 287-300.

Elmeer K, Sarwath H, Malek J, Baum M & Hamwieh A. 2011. New microsatellite markers for assessment of genetic diversity in date palm (*Phoenix dactylifera* L.). *Biotechnology* 1: 91-97.

Faleiro JR, El-Shafie HAF, Ajlan AM & Sallam AA. 2014. Screening date palm cultivars for resistance to red palm weevil, *Rhynchophorus ferrugineus* (Coleoptera: Curculionidae). *Florida Entomologist*, 97(4):1529-36.

Gogi MD, Ashfaq M, Arif MJ, Sarfraz RM & Nawab NN. 2010. Investigating phenotypic structures and allelochemical compounds of the fruits of *Momordica charantia* L. genotypes as sources of resistance against *Bactrocera cucurbitae* (Coquillett) (Diptera, Tephritidae). *Crop Protec* 29: 884-90.

Haldhar SM & Maheshwari SK. 2021. Pest management in dryland horticultural crops. Biotech Books, pp-473; ISBN: 978-81-7622-491-8.

Haldhar SM, Berwal MK, Samadia DK, Kumar R, Gora JS & Choudhary S. 2018d. Biochemical basis of plant-insect interaction in arid horticulture crops: a scientific review. *Journal of Agriculture and Ecology* 6: 1-16.

Haldhar SM, Bhargava R, Choudhary BR, Pal G & Kumar S. 2013a. Allelochemical resistance traits of muskmelon (*Cucumis melo*) against the fruit fly (*Bactrocera cucurbitae*) in a hot arid region of India. *Phytoparas.* 41: 473-81.

Haldhar SM, Bhargava R, Krishna H, Berwal MK and Saroj PL. 2018a. Bottom-up effects of different host plant resistance cultivars on ber (*Ziziphus mauritiana*)-fruit fly (*Carpomyia vesuviana*) interactions. *Crop Protec.* 106: 117-24.

Haldhar SM, Bhargava R, Singh RS, Samadia DK, Jat GC & Singh D. 2019. Do bottom-up effects (resistance) of different accessions of Indian cherry (*Cordia myxa* L.) help against tingid bug [*Dictyla cheriani* (Drake)] attack? *Journal of Agriculture and Ecology* 8: 84-103

Haldhar SM, Choudhary BR, Bhargava R & Gurjar K. 2015b. Host plant resistance (HPR) traits of ridge gourd (*Luffa acutangula* (Roxb.) L. against melon fruit fly, (*Bactrocera cucurbitae* (Coquillett)) in hot arid region of India. *Horti Scientia.* 194: 168-74.

Haldhar SM, Choudhary BR, Bhargava R & Meena SR. 2015a. Antixenotic and allelochemical resistance traits of watermelon against *Bactrocera cucurbitae* in a hot arid region of India. *Florida Entomol.* 98: 827-34.

Haldhar SM, Choudhary BR, Bhargava R & Sharma SK. 2013c. Screening of ridge gourd varieties/ genotypes (*Luffa acutangula*) for resistance fruit fly (*Bactrocera cucurbitae*) in hot arid region of Rajasthan. *Indian J Arid Horticul.* 8(1-2): 21-24.

Haldhar SM, Deshwal HL, Jat GC, Berwal MK & Singh D. 2016a. Pest scenario of ber (*Ziziphus mauritiana*Lam.) in arid regions of Rajasthan, a review. *J Agricul Ecol.* 1: 10-21.

Haldhar SM, Karuppaiah V, Sharma SK, Singh D. 2012. Insect-pests of ber (*Ziziphus mauritiana* Lam.) as influenced by abiotic factors in arid region of Rajasthan. Global conference on "*Horticulture for food, nutrition and livelihood options*" organised by ASM foundation, New Delhi and OUAT, Bhubaneswar. Odisha during 27-31, May, 2012.

Haldhar SM, Kumar R, Corrado G, Berwal MK, Gora JS, Thaochan N, Samadia DK, Hussain T, Rouphael Y, Kumar P & Basile B. 2022. A field screening of a pomegranate (*Punica granatum*) ex-situ germplasm collection for resistance against the false spider mite (*Tenuipalpus punicae*). *Agriculture* 12: 1686. https://doi.org/10.3390/agriculture12101686.

Haldhar SM, Samadia DK, Bhargava R & Singh D. 2016b. Screening of snapmelon (*Cucumis melo* var. momordica) genotypes for resistance against fruit fly (*Bactrocera cucurbitae* (Coquillett)) in hot arid region of Rajasthan. *Internat J Horti Canada* 6(19): 1-7.

Haldhar SM, Samadia DK, Bhargava R & Singh D. 2017a. Host plant genotypes determine bottom-up effect of *Cucumis melo var. callosus* against melon fruit fly. *Crop Protec.* 98: 157-65.

Haldhar SM, Samadia DK, Bhargava R, Choudhary BR & Singh D. 2018b. Host plant accessions determine bottom-up effect of snapmelon (*Cucumis melo* var. *momordica*) against melon fly (*Bactrocera cucurbitae* (Coquillett)). *Breeding Scien.* 68: 499-507.

Haldhar SM, Sharma SK, Bhargava R, Singh RS, Sharma BD & Singh D. 2013b. Insect pests of arid fruit crops, practical manual for field identification and damage symptoms. CIAH/ Tech./ Pub. No. 42 pp. 1-53.

Haldhar SM, Singh AK, Gora JS, Berwal MK & Hussain T. 2023b. Plant-insect interactions in different genotypes of ber (*Ziziphus mauritiana* L) against fruit borer (*Meridarchis scyrodes* Meyrick). *Journal of Agriculture and Ecology* 16: 78-87; https://doi.org/10.58628/JAE-2316-216.

Haldhar SM, Singh AK, Kumar K & Sarolia DK. 2018c. Antixenotic and allelochemical resistance traits of ber (*Ziziphus mauritiana*) against stone weevil, *Aubeus himalayanus* in hot arid region of India. *Indian Journal of Arid Horticulture* 13: 50-58.

Haldhar SM, Singh RS & Singh D. 2021. HPSI of date palm (*Phoenix dactylifera* L.) varieties/ genotypes against fruit borers, *Batrachedra amydraula* and *Arenipses sabella* in hot arid region. *Journal of Agriculture and Ecology* 11: 26-36.

Haldhar SM, Singh S, Singh AK, Bhargava R, Singh RS & Singh D. 2016c. Host Plant Resistance (HPR) study in Jamun (*Syzygium cumini*) against fruit borers, *Meridarchis scyrodes* Meyrick and *Dudua aprobola* (Meyrick) in semi-arid region. International Conference on Entomology organized by Punjabi University, Patiala during 03 to 05, December 2016

Haldhar SM, Sinha B, Choudhary BR, Singh D, Konsam J and Thaochan N. 2023a. Plant-insect interaction in underutilized horticultural crops for sustainable production. *Journal of Agriculture and Ecology* 17: 1–13. DOI:https://doi.org/10.58628/JAE-2316-301.

Han P, Desneux N, Michel T, Le Bot J, Seassau A, Wajnberg E, Amiens-Desneux E and Anne-Violette Lavoir AV. 2016. Does plant cultivar difference modify the bottom-up effects of resource limitation on plant-insect herbivore interactions? *J. Chem. Ecol.*, 42: 1293-1303.

Hosagoudar IM, Nandihalli BS, Lingappa S & Patil DR. 1999. Incidence of ber fruit borer (*Meridarchis scyrodes* Meyrick) and fruit fly (*Carpomyia vesuviana* Costa) on different varieties. *Pest Management in Horticulture Ecosystem* 5(2): 82-85.

Ismail HI, Chan KW, Mariod AA & Ismail M. 2010. Phenolic content and antioxidant activity of cantaloupe (*Cucumis melo*) methanolic extracts. *Food Chemist*. 118: 120–27.

Johnson DV, Al-Khayri JM & Jain SM. 2013. Seedling date palms (*Phoenix dactylifera* L.) as genetic resources. *Emirates Journal of Food and Agriculture* 25(11): 809-30.

Johnson MT & Agrawal AA. 2005. Plant genotype and environment interact to shape a diverse arthropod community on evening primrose (*Oenothera biennis*). *Ecology* 86: 874-85.

Kavitha Z & Savithri P. 2002. Documentation of insect pests on *ber*. *South Indian Horticulture* 50(1/3): 223-25.

Nidhi CN & Haldhar SM. 2021. Pest management in dryland horticultural crops: book review. *Journal of Agriculture and Ecology* 11: 78-82.

Shahid I, Muhammad A, Muhammad AK & Nazir J. 2014. Cultivar variation of cauliflower against cabbage butterfly *Pieris brassicae* (L.) Pieridae: Lepidoptera. *Pakishtan Journal of Agriculture Science* 51(2): 325-29.

Sharma HC Sujana G & Rao DM. 2009. Morphological and chemical components of resistance to pod borer, *Helicoverpa armigera* in wild relatives of pigeonpea. *Arthropod Plant Interac* 3: 151-61.

Sobhani M, Goldansaz SH, Hatami B & Hosseini SA. 2015. A field screening of pomegranate cultivars for resistance to the carob moth, *Ectomyelois ceratoniae*, and compatibility with its larval parasitoids. *Int. J. Pest Manag*. 61: 346–52.

Tews J, Brose U, Grimm V, Tielborger K, Wichmann MC, Schwager M *et al.,* 2004. Animal species diversity is driven by habitat heterogeneity/diversity: the importance of keystone structures. *J Biogeogr.* 31: 79–92.

Vir O, Roy S, Tyagi RK, Singh AK & Meghwal P R. 2015. *Tikdi*-a fruit fly resistant land race of ber (*Ziziphus mauritiana* Lamarck). *Indian Journal of Plant Genetic Resources* 28(2): 252-55.

Visschers IGS, Peters JL, Timmermans LLH, Edwards E, Ferrater JB, Balatero, CH, Stratongjun M, Bleeker PM, Herwijnen Z, Glawe, GA, Bruin J, Vandam, NM & Macel M. 2019. Resistance to three thrips species in *Capsicum* spp. depends on site conditions and geographic regions. *Journal of Applied Entomology* 143: 929-41.

War AR, Paulraj MG, Ahmad T, Buhroo AA, Hussain B, Ignacimuthu S & Sharma HC. 2012. Mechanisms of plant defense against insect herbivores. *Plant Sign Behav.* 7: 1306-20.

CHAPTER 22

# NOVEL PESTS OF VEGETABLE CROPS: CURRENT STATUS AND WAY FORWARD

**A.B. Rai and Jaydeep Halder**

*Division of Crop Protection, ICAR-Indian Institute of Vegetable Research, Varanasi, Uttar Pradesh, India*

*Corresponding author: abraiiivr@gmail.com*

---

Threats to food security posed by invasive and emerging insect pests are among the worst problems for mankind, and they have got progressively worse in recent years as a result of climate change and global warming. These days, sucking insect pests are a considerably bigger concern in agricultural and horticultural crops, especially vegetables, than in previous years due to rising temperatures (Rathee and Dalal, 2018). Increased global trade in agriculture without much quarantine has enhanced the chances of introduction of exotic pests. Chilli black thrips (*Thrips parvispinus* Karny), tomato pin worm (*Tuta absoluta* (Meyrick)), Solenopsis mealy bug (*Phenacoccus solenopsis* (Tinsley)), Western flower thrips (*Frankliniella occidentalis* (Pergande) and more recently China-3 race of whitefly (*Bemisia tabaci* (Gennadius) are some recent examples of invasive alien pests noted infesting vegetable crops in India (Ananthakrishnan, 2009; Sridhar *et al.*, 2014; Halder and Rai, 2021). These invasive pests cause more harm in absence of natural enemies in new habitat. Similarly, some newly emerging indigenous pests namely Mirid bugs (*Nesidiocoris cruentatus* (Ballard) and *Metacanthus pulchellus* Dallas) on bottle gourd; melon weevil (*Acythopeus curvirostris citrulli* (Marshall) on gourd crops; White plume moth (*Sphenarches caffer* (Zeller) on bottle gourd; Spotted pod borer (*Maruca vitrata* Fab.) on cowpea; gall midge (*Asphondylia capparis* (Rubsaman) on brinjal and chilli; Hadda beetle (*Henosepilachna vigintioctopunctata* (Fabricius) and *Epilachna dodecastigma* (Wiedemann) on cowpea and bitter gourd causes serious problem. Mites belonging to the Tetranychidae, Tarsonemidae and Eriophyidae family on beans, brinjal, cucurbits, okra, tomato and chilli have intensified their incidence in eastern part of the country (Rai *et al.*, 2007; Singh and Raghuraman, 2011; Rai *et al.*, 2014, Halder and Srinivasan, 2011, Halder *et al.*, 2011; Maurice and Ramteke, 2012; Jamwal *et al.*, 2013; Puri and Mote, 2004; Patel *et al.*, 1993; Halder and Rai, 2021) which are adding to heavy loss of crops (Table 1). At present various pests cause losses in the range of 10-30 per cent in vegetables depending upon the infestation severity in time and space (Shivlingaswamy *et al.*, 2002; Rai *et al.*, 2014a,b).

Many pests have expanded their host horizon, developed pesticide resistance/ resurgence and emerged as major pests. In vegetable cultivation, large scale cultural operations, short cropping periods as well as scheduled based and frequent chemical interventions do not allow the natural enemies to grow to their potential besides causing several other problems like environmental pollution, pesticide residue in harvested products and increased cost of production (Banik and Halder, 2013; Halder *et al.*, 2019(a); Roy *et al.*, 2017). As majority of natural enemies (parasitoids and predators) active under natural environment are not amenable to mass multiplication and release, their conservation through various means should be given focus of main attention to save agro-environment. This can be effectively achieved through rationalization of pesticide use including new generation highly degradable green chemistry safer molecules in agriculture that require 10-15 times lower dose than conventional insecticides in case of sucking pests and 4-53 times less dose for borer pests (Kodandaram *et al.*, 2010) as well as integration of several other ecofriendly approaches with major focus on biological based pesticides which is becoming so pertinent with the advent of environmentally conscious organic and natural farming system in the present context of human health. The information generated on important novel insect pests (indigenous and exotic) infesting vegetable crops has been elaborated in the light of their status, management and future direction.

**Table 1. Emerging insect pests (exotic and indigenous) in vegetable crops in India**

| Pests | Crops |
|---|---|
| **Exotic insect pests** | |
| South American Tomato pinworm, *Tuta absoluta* (Meyrick) (Lepidoptera: Gelechiidae) | Tomato |
| Solenopsis mealy bugs *Phenacoccus solenopsis* Tinsley (Hemiptera: Pseudococcidae) | Okra, tomato, brinjal, chilli, cucurbits |
| Western flower thrips (WFT), *Frankliniella occidentalis* (Pergande) (Thysanoptera: Thripidae) | Tomato |
| China3-Whitefly, *Bemisia tabaci* (Gennadius) (Hemiptera: Aleyrodidae) | Okra, crucifers, cucurbits and many solanaceous crops |
| Black chilli thrips, *Thrips parvispinus* Karny (Thysanoptera: Thripidae) | Chilli |
| Fall Armyworm, *Spodoptera frugiperda* (J E Smith) (Lepidoptera: Noctuidae) | May infest cabbage, soybean |
| **Indigenous pest** | |
| Chilli gall midge, *Asphondylia capparis* Rubsaman (Diptera: Cecidomyiidae:) | 50-60% fruit damage in chilli and brinjal |

| | |
|---|---|
| Spotted pod borer, *Maruca vitrata* Fab. ( Lepidoptera : Crambidae) | Up to 42% pod damage in cowpea |
| Hadda beetles, *Henosepilachna vigitioctopunctata* Fab and *Epilachna dodecastigma* (Wied.) (Coleoptera: Coccinellidae) | Up to 80% leaf damage in cowpea |
| White plume moth, *Sphenarches caffer* (Zeller) (Lepidoptera: Pterophoridae), | Minor pest in bottle gourd but major in Indian bean |
| Mirid bugs, *Nesidiocoris cruentatus* (Ballard) and *Metacanthus pulchellus* Dallas (Hemiptera: Miridae) | 70-80% fruit damage, 30% shoot damage in bottle gourd) |
| Spider mite (*Tetranichus cinnabarinus* (Boisduval), *T. urticae* C. L. Koch) | Okra, brinjal, tomato, cucurbits |
| Cucumber moth, *Diaphania indica* Saunders (Lepidoptera: Crambidae) | 78% fruit damage, 36% shoot damage (bitter gourd) |
| Melon weevil, *Acythopeus curvirostris citrulli* (Marshall) (Coleoptera: Curculionidae) | 70-80% fruit damage, 30% vine damage (sponge gourd) |
| Drumstick fruit and seed borer / Moringa Leaf caterpillar, *Noorda blitealis* (Walker) (Lepidoptera: Crambidae) | 30-40% fruit damage in Moringa |
| Tortoise beetle, *Cassida circumdata* Herbst (Coleoptera: Chrysomelidae) | Up to 50% leaf damage |

**Source: Modified from Rai *et al.*, 2014a,b; Halder and Rai, 2021**

## Reasons for increasing pest status in vegetables

Advent of global warming, climate change, increasing connectivity due to globalization for agricultural trade and commerce, human movement, quarantine failure, change in cropping pattern, introduction of high-yielding varieties (HYVs), intensive mono-cropping, off-season cultivation, injudicious use of insecticides leading to pest resistance / resurgence. The adverse effect on natural enemies, non-target organisms and environment are noted to be the important elements for increasing pests status in time and space under vegetable ecosystem. Therefore, a search for alternatives to chemical pesticides is of high priority. Besides many pests have extended their host horizon which is a cause of concern to vegetable growers. In comparison to other pests, the problem of sucking insect pests has been worse over the past few years due to rising temperatures.

## Emerging Threats

### Invasive Alien Species

These species are source of complex and far-reaching challenges threatening agro-ecosystems worldwide. International trade and movement of planting materials from one country to another breach the existing framework of biogeographical boundaries paving ways for the widespread and rapid spread of alien invasive insect pests across the world. India harboured around 173 invasive species including 47 invasive species of agricultural ecosystem of which 23 are invasive alien insects (Viraktamath, 2002; Singh *et al.*, 2020). Due to the lack of their natural enemies these alien species are more serious in new areas and cause many economic, and ecological impacts facilitating subsequent invasions. Some of the important invasive alien insect species introduced in India in the last couple of years infesting vegetable crops are discussed.

**American Pinworm, *Tuta absoluta* (Meyrick):** It is a serious invasive alien insect pest known as South American tomato pinworm or tomato pin worm. This pest has been classified as the most serious threat for tomato production in protected and open fields worldwide. The pest is native to Andes Region of South America. For over 30 years it was a major problem for tomato production in Central America (Garcia and Espul, 1982) and has recently spread to several parts of Europe, North Africa, Middle East and now introduced to Indian subcontinent in 2014 infesting tomato leaves, flowers, terminal shoots and fruits in Pune and subsequently in major tomato growing areas of Maharastra (>50% plant affected) (Shashank *et al.*, 2015) and Karnataka (with 87 % infestation) (Sridhar *et al.*, 2014) as well as Tamil Nadu, Himachal Pradesh, Telangana, Andhra Pradesh and Uttarakhand. In Varanasi region, Uttar Pradesh it was first observed on tomato during January, 2017 (Halder *et al.*, 2019(b)). Though the pest's primary host is tomato, it has also been noted to feed on other solanaceous vegetables like potato, brinjal, pepper, common bean including solanaceous weeds like *Solanum nigrum* and *Datura stramonium* (Pereyra and Sanchez, 2006). Due to its aggressive nature, crop destruction potential and host range it can become a key pest of vegetable crops in India. Potential yield loss (both quality and quantity) is significant and if the pest is not managed in time, can reach up to 100% in tomato and is considered a key pest of greenhouse and open-field tomato (Arturo *et al.*, 2012). One of the possible pathways for a long-distance dissemination of *T. absoluta* could be through packaging materials (boxes) from infested countries (EPPO, 2005 and 2010). It feeds on leaves, buds, calyces, young green fruit/red ripe fruit. On leaves, the larvae mine in the mesophyll tissue, forming irregular blotches or bigger zigzag papery gallery than that of *Liriomyza trifoli*. Infested fruits show minute holes forming galleries representing open areas for invasion/entry by secondary pathogens and consequent fruit rot. Eggs are laid singly on under surface of leaves including stems and sepals. Pupation takes place in soil or on plant parts such as leaves and stem. Adult moths are silvery brown with mottled wings. More than 60% plants were observed affected by the pest and number of larvae varied from 5-17 per infested plant (Halder *et al.*, 2019(b)). The pest is multivoltine having nearly 12 generations per year. Thermal requirement for the pest is estimated as 416.7 degree days to complete life-cycle with 8°C lower temperature threshold and 37.3°C upper temperature threshold (Krechmer and Forester, 2015). The rapid growth, potential natural dispersal and resistance to insecticides render this pest as the most serious threat for tomato production systems.

**Management:** Use of insect-proof net to cover nursery, use of pest-free seedlings, and avoidance of solanaceous crops after tomato and destruction of affected plants and fruits are suggested to minimize the pest infestation. More than 70 predators and parasitoids have been reported associated with this pest worldwide and some of the Eulophids, Braconids and Mirids have encouraging prospective to be incorporated in the management strategies for the pest in newly invaded areas. A mirid bug, *Nesidiocoris tenuis* (Reuter) found predating on eggs and early instars of *T. absoluta*. Hymenopteran parasitoid, *Neochrysocharis formosa* (Westwood) has also been recovered from the infested leaves but, its association with *T. absoluta* remained uncertain and needs further confirmation as the serpentine leaf miner, *Liriomyza trifolii* (Burgess) was also present on the leaves (Sharma and Gavkare, 2017). A detailed examination of numerous natural enemies attacking *T. absoluta* in India would be needed. *Beuvaria bassiana, Bacilus thuringiensis*, Abamectin, Azadiractin, egg parasitoid, *Trichogramma achaeae* are recommended for pest management in Azores Island (Oliveira *et al.*, 2017). Use of yellow sticky traps (40-50 traps/ha) and Delta sex- pheromone trap for monitoring (2-4 traps/ha) and for mass trapping (20-25 traps/ha) in open fields is suggested for pest suppression. In both cases, the pin worm lure should be changed 4-6 weeks. Luminous traps with water, with limed bucket (Advantage low cost capturing both male and female) and light trap @1/150 m2 + 1 pheromone trap/300m2 + Spinosad or Flubendamide are also suggested for pest suppression.

**Western flower thrips (WFT), *Frankliniella occidentalis* (Pergande)**: It is an American origin cosmopolitan, polyphagous pest and rated as one of the most destructive thrips of the world. It punches a hole into the host plants and then sucks the fluids from the cells. Apart from direct feeding, it is a highly efficient vector of different plant tospoviruses including the tomato spotted wilt virus (TSWV) causing high economic loss worldwide. It is known to transmit as many as five species of tospoviruses. In India, Tyagi and Kumar (2015) have reported it for the first time from Bangalore, Karnataka causing huge economic losses on tomato and subsequently on *Chrysanthemum* causing necrosis disease (Suganthy *et al.*, 2016). It is highly polyphagous pest species in Thysanoptera, causing direct damage on a variety of hosts in 65 families including groundnut, cotton, tobacco, alfalfa, vegetables (cucurbits, cabbage, eggplant, tomato, carrot, peas, beans, capsicum, wild radish etc.), fruits (grapes, apple, peach, strawberry, watermelon plum, apricots, artichoke etc.) and ornamentals (gladiolus, carnations, chrysanthemum, gerbera, hibiscus, geranium, roses, orchids, etc.). The insect completes its life-cycle in around 2-3 weeks, passing through egg, nympal stages (I and II), pre-pupal, pupal and adult stages. Female produces 60-80 eggs in her lifetime.

**Management:** it is a typical pest of greenhouse and covered crops, which opens up many opportunities for alternative ways to control the pest. Blue sticky roller traps can be employed to catch flying adult thrips, anti-thrips screens reduce thrips immigration into the greenhouses. Release of predators *viz., Amblyseius cucumeris, A. degenerans, Orius* sp., *Neoseiulus cucumeris* is reported for its management, entomopathogenic nematodes and fungi for example *Steinernema, Beauveria, Paecilimyces* sp. can be used as foliar sprays in flowering crops against adult thrips under high humid conditions, use of beneficial-compatible compounds or those with low foliar residual effect to limit long term effects on predators and parasites, improved insecticide formulations and the use of adjuvants may be a key factor for successful control.

**Solonepsis mealybug, *Phenacoccus solenopsis* (Tinsley):** It originated from USA and reported to damage cotton and other crops of 14 families. In India during 2007, the pest was recorded to be pre-dominant one, comprising 95% of samples from 47 locations representing 9 cotton growing states of India (Punjab, Rajasthan, Maharashtra, Gujarat) (Jhala *et al.*, 2008). In Varanasi region, it was first recorded during 2010-11 on tomato and recently infesting Brinjal, Chilies, Pointed Gourd, Pumpkin, Okra, Tomato, Chilli and weeds like *Parthenium* (Halder *et al.*, 2013). Its peak period of infestation was recorded during December to April in vegetable ecosystem. Both nymph and adult stages suck sap from growing points, plants lose vitality, remain stunted, leaves turn yellow, dry up and finally death of plants. It excretes honey dew while feeding which favours the development of black sooty-mould on leaves retarding the photosynthetic activity of plants (Halder *et al.*, 2015(b)). They often attended by the ants which cause for their rapid spreading. The mean developmental period from immature crawler to adult stage varied from 13.2±1.8 days under optimum conditions (23.3-30.2 C temperature and 40.5-92.5% relative humidity). The pest was noted was more serious in vegetable crops grown under protected condition. Some of the characteristic features of the pest for its upsurge are due to quarantine failure, having protective white waxy coating, high reproductive rate and ability to hide in cracks and crevices and its movement aided through '*balooning*', mechanical and ants (Halder *et al.*, 2015(b)).

**Management:** Field removal of weeds especially *Parthenium* and *Xanthium* the most suitable hosts for mealybugs, Neem Seed Kernel Extract (NSKE 5%) 50 ml/l, spray of Fish Oil Rosin Soap (FORS), *Leccanicillium lecanii* ( 2x108 CFU/ml) @ 2.5 g/ l+ neem oil @ 2.5 ml/lit (1:1 ratio), inoculative release of adult *Zygogramma bicolorata* beetle @ 500-1000 beetles /ha on *Parthenium* grass and need based spray of CIBRC approved chemicals will be beneficial.

**Black thrips, *Thrips parvispinus* (Karny):** Cosmopolitan species of quarantine importance is reported from Thailand to Australia (Mound and Collins, 2000). In Last couple of decades there has been drastic expansion in geographical distribution covering France, Greece, Hawaii, Mauritius, Spain, Tanzania, Netherland, Indonesia (Johari *et al.*, 2014; NPPO, 2019). It is a polyphagous pest noted to infest beans, eggplant, papaya, pepper, potato, shallot and strawberry (NPPO, 2019). Mounds and Collins (2000) reported its occurrence in India. Subsequently in India, it was first reported on papaya (*Carica papaya*) in Bengaluru, Karnataka. Haplotyping data suggested that the Indonesia may be a probable source of invasion of this pest to India (Tyagi *et al.*, 2015). In addition, it damages on ornamental plants like Anthurium, Chrysanthemum, Dahlia, Dipladenia, Gardenia, and Ficus (Nagaraju *et al.*, 2021; Rachana *et al.*, 2021). It has spread rapidly, affecting chilli crop in Telangana and Andhra Pradesh and also seen in Tamil Nadu, Karnataka, Chhattisgarh, Kerala and more recently recorded infesting chilli cops in Uttar Pradesh. About 90-95% flowers (chilli) were badly affected (18-20 thrips/flower) in Andhra Pradesh,

Chhattisgarh. Eggs are inserted into leaves and flowers and incubation period is 4-5 days. It passes through five stages and completes its life-cycle in 13-15 days (Thorat *et al.*, 2022). Though the pest is not currently reported a vector of Tospoviruses, it may likely acquire viruliferous trait. Therefore, it is imperative to strengthen domestic quarantine mechanisms to check the spread of this nefarious pest to the rest of India.

**Management:** Domestic quarantine mechanism to be put in place, constant exhaustive monitoring for its infestation in new areas, complete destruction of the infested plants in the specific areas to avoid further spread in new areas, use of healthy and pest free seedlings, uprooting off-seasonal host weeds (*Parthenium hysterophorus* and *Abutilon* spp) from field and installation of blue sticky traps 25-35 per acre for mass trapping immediately after transplanting, *Pseudomonas fluorescence*-BAIRPFDWD @ 20 g/l or *Bacillus albus*-NBAIR-BATP @ 20 g/l spray focusing on flowers and fruits, use of neem oil, pongamia oil or soap solution in heavily infested sites are suggested for pest management (Murai *et al.*, 2009; Thorat *et al.*, 2022). Using kairomones (volatile chemicals from host plants) increases trapping of thrips species up to six-fold can be advantageously utilized (Nielsen *et al.*, 2010),

**China3-Whitefly, *Bemisia tabaci* (Gennadius):** In vegetables, okra, crucifers, cucurbits and solanaceous crops are major hosts of whitefly. Diversity of whitefly biotypes on vegetable crops in Varanasi using mtCOI gene revealed the presence of most predominant biotype Asia I and Asia II 5. China-3 was also noted to be dominant and widespread in the region, and in future and it may replace the indigenous species. With regards to infection frequencies of endosymbioants in different genetic groups of whitefly, *Wolbachia* infection was noted to be maximum. Moreover, *Arsenophonus* infection was maximum in Asia II 5 where as China 3 harboured highest *Wolbachia, Cardinium* and *Rickettsia* infection (Anonymous, 2015). The silver leaf whitefly or 'B' biotype of whitefly was identified for the first time in India at Kolar district, Karnataka in 1999. This whitefly has wider host range, rapid reproduction ability inflicting more damage and more resistance to insecticides. In recent past, high incidence caused failure of the tomato crop in Bangalore, Karnataka, through direct damage to plants by sap feeding on cell sap and indirectly transmitting tomato leaf curl virus. The silver leaf whitefly could spread to affect a wider range of crops, including many vegetables, tobacco and cotton (Ananthakrishnan, 2009). Whitefly lays eggs in circular group, on undersides of leaves by a pedicel inserted into a fine slit made by the female. On hatching, first instar larva or crawler moves to a suitable feeding site on lower leaf surface where it moults and becomes stationary throughout the remaining stages, first three larval stages last 2-4 days, fourth larval stage is called as puparium which lasts for about 6 days. Adult emerges through a 'T'-shaped rupture in the puparium. Female may live up to 60 days, while male lives shorter (9 to 17 days). It passes about 11 to 15 generations in a year. Besides direct damage, the pest is an important vector of a wide array of plant virus diseases like YLCV, BGMV, ACMV etc. Infected plants exhibit any one or a combination of symptoms like vein yellowing, inter-vein yellowing, leaf yellowing, yellow blotching of leaves, yellow mosaic of leaves, leaf curling, vein thickening, leaf enations, leaf cupping, plant stunting.

**Management:** Yellow sticky trap for monitor at early stage and also for mass trapping of this pest is recommended. Removal of weed hosts is found to minimize both the incidence of whiteflies and associated viral diseases. Conservation and release of parasitoids like *Encarsia brevivena, Eretmocerus corni, Eretmocerus mundus*, predators like *Chrysoperla zastrowi sillemi, Mallada boninensis, Coccinella septempunctata* and application of entomopathogens like *Beauveria bassiana, Lecanicillium lecanii* and *Paecilomyces farinosus* reported to be promising for the management of this pest. Protection of seedlings in nursery by nylon net (40 mesh) and/or seed treatment with Imidacloprid 70% WS @ 2.5 g per kg of seed provides good protection from the pest for 25-30 days. Need based application of systemic insecticides like Imidacloprid 17.8% SL@ 0.35 ml/ l water or Thiomethoxam 25% WG @ 0.33 g/ l water or Dimethoate 30% EC @ 2 ml/l water was effective after transplanting in reducing the whitefly population (Rai *et al.*, 2014b). Ameta and Kumar (2005) reported Spiromesifen 240 SC @ 500 ml/ha to be highly effective in controlling whitefly in cotton.

**Fall Armyworm (FAW), *Spodoptera frugiperda* (JE Smith):** FAO proclaimed this pest in 2021 as most destructive pest across the globe and ranked 15 of globally most resistant arthropods. The pest is native to North America, invaded Africa in 2016, Asia from 2018, Australia and Europe (Canary Island Spain) in 2020 (Acharya *et al.*, 2021; Kumar *et al.*, 2022). There are two species of *S. frugiperda* each associated with different host plants: one with rice (and Bermuda grass) and other with corn (and cotton, sorghum). The two occur sympatric ally in USA and exhibit high level of reproductive isolation and are physiologically adapted to their host plants (Pashley, 1986). It is a polyphagous pest that feeds on more than 80 different plants. It is predominantly a maize pest, and in the lack of maize, it will seek out sorghum; in the absence of both hosts, it will attack other crops.

The pest causes serious leaf feeding damage as well as direct injury during its larval stages (4-6) to growing points and cobs. The larva has predominant white inverted Y-shaped suture on its head between the eyes. The pest has specialized mechanism to detoxify xenobiotics and chemical insecticides and developed resistant to many classes of insecticides and Bt protein due to continuous selection pressure and frequent application of insecticides (Wan *et al.*, 2021). It prefers to lay eggs in group (100-150 eggs) on maize compared to other vegetable crops including cabbage, soybean and showed reluctance to tomato. It may not move to other crops for oviposition under natural condition but the development of offspring have been noted to be supported by vegetable crops like cabbage and soybean and provide high benefit for offspring performance exceeding the benefit of maize-based diet (Paola *et al.*, 2021). This envisages that the pest may not move to other crops for oviposition under natural condition but offspring will be supported by vegetable crops once they adapt to them. Migrate to vegetables in the absence of preferred hosts, necessitating constant and thorough surveillance to prevent spread. Shylesha *et al.*, (2018) reported on its natural enemies found in Karnataka, India.

**Management:** Maintain fields weed free and follow balanced fertilizer, plant Napier grass in border rows to act as FAW trap crop and create plant diversity by inter cropping: Maize + pigeon pea/black gram /green gram. Microbial biopesticides and botanicals (Neem oil, *Bacillus thuringiensis*, pyrethrin, and diatomaceous earth), newer molecules chlorantraniliprole 18.5% SC, Emamectin benzoate 5% SG, Indoxacarb 14.5% SC, Flumendamide 480 SC, may be taken advantage for the control (Deshmukh *et al.*, 2020).

## Emerging indigenous pests

**Mirid bugs, *Nesidiocoris cruentatus* (Ballard), *Metacanthus pulchellus* (Dallas):** Earlier, only one species *i.e. M. pulchellus* was on record as a pest of bottle gourd from eastern part of India (Dhiman, 1991). In addition to above, another species, *Nesidiocoris cruentatus* was for the first time recorded concurrently occurring as serious and emerging pest of bottle gourd from Varanasi, Uttar Pradesh, India during 2016 causing about 70-80% fruit damage and 30% shoot damage by these bugs (Halder *et al.*, 2021). Infested leaves also showed minute puncture spots with yellow hallows. The damage was more conspicuous with the presence of numerous brown puncture spots on young fruits, the rind along with sap oozing at puncture site were the common symptoms produced by these sucking pests. Affected fruits therefore had reduced market value. As regards the species composition, *N. cruentatus* was dominant contributing 68.63% of the total mirid bug population. Major active periods were November and December (Halder *et al.*, 2021).

**Management:** Three sprays of entomopathogen fungus, *Lecanicilium lecanii* (2.5 g/L water) + Neem oil (2.5 ml/L water) in 1:1 ratio at 15 days interval (LT50 value: 41.85 h as against 56.59 h in *L. lecanii* and 50.31 h in neem oil alone) proved effective against mirid bugs (Halder *et al.*, 2021).

**Melon weevil, *Acythopeus curvirostris citrulli* (Marshall):** An emerging serious borer pest of gourd crops was first record from sponge gourd in the world from Varanasi, India causing 70-80% fruit and 20-30% shoot damage during *Kharif* season (Halder *et al.*, 2016(a)). Eggs are laid in small batches on tender fruits. Grubs feed on soft, tender fruit pulp and continue till pupation inside fruits. Cocoons are hard blackish in colour made up of fibrous materials of the fruits and larval excreta. Affected fruits show many brown puncture spots. Initially feeding by the weevil causes secretion of yellowish white fluid later it turns in to brown gummy encrustations. Due to its feeding, affected fruits rot with no seed formation. Adults emerge from dry fruits by making small minute holes. Infestation started from third week of August with a peak during mid-October. Weevils also punctured on vines leading to their gradual drying bearing fruits and flowers and consequent profuse losses (Halder *et al.*, 2016(a), Halder and Rai, 2021).

**Management:** Certain inherent characteristics of the pest such as being an internal feeder, having long hibernation during winter season, over lapping generations, rapid multiplication, and lack of information on biocontrol agents are the critical issues need to be considered while developing management strategies.

According to Halder *et al.*, 2016(a), the summer crops and early sown sponge gourd and ridge gourd crops suffers less damage compared to rainy season than the late sown crops. They also observed significantly lowest fruit damage (43.48%) in the sponge gourd grown in raised bed system than the trailing system (80.73%). Among the biopesticides, *Lecanicillium lecanii* (5 g/l water) was found effective (LT50= 87.84 h), followed by *Metarhizium anisopliae* (LT50= 101.08 h) and they showed synergistic action with neem oil with half of their recommended doses (Halder *et al.*, 2016(a)).

**White Plume moth, *Sphenarches caffer* (Zeller):** It is a well-known major pest of lablab, beans etc. in South India (Sujithra *et al.*, 2010). This pest was reported hundred years back from Pusa, Bihar. Recently, according to Halder *et al.*, 2014 in eastern part of the country it was recorded as foliage feeder in bottle gourd by scrapping the chlorophyll thereby reducing the photosynthetic activity of the plants in and around Varanasi region and also damages buds with characteristic black excreta inside it. During the peak summer months of May – June when atmospheric temperature was around 45 C in Varanasi its incidence was also recorded. Interestingly there was no pest incidence from mid-October onwards when Rabi season bottle gourd available in vegetative stage and again started from May-June. However, in Varanasi region incidence of this plume moth is restricted to bottle gourd as foliage feeder and also damage emerging buds with black excreta inside, a characteristic symptom (Halder *et al.*, 2014). An endoparasitoid, *Apanteles paludicole* is found to be highly active in the region parasitizing the pest up to 40.91% besides the involvement of another chalcid pupal parasitoid, *Tropimeris monodon*. This appears to be the main reason that this pest has not acquired a major status in eastern part of India (Halder *et al.*, 2014).

**Management:** Conservation of solitary larval braconid doparasitoid, *Apanteles paludicole* and chalcid pupal parasitoid, *Tropimeris monodon* are beneficial. Need based application of *Bacillus thuringiensis* @ 2 g/l of water is able to control this pest (Halder *et al.*, 2014).

**Hadda beetle, *Henosepilachna vigintioctopunctata* (Fabricius) and *Epilachna dodecastigma* (Wiedemann):** These are well-known pests of eggplant and emerging as serious foliage feeder on cowpea in many parts of the country particularly eastern Uttar Pradesh and Bihar. More than 80% leaves were infested by the grubs and adults of this beetle on cowpea during summer season of 2009-10 in Varanasi region (Halder *et al.*, 2011; Kodandaram *et al.*, 2014). Its serious incidence have also been observed on bitter gourd, Ivy gourd (*Coccinia grandis*) (Maurice and Ramteke, 2012) at Allahabad in UP; on *M. charantia* in Jammu (Jamwal *et al.*, 2013) and on wild bitter gourd (*Momordica dioica*) in Maharastra (Deshmukh *et al.*, 2012).

**Management:** Conservation of larval-pupal endoparasitoid, *Pediobius foveolatus* (Eulophidae: Hymenoptera) which caused up to 36.6% reduction in beetle population in and around Varanasi, Uttar Pradesh (Halder *et al.*, 2011). Islam *et al.*, 2011 observed that crude aqueous extracts of leaves *Ricinus communis* had highest larvicidal toxicity (LC50=18.40%) besides significant reduction in both oviposition and egg-hatch, prolonged larval duration ($P<0.001$), and inhibited pupae formation and adult emergence. Amongst the entomopathogenic fungi, *Beauveria bassiana* @ 3 g/l water effectively controls the population of *Epilachna* beetle (74.91%) (Vishwakarma *et al.*, 2011) and *Metarhizium anisopliae* @ 5 g/l also gave good control.

**Spotted pod borer, *Maruca vitrata* (Fab.)**: It has emerged as a predominant pest in recent years in all pigeon pea and cowpea growing areas in almost all the states during Rabi and pre-*Kharif* season covering southern, northern and central parts of India causing up to 42% damage in case of cowpea. Apart from damaging pods, the larvae also feed on flower, buds and sometime stem of the plants (Halder and Srinivasan, 2011; 2012). Some of the attribute like occurrence of over lapping generations, short life-cycle, wide host range and protective web forming nature of this pest are main factors for its cosmopolitan distribution.

**Management:** Clean cultivation and removal of the web reduce the pest load from the field. Conservation of bioagents like *Apanteles* sp is effective. Need based spraying of azadirachtin @ 0.005 % or NSKE 4% or *Bacillus thuringiensis* var Kurstaki @ 1 kg/ha is recommended (Rai *et al.*, 2014). Mittal and Ujagir, 2005 also found Spinosad 90 g ai/ha to be effective against pod borer complex of pigeon pea including *M. vitrata.*

**Phytophagous Mites:** With change in climatic scenario, spider mite species (Tetranychids) notably *Tetranychus neocaledonicus* Andre, *T. urticae* (Koch.) *T. cinnabarinus* (Boisduval), tarsonemid mite, *Polyphagotarsonemus latus* (Banks) and eriophyid mites have intensified their damage potential on many of the vegetable crops like okra, brinjal, chilli, tomato, cowpea, beans, cucurbits etc mainly during post-monsoon period and dry period. Spider mite, *T. urticae* has commonly been found associated with plants like ladies finger, brinjal, beans, cow pea, rajmah and cucurbits whereas *T. cinnabarinus* mainly with cucurbits but also found on other crops including brinjal, beans, onion, peas, tomato, sweet potato and various ornamental crops (Puri and Mote, 2004). *P. latus* is a polyphagous pest is noted to be a serious on chilli wherever it is grown across the world. Infested leaves curl downwards and turn reddish and plants become rosette in appearance which is also called as '*Murda*' disease of chilli. The dangerous diffusion of this broad mite on vegetables and floral crops especially in green houses has also been reported. At cellular level, the typical symptoms of mite can be recognized by thickening the lower epidermal cells, elongation of the palisade cell layer, shrinkage and deformation of parenchymatous tissues (Rai *et al.*, 2007). Mites are small and microscopic creature, difficult to observe and initial infestation, the damage symptoms appear to be of physiological deficiency. Red spider mite can spread through air current *i.e.*, called ballooning.

**Management:** Spider mite infestation is more severe in summer season also in drought prone areas, so frequent irrigation reduces its population build up. During initial stage of broad mite (*P. latus*) infestation, destruction of severely attacked plant part mechanically reduces mite population. Conservation and release of native predatory mites also can minimize the mite populations from the field. *Amblyseius* spp. is an efficient predatory mite, which regulates the pest under field conditions. Using botanicals like Kochea and Calotropis leaf extract showed strong acaricidal action under field conditions (Rai *et al.*, 2014(b)). Under severe infestation, Propargite 57% EC @ 3 ml/l (Singh and Singh, 2004) or need based application of Fenazaquin 10% EC @ 2 ml/l or Spiromesifen 22.9% SC @ 0.75 ml/l or Chlorfenapyr 10% SC @ 1.25 ml/l of water should was found effective against yellow mites in chilli (Halder *et al.*, 2015(b)). The integrated module comprising seedling root dip with Imidacloprid 17.8% SL @1 ml/ l, spray of Buprofezin 25% SC @ 1 ml/l at 25 days after transplanting (DAT), Fipronil 5% SC @ 2 ml/l at 35 DAT, *Lecanicillium lecanii* @ 5 g/l at 45 DAT, Chlorfenapyr 10% SC @ 1.25 ml/l at 55 DAT, Neem oil @ 0.5% at 65 DAT gave yellow mite reduction over control in chilli was highest in both Kashi Anmol (75.48%) and Kashi Gaurav (72.90%) (Halder *et al.*, 2016(b))

**Tortoise beetle:** In many regions of India, water spinach (*Ipomoea aquatica* Forsk), an aquatic and/or semi-aquatic vegetable that grows both wild and under cultivation, is becoming more and more problematic due to the tortoise beetle [*Cassida circumdata* Herbst (Coleoptera: Chrysomelidae: Cassidinae)]. GreSSitt (1952) observed that *Ipomoea palmata, I. batatas, I. aquatica, I. cairica,* and *I. digitata* were among the Convolvulaceae plant species that *C. circumdata* fed on. Close observation showed that the water spinach leaves skeletonize as a result of the early instar grubs removing the chlorophyll portion of the leaves. On the leaves, later instars produce tiny, asymmetrical shot holes and notches. On a single leaf, there were numerous of these tiny holes (2 to 39 with an average of 11.37). Often, black excrement was visible on the upper surface of the leaves. Reduced photosynthetic activity and reduced market prices were associated with the affected leaves (Halder and Rai, 2020). This leaf feeder was found to infest a maximum of 50% of the leaves.

**Management**: Collection and destruction of the grubs, adults and severely infested plant parts is advisable.

## Future Emphasis on Pest Management Programme for Sustainable Development Goal

Successful Integrated pest management programme relies chiefly upon the follow-ups of its key components like avoidance of pest problem through physical, mechanical, cultural and biological means; sampling and early pest detection and curative measures. In case of invasive alien pests, the process of its management should be followed at three level of invasion *viz.*, if the pest has not been introduced then preventive measure to avoid entry like pest risk analysis (PRA), quarantine and early data base monitoring (early pest detection); when the pest has been introduced but not spread to nearby areas then post-quarantine such as rejection of consignment from which the pest has introduced and eradication by means of fumigation of consignment lot. In case of introduced insects which has established in the area then various curative measures like cultural, biological and chemical means of management are adopted for successful pest management. In vegetable cultivation, scheduled based and frequent chemical interventions do not allow the natural enemies to grow to their potential besides causing several other problems like environmental pollution, pesticide residue in harvested products and increased cost of production. Vegetable ecosystem is endowed with a large complex of natural enemies attacking different stages of insect pests to varying extent (Table 2 and 3). However, most of these natural enemies are under exploited in vegetables and also not amenable to mass multiplication and release except a few promising ones (*Trichogramma, Chrysoperla,* NPV, *Bacillus thuringiensis, Lecanicillium, Beauveria, Metarhizium, Nomuraea*) which need to be aggressively advocated and implemented at large scale for adoption by end users which is still very low. In view of the occurrence of large complex of natural enemies as our major strength as evident from the tables (2 and 3), their conservation through various means need to be given emphasis for enhancing their activities in natural environment as many of them not amenable for mass multiplication, rationalizing pesticide use and integration of several other ecofriendly approaches in vogue with major focus on biological based pesticides which is becoming so pertinent with the advent of environmentally conscious organic and natural farming system. Moreover, availability of biological control agents is one of the major limiting factors in India needs revival of policy framework to ensure accessibility to agrarian. Technology for mass production and application of bioagents has been developed in few cases but needs refinement with a major focus on the quality of bioagents and suitable storage and packaging technique.

## Enhancing Natural Enemy Population:

Monoculture environments are usually very poor environments for natural enemies hence maintaining successions of crop diversity, crop cover, standing crop, crop residue and altering harvest and/or cultivation practices to maintain 'refuge strips' are highly beneficial and suggested for increased activities of natural enemies. This is rather the environment or existing practices modified in a way to protect and enhance specific enemies or other organisms to reduce the impact of pests. It is now widely recognized that plant characteristics (leaf trichomes, cuticular waxes, volatiles etc.) can positively influence the attractiveness and suitability of a plant to entomophages and herbivores both. These plant characteristics may be advantageously utilized combining with other biological control techniques that refine natural enemy fitness. With this consideration, the occurrence of natural enemies in vegetable ecosystem and increasing their efficiency through ecological engineering and several other safer means are highlighted for consideration to achieve safe and sustainable vegetable production.

**Composition of important natural enemies associated with vegetable pests:** Predators like coccinellid beetles, rove beetles, preying mantids, green lace wing, syrphids, and spiders which feed on aphids, whiteflies, leafhoppers, thrips, mites. Several birds eg. Fork-tailed Drongo *Dicrurus adsimilis* (Bechstein), Indian Mynah *Acridotheres tristis* (Linnaeus) and Egrets are efficient feeders on caterpillar pests. Some of the parasitoids like *Trichogramma* (on *Helicoverpa* and *Earias*), *Cotesia plutellae* (on diamond back moth) and *Trathala flavoorbitalis* (on *Leucinodes orbonalis*) help in bringing down the pest population significantly. Several important natural enemies (parasitoid and predators) active against some of the emerging pests of different vegetable crops are depicted (Table 2 and 3) (Halder *et al.*, 2018; Rai and Halder, 2023). This envisages for conservations of those effective natural enemies under field condition (Table 3) through various approaches involving ecological engineering (habitat management) and others to enhance efficiency and efficacy of natural enemies avoiding all sort of steps harmful to safeguard and sustain these living useful fauna.

**Table 2. Important parasitoids and predators associated with emerging insect pests of vegetable crops**

| Pest | Parasitoid | Predator |
|---|---|---|
| **Crucifers** | | |
| Aphids, *Myzus persicae, Brevicoryne brassicae, Lipaphis erysimi* | *Diaeretiella rapae* (Hymenoptera: Braconidae) | *Chrysoperla zastrowi sillemi* (Neuroptera: Chrysopidae), *Coccinella septempunctata* (Coleoptera: Coccinellidae), *Episyrphus balteatus* (Diptera: Syrphidae) - predator |
| **Tomato** | | |
| *Bemisia tabaci* | *Encarsia brevivena, E. Formosa, Eretmocerus corni* and *E. mundus* (Hymenoptera: Aphelinidae); *Chrysocharis pentheus* (Hymenoptera: Eulophidae) | *Chrysoperla zastrowi sillemi* (Neuroptera: Chrysopidae), *Mallada boninensis* (Neuroptera: Chrysopidae:), *Dicyphus Hesperus* (Heteroptera: Miridae), *Coccinella septempunctata* (Coleoptera: Coccinellidae) |
| **Okra** | | |
| Leafhopper, *Amrasca biguttula biguttula* | *Anagrus empoascae, A. flaveolus Gonatocerus ater* and *Stethynium triclavatum*, (Hymenoptera: Mymaridae) as well as *Oligosita sp.* (Hymenoptera: Trichogrammatidae) | *Cheilomenes sexmaculata* (Coleoptera: Coccinellidae), *Chrysoperla zastrowi sillemi* (Chrysopidae: Neuroptera), *Mallada boninensis* (Chrysopidae: Neuroptera), *Paederus fuscipes* (Coleoptera: Staphylinidae) |
| *Aphis spp.* | | *Coccinella septempuctata* (Coleoptera: Coccinellidae), *Chrysoperla zastrowi sillemi* (Chrysopidae: Neuroptera), |
| Red Spider mites: *Tetranychus spp* | | *Amblyseius alstoniae, A. tetranichivorous, A. longispinosus, A. fallacis, A. multidentatus* (Phytoseiidae: Acarina), *Stethorus punctillum, S. pauperculus* (Coleoptera: Coccinellidae), *Scirtothrips indicus* (Thysanoptera: Thripidae), Anthocorid bug |

| Pest | Parasitoid | Predator |
|---|---|---|
| **Brinjal** | | |
| Brinjal Stem borer, *Euzophera perticella* | *Pristomerus testaceus, P. euzopherae* | |
| Hadda beetle, *Epilachna vigintioctopunctata* | *Pediobius foveolatus*, (Hymenoptera: Eulophidae), *Uga menoni* (Hymenoptera: Chalcididae) | |
| Gall midge *Asphondylia sp.* | *Eurytoma* sp. (Hymenoptera, Eurytomidae) | |
| Mealybug *Phenacoccus solenopsis* | *Aenasius arizonensis* (=*Aenasius bombawalei*) (Hymenoptera: Encyrtidae) | |
| **Chilli** | | |
| *Scirtothrips dorsalis* | | *Franklinothrips megalops, Mymarothrips garuda* (Thysanoptera: Aeolothripidae), *Scolothrips indicus* (Thysanoptera: Thripidae), *Orius maxidentex* (Hemiptera: Anthocoridae), *Geocoris ochropterus* (Hemiptera: Lygaeidae ) |
| Broad mite or yellow mite, *Polyphagotarsonemus latus* | | *Amblyseius sp* and *A. ovalis* Evans (Acari: Phytoseiidae) |
| **Cucurbits** | | |
| Leaf eating caterpillar *Diaphania indica* | *Apanteles taragamae* and *A. machaeralis* (Hymenoptera: Braconidae), *Elasmus brevicornis* (Hymenoptera: Elasmidae), *Trathala flavo-orbitalis* (Hymenoptera: Ichneumonidae) | |
| **Legume vegetable** | | |
| Cowpea Pod borer, *Maruca vitrata* | *Cedria paradoxa* (Hymenoptera: Braconidae), *Phanerotoma hendecasiella, Apanteles* sp (Hymenoptera: Braconidae) | |

| Pest | Parasitoid | Predator |
|---|---|---|
| Bean Aphids: *Aphis craccivora* | *Trioxys* (*Binodoxys*) *indicus* (Hymenoptera: Braconidae) | *Cheilomenes sexmaculata, C. septempuncta* (Coccinellidae: Coleoptera), *Episyrphus balteatus, Ischiodon scutellaris* (*Diptera: Syrphidae*) |

**Source: Modified from Halder** *et al.*, **2018; Rai and Halder, 2023**

**Habitat management for sustenance and amplification of natural enemies:** The access to flowering resources is required to the natural enemies that belong to a broad range of order such as Hymenoptera, Diptera, Coleoptera, Heteroptera, Neuroptera, Araneae and Acari. Important flowering plants for direct provision of resources which attract natural enemies are given (Table 3) (Wackers *et al.*, 2005; Reddy, 2016). The experimental studies have also shown that provision of floral resources, food sprays (sugar-water or protein sprays to attract / maintain natural enemies) and shelter habitats (avoiding destruction of nests of social wasps) can increase natural enemy abundance, diversity and fitness which leads to decreased pest damage, increased crop yield and greater profit for farmers (Heimpel and Jervis, 2005). Classical biological control is the modification of the environment or existing practices to protect and enhance specific enemies or other organisms to reduce the effect of pests (Eilenberg *et al.*, 2001). It does not involve introducing exotic organisms into the environment, thus avoiding some major environmental and legal concerns associated with classical biocontrol. It is now widely recognized that plant characteristics can positively influence the attractiveness and suitability of a plant to entomophages. Leaf trichomes, cuticular waxes on plants, volatiles emit from plants etc. can interfere with both herbivores and their entomophages. Exploitation of such volatiles may be particularly effective if they are combined with other classical biological control techniques that improve natural enemy fitness (Reddy, 2016).

**Provisioning pollen and nectar resources:** The non-prey food, like pollen and nectar, is needed by most of the natural enemies (predators and parasitoids) as their food so as to increase in their activity, longevity and fecundity, without providing resources to pest insects. The best nectar sources for small parasitoids and larger predators include plants having very small flowers. Adult hoverflies (aphid predators) use pollen essential for eggs to mature and produce young ones. Lacy phacelia (*Phacelia tanacetifolia*) and Nasturtium (*Tropaeolum majus*) provide resources for natural enemies without enhancing moth pests. The integration of non-prey food along with attraction of natural enemies in an attract-and-reward approach has potential for enhancing biological control synergy (Hogg *et al.*, 2011).

**Corridors for natural enemies:** The strip highways of habitat for multiplication and dispersal to natural enemies (predators and parasitoids) can be provided by growing different flowering plants as strips across the main crops every after 50-100 meter. The bean aphids (*Brevicoryne brassicae*) on cabbage and sugar beets can be managed by planting tansy leaf (*Phacelia tenacetiolia*) every twenty to thirty rows to enhance the population of syrphid predators. Likewise, the cabbage aphids (*Brevicoryne brassicae*) can be controlled by planting strips of tansy leaf and buck-wheat in Swiss cabbage fields which enhanced small parasitic wasp population (White *et al.*, 1995). The aphids on lettuce and cruciferous crop fields can be managed by planting strips of alyssum every 50-1oom in order to attract syrphid flies (Reddy, 2016).

**Artificial Food Supplements:** In absence of abundant prey, artificial food supplements containing yeast, whey proteins (casein hydrolysate), and sugars-water may be provided in order to attract predators such as lady bird beetles, syrphid flies, and lacewing bugs which is required for egg production. Some of the artificial foods available from suppliers of natural enemies include Wheast, Bug Pro, and Bug Chow (Reddy, 2016).

**Kairomones:** Tricosane is known to be used as pre-conditioning and reinforcing agent that increases the predation capacity in case of Chrysopid. Similarly, alpha-Tryptophan is applied on the crop as ovipositional attractant for Chrysopid, *Trichogramma*, Coccinellid which increases the rate of oviposition in these bioagents.

**Table 3. Habitat manipulation through maintaining the plants in and around the main crop which are attractive to natural enemies can be utilized in pest suppression.**

| Plant | Beneficial insects attracted |
|---|---|
| Caraway | Lacewings, hoverflies, spiders, parasitic wasps |
| Common knotweed | Big-eyed bugs, hoverflies, parasitic wasps |
| Cowpea | Parasitic wasps, lady bird beetle |
| Crimson clover | Minute pirate bugs, big-eyed bugs, lady beetles |
| buckwheat | Hoverflies, wasps, tachinid flies, lacewings, lady beetles |
| Hairy vetch | Lady beetles, minute pirate bugs, predatory wasps |
| Queen Anne's lace | Lacewings, predatory wasps, minute pirate bugs, tachinid flies |
| Soap bark tree | Hoverflies, green lacewings, brown lacewings |
| Sweet alysuum | Tachinid flies, hoverflies, chalcids |
| Sweet fennel | Parasitic wasps, predatory wasps |
| Toothpick ammi | Hoverflies, minute pirate bugs, flower beetle, tachinid flies |
| Tansy (Basil) | Parasitic wasps, lady beetles, insidious flower bugs, lacewings |
| White sweet clover | Tachinid flies, bees, predatory wasps |
| Wild buckwheat | Hoverflies, minute pirate bugs, tachinid flies |

**Source: Reddy, 2016**

## The Way Forward .......

With the advent of information explosion, the consumer awareness is growing with respect to the right to have safe food. Agricultural Practices integrating all possible non-chemical methods with a major focus on effective biological based control method (although recommended in many cases but large-scale adoption by the end-users is very less) need to be devised, advocated and sincerely followed in a way that protects human health and our environment. In this endeavours, important aspects in integrated pest management programme such as correct identification of emerging pest species, ecological study of invasive alien species along with careful tracking of geographical distribution, phytosanitary regulations to prevent or limit the introduction of risky insect pests, developing weather based pest forecasting; intensive and extensive search for effective natural enemies of emerging insect pests for their conservation and augmentation. Incorporation of effective microorganisms, botanicals, PGPR, pheromones, kairomones, trap crops, habitat manipulation and CIBRC approved new generation highly degradable safer and effective molecules (with PHI, MRL) for integration in developing area-wide suitable integrated pest management approaches/modules. Use logical and convergent use of the technological advancements including information technology are some of the focused areas need impetus to deal with emergent pest situation in present context of environmental changes and global warming. Role of public-private partnership in production, distribution and quality control of different components of IPM/ BIPM are imperative and brought to the fore in time.

## References

Acharya R, Akintola AA, Malekera MJR, Patrick K, Benedictor K, Mutimbu MK, Shrestha YK, Jahan SM, Hemayet, HTX, Dao HT, Park JH, Nam M, Lee SJ, Kim SM, Kim I, Hwang HS and Lee KY. 2021. Genetic Relationship of Fall Armyworm (*Spodoptera frugiperda*) Populations That Invaded Africa and Asia. *Insects*, 12(5): 439. https://doi.org/10.3390/insects12050439

Ameta, OP and Kumar A. 2005. Field evaluation of newer insecticides for the management of white fly *Bemisia tabaci* in cotton. *Indian Journal of Applied Entomology*, 19(1): 58-60.

Ananthakrishnan T. 2009. Invasive insects in Indian agriculture, forestry and medicine. *Current Science*, 97(3): 300-303.

Anonymous. 2015. IIVR Annual Report 2014-15, published by Indian Institute of Vegetable Research, Varanasi, Uttar Pradesh-221305, pp: 43.

Aturo C, Salvatore D and Gravino D. 2012. Integrated control in protected crops Mediterranean climate (In Congress on Insect Science. April, 16-27, INSAIS). IOBC-WPRS Bulletin, 80: 319-324.

Banik S and Halder J. 2013. Protection of Crop Plants with Newer Chemical Molecules. *In*. Current Concepts in Crop Protection. Edited by Susanta Banik, ISBN: 9789380012421, Published by Studium Press (India) Pvt. Ltd., New Delhi-110002, India, pp: 307-349.

Deshmukh PS, Chougale AK, Shahasane SS, Desai SS and Gaikwad SG. 2012. Studies on biology of Hadda beetle, *Epilachna vigintioctopunctata* (Coleptera, Coccinillidae): a serious pest of wild better gourd, *Momordica dioica*. *Trends in Life Sciences*, 1(3):46-48.

Deshmukh S, Pavithra HB, Kalleshwaraswamy CM, Shivanna BK, Maruthi MS and Mota-Sanchez D. 2020. Field efficacy of insecticides for management of invasive fall armyworm, *Spodoptera frugiperda* (JE Smith) (Lepidoptera: Noctuidae) on maize in India. *Florida Entomologist*, 103(2): 221-227. http://doi.org/10.1653/024.103.0211

Dhiman SC. 1991. *Hypopharynx* of *Metacanthus pulchellus* (Heteroptera: Berytidae), a pest of bottle-gourd. *Bioved*, 2(2): 155-158.

Eilenberg J, Hajek A and Lomer. C. 2001. Suggestions for unifying the terminology in biological control. *Bio Control*, 46: 387-400.

EPPO, 2005. Data sheets on quarantine pests: *Tuta absoluta*. *EPPO Bulletin*, 35: 434-435.

EPPO, 2010. First report of *Tuta absoluta* in Hungary. *EPPO Reporting Service*, 3(52): 2.

Garcia MF and Espul JC. 1982. Bioecology of the tomato moth (*Scrobipalpula absoluta*) in Mendoza, Argentina Republic. *Revista de investigaciones Agropecuarias*. 17:135-146.

Gres Sitt JL. 1952. The tortoise beetles of China (Chrysomelidae: Cassidinae). *Proceedings of California Academy of Science*, 27: 433-592.

Halder H and Rai AB. 2020. Bionomics of tortoise beetle, *Cassida circumdata* Herbst: An emerging pest of water. *Pest Management in Horticultural Ecosystems*, 26(1): 25-28. https://doi:10.5958/0974-4541.2020.00004.1

Halder J and Rai AB. 2021. Emergence of new insect pests on vegetables during the last decade: a case study. *Current Horticulture*, 9(1): 20–26.

Halder J and Srinivasan S. 2011. Varietal screening and role of morphological factors on distribution and abundance on spotted pod borer, *Maruca vitrata* (Geyer) on cowpea. *Annals of Plant Protection Sciences*, 19: 71-74.

Halder J and Srinivasan S. 2012. Relative infestation of *Maruca vitrata* on different host plants. *Annals of Plant Protection Sciences*, 20(1): 205-269.

Halder J, Kodandaram MH, Rai AB and Kumar R. 2016(b). Impact of different pest management modules against the major sucking pests complex of chilli (*Capsicum annuum*). *Indian Journal of Agricultural Sciences*, 86(9): 792-795.

Halder J, Kodandaram MH, Rai AB and Singh B. 2015(b). Bio-efficacy of some newer acaro-insecticides against yellow mite (*Polyphagotarsonemus latus* (Banks)) and thrips (*Scirtothrips dorsalis* Hood) in Chilli. *Pesticide Research Journal*, 27(2): 171-174.

Halder J, Kushwaha D, Deb D and Rai AB. 2019(a). Abundance and distribution of sucking pests complex of okra in relation to meteorological parameters. *Journal of Agrometeorology*, 21(2): 227-229.

Halder J, Kushwaha D, Deb Dibyendu, Dey Debjani and Rai A B. 2021. Mirid bugs as an emerging threat to bottle gourd cultivation: Dynamics and management. *Indian Journal of Agricultural Sciences*, 91(5): 723–728.

Halder J, Kushwaha D, Rai A B, Dey D, Chaubey T and Singh B. 2016(a). First record of *Acythopeus curvirostris citrulli* (Marshall) (Coleoptera: Curculionidae) on sponge gourd, *Luffa cylindrica* (Linn.), its bionomics, diurnal activity and ecofriendly management. *Vegetable Science*, 43(2): 190-197.

Halder J, Kushwaha D, Rai AB and Singh B. 2019(b). Biology and Biorational Management of *Tuta absoluta* (Meyrick) (Lepidoptera: Gelechiidae): A Global Challenge to Tomato Production. *Proceedings of Zoological Society*, 72(2): 107-110. https://doi.10.1007/s12595-017-0232-0

Halder J, Rai AB and Dey D. 2015(a). Occurrence of *Phenococcus solenopsis* (Tinsley) in vegetable ecosystem and host-mediated effects on its dominant parasitoid, *Aenasius bambawalei* Hayat. *Vegetable Science*, 42(2): 30-33.

Halder J, Rai AB and Kodandaram MH. 2013. Compatibility of neem oil and different entomopathogens for the management of major vegetable sucking pests. *National Academy Science Letters*, 36(1): 19-25. https://doi.10.1007/s40009-012-0091-1

Halder J, Rai AB, Dey D and Kodandaram MH. 2014. Is *Apanteles paludicole* Cameron, a Limiting Biotic Factor for Minor Pest Status of *Sphenarches caffer* (Zeller)? *Journal of Biological Control*, 28(2): 119–121.

Halder J, Rai AB, Dey D and Singh B. 2018. Abundance of important parasitoids in the vegetable ecosystem and their prospects in integrated pest management. *Journal of Entomology and Zoology Studies*, 6(4): 762-769.

Halder J, Rai AB, Kodandaram MH and Dey D. 2011. *Pediobius foveolatus* (Crowford): A promising bioagent against *Epilachna vigitionctopunctata*. National conference on horti-business linking farmers with market. Held at Dehradun, Uttarakhand, India, 28-31 May. p. 133.

Heimpel GE and Jervis MA 2005. Does floral nectar improve biological control by parasitoids? *In*: Wackers, F.L., van Rijn, P.C.J. and Bruin, J. (Eds.), Plant-provided Food for Carnivorous Insects: A Protective Mutualism and its Applications. Cambridge University Press, Cambridge, UK. 267-304.

Hogg BN, Bugg RL and Dane KM. 2011. Attractiveness of common insectry and and harvestable floral resources to beneficial insects. *Biological Control*, 56: 76-84.

Islam K, Islam MS and Ferdousi Z. 2011. Control of *Epilachna vigintioctopunctata* Fab. (Coleoptera: Coccinellidae) using some indigenous plant extracts. *Journal of Life and Earth Sciences*, 6: 75-80.

Jamwal VVS, Ahmad H, Sharma D, Srivastava K and Kumar V. 2013. Comparative biological and morphometric attributes of *Epilachna vigintioctopunctata* Fabr. on brinjal and bitter gourd. *Vegetos*, 26(2): 426-437.

Jhala RC, Bharpoda TM and Patel MG 2008. *Phenacoccus solenopsis* Tinsley (Hemiptera: Pseudococcidae), the mealy bug species recorded first time on cotton and its alternate host plants in Gujarat, India. *Uttar Pradesh Journal of Zoology*. 28(3): 403-406.

Johari A, Herlinda S, Pujiastuti Y, Irsan C and Sartiami D. 2014. Morphological and Genetic Variation of *Thrips parvispinus* (Thysanoptera: Thripidae) in Chilli Plantation (*Capsicum annuum* L.) in the Lowland and Highland of Jambi Province, Indonesia. *American Journal of Bioscience*, 2(6-1): 17-21. https://doi:10.11648/j.ajbio.s.2014020601.14

Kodandaram MH, Halder J and Rai AB. 2014. New insecticide molecules and entomopathogens against hadda beetle, *Henosepilachna vigitioctopunctata* infesting vegetable cowpea. *Indian Journal of Plant Protection*, 42(4): 333-337.

Kodandaram MH, Rai AB and Halder J. 2010. Novel insecticides for management of insect pests in vegetable crops: A review. *Vegetable Science*, 37: 109-123.

Krechemer F da Silva and Forester LA. 2015. *Tuta absoluta* (Lepidoptera: Gelechiidae): Thermal requirements and effect of temperature on survival, reproduction and longevity. *European Journal of Entomology*, 112(4): 658-663. https://doi:10.14411/eje.2015.103

Kumar RM, Gadratagi BG, Paramesh V, Kumar P, Madivalar Y, Narayanappa N and Ullah F 2022. Sustainable management of invasive fall armyworm, *Spodoptera frugiperda*. *Agronomy* (special issue: Ecological management of pests), 12(9): 2150; http://doi.org/10.3390/agronomy12092150

Maurice NG and Ramteke PW. 2012. Development of *Epilachna vigintioctopunctata* (Fabricius) (Coleoptera: Coccinellidae) on essential and alternative foods. *International Journal of Advance Pharmaceutical and Biological Sciences*, 2(3): 199-204.

Mittal V and Ujagir R. 2005. Evaluation of naturalyte spinosad against pod borer complex in early pigeon pea. *Indian Journal of Plant Protection*, 33(2): 211-215.

Mound LA and Collins DW. 2000. A South East Asian pest species newly recorded from Europe: *Thrips parvispinus* (Thysanoptera: Thripidae), its confused identity and potential quarantine significance. *European Journal of Entomology*, 97(2): 197200. https://doi:10.14411/eje.2000.037

Murai T, Watanabe H, Toriumi W, Adati T and Okajima S. 2009. Damage to vegetable crops by *Thrips parvispinus* Karny (Thysanoptera: Thripidae) and preliminary studies on biology and control. *Journal of Science*, 10: 166.

Nagraju DK, Vivek U, Ranjith M, Sriharsha RG, Ramesh C, Verma OP and Prakash R. 2021. Occurrence of *Thrips parvispinus* (Karny) (Thripidae: Thysanoptera) in major chilli (*Capsicum annum*) growing areas of Karnataka. *Insect Environment*, 24(2): 523-532.

Nielsen MC, Worner S, Chapman RB, DeKogel WJ, Perry NB, Sansom C, Murai T, and Muvea A. 2010. Optimising the use of allelochemicals for thrips pest management, p.324. In 26th Annual Meeting International Society of Chemical Ecology Conference Book of Abstract. 31 July to 4 August 2010: http://www.atout=org.com/isce2010/docs/book isce2010.pdf.

NPPO. 2019. The Netherlands plant Protection Organization. *Thrips parvispinus*. Quick scan, 2019, of climate change for invasive species. *Conservation Biology*, 22: 534-543.

Oliveira LI, Durao AC, Fontes J, Raja IS and Tavares J. 2017. Potential of *Trichogramma achaeae* (Hymenoptera: Trichogrammatidae) in biological control of *Tuta absoluta* (Lepidoptera: Gelechiidae) in Azorean greenhouse tomato crops. *Journal of Economic Entomology*, 110 (5): 2010-2015. https://doi:10.10.1093/jee/tox197

Paola SC, Chuang WP, Lin MY, Chiang MY and Srinivasan R. 2021. Oviposition preference not necessarily predicts offspring performance in the fall armyworm, *Spodoptera frugiperda* (Lepidoptera: Noctuidae) on vegetable crops. *Scientific Reports*, 11: 15885. https://doi.org/10.1038/s41598-021-97184-9

Pashley DP. 1986. Host associated genetic differentiation in fall armyworm (Lepidoptera: Noctuidae): a sibling species complex? *Annals of the Entomological Society of America*, 79(6): 898-904. http://doi.org/10.1093/aesa/79.6.898

Patel CB, Shah AH and Rai AB. 1993. Problems of Mite Pests of Crop in Gujarat and their Management. Technology Bulletin No. 1: Gujarat Agricultural University, Navsari, Gujarat, pp.24.

Pereyra PC and Sanchez N. 2006. Effect of two Solanaceous plants on developmental and population parameters of the tomato leaf miner, *Tuta absoluta* (Meyrick) (Lepidoptera: Gelechiidae). *Neotropical Entomology*, 35: 671-676.

Puri SN and Mote UN. 2004. Emerging pest problems in India and critical issues in their management: an overview. In: Proceedings on the National Symposium on Frontier Areas of Entomological Research (Eds., Subrahmanyam B, Ramamurthy VV and Singh VS), 5-7November, 2003, Division of Entomology, IARI, New Delhi. pp. 13-24.

Rachana RR, Roselin P, Amutha M, Sireesha K and Reddy NG. 2021. Invasive pest, *Thrips parvispinus* (Karny) (Thysanoptera: Thripidae) a looming threat to Indian agriculture. *Current Science*, 122(2): 211-213.https://doi.10.18520/cs/v122/i2/211-213.

Rai AB and Halder J. 2023. Parasitoid and predator for ecofriendly management of insect pests in vegetable crops (Chapter-02); 07-25. *In*: Identification and diagnosis of pest and diseases of horticultural crops and their management (Eds., A. K. Pandey and P. Bhumita), Training Mannual – 03 (CAU/COH/Bermiok/2023/TM-03), College of Horticulture, Central Agricultural University, Sikkim.

Rai AB, Halder J and Kodandaram MH. 2014a. Emerging insect pest problems in vegetable crops and their management in India: An appraisal. *Pest Management in Horticultural Ecosystems*, 20(2): 113-122.

Rai AB, Loganathan M, Halder J, Venkataravanappa V and Naik PS. 2014b. Eco-friendly Approaches for Sustainable Management of Vegetable Pests. IIVR Technical Bulletin No. 53, Indian Institute of Vegetable Research, Varanasi, Uttar Pradesh, pp. 104.

Rai AB, Satpathy S, Gracy RG, Swamy TMS and Rai M. 2007. Yellow mite (*Polyphagotarsonemus latus* Banks) menace in chilli crop. *Vegetable Science*, 34(1): 1-13.

Rathee M and Dalal P. 2018. Emerging insect pests in Indian agriculture. *Indian Journal of Entomology*, 80(2): 267-281. https://doi:10.5958/0974-8172.2018.00043.3

Reddy PP. 2016. Sustainable intensification of crop production (Chapter 20: Habitat management, 305-322), Springer Nature, Singapur Pvt Ltd: pp: 405. https://doi:10.1007/978-981-10-2702-4

Roy S, Halder J, Singh S, Rai A B, Prasad R N and Singh B. 2017. Do vegetable growers really follow the scientific plant protection measures? An empirical study from eastern Uttar Pradesh and Bihar. *Indian Journal of Agricultural Sciences*. 87(12): 1668–1672.

Sharma PL and Gavkare O. 2017 New distributional record of invasive pest *Tuta absoluta* (Meyrick) in North-Western Himalayan region of India. *National Academy Science Letters*, 40(3): 217–220. https://doi:10.1007/s40009-016-0526-1

Shashank PR, Chandrashekhar K, Meshram NM and Shreedevi K. 2015. Occurrence of *Tuta absoluta* (Lepidoptera: Gelechiidae) an invasive pest from India. *Indian Journal of Entomology*, 77(4):323-329.

Shivalingaswamy TM, Satpathy S and Banerjee MK. 2002. Estimation of crop losses due to insect pests in vegetables. In: Resource management in plant protection, published by Plant Protection Association of India, NPPTI Campus, Rajendranagar, Hyderabad. pp. 24-31.

Shylesha AN, Jalali SK, Gupta AN, Varshney RI, Venkatesan T and Shetty PR, 2018. Studies on new invasive pest *Spodoptera frugiperda* (JE Smith) (Lepidoptera: Noctuidae) and its natural enemies. *Journal of Biological Control*, 32(3): 145-151.

Singh J and Raghuraman M. 2011. Emerging scenario of important mite pests in North India. *Zoosymposia*, 6: 170-179.

Singh J and Singh RN. 2004. Major harmful mites of vegetables and other crops and their control. Technical Bulletin. Courtesy: Dhanuka Group, New Delhi., pp. 5.

Singh S, Sharma JH, Udikeri A and Ansari H. 2020. Invasive Insects in India, In: Invasive Species - Introduction Pathways, Economic Impact, and Possible Management Options (eds:Hamadttu El-Shafie). https://doi:10.5772/intechopen.91986

Sridhar V, Chakravarthy AK, Asokan R, Vinesh LS, Rebijith KB, Vennila S. 2014. New record of the invasive South American tomato leaf miner, *Tuta absoluta* (Meyrick) (Lepidoptera: Gelechiidae) in India. *Pest Management in Horticultural Ecosystem*, 20(2): 148-154.

Suganthy M, Rajeswari S, Senthilraja C, Nakkeeran S, Malathi VG, Ramaraju K and Renukadevi P. 2016. New records of western flower thrips, *Frankliniella occidentalis* (Pergande) (Thysanoptera; Thripidae) in South India. *International Journal of Environment, Agriculture and Biotechnology*, 1(4). http://dx.doi.org/10.22161/ijeab/1.4.33

Sujithra, M, Srinivasan S and Muralikrishna T. 2010. Bionomics and seasonal occurrence of plume moth, *Sphenarches caffer* of field bean. *Annals of Plant Protection Sciences*. 18(1): 241-243.

Thorat SS, Sisodia DB and Gangawar. 2022. Invasive Thrips, *Thrips parvispinus* (Karny) an invasive threat: A review. *Environment and Ecology*, 40(4A): 2170-2175.

Tyagi K and Kumar V. 2015. First report of western flower thrips, *Frankliniella occidentalis* (Pergande) (Thripidae: Thysanoptera) from India-a potential havoc to Indian agriculture. *Halteres*, 6: 1-3.

Tyagi K, Singha D and Chakraborty R. 2015. Morphological and DNA barcoding evidence for invasive pest thrips, Trips parvispinus (Thripidae: Thysanoptera), newly recorded from India. *Journal of Insect Science*, (1): 105.

Viraktamath CA. 2002. Alien invasive insects and mite pests and weeds in India and their management. Microensica Supplementia, 6: 67-83.

Vishwakarma R, Prasad PH, Ghatak SS and Mondal S. 2011. Bio-efficacy of plant extracts and entomopathogenic fungi against Epilachna beetle, *Henosepilachna vigintioctopunctata* (Fabricius) infesting bottle gourd. *Journal of Insect Science*, 24(1): 65-70.

Wackers FL, Rijin PCJ and Bruin J. 2005. Plant-provoded food for carnivorous insects: a protective mutualism and its applications. Cambridge University Press, Cambridge

Wan J, Hung C, Li CY, Zhou HX, Ren YL, Li ZY, Xing LS, Zhang B, Qiao X, Liu B, Liu CH, Xi Y, Liu WX, Wang WN, Qian WG, Mckirdy S and Wan FH. 2021. *Journal of Integrative Agriculture*, 20(3): 646-663.

White AJ, Wratten SD, Berry NA and Weighmann U. 1995. Habitat manipulation to enhance biological control of Brassica pests by hover flies (Diptera: Syrphidae). *Journal of Economic Entomology*, 88: 1171-1176.

CHAPTER 23

# RECENT TRENDS IN BIOLOGICAL DISEASE MANAGEMENT IN HORTICULTURAL CROPS

**Bhavya Mishra and Arvind M**

*Department of Plant Pathology, SKNAU - Rajasthan Agricultural Research Institute (RARI), Durgapura, Rajasthan*

Horticultural crops contribute significantly to India's GDP, constituting approximately 30% from a marginal 13.1% of the gross cropped area. Despite the sector's rapid growth, various biotic and abiotic constraints hinder production, causing economic yield losses. Pathogens, such as fungi, bacteria, and viruses, contribute to major losses during cultivation, harvesting, and storage. Chemical control measures are common but pose environmental and health risks. Biological management thus offers a safer alternative. This article explores both age-old and emerging biocontrol agents, their mechanisms, and strategies to enhance biocontrol potential against diverse pathogens. The mechanisms include competition, lytic enzyme production, hyperparasitism, antibiosis, and induced immunity through priming. Furthermore, it delves into the evolving field of combining nanotechnology with biological control, offering a synergistic approach for enhanced efficiency in managing plant diseases. Targeted use of substrates to sustain beneficial biocontrol microbes in the crop environment is also explored as a proactive strategy for disease management, aligning with sustainable agricultural practices. Some recent approaches like phage cocktails, antagonistic yeasts, genetically modified biocontrol agents, potential of mycoviruses etc. have also been addressed. In conclusion, biological control, with its multifaceted strategies and emerging innovations, stands as a promising and environmentally friendly alternative for integrated disease management in horticultural crops.

Horticultural crops contribute to approximately 30 per cent of the India's GDP and 33 per cent of the agriculture sector's Gross Value Added (GVA). In addition to guaranteeing the country's nutritional security, it offers farmers more money, varied farm operations, and alternative rural employment alternatives. Recently, the total horticultural production surpassed the total production of food grains reflecting the bright future of the sector. Horticulture covers 18 per cent of the net cultivated area of India and horticultural production of 333.25 million MT has been achieved from an area of 27.5 million ha in 2021-22. India is the largest producer of mango, banana, guava, papaya, lemon & lime, okra in the world and the second largest producer of crops like potato, tomato, onion, cabbage, cauliflower and brinjal in the world showcasing its strength in world horticulture sector. India exported worth 6510.53 million US Dollar of horticultural produce despite the pandemic in 2020-21 and today it is the 5th largest and fastest growing vegetable seed market in the world worth Rs. 6 billion.

Horticultural crops are very important because of their nutritive value as well as important gene repositories that can withstand a variety of stresses. The challenge of further boosting horticulture crop productivity and output stems from the growing demand for horticultural produce brought on by increased health consciousness, rising incomes, increased demand for exports, and growing population. Despite the fact that the horticultural sector is blooming at a massive pace, the production is severely curtailed by various biotic and abiotic constraints which contributes to the economic yield losses every year (Anonymous, 2023). Among the biotic challenges, crops attacked by diverse groups of pathogens such as fungi, bacteria and viruses are responsible for the major losses during cultivation, harvesting and storage phase.

Farmers predominantly rely on chemical control measures to combat agricultural challenges, posing adverse effects on both the environment and human health. The use of these chemicals contributes to the development of resistance among more and more pathogens. In addition to their harmful effects on beneficial soil microflora, the issue of pesticide residue in edible food is becoming more and more prevalent, particularly in horticultural crops. Moreover, chemical management is expensive and difficult for resource-constrained farmers to procure and apply. Considering these factors, the preference leans towards biological methods and the development of resistant crop varieties. However, resistance breeding proves challenging in many horticultural crops due to the scarcity of resistant stock, the complexity of the breeding process, breakdown of resistance in intensive monoculture, and the extensive time required for the final outcome.

Biological management emerges as an environmentally safe, affordable and easily applicable alternative. This approach primarily involves the application of biocontrol fungi and bacteria to the plant biosystem or ecosystem, preventing the harmful accumulation of plant pathogens. Bioinoculants play a crucial role by scavenging essential nutrients in the soil, competing with pathogens for space and plant receptors. Additionally, these bioagents induce a systemic response known as induced systemic resistance (ISR) through the release of antimicrobial substances like reactive oxygen species (ROS) and phytoalexins. ISR proves effective against biotrophs, herbivores, and wounds, relying on the jasmonic acid-ethylene (JA-ET) signaling pathway (Burketova *et al.*, 2015). This priming mechanism enhances the preparedness of plants, enabling faster and stronger responses to pathogen attacks.

The historical use of bioagents like *Trichoderma harzianum, T. viride, Pseudomonas fluorescens* and *Bacillus subtilis* dates back to time immemorial. However, a recent shift in focus has directed attention towards indigenous species, including non-pathogenic *Fusarium spp., Bacillus velezensis, Chaetomium globosum,* and biocontrol yeast species such as *Aureobasidium pullulans, Rhodosporidium kratochvilovae, Leucosporidium scottii,* and *Cryptococcus laurentii* (Palmeiri *et al.*, 2022). This study explores both traditional and newly recognized biocontrol agents, delving into the associated mechanisms and strategies influencing the biocontrol potential against a diverse range of pathogens.

## Biological Control

The term "biological control," coined by Harry Scott Smith of the University of California, originally focused on the suppression of insect populations through the actions of native and introduced enemies (Smith, 1908). Baker and Cook (1974) later broadened the term, defining it as the reduction of inoculum or disease-producing activities of a pathogen or parasite, whether in active or dormant form, by one or more organisms. This reduction is achieved naturally or through the manipulation of the environment, host, or antagonist. The definition has undergone multiple modifications by various authors, reflecting the evolving scope and perspectives of biological control. Generally, biocontrol employs two approaches: importation (the silver bullet approach) and augmentation (the black box approach). Importation involves introducing bioagents foreign to a specific location to suppress pest populations. In contrast, augmentation supports naturally occurring bioagents through mass culture and periodic release. Augmentation proves more successful than importation, as it eliminates the need for bioagents to acclimate to new environments and is less influenced by ecological characteristics such as host genetics, physiology, and climatological circumstances (Marian and Shimizu, 2019). Despite these advantages, most biocontrol applications are confined to greenhouse crops, where environmental conditions can be strictly monitored to meet the specific needs of the bioagents. To address environmental instability, an alternative approach involves applying a consortium of microbes with collaborative properties, deviating from reliance on a single organism should be used.

## Biocontrol agents

### Fungal Biocontrol

The *Trichoderma* species, ectomycorrhizas, arbuscular mycorrhizas (AMF), yeasts, and endophytes are a few of the well-known fungal biocontrol agents. Hypovirulence-associated mycoviruses can be used by even nonvirulent strains of some pathogens to act as biocontrol fungi. Better biotechnological and genetic advancements allow for the introduction of advantageous fungal genes into host plant genomes as well as the interruption or overexpression of these genes to enhance biocontrol capacity (Ghorbanpour *et al.*, 2018). The genus *Trichoderma*, which includes 25 species that have been employed as biocontrol agents against various plant fungal diseases, has been found to have the highest potential for biocontrol. Furthermore, nine additional genera, which have five or more known antagonistic species: *Alternaria, Aspergillus, Candida, Fusarium, Penicillium, Pichia, Pythium, Talaromyces,* and *Verticillium.* The majority of plant growth-promoting fungi (PGPF), which include *Trichoderma, Penicillium, Aspergillus,* and *Fusarium* spp., are known to stimulate the plant immune responses in response to enemy attack. They are also regarded as one of the safest methods for promoting plant growth and inducing systemic resistance (ISR) (Thambugala *et al.*, 2020).

## Plant Growth-Promoting Rhizobacteria (PGPR)

They are the most common biological control agents that have extensively been used to control disease and to simultaneously enhance the yield of crops. Among PGPR, species of *Arthrobacter, Azotobacter, Bacillus, Enterobacter, Erwinia, Flavobacterium, Actinobacteria, Pseudomonas, Pasteuria, Rhizobium, Bradyrhizobium, Serratia, Xanthomonas* etc. have been identified as the most common biocontrol agents. Of these, Bacillus and Pseudomonas spp. are widely used as biocontrol agent for disease suppression (Shahid *et al.*, 2017). Numerous rhizospheric bacteria and fungi, including *Xenorhabdus bovienii, Bacillus thuringiensis, Pasturia penetrans, Pseudomonas fluorescens, B. amyloliquefaciens, Paecilomyces lillacinus,* and *Trichoderma harzianum,* have been reported to effectively control root-knot nematodes. All of these microorganisms control root-knot nematodes in various ways, including hatching inhibition, killing root-knot nematode females and juveniles (J2), and reducing nematode density in soil (Liu *et al.*, 2020).

## Plant Virus and Biocontrol

Plant virus strains that cause mild symptoms have been used to cross-protect against strains that cause more severe symptoms, and pathogen-derived transgenic resistance techniques make use of this characteristic. Qu *et al.* (2021) have elucidated the effects of a single-stranded DNA virus, *Sclerotinia sclerotiorum* hypovirulence-associated DNA virus 1 (SsHADV-1) that infects the fungus *Sclerotinia sclerotiorum,* a disease causative agent of many crops.

## Arbuscular Mycorrhizal Fungi (AMF) Biocontrol

AMF provides protection against several fungal infections from genera like *Macrophomina, Botrytis, Rhizoctonia, Fusarium, Cylindrocladium, Sclerotium, Verticillium, Alternaria, Erysiphe, Gaeumannomyces,* and *Macetotrichum.* However, they don't provide much defense against a wide range of bacterial and viral infections.

## Phage-Based Biocontrol

For a very long time, phages have been employed as biocontrol agents against bacterial infections. In 1924, Mallmann and Hemstreet conducted the first investigation showcasing their capacity for biocontrol. They isolated *Xanthomonas campestris pv. campestris* from the cabbage-rot infected plant tissues. Subsequent investigations revealed that phages might prevent soft-rot in carrots (*Pectobacterium carotovorum* subsp. *carotovorum*) and tomato bacterial spots (*X. campestris pv. vesicatoria*) as well as *Pectobacterium atrosepticum* in potato slices. The goal of more recent research on the application of phage biocontrol has been to increase their durability in field settings. Investigating the application of phage cocktails and systemic acquired resistance activator in the treatment of citrus bacterial canker and citrus bacterial spot caused by *X. citri* subsp. *citri* and *Xanthomonas axonopodis* pv. *citrumelo* respectively showed positive results in field experiments.

## Natural Compounds against Plant Diseases

Several bioactive compounds that fall into the phenolic, terpenoid, or alkaloid groups have been found to be useful as biopesticides in organic farming, including chitin, laminarin, allicin, terpenes, chitosan, naringin, and carrageenans. Under field conditions, allicin, derived from garlic, demonstrates antibacterial and antifungal effects. Garlic juice inhibits the growth of a number of bacteria viz. *Pseudomonas, Agrobacterium, Xanthomonas*, and *Erwinia* and fungi *Cercospora arachidicola, Botrytis cinerea, Rhizoctonia solani, Alternaria alternata, Fusarium moniliforme, Colletotrichum coccodes* etc. Naringin, a bioactive molecule found in seeds and pulp of grapefruit, is effective against fusariosis, alternariosis, and gray mold infections in soybean, ornamental plants, and vegetables such as potato. Tea tree oil is particularly effective against downy mildew of lettuce. Chitin (a polysaccharide found in the fungal cell wall and exoskeleton of crustaceans and insects) has antifungal influence against soil-borne pathogenic fungi.

## Algal and Cyanobacterial Biocontrol

Use of extracts from the algae *Sargassum filipendula, Ulva lactuca, Caulerpa sertularioides, Padina gymnospora* and *Sargassum liebmannii* reduces symptoms of fungal infection on tomato produced by *Alternaria solani* and *Xanthomonas campestris pv. vesicatoria*. The activity of polyphenol oxidase and peroxidase enzymes important in plant defense in tomatoes was also shown to improve when extracts from *Cystoseira myriophylloides, Laminaria digitata* and *Fucus spiralis* were utilized against *Verticillium dahliae* wilt. Cyanobacteria has also been tested against plant pathogens. In tomato, application of *Nostoc linckia* in soil against *Fusarium oxysporum f. sp. lycopersici* decreased wilt. Usage of cyanobacteria, *Anabaena* sp. on zucchini cotyledons infected with powdery mildew (*Podosphaera xanthii*) resulted in enhanced enzymatic activity of peroxidases, endochitinase.

# Mechanism of biocontrol action

Biocontrol agents have the ability to combat pathogens directly or indirectly. Parasitism, release of harmful metabolites such lytic enzymes, antibiotics, volatile chemicals, siderophores, etc., competition for nutrients, and niche exclusion are some of the ways that direct antagonism manifests itself. Conversely, indirect antagonism occurs when a biocontrol agent induces systemic resistance in plants, or when plant growth is promoted. While bioagents employ numerous strategies to combat disease-causing agents, the extensively studied mechanisms encompass competition, lytic enzyme production, hyperparasitism, antibiosis and induced resistance.

## Competition

Obligate pathogens are solely dependent on living infected host, whereas, majority of the plant pathogens especially those belonging to the category of necrotrophs, kill and subsequently invade tissues of host plants and utilize the available nutrients. Once necrosis has been induced by the pathogen, non-pathogenic saprophytes come into picture which results in competition for space and nutrients. The root surface and the adjoining rhizosphere acts as a major carbon sink (Rovira, 1965). This region attracts a great plethora of beneficial and pathogenic microorganisms as a result of its nutrient rich nature especially iron. The bioagents reach root surfaces by active movement facilitated by chemical exudates such as organic acids, amino acids, and specific sugars and produce low-molecular-weight siderophores to competitively acquire ferric ion depriving the pathogenic fungi of iron element. The suppression of *Botrytis cinerea* infection through competition for niches and nutrients has been illustrated in several studies (Rabosto *et al.*, 2006). Apart from this, the rapid colonization of the fruit wounds has been advocated as one of the key strategy of antagonists against the post-harvest pathogens (Jamalizadeh *et al.*, 2011). Compared to root, foliar surfaces have dearth of nutrients and efficient colonization of phyllosphere surfaces by PGPB gives competitive disadvantage to the foliar pathogens. These bacteria have also been observed to compete for nutrients near the spore, resulting in inhibition of spore germination of *B. cineria* (Brodie and Blakeman, 1975). Francesco *et al.* (2017) reported that the antagonism of *A. pullulans* strains against peach brown rot *pathogen Monilinia laxa* was associated with the depletion of asparagine by the bioagent which resulted in inhibition of conidia and mycelial development. Guijarro *et al.* (2017) and Poppe *et al.* (2003) demonstrated that the effective control of *Monilinia* spp. and *Penicillium* spp. in fruit crops using *Penicillum frequentans* strain 909 and *Pantoea agglomerans* CPA-2 was based on competition for nutrients and space. On the other hand, competition for carbohydrates in the carbohydrate rich wound environment along with competition for the limited amounts of amino acids was driving force of antagonism in case of *Pichia guilliermondii* against *Penicillium digitatum, P. expansum, B. cinerea,* or *Colletotrichum* spp. in wounds of different fruits (Spadaro and Droby, 2016).

## Antibiosis

Antibiosis, a prevalent strategy in biocontrol activity, involves the production of a chemically diverse group of low molecular weight compounds or antibiotics by microorganisms, impairing the growth and metabolic activities of other microorganisms. The majority of these antimicrobial metabolites in the environment remain unknown, with characterized ones primarily attributed to actinobacteria, fungi, and bacteria in decreasing order (Berdy, 2005). Various biocontrol bacteria, such as *Agrobacterium, Bacillus, Pantoea, Pseudomonas, Stenotrophomonas, Burkholderia* sp., Streptomyces, and fungal antagonists like *Trichoderma* and closely associated *Gliocladium*, are known for producing antibiotics such as meropenem, moxalactam, iturin, surfactin, fengycin, DAPG, pyrrolnitrin, phenazine, maltophilin, lipopeptides, cepaciamides A and B, cepacidines, siderophores, altericidin, pyrrolnitrin, glidobactins, 6-PAP, gliovirin, gliotoxin, and viridin (Ghorbanpour *et al.*, 2018).

Cyclic lipopeptides (LPs), a crucial component of biological control, are classified into three major families: surfactin, iturin, and fengycin. While surfactin exhibits antibactericidal, antiviral, and antimycoplasma activities, iturin and fengycin lack antibacterial activity and instead demonstrate antifungal activity against pathogenic fungi (Ongena and Jaques, 2008). These LPs bind to lipid bilayers in pathogen cells, inducing permeability changes (fengycin and iturin) and structural damage (surfactin). Additionally, these compounds act as elicitors, stimulating plant immune-related responses (Cawoy *et al.*, 2014).

Antibiotic production is closely linked to the overall metabolic condition of the cell, influenced by factors such as nutrient availability, major and minor minerals, carbon source type, pH, temperature, and trace element availability. Although many antagonistic strains produce a variety of antimicrobial metabolites, the conditions favoring one compound may not favor another. In some cases, a single antibiotic may prove effective against a multitude of pathogens on the same crop. For instance, tetramycin 0.3% aqueous solutions exhibited antibacterial and antifungal activity against eight kiwi fruit pathogens in a study conducted by Wang *et al.* (2021). Thus, the diverse arsenal of biocontrol strains enables antagonists to achieve their ultimate objective of pathogen suppression under a wide range of environmental conditions.

## Hyperparasitism

Hyperparasitism is a direct competitive interaction wherein two organisms engage, with one organism, typically a parasite or plant pathogen, extracting nutrients from another parasite. This interaction is commonly observed among fungi and occasionally among bacteria, showcasing a unique ability of these organisms to utilize the cytoplasm of others as a nutrient source. A prominent example is *Bdellovibrio bacteriovorus*, initially reported to parasitize *Burkholderia cepacia* in liquid cocultures and subsequently demonstrated to prey on a wide range of pathogens including *Agrobacterium tumefaciens*, *Xanthomonas vesicatoria*, *X. campestris* pv. *campestris*, *Erwinia carotovora* pv. *carotovora*, *P. syringae* pv. *glycinea*, P. *syringae* pv. *tomato*, *P. marginalis*, and *Erwinia herbicola* (McNeely *et al.*, 2017).

Mycoparasitism can be broadly categorized into biotrophic and necrotrophic interactions based on lifestyle. In biotrophic interactions, hyperparasites solely rely on living hosts, extracting nutrients via haustoria without causing host cell death. Although this species-specific interaction plays a vital role in disease suppressiveness within ecosystems, its commercial exploitation is limited due to the necessity of a fungal substrate for mass propagation. Conversely, necrotrophic hyperparasites obtain nutrients from both dead host cells and available organic matter, making them more suitable for commercial use, as they can be mass-produced on artificial media.

The primary mechanisms of parasitism involve the secretion of cell wall-degrading enzymes (CWDEs) such as chitinases, -1,3-glucanases, proteases, cellulases, and certain secondary metabolites into the host cell. This process leads to the disruption of the cell wall and subsequent disorganization of the cytoplasm. Screening for hyperparasites involves the observation of structures like sclerotia, microsclerotia, uredinospores, and conidia, necessitating meticulous microscopical assessments of the viability of invaded host structures. This process is laborious and demands considerable patience. Alternative screening methodologies based on the activity of fungal CWDEs under in vitro conditions have limitations, as measuring the constitutive level of a single enzyme in the absence of host interaction may not accurately represent the complex interplay between bioagents and pathogens. Additionally, isolates selected for high constitutive enzyme production may not perform well in competitive environments due to the specificity, temporal and spatial restrictions, and potential rapid denaturation of microbial activities. As of now, there is no reported development of resistance by plant pathogens against hyperparasitism, although pathogens can inhibit the synthesis of enzymes required by antagonists, thereby evading hyperparasitic interactions.

**Production of lytic enzymes**

In addition to antibiotics, some biocontrol strains produce lytic enzymes such as chitinases, glucanases, cellulases, proteases, cell wall hydrolases and lipases that can interfere with the metabolic activities and/ directly suppress the plant pathogens (Compant *et al.*, 2005). These lytic enzymes are not constitutive in nature and their production is triggered by cascade signaling after recognition of host surface compounds such as lectins, diffusible host-released secondary metabolites and signaling molecules such as MAP Kinase, cyclic AMP and G-protein. This recognition process triggers the expression of the "molecular weapons" such as cell wall degrading enzymes. Apart from common ones, some peculiar enzymes capable of degrading oxalic acid into oxalate detoxify virulence factors of *B. cinerea* in various crop plants (Schoonbeek *et al.*, 2007). The bioagents namely, *Serratia plymuthica* C48, *Serratia marcescens*, *Paenibacillus* sp. and *Streptomyces* sp. produce extracellular chitinase which inhibits spore germination and germ-tube elongation in *B. cinerea*, *Sclerotium rolfsii* and *F. oxysporum* f. sp. *cucumerinum* (Compant *et al.*, 2005). Abo-Zaid *et al.* (2020) applied strain of *Streptomyces cellulosae* Isolate Actino 48 against *Tobacco mosaic virus* in tomato plant and observed the increased activity of defense related enzyme PAL and PR-1 gene compared to control. Yulang *et al.* (2019) reported that the pre- inoculation of soil with *Streptomyces pactum* Act12 agent enhanced the production of peroxidase, chitinase and genes related to plant systemic resistance (*PR-1* and *SIPI-II*) in response to tomato yellow leaf curl disease under pot and field conditions. The Act12 application also increased the diversity of bacteria and the abundance of beneficial microbes in the root-zone. Ghule *et al.* (2021) evaluated vitro and in vivo antifungal activity of chitosan against *F. solani* causing root rot in fenugreek. Chitosan treatments ranging from 0.1 to 2.0 g/L were tested against *F. solani* on PDA. In pot and field studies, fenugreek seeds treated with chitosan at 2.0 g/L greatly reduced root rot disease severity and also enhanced yield parameters. The activity of defense enzymes, such as chitinase, -1, 3- glucanase and total phenol were also increased after treatment.

## Induced Immunity and Priming

Induced immunity is defined as pre-treatment of plants by chemical or non-chemical elicitors to trigger defense response either through microbe/ pathogen-associated molecular patterns (PAMP) Triggered Immunity (PTI) or Effector Triggered Immunity (ETI) against wide spectrum of plant pathogens. The main and evolutionary layer of this inducible defense is based on the recognition of signals such as PAMPs and damage-associated molecular patterns (DAMPs) by the plant receptors. PAMPs are conserved molecular structures essential for the survival of the microorganism (Boller and Felix, 2009) and it includes flagellin from bacterial flagella, chitin or glucan present in fungal cell walls. On the other hand, DAMPs or danger signals such as oligo galacturonides are triggered mainly due to herbivore or insect feeding and is recognized by plasma-membrane localized pattern recognition receptors (PRRs) (Dodds and Rathjen, 2010). The recognition of these PAMPs and DAMPs induce a variety of defense responses commonly referred to as PTI. During this, PRRs trigger a cascade of signaling events, leading to depolarization of plasma membrane, production of ROS, nitric oxide (NO) and activation of mitogen activated protein kinases (MAPKs) (Boller and Felix, 2009). This response results in reduced disease severity and damage done to plant by the pathogens. The second type of immunity ETI is associated with hypersensitive response (HR) resulting to cell death at the site of infection (Dodds and Rathjen, 2010). These defense alerts are further echoed and transferred from the site of infection by a system of mobile signals located distally by a phenomenon known as systemic acquired resistance (SAR). The phenomenon of priming is an important component of SAR. The phenomenon of induced resistance directly depends upon induction and priming (Walters *et al.*, 2013). It develops during the primed phase in the plants with some changes observed at transcriptional, metabolic, and cellular level. After these changes, plant enters post primed state in which primed cells have enhanced defensive capabilities to fight against the intruding microorganism. Priming plays a vital role in development of memory and also minimizes the energy cost of plant during immune responses (Mauch-Mani *et al.*, 2017). Primed tissues are in alert state that enables them to confront both biotic and abiotic stresses rapidly and efficiently. Priming is associated with synthesis and accumulation of defense signaling proteins and histone modifications brought about by acetylation or methylation at promoter site of defense genes (Jaskiewicz *et al.*, 2011). The induced immunity can be achieved by using resistance inducers through chemical and non-chemical methods. The chemical immunity inducers include SA and JA analogues, microbial elicitors and several natural compounds whereas biocontrol agents, non-pathogenic species, biochar and compost are used to induce resistance through non-chemical or biological methods (Burketova *et al.*, 2015). Based on the type of resistance inducers and the signaling pathway involved, the induced immunity can be categorized into two forms- SAR and induced systemic resistance (ISR). Contrary to the SAR which is triggered after sensitization of infection molecule ISR is triggered by the PGPRs and plant growth promoting fungi (PGPFs) prior to root colonization. These signaling events modulate transcription factor (TF) activities leading to the defense related transcriptional restructuring. Defense gene activation results in the accumulation of different enzymes and stress-specific metabolites such as -1,3-glucanases, peroxidases, phytoalexins, lignin and chitinase. The fine tuning between these signaling molecules leads to the activation of these above-mentioned immunity and allow directing immune responses against the specific intruder (Robert- Seilaniantz *et al.*, 2011).

## Emerging Biocontrol Strategies

a. **Microbial Volatilome**: Among the several microbiological strategies used by BCAs, the production of volatile organic compounds (VOCs) is a method that is helpful in situations where the straightforward association between the pathogen and its competitor is not possible. All living forms synthesize VOCs and these can be exploited for usage in biocontrol of plant pathogens like bacteria, oomycetes, and fungi. VOCs are a sustainable preference for synthetic fungicides due to their ease of application, low residue deposition in the environment and on crops, and their biocontrol efficacy.

b. **Microbiome-Based Solutions**: Novel biocontrol agents could potentially be sourced from endophytes. Fungi and bacteria have primarily been employed for biocontrol up to this point. It is unknown how archaea affect plants or whether they have the potential to be used for biocontrol, despite their recent recognition as a component of the plant microbiome. In order to attain the intended microbiome outcomes, microbial consortia and biocontrol agents might be utilized to enhance crop biodiversity through microbiome engineering (Erlacher *et al.*, 2014). In this context, a pool of chosen biocontrol agents can be organized into crop-specific biological consortia. Targeted and preemptive biocontrol strategies are made possible by adopting a comprehensive strategy and utilizing microbiome-based approaches.

c. **Antagonistic Yeasts**: Yeast, classified as unicellular fungi, thrives in diverse environments and is easily cultivable. Employing phage-based competitive strategies, mycoparasitism, and the production of volatile compounds, enzymes, and toxins, yeast competes with other organisms for nutrients. Additionally, yeast enhances the immune response of host plants, triggering the secretion of defense molecules. Noteworthy yeast species, including *Candida oleophila, Aureobasidium pullulans, Metschnikowia fructicola, Cryptococcus albidus, Saccharomyces cerevisiae, Rhodosporidium kratochvilovae, Leucosporidium scottii* and *Cryptococcus laurentii*, exhibit significant potential for serving as commercial biocontrol agents. Di Canito *et al.* (2021) suggests that Saccharomyces and non-Saccharomyces yeasts as potential antagonists against phytopathogenic fungi of the genera Penicillium and Aspergillus and the species *Botrytis cinerea* on table grapes, wine grapes, and raisins.

d. **Genetically Modified Biocontrol Agents**: Techniques for genetic engineering of all organisms can be applied to increase the effectiveness of BCAs. E.g. *Rhizoctonia solani* infection in beans can be brought under control by transferring a gene coding for the enzyme chitinase from *Serratia* to a *Pseudomonas* endophyte. Genome shuffling was used by Clermont *et al.* (2011) to produce improved *Streptomyces melanosporofaciens* EF76 biocontrol strains. After two rounds of genome shuffles, four strains with enhanced antagonistic activity against the potato diseases *Phytophthora infestans* and *Streptomyces scabies* were identified.

e. **Phage Cocktail**: To guarantee the survival of phage combinations during long-term storage under ambient conditions, more sophisticated protective formulations are required. Better defense against the phytopathogens in that environment can be achieved by utilizing phages from the phyllosphere that are already in existence. To boost phage persistence in the phyllosphere, light-absorbing chemicals and/or protective formulations could be added to phages that have evolved to resist UV-induced damage. It is also possible to develop synthetic phage cocktails with unique host ranges.

f. **Use of Mycoviruses**: After it was discovered that mycoviruses cause their hosts to become less virulent, or hypovirulent, there was a lot of interest in characterizing the viruses that come from phytopathogenic fungus that could be used as biocontrol agents. Many plant pathogenic fungus have mycoviruses, which lessen the virulence of their fungal host. However, commercial mycovirus biocontrol techniques for these phytopathogenic fungi have been challenging to develop due to the difficulty of transmitting these mycoviruses between vegetatively incompatible groups.

g. **Combining nanotechnology and biological control**: Biological control demonstrates efficacy, but the synergistic integration of nanotechnology significantly enhances its efficiency. This convergence of biological control and nanotechnology not only contributes to enhanced food security and productivity but also represents a pivotal step toward advancing societal well-being (Eltarahony *et al.*, 2018). The focal point of nanoscience research revolves around the preparation and utilization of metal nanoparticles (NPs), underscoring their pivotal role in driving advancements in nanotechnology. Green chemistry or biological approach employs plants, fungi, bacteria, algae and actinomycetes to synthesise the NPs. The use of microorganisms for the synthesis of nanoparticles (NPs) is a fairly new and emerging area of research with considerable potential to control plant diseases. Microorganisms offer several advantages like maneuverability for desired result using biotechnology, ease of handling, cheap production technology availability, easy scale up process, high efficiency and eco-friendliness. Easy downstream processing, easy mass production and precipitation of NPs externally without producing cellular contaminants have drifted the research towards fungi from bacteria for their use in synthesis of NPs. However, being a eukaryote, fungi are less amenable to genetic alteration compared to prokaryotes. Among the metal NPs, silver NPs are most commonly used in agriculture both in ionic or nanoparticle forms. AgNPs are widely used in bio-labeling, sensors, and possess inhibitory effects on growth of bacteria, virus and eukaryotic microorganisms. Antifungal properties and ability to deactivate enzymes of microorganisms possessed by silver NPs make them an apt candidate for the plant disease management. The disease control properties of silver NPs have been well studied against *B. cinerea, Cladosporium cucumerinum, Corynespora cassiicola, Cylindrocarpon destructans, F. oxysporum* f.sp. *cucumerinum, Glomerella cingulata, Monosporascus cannonballus, Pythium aphenidermatum, Alternaria alternata, A. brassicicola, A. solani, Didymella bryoniae, F. oxysporum* f.sp. *lycopersici, F. solani, Stemphylium lycopersici, B. megaterium, P. syringae, Burkholderia glumae*, and *X. oryzae.*

h. **Use of substrates to maintain beneficial biocontrol microbes**: The application of substrates to sustain beneficial biocontrol microbes in the vicinity of crops is a proactive strategy in the management of plant diseases. By incorporating specific substrates, such as organic matter or bioactive compounds, into the soil or onto plant surfaces, the goal is to create an environment that fosters the proliferation and activity of these advantageous microbes. These substrates serve as a source of nutrition and support the growth of biocontrol microbes, facilitating their establishment and persistence in the rhizosphere or on plant surfaces. The enhanced presence of these beneficial microbes can help suppress the growth and spread of plant pathogens through various mechanisms, including competition for nutrients, production of antimicrobial compounds, and induction of systemic resistance in plants. This targeted use of substrates not only provides a conducive habitat for biocontrol microbes but also contributes to the overall balance of the soil microbiome. The approach aligns with sustainable agriculture practices by reducing the reliance on chemical interventions and promoting a more resilient and disease-resistant crop environment.

## Conclusion

In conclusion, biological control has emerged as a multifaceted and dynamic strategy for managing plant diseases, providing an environmentally friendly alternative to conventional chemical interventions. The use of biocontrol agents, particularly microorganisms like PGPRs and PGPFs, has shown promise in suppressing plant diseases and promoting overall plant health. The mechanisms employed by these agents, including competition, lytic enzyme production, hyperparasitism, antibiosis, and induced immunity, contribute to their efficacy in curbing pathogenic threats. Notably, the concept of induced immunity and priming introduces a proactive element to disease management, enhancing the plant's ability to defend itself. Despite these advancements, challenges persist, such as the limited commercial adoption of biological control due to factors like inconsistent efficacy, climate dependency, and genotype specificity. Biocontrol is also frequently more difficult to manage than physical and chemical methods because it requires on knowledge of biological interactions at the ecosystem, organism, cellular, and molecular levels. It is also likely to be less remarkable than most physical or chemical controls. However, the integration of modern technology and the exploration of the several emerging strategies could further enhance the efficiency of biological control. While biological control faces hurdles, ongoing research and innovative approaches offer hope for its continued evolution and broader acceptance as an integral component of integrated disease management which will be more stable and longer lasting.

## References


Abo-Zaid GA, Matar SM and Abdelkhalek A. 2020. Induction of plant resistance against *Tobacco Mosaic Virus* using the biocontrol agent *Streptomyces cellulosae* isolate actino 48. *Agronomy*, 10: 1620.

Anonymous. 2023. Challenges associated with Indian agricultural and horticultural sector retrieved from https://economictimes.indiatimes.com/news/economy/agriculture/indias-horticulture-sector-holds-untapped-potential-despite-challenges/articleshow/ 99763533.cms

Baker KF and Cook RJ. 1974. Biological control of plant pathogens. WH Freeman and Company, New York, US.

Bérdy J. 2005. Bioactive microbial metabolites. *Journal of Antibiotics*, 58: 1–26.

Boller T and Felix G A. 2009. Renaissance of elicitors: perception of microbe-associated molecular patterns and danger signals by pattern-recognition receptors. *Annual Review of Plant Biology*, 60: 379–406.

Brodie IDS and Blakeman JP. 1975. Competition for carbon compounds by a leaf surface bacterium and conidia of *Botrytis cinerea*. *Physiological Plant Pathology*, 6: 125–136.

Burketova L, Trda L, Ott P G and Valentova O. 2015. Bio-based resistance inducers for suitable plant protection against pathogens. *Biotechnology Advances*, 33: 994-1004.

Cawoy H, Mariutto M, Henry G, Fisher C, Vasilyeva N, Thonart P, Dommes J and Ongena M. 2014. Plant defense stimulation by natural isolates of *Bacillus* depends on efficient surfactin production. *Molecular Plant-Microbe Interactions*, 27(2): 87-100.

Clermont N, Lerat S and Beaulieu C. 2011. Genome shuffling enhances biocontrol abilities of Streptomyces strains against two potato pathogens. *Journal of applied microbiology*, 111(3): 671-682.

Compant S, Duffy B, Nowak J, Clement C and Barka EA. 2005. Use of plant growth-promoting bacteria for biocontrol of plant diseases: principles, mechanisms of action, and future prospects. *Applied and Environmental Microbiology*, 71: 4951–4959.

Dodds P N and Rathjen J P. 2010. Plant immunity: towards an integrated view of plant- pathogen interactions. *Nature Reviews Genetics*, 11: 539–548.

Di Canito A, Mateo-Vargas MA, Mazzieri M. 2021. The Role of Yeasts as Biocontrol Agents for Pathogenic Fungi on Postharvest Grapes: A Review. *Foods*, 10: 1650.

Elad Y. 1996. Mechanisms involved in the biological control of *Botrytis cinerea* incited diseases. *European Journal of Plant Pathology*, 102: 719–732.

Eltarahony M, Zaki S, ElKady M, Abd-El-Haleem D. 2018. Biosynthesis, characterization of some combined nanoparticles, and its biocide potency against a broad spectrum of pathogens. *Journal of Nanomater*, 2018: 1-16.

Erlacher A, Cardinale M, Grosch R, Grube M and Berg G. 2014. The impact of the pathogen Rhizoctonia solani and its beneficial counterpart *Bacillus amyloliquefaciens* on the indigenous lettuce microbiome. *Frontiers in Microbiology*, 5: 175.

Francesco A, Di Foggia M, Corbetta M, Baldo D, Ratti C and Baraldi E. 2021. Biocontrol activity and plant growth promotion exerted by *Aureobasidium pullulans* strains. *Journal of Plant Growth Regulation*, 40: 1233-1244.

Ghorbanpour M, Omidvari M, Abbaszadeh-Dahaji P, Omidvar R and Kariman K. 2018. Mechanisms underlying the protective effects of beneficial fungi against plant diseases. *Biological Control*, 117: 147–157.

Guijarro B, Casals C, Teixidó N, Larena I, Melgarejo P and De Cal A. 2020. Balance between resilient fruit surface microbial community and population of *Monilinia* spp. after biopesticide field applications of *Penicillium frequentans*. *International Journal of Food Microbiology*, 333: 108788.

Jamalizadeh M, Etebarian H, Aminian H and Alizadeh A. 2011. A review of mechanisms of action of biological control organisms against post-harvest fruits spoilage. *Bulletin OEPP/EPPO*, 41: 65–71.

Jaskiewicz M, Conrath U and Peterhansel C. 2011. Chromatin modification acts as a memory for systemic acquired resistance in the plant stress response. *EMBO Reporter*, 12: 50–55.

Liu G, Lin X, Xu S, Liu G, Liu F and Mu W. 2020. Screening, identification and application of soil bacteria with nematicidal activity against root-knot nematode (*Meloidogyne incognita*) on tomato. *Pest Management Science*, 76(6): 2217-2224.

Marian M and Shimizu M. 2019. Improving performance of microbial biocontrol agents against plant diseases. *Journal of General Plant Pathology*, 85: 329–336.

Mauch-Mani B, Baccelli I, Luna E and Flors V. 2017. Defense priming: an adaptive part of induced resistance. *Annual Review of Plant Biology*, 68: 485–512.

McNeely D, Chanyi RM, Dooley JS, Moore JE and Koval SF. 2017. Biocontrol of *Burkholderia cepacia* complex bacteria and bacterial phytopathogens by *Bdellovibrio bacteriovorus*. *Canadian Journal of Microbiology*, 63(4): 350-358.

Ongena M and Jacques P. 2008. Bacillus lipopeptides: versatile weapons for plant disease biocontrol. *Trends in microbiology*, 16(3): 115-125.

Palmieri D, Ianiri G, Del Grosso C, Barone G, De Curtis F, Castoria R and Lima G. 2022. Advances and perspectives in the use of biocontrol agents against fungal plant diseases. *Horticulturae*, 8(7): 577.

Poppe L, Vanhoutte S and Höfte M. 2003. Modes of action of *Pantoea agglomerans* CPA-2, an antagonist of postharvest pathogens on fruits. *European Journal of Plant Pathology*, 109: 963-973.

Qu Z, Fu Y, Lin Y, Zhao Z, Zhang X, Cheng J, Xie J, Chen T, Li B and Jiang D. 2021. Transcriptional Responses of *Sclerotinia sclerotiorum* to the Infection by SsHADV-1. *Journal of Fungi*, 7(7): 493.

Rabosto X, Garrau M, Paz A, Boido E, Dellacassa E and Carrau F. 2006. Grapes and vineyard soils as source of microorganisms for biological control of *Botrytis cinerea*. *American Journal of Enology and Viticulture*, 57: 332–338.

Robert-Seilaniantz A, Grant M and Jones J D. 2011. Hormone crosstalk in plant disease and defense: more than just jasmonate-salicylate antagonism. *Annual Review of Phytopathology*, 49: 317–343.

Rovira A D. 1965. Interactions between plant roots and soil microorganisms. *Annual Review of Microbiology*, 19:241–266.

Schoonbeek HJ, Jacquat-Bovet AC, Mascher F and Mé- traux JP. 2007. Oxalate-degrading bacteria can protect *Arabidopsis thaliana* and crop plants against *Botrytis cinerea*. *Molecular Plant-Microbe Interactions*, 20: 1535–1544.

Shahid M, Zaidi A, Khan MS, Rizvi A, Saif S and Ahmed B. 2017. Recent advances in management strategies of vegetable diseases. *Microbial strategies for vegetable production*, 197-226.

Smith HS. 1908. On some phases of insect control by the biological method. *Journal of Economic Entomology*, 12: 288.

Spadaro D and Droby S. 2016. Development of biocontrol products for postharvest diseases of fruit: the importance of elucidating the mechanisms of action of yeast antagonists. *Trends in Food Science and* Technology, 47: 39–49.

Thambugala KM, Daranagama DA, Phillips AJ, Kannangara SD and Promputtha I. 2020. Fungi vs. fungi in biocontrol: An overview of fungal antagonists applied against fungal plant pathogens. *Frontiers in cellular and infection microbiology*, 10: 604923.

Walters DR, Ratsep J and Havis ND. 2013. Controlling crop disease using induced resistance: challenges for the future. *Journal of Experimental Botany*, 64: 1263-1280.

Wang Q, Zhang C, Long Y, Wu X, Su Y, Lei Y and Ai Q. 2021. Bioactivity and control efficacy of the novel antibiotic tetramycin against various kiwifruit diseases. *Antibiotics*, 10(3): 289.

Yulang L, Qiao G, Yunzhou L, Yifan S, Quanhong X and Hangxian L. 2019. *Streptomyces pactum* Act12 controls tomato yellow leaf curl virus disease and alters rhizosphere microbial communities. *Biology and Fertility of Soils*, 55: 149-169.

## CHAPTER 24

# INDIAN BEAN FOR LIVELIHOOD AND NUTRITIONAL SECURITY IN SEMI-ARID AREAS

**Gangadhara K[1], L. P. Yadav[1], A.K. Singh[1], Anil[1], V.V. Appa Rao[1], Anand Sahil, P. Ravat[1], Manjunath, P.P[2]. and Ramesh K.B[3]**

[1]*ICAR-Central Horticultural Experiment Station (CIAH), Vejalpur, Godhra, Gujarat*
[2]*ICAR-Indian Institute of Groundnut Research, Junagadh, Gujarat*
[3]*ICAR-Indian Institute of Vegetable Research (Varanasi), RS-Sargatia, UP*

## Introduction

The Indian bean or Dolichos bean is one of the most ancient crop among cultivated plants and an important legume as well as vegetable crop. It is commonly called as dolichos bean, Indian bean, hyacinth bean, lablab bean, country bean, Egyptian bean, Tonga bean, and field bean. In Indian languages, it is called by different names viz., Sem-Hindi, Sim- Bengali, Valor- Gujarati, Val papadi-Marathi, Moang-Kashmiri, Avare or Chapparada Avare-- Kannada, Avara- Malayalam, Avarai or Motchai- Tamil, Chikkudu- Telugu and semi- Oriya. This multiplicity of names of this crop indicates the wide range of forms available globally and the fact that it has long been cultivated for human food and as a green manure. Dolichos came from Greek language which means long pod and lablab is Arabic or Egyptian word which means dull rattle sound (Sharp knocking sound) of the seed inside the dry pod (Dhaliwal, 2017). Dolichos bean is vigorous in growth having a trailing, twining nature annually propagated through seed.

## Origin and Distribution

Dolichos bean originated in India as a wild form. Further it was distributed from India to other countries of South Asia, Southeast Asia, Africa and other tropical countries (Vishnu Swarup, 2012). It is cultivated mainly in the tropical regions of Asia, Africa, and America. In India, it is majorly grown in the states of Karnataka,

Kerala, Tamil Nadu, Andhra Pradesh, Telangana, Maharashtra, Gujarat, Madhya Pradesh, Uttar Pradesh, West Bengal and North eastern states.

## Botany and Taxonomy

Dolichos bean, *Lablab purpureus* L.(2n=2x=22, 24), belongs to the family Fabaceae (Leguminosae). It is perennial herbaceous plant often cultivated as an annual. Leaves are alternate and trifoliate. Flowers are borne in axillary racemes, typically papilionaceous and are self- pollinated. Flowering takes place under short day periods irrespective of planting time. Anthesis occurs from 9 a.m. to 5. p.m. (Pokle and Deshmukh, 1971). Anther dehiscence is from 5.0 a.m. to 2.0 p.m. and stigma is receptive on the day of anthesis.

Two types of *Dolichos* bean are recognized (Prabhakara *et al.*,2018), i.e., *Lablab purpureus* var. *typicus*, which is a garden type with soft edible pods having less fibre in their pod walls. The second type is *Lablab purpureus* var. *lignosus*, which is a field bean grown for dry seeds generally used as a pulse. Its pods have a characteristic aroma and the pod walls have high fibre content. Both the varieties are cross compatible.

## Nutritive value

Dolichos bean is an important source of proteins, minerals and dietary fibre. The pods have a strong flavour. Some varieties have purple colour of the pods, which disappears after cooking. According to Davis *et al.*, (2000) per 100 g of edible green pod contain moisture 86.1 g, carbohydrates 6.7 g, protein 3.8 g, fibre 1.8 g, minerals 0.9 g (calcium 210 mg, phosphorus 68 mg, sodium 55.4 mg, potassium 74 mg, sulphur 40 mg, magnesium 34 mg, iron 1.7 mg) and fat 0.7 g. The vitamin composition is: vitamin C 9 mg, vitamin A 312 IU, riboflavin 0.06 mg, thiamin 0.1 mg and nicotinic acid 0.7 mg. According to Venkatachalam *et al.* (2002) on dry weight basis, it accounts 30% protein. Approximately Albumin (20%), globulin (48%), prolamin (1%) and glutelin (31%) accounted for the total seed proteins.

## Uses

The use of dolichos bean in different forms mainly depends upon the location to location or region to region where it is grown. However, for vegetable purposes, it is primarily grown for its green pods as well as tender seeds which are consumed as cooked or fried vegetables and the mature dry seeds are consumed as a pulse. In some regions, young tender leaves are also used in salads and they prepare potherb like spinach from the relatively matured leaves. The mature seeds, however, contain anti-nutritional factors such as tannins, phytate and tryps in inhibitors. Removing seed coat, soaking, and proper cooking minimize the activity of these compounds (Ariina *et al.*, 2021). Seed sprouts are consumed as those of other beans. Besides its cultivation as a vegetable crop, it is also grown as a green manure and fodder crop.

**Varieties:** A brief description of improved varieties is given below.

| Institute | Varieties | Special features |
|---|---|---|
| Indian Institute of Horticultural Research (IIHR), Bengaluru (Karnataka). | Arka Jay | Bush type, photo-insensitive. Pods long, light green slightly curved, without parchment. Vegetable type with excellent cooking qualities. Tolerant to low moisture stress. It is developed by selection at $F_7$ from Hebbal Avare-3 (Bush) x IIHR 99(Pole-recurrent parent).<br>Pod Yield: 12 t/ha in 90 days. |
| | Arka Vijay | Bush-type, plants are dwarf, bushy, erect, and photo-insensitive. Pods are short, dark green. Seeds bold. Pods with characteristic aroma and without parchment. Tolerant to low moisture stress. Pod Yield: 12 t/ha in 90 days. Developed by Pedigree method of selection from $F_7$ generation involving Hebbal Avare x IIHR93.Pod Yield: 12 t/ha in 90 days. |
| | Arka Soumya | Bush-type, plants are medium tall and photo-insensitive. Pods are slender, wavy, medium long, and ready for harvest in 55 days. Suitable for Andhra Pradesh. Developed by the Pedigree method of selection from $F_7$ generation involving (Arka Jay X Arka Vijay) X Konkan Bhushan). Pod Yield: 19.0 t/ha in 75 days |
| | Arka Sambhram | Bush-type, plants are medium tall and photo-insensitive. Pods are flat, light green, medium long, and ready for harvest in 55 days. Suitable for Tamil Nadu. Developed by the Pedigree method of selection from $F_7$ generation involving (Arka Jay X Arka Vijay) X Konkan Bhushan). Pod Yield: 19-20 t/ha in 75 days. |
| | Arka Amogh | Bush-type, plants are medium tall and photo-insensitive. Pods are wavy, green, medium long, and ready for harvest in 55 days. Suitable for Maharashtra. Developed by the Pedigree method of selection from $F_7$ generation involving (Arka Jay X Arka Vijay) X Konkan Bhushan). Pod Yield: 19-20 t/ha in 75 days. |
| | Arka Swagath | Pole type and photo-insensitive variety and suitable for round-the-year cultivation. Pods are light green, medium long, and suitable for Karnataka. Developed by Pureline selection from IC 556736. Pod Yield: 26.0 t/ha in 120 days. |
| | Arka Prasidhi | Pole type and photo-insensitive variety. Pods are dark green, long, flat, and slightly curved. Resistant to rust. Suitable for south Indian states. Developed by the Pedigree method of selection from $F_7$ generation involving (IC 556824 IPS-2 X Arka Swagath). Pod Yield: 37.0 t/ha in 120 days. |

| Institute | Varieties | Special features |
|---|---|---|
| | Arka Bhavani | Pole type and photo-insensitive variety. Pods are slender, wavy, and dark green colored. Suitable for cultivation in Andhra Pradesh. Developed by the Pedigree method of selection from $F_7$ generation involving (IIHR 178 X Arka Swagath).<br>Pod Yield: 32.0 t/ha in 120 days. |
| | Arka Visthar | Pole type and photo-insensitive variety. Pods are long, thick, very broad, and dark green colored. Suitable for cultivation in Tamil Nadu and North Eastern states. Developed by the Pedigree method of selection from $F_7$ generation involving (IIHR 178 X Arka Swagath).<br>Pod Yield 37.0 t/ha in 120 days. |
| | Arka Pradhan | Pole type and photo-insensitive variety. Pods are green in colour, smooth, and shiny with an undulating surface. Suitable for cultivation in Maharashtra. Developed by the Pedigree method of selection from $F_7$ generation involving (IC 556824 IPS-2 X Arka Swagath).<br>Pod Yield: 35.0 t/ ha in 120 days. |
| | Arka Krishna | Pole type and photo-insensitive and early variety. Pods are borne in clusters and dark green colored. Suitable for Karnataka. Developed by the Pedigree method of selection from $F_7$ generation involving (IIHR 178 X Arka Swagath).<br>Pod Yield: 30.0 t/ha in 120 days. |
| | Arka Adarsh | Pole type and photo-insensitive and early variety. Pods are medium long, slightly thick, broad, and dark green colored. Suitable for Karnataka. Developed by the Pedigree method of selection from F7 generation involving (IIHR 178 X Arka Swagath). Pod Yield: 41.0 t/ha in 120 days. |
| | Arka Supriya | The pods are light green, flat and long, and bearing in clusters. It is Photo-sensitive and suitable for Rabi season only. Yield potential of 272.1 q/ha. |
| UAS, Bengaluru (Karnataka) | Hebbal Avare-1 | Bush and photo-insensitive variety. The pods are small and soft. Yield 0.8 t/ha in 90-100 days. |
| | Hebbal Avare-3 | Bush and photo-insensitive variety. Flowers white. Pods green, 2-3 seeded. Seeds are brown, round, and small. Yield 8-10 t/ha in 90-100 days. |
| | Hebbal Avare-4 | Bush and photo-insensitive variety. Pods are soft and harvested in 5 pickings. Yield 6t/ha. |

| Institute | Varieties | Special features |
|---|---|---|
| IARI, New Delhi | Pusa Early Prolific | Pole type. It is developed through selection from local types. Pods are flat, green, narrow, cycle shaped. Pod length 9.3 cm, width 1.5 cm, weight 3.5 g. Yield 14 t/ha in 200-215 days. |
| | Pusa Sem-2 | Pole type. It is developed through selection. Pods are semi-flat, dark green, fleshy and stringless. Pod length 15-17 cm. Yield 13-22 t/ha in 200-215 days. Tolerant to anthracnose, yellow bean mosaic virus, aphids, pod borers, and frost. |
| | Pusa Sem-3 | Pole type. It is developed through the selection of Pods flat, green, fleshy, and stringless. Pod length 15 cm. Yield 17-27 t/ha in 200-215 days. Tolerant to anthracnose, yellow bean mosaic virus, aphids, pod borers, and frost. |
| CIAH, Bikaner, Rajasthan | Thar Kartiki | Pole type. Days to flowering, DAS is 70-75 after 5.94. The first harvest is 90-95 DAS. Pod yield is 1.75 kg/plant and yield 110 q/ha. |
| | Thar Maghi | Pole type. It is high-yielding, stable, and found to be the most potential for commercial cultivation. For the first marketable harvesting, it took 90-95 days in comparison to AHDB-3 (95-100 days). The average pod yield is 1.77 kg/plant and the tender pod yield is 113.25 q/ha. |
| Central Horticultural Experimental Station (CHES), Vejalpur, Godhra, Gujarat. (Regional station of ICAR-CIAH, Bikaner). | Thar Ganga | Pole type. It is developed through selection from local types. The pods are attractive long green with an average pod length of 17.5cm and pod girth of 5.21cm with a pod weight of 15.2g. The first harvest is at 98 to 110 days after sowing. Pod yield: 8-10 kg/plant of fresh pod with a yield potential of 60-70 t/ha. It is performing well under rainfed semi-arid conditions and found moderately resistant to dolichos bean yellow mosaic virus disease under field conditions. |
| | Thar Kiran | Pole type. It is a unique variety of dolichos beans developed through selection from local types. It has a purple pod colour high yielding and rich in anthocyanins (190 mg/100g). The average pod length is 10.8cm, pod girth is 5.33cm and pod weight is 8.4g. The first harvest is 100 to 110 days after sowing. Pod yield: 7-9 kg/plant of fresh pod with a yield potential of 60 t/ha. It performs well under rainfed semi-arid conditions and is resistant to dolichos bean yellow mosaic virus disease under field conditions. |

| Institute | Varieties | Special features |
|---|---|---|
| | Thar Vinaya | It is a pole-type variety having attractive long, light green colour pods. It is a pole-type Indian bean variety with cluster pod bearing in nature. The plants have a climbing growth habit and grow up to 4.0 - 4.6m. The plant requires stacking or support for higher yield and better-quality pod production. Pods are long (14 cm), girth (4.20 cm) with pod weight (8.5g). It takes 90-91 days for the first flowering & 102-105 days after sowing for the first harvest. Yield total of 1000-1265 fresh pods/ plant with yield 6.5-7.5 kg/plant and 50t/ha of fresh tender pods. |
| | Thar Lakshmi | It is a Pole type variety having long flat and medium-sized green colour pods. Length of pod 14 cm, girth 5.5 cm, and weight 10 g. The first harvest of fresh pods starts 90-95 days after sowing. Total of 800-1240 pods/plant with an average yield of 6.5-6.7 kg/plant and 45-50t/ha of fresh pods. |
| KKVP, Dapoli (Maharashtra) | Wal Konkan 1 | Bushy, photo-insensitive, resistant to yellow mosaic virus. Yield 9-10 t/ha in 110-115 days. |
| | Konkan Bhushan (DPLD 1) | Bushy, photo-insensitive, resistant to yellow mosaic virus. Yield 9-10 t/ha in 110-115 days. |
| DPKV, Akola (Maharashtra) | Dasarawal | Pole type. The pods are dirty green with a purple tinge at both borders. Pod length 7.8 cm, width 2.0 cm, weight 3.2 g. Yield 7-8 t/ha. |
| | Deepaliwal | Pole type. Pods extra long (18.4 cm), white, not smooth due to bulging at each seed. Pod width 2.7 cm, weight 1.5 g. Yield 6-8 t/ha in 200-210 days. |
| Tamil Nadu Agricultural University, Coimbatore | CO.1 | Pole type. Pods are green, and fleshy with slow fiber development. Pod weight 9.7 g. Yield 18-20 t/ha in 160-180 days. |
| | CO.2 | Pole type. Pods, flat, green with purple margin. Pod length 9.3 cm, width 2.1 cm, weight 6.17 g. Yield 11.8 t/ha in 210-220 days. |
| | CO.3 | Pole type. Pods fleshy, green with purple tinge. Pod length 10.6 cm, width 4.8 cm, weight 11.77 g. Seeds black. Yield 10.0 t/ha in 230 days. |
| | CO.4 | Pole type. Pods deep purple throughout and fleshy. Pod length 10.2 cm, width 3.3 cm, weight 7.43 g. Seeds black. Yield 13.5 t/ha in 215-220 days. |
| | CO.5 | Pole type. Pods long, narrow, light green to white in colour, tubular, curved with serrated margin. Pod length 13.4 cm, breadth 1.5 cm, weight 5.26 g. Seeds chocolate brown. Yield 6-7 t/ha in 235 days. |

| Institute | Varieties | Special features |
|---|---|---|
| | CO.6 | Bush variety. Selected from DL 3169 x CO.5. Pods slightly curved and bloated. Yield 12 t/ha in 240 days. |
| | CO.7 | Bush variety. Selected form DL 3169 x CO.5. Pods long, succulent, flat, greenish white and broad. Yield 12 t/ha in 240 days. |
| | CO.8 | Bush variety. Pods green tubular and fleshy. Yield 6-8 t/ha in 120 days. |
| | CO.9 | Bush variety. Pods and grains are used. Yield 7-8 t/ ha in 120 days. |
| | CO.10 | Bush variety. Induced mutant from CO.6 by gamma ray (24 krad). Pods greenish white, tubular and curved. Yield 5-6 t/ha in 120 days. |
| | CO.11 | Bush variety. Hybrid derivative of CO.9 x a pandal type. Compact plant type. Pods flat and light green with purple margin. Yield 9-10 t/ha. |
| | CO.12 | Bush variety. Hybrid derivative of CO.9 x CO.4. Pods deep purple. Yield 10-12 t/ha in 110 days. |
| | CO.13 | Bush variety. Hybrid derivative of Co.9 x a training type. Pods long green. Yield 10t/ha in 110-120 days |
| CSAUA&T, Kanpur, UP. | Rajani | Pole type. Pods narrow oval in cross section, shining green. Pod length 10.4 cm, width 1.2 cm, weight 1.78 g. Yield 7-8 t/ha in 200-210 days. |
| | KDB 403 | Pole type. Pods long, narrow, shining green. Pod length 12.9 cm, width 1.2 cm, weight 2.0 g. Yield 5-6 t/ha in 180-210 days. |
| | KDB 405 | Pole type. Pods medium long, narrow, dark green band in the middle and boarders light green. Pod length 9.6 cm, width 1.3 cm, weight 1.1 g. Yield 3-4 t/ha in 180-200 days. |
| JNKV, Jabalpur, MP. | JDL. 79 | Pole type. Pods flat, broad, whitish green with parrot green boarder along the line of seed attachment. Pod length 11.8 cm, width 3.6 cm, weight 1.5 g. Yield 5.6 t/ha in 200 days. |
| | JDL 53 | Pole type. Pods flat, small, narrow, dull whitish green with purple tinge along the border. Pod length 7.2 cm, width 1.8 cm, weight 3.75 g. Yield 10-12 t/ha in 200-220 days. |

| Institute | Varieties | Special features |
|---|---|---|
| IIVR, Varanasi, UP. | Kashi Khushal (VRSEM-3) | Semi-pole type variety has shiny dark green, slightly curved glabrous fruit containing 4-5 coffee colour seed. Tolerant to high temperature and DYMV. Variety is rich in protein 635.6 mg/g low in total sugar and phenol content. Yield 35-38 t/ha. Recommended for cultivation in Uttar Pradesh, vide gazette notification number S.O. 692(E), dated 05.02.2019. |
| | Kashi Sheetal (VRSEM-11) | Semi-pole type, tolerant to low temperature and DYMV. It can give yield of 18-20 t/ha. Variety is rich in protein 590.8mg/g and low in total sugar 0.621g/100g. Yield 182.06. Recommended for cultivation in Uttar Pradesh, vide gazette notification number S.O. 692(E), dated 05.02.2019. |
| | Kashi Harittima | Suitable for sowing in Kharif season. High yielding and good pod quality (green, tender and parchment free). It is moderately resistant to Dolichos Yellow Mosaic Virus diseases under field condition. Moderately tolerant to jassid, aphid and pod borer under field condition |

Source: Gopalakrishnan (2007), Singh and Nath (2011), Dhaliwal, (2017), Chakraborty and Parthasarathy, (2003), https://www.iihr.res.in , https://iivr.icar.gov.in and http://iari.res.in

## Cultivation

### Soil and climate

Dolichos bean is a cool season crop. It grows well under warm, humid conditions at temperatures ranging from 18-35 C and is fairly tolerant to high temperatures. The bean is drought-hardy and is grown in wide areas under diverse climatic conditions such as arid, semiarid, Subtropical and humid regions, lowlands and uplands. pH varying from 5.3 to 6.0 is suitable. The varieties suitable for cultivation under long-day and short-day conditions are different, however, some photo-insensitive varieties, like Arka Jay and Arka Vijay can be grown throughout the year (Prabhakara *et al.*,2018).

The soil should be plowed to a depth of 25 cm to break the clods and remove the weeds completely. Shallow cultivation is practiced to bring the soil into fine tilth. After that, the field should be divided into plots or ridges and furrows of convenient size (Prabhakara *et al.*,2018).

### Sowing

The seeds are dibbled or drilled behind the plough. The climbing types are sown near houses and allowed to climb on the rooftops. The seed rate for this crop is 50-60kg/ha (bush) and 5-6 kg/ha (pole type) with a spacing of 30x60 (bush type), 75cmx1.5 m, or 1x1.5 m (pole type) respectively.

## Nutrient requirement

Although dolichos bean is a legume crop, it responds well to the application of fertilizers. About 20-25t/ha of well-composted FYM should be applied to the soil at the time of land preparation. Application of 20:60:60 kg/ha of the N (Nitrogen), P (Phosphorus) and K (Potassium) is recommended (Prabhakar *et al.*,2018). Half of N along with the entire dose of P and K fertilizer should be applied at sowing time. The remaining half dose of N should be top-dressed 30 days after sowing.

## Weed control

Weeds may be controlled mechanically or by using weedicides. Shallow cultivation during the early stages of crop is necessary to control the weeds. Pre-sowing application of Fluchloralin at the rate of 2litre/ha checks the weed growth for 20-25 days (Prabhakara *et al.*, 2018; Gopalakrishnan, 2007). At the later stages of the crop growth, weeds are kept under check due to the thick canopy of the crop as the dolichos bean grows vigorously.

## Staking or support

The Pole type of dolichos bean varieties needs support since the plants have a twinning growth habit. After 20-30 days after sowing, the supporting structure should be made with the help of GI-iron wires supported by bamboo poles or cement poles. All the plant vines should be twined with gunny threads or any other plastic threads and allowed to twine and spread on the bamboo stakes for better growth and fruit set. Over-crowding of plant vines due to non-staking affects the formation of pods and yields adversely. Staking improves plant spread and photosynthetic activity. As a result, there is a higher yield due to a higher number of pods per plant.

## Water management

Dolichos bean is a hardy crop that comes up well under rainfed conditions. Flowering and pod development periods are the critical stages. based on the atmospheric conditions, five to six protective irrigation may be given. Light irrigation is given when required. For higher yield, the crop should be irrigated regularly at 7 -10 days intervals (Prabhakara *et al.*, 2018).

## Harvesting

The fresh pods are harvested when they are at the tender and succulent stage. Fully grown immature pods are harvested. Avoid delaying the harvest, otherwise, the pods may become fibrous. In bush type, the crop is ready for harvest two months after sowing and only 2-3 picking are obtained. In pole type, in general, it takes 3 months for the first harvest with 9-10 picking at 7-10 days intervals. On average, 30-40 tonnes/ha of fresh pods are expected from the improved varieties.

## Plant Protection

**Powdery mildew (*Leveillula taurica var. macrospora*)**: The infection on leaves appears initially as small darkened areas, later becoming white powdery growth that spreads to stems and other parts of the plant. In severe infections, the leaves become chlorotic and fall off. Pods are reduced in size and distorted in shape. Plant growth is reduced due to defoliation.

Spray with 0.5% wettable sulfur or with benlate or Bavistin 0.15%. If needed, repeat the spray at 14-day intervals.

**Dolichos bean yellow mosaic virus disease (DYMVD):** It is a viral disease and spreads by the whitefly (*Bemisia tabaci*). The Leaves of diseased plants are reduced in size and develop bright yellow patches interspersed with green areas. In severe cases, the entire leaf looks golden yellow in colour.

Remove and destroy the infected plants. Control the insect vector by spraying of Diafenthiuron 50WP @1.5g/litre, Imidacloprid (17.8% SL) @ 1ml/litre or Acetamiprid 2ml/litre of water. Repeat the spray at 10-15days intervals based on the insect vector population.

**Leaf spot (*Cercospora dolichii*):** The affected leaves have circular to angular spots with grey centre and reddish border. The spots gradually cover the entire leaf.

Destruction of diseased plant materials and proper crop rotation has to be done. Spraying with benomyl @ 0.2% or Dithane M-45or Blitox @ 0.2% or Thiram @0.2%.

**Ashy stem blight (*Macrophomina phaseoli*):** Black sunken cankers on cotyledons and stem of young seedlings which results in damping off and death of seedlings. Yellowing and drooping of foliage in older plants. The disease is seed borne. Warm and humid weather are favorable for the disease infection.

Destruction of diseased plant debris left from previous crop and follow phytosanitary measures. Use disease free seeds. Seed treatment with Thiram (2g/kg). Avoid excessive soil moisture. Use organic manure preferred.

**Rust (*Uromyces phaseolivignae*)**: The disease mostly attacks leaves and rarely stem and petioles. Rust pustules appear in the form of minute, slightly raised spots which enlarge to form reddish brown sori. These sori in advanced stage turn dark brown to black with formation of teliospores. Severely affected leaves turn yellow and may fall off.

To manage the disease, destroy infected plant debris and follow crop rotation and wider plant spacing. Spray the crop with Indofil M-45 @ 2.5 g per litre of water.

**Anthracnose (*Colletotrichum lindemuthianum*):** This is a seed-borne fungal disease that attacks leaves, stems, and pods. Infected spots show dark-brown sunken spots with raised reddish or yellowish margins. Tissues underneath the necrotic spots are thin and papery and produce abundant pinkish-slimy masses in humid conditions.

Use disease-free seed and follow crop rotations. Treat the seed with Captan/ Captaf or Thiram @ 2-3 g per kilogram of seed. Spray the crop with Indofil M-45 @ 2g/l litre of water or copper oxychloride @ 0.25 percent.

**Bacterial blight (*Xanthomonas campestris pv. phaseoli*):** Bacterial blight is a seed-borne disease. The affected leaves show light yellow, irregular to circular spots with necrotic brown centers which later on change to straw colour. Dark green water-soaked lesions of variable shapes and sizes appear on pods, which later on turn, yellow and dry.

Uprooting of the diseased plant and also, avoiding irrigation through diseased plant to protect the healthy plants from further infection. Use disease-free seeds and grow resistant varieties. The disease spreads rapidly when moisture is present hence, avoid overhead watering. Drenching with Copper Oxychloride (COC) @ 4-5g/l of water along with Streptomycin (150ppm) is effectively control the disease.

## Major Insect Pests and their Management Practices

**Bean Aphid (*Aphis craccivora*):** It Suck the plant sap which results in stunted plant growth, curling of leaves, twisting of twigs and developing fruits and sometimes shedding of flowers. Aphids excrete honeydew, a sugary substance that causes sticky, shiny leaves and black sooty mold growth and it spreads plant diseases (a large number of viruses are vectored by aphids).

Use the yellow sticky traps (10 numbers/ha). Spraying of neem oil @ 3g/l of water followed by the systemic insecticides such as., Dimethoate 30% EC (Rogor) 2ml/litre or Imidacloprid (17.8%SL) @1ml/litre of water. If required, repeat the spray at fortnightly intervals.

**Whitefly (*Bemisia tabaci*):** Whiteflies feed on the sap of cowpea plants using their piercing-sucking mouthparts. Whiteflies excrete a sticky substance known as honeydew, which can lead to the accumulation of honeydew which promote the growth of sooty mold, cover leaves and reduce photosynthesis. It is a vector for several plant viruses. The infected leaves may turn yellow and curl downward, indicating stress from feeding or virus infection. Infested plants may produce fewer pods, and the quality of the pods can be compromised, affecting harvest and yield.

Remove and destroy the infected plants. Control the insect vector by spraying Diafenthiuron 50WP @1.5g/litre, Imidacloprid (17.8% SL) @ 1ml/litre, or Acetamiprid 2ml/litre of water. Repeat the spray at 10-15 days intervals based on the insect vector population.

## Integrated Pest Management for Sucking Pests Complex

(Arora *et al.*, 2017 and Rai *et al.*, 2014)

### Cultural control:

- Irrigation must be given at critical stages of the crop growth
- Growing plants like sunflower and coriander are helpful for the attraction of natural enemies

### Physical control:

- Use yellow sticky traps for aphids, whiteflies, and jassids @ 4-5 traps/acre.
- Installation of light trap @ 1/acre and operate between 6 pm and 10 pm to trap the nymphs and adult stages of the pests

### Mechanical control:

- Shaking the infested plants over the vessels of oil and water or oily cloth gives the most effective control
- Collection and destruction of eggs, and larvae
- Collect and destroy insect-infested plant parts

### Biological control:

- Conserve predators such as ladybird beetles (*Menochilus sexmaculata, Hippodamia variegata,* Brumus *suturalis Cocciniella septumpunctata,* and *Cheilomones vicina*). Adult beetle feed @10 to15 adults/day.
- Release of natural enemies of bean aphid: Parasitoids: *Diaeretiella rapae, Aphelinus abdominalis,* etc., Parasitoids: *Encarsia formosa, Chrysocharis pentheus, Eretmocerus* spp. etc.
- Spray of NSKE 5% or neem oil 5%.

### Chemical control

- systemic insecticides such as., Dimethoate 30% EC (Rogor) 2ml/litre or Imidacloprid (17.8%SL) @ 1ml/litre of water or Diafenthiuron 50WP @1.5g/litre. If required, repeat the spray at fortnightly intervals.

**Pod borer (*Helicoverpa armigera* and *Adisura atkinsoni*)**: It is a polyphagous pest and is an important pest on dolichos bean, cowpea and other pulses. Caterpillar first feeds on leaves, later bores into pods and feeds on seeds thereby, reducing the marketable yields drastically. Larva is seen feeding with the head alone thrust inside the parts and the rest of the body hanging out. Boreholes on pods, absence of seeds on pods and defoliation in early stages are the symptoms of attack.

Both species of pod borer are effectively controlled by spraying of NSKE 5g/litre and the use of chemical insecticides like spinosad 45 EC @ 0.4 ml/litre or Emamectin Benzoate @ 0.4 g/litre reduces the incidence of pod borer effectively.

### Spotted pod borer (*Maruca testulalis*)

Adults prefer to lay eggs on flower buds, flowers, terminal shoots and tender pods. The larvae feed from inside a webbed mass of leaves, flowers, flower buds and pods. This concealed feeding habit protects the larvae from natural enemies and insecticides and causing severe damage throughout the reproductive cycle of the crop.

Use pheromone traps, neem oil 3g/l, NSKE 5g/l, use any of these chemical insecticides like chlorpyriphos @ 0.05%, Novaluron 10% E.C @ 1.0 ml/l, Profenophos 50% EC @ 2.0 ml/litre, Flubendiamide @ 0.2 ml/litre and Emamectin Benzoate @ 0.4 g/litre of water were effective against spotted pod borer.

## Integrated Pest Management

(Arora *et al.*, 2017 and Rai *et al.*, 2014)

### Cultural control:

- Summer deep ploughing by exposing the larval and pupal stages of the insect pests to UV radiation or predated by the birds
- Early sowing, short-duration varieties
- Avoid closer plant spacing.

### Physical control:

- Setting light traps (1 light trap/5 acre) to kill the moth population.

### Mechanical control:

- Collection and destruction of grown-up larva
- Installation of Pheromone traps for *Helicoverpa armigera* 12/ha

### Biological control:

- Release of natural enemies viz., Parasitoids: *Trichogramma* spp, *Bracon pectoralis* and *Phanerotoma planifrons*, etc. Predators: Lacewing, ladybird beetle, spider, fire ant, dragonfly, robber fly, reduviid bug, preying mantid, black drongo, wasp, common mynah, etc.
- Use of botanical insecticide: Azadirachtin 0.03%WSP 2500-5000 g/ha
- Usage of biopesticides:
    a. Ha NPV 3 x1012 POB/ha in 0.1% teepol
    b. *Bacillus thuringiensis* serovar kurstaki (3a,3b,3c) 5%WP 1000-1250 g/ha
    c. Application of NPV 250 LE /ha with teepol 0.1% and Jaggery 0.5% thrice at 10 – 15 days intervals commencing from the flowering stage.

### Chemical control:

- Once the pests cross or tend to cross the ETL i.e., 10% pod damage. Application of insecticides is recommended viz.,
- Emamectin benzoate 5%SG 220 g/ha

- Indoxacarb 15.8%SC 333 ml/ha
- Spinosad 45%SC 125-162 ml/ha

## References

Ariina, S.M.M., Kanaujia, S.P., Alila, P., Sebastian, K.S. and Bier, K. 2021. Nutritional & Anti-Nutritional Profile of Indian bean: A Mini Review. *Just Agriculture*, 1(9): 2582-8223

Arora, R., & Dhawan, A. K. (2017). Integrated pest management: concept and approaches. *Theory and Practice of Integrated Pest Management, 1.*

Birari., S.P.and Shankar., S.L. 1992. Genetic diversity in Lablab bean. *J. Maharashtra Agril.Univ.* 17(2):257-260

Chakroborti, A.K. and Parthasarathy, V.A. 2003.Lablab Bean. "Vegetable crops Volume-2" Bose,T.K, Kabir, J, Maity, T.K, Parthasarathy, V.A and Som, M.G (eds), , Naya Udyog, Kolkata. pp:263-272.

Davis, D.W., Oelke, E.A., Oplinger, E.S., Doll, J.D., Hanson, C.V. and Putnam, D.H.2000. Alternative field crops manual. http://www.hort.purdue.edu/newcrop/afem/cowpea.html.

Dhaliwal, M.S.2017.Legume vegetable.pp:228-278. In Book: Hand Book of Vegetable Crops. Kalyani publishers, New Delhi.

Gaangadhara, K., Yadav,L.P., Apparao, V.V., Singh, A.K., Verma, A.K., Selvakumar, R and Gograj, S.J.2023. Genetic variability and principal component analysis in Indian bean (*Lablab purpureus* var. *Typicus*L.) genotypes under rainfed conditions of western India. Genet resour Crop Evol. https://doi.org/10.1007/s10722-023-01702-9

Gangadhara K,Lalu Prasad Yadav, V.V. Appa Rao and A.K. Singh. 2023b. Thar Kiran: Anthocyanin Rich High Yielding Variety of Indian bean. *Agri Tech today* **1**(3): 80-82.

Gangadhara.K,L.P.Yadav, V.V.Appa Rao, A.K.Singh, A.K.Verma, Vikas Yadav and Anita Meena. 2023a. Genetic diversity through $D^2$ and cluster analysis for structural and economic traits in yardlong bean (*Vigna unguiculata var sesquipedalis* L.) under rainfed semi-arid conditions of Gujarat. *Agril. Mecha. in Asia, Africa, L. America.***54** (08):15299-308.

Gopalakrishnan, T.R.2007.Vegetable crops.pp:1-354. In K.V. Peter (ed), *"Horticulture Science Series-4"*, New India Publishing House, New Delhi.

Ng, N.Q. and Marechal, R. 1985. Cowpea taxonomy, origin and germplasm. Cowpea Research, Production and Utilization. Wiley, New York, pp 11–21

Parthasarathy, V.A. and Chakroborti, A.K. 2003. Cowpea. "Vegetable crops Volume-2" Bose,T.K, Kabir, J, Maity, T.K, Parthasarathy, V.A and Som, M.G (eds), , Naya Udyog, Kolkata. pp:229-250.

Pokle, Y.S. and Deshmukh, N.N.1971. Anthesis and pollination in garden bean (*Dolichos lablab* L.). Nagpur College of Agriculture Magazine. **44**:17.

Prabhakara, B.S., Naik, L.B., Mohanm, N. 2018. Indian bean (*Dolichos purpureus* L).pp: 214-218. "Textbook of vegetables, Tuber Crops and Spices). Directorate of Knowledge Management in Agriculture, Indian Council of Agricultural Research, New Delhi.

Rai, A. B., Halder, J., and Kodandaram, M. H. (2014). Emerging insect pest problems in vegetable crops and their management in India: An appraisal. *Pest Management in Horticultural Ecosystems*, **20**(2): 113-122.

Steele,W.M. 1976. Evolution of Crop Plants, Ed.N.W.Simmonds, Longman, London, New York.

Venkatachalam, M., Kshirsagar, H.H., Tiwari, R. and Sathe, S.K. 2002. Annual meeting and food Expo- Anaheim, California, Session 30C, Food Chemistry: Protiens.

Vishnu Swarup.2012.*Vegetable Science and Technology in India*. pp:1-656.Kalyani publishers, New Delhi.

Web sites: https://www.iihr.res.in , https://iivr.icar.gov.in, http://iari.res.in

CHAPTER 25

# VEGETABLE COWPEA FOR LIVELIHOOD AND NUTRITIONAL SECURITY IN SEMI-ARID AREAS

**Gangadhara K[1], L. P. Yadav[1], A.K. Singh[1], Anil[1], V.V. Appa Rao[1], Anand Sahil, P. Ravat[1], Manjunath, P.P.[2] and Ramesh K.B.[3]**

[1]*ICAR-Central Horticultural Experiment Station (CIAH), Vejalpur, Godhra, Gujarat*
[2]*ICAR-Indian Institute of Groundnut Research, Junagadh, Gujarat*
[3]*ICAR-Indian Institute of Vegetable Research (Varanasi), RS-Sargatia, UP*
*Corresponding author: aksbicar@gmail.com*

---

Cowpea (*Vigna unguiculata* L. 2n=2x=22) is a warm season, annual and herbaceous legume known by various names such as black-eye bean, China pea, southern pea and lobia. It is one of the most ancient human foods and has probably been cultivated since Neolithic times (Dhaliwal, 2017). *Vigna unguiculata* ssp. *unguiculata* (dual purpose type) is widely distributed. Other cultivated species of cowpea is *Vigna unguiculata* var. *sesquipedalis,* known as yard long bean or vegetable cowpea is used as vegetable purpose. Cowpea is an important source of protein in human diet. In African countries it is more commonly used as pulse. In India it is used as both pulse as well as vegetable (Vishnu Swaroop, 2012). Due to its nutritive value and soil improving properties, it is used as a fodder, green manure and cover crop. Being a legume crop, cowpea suits well as inter crop with many other crops.

**Origin and Distribution:** Cowpeahas been used since from neolithic times. It originated in Africa. Its domestication was probably in Ethopia, Central Africa, South Africa and West Africa. Further it was moved into India about 2000 years ago and southeast Asia(2300 BC). It was introduced in Southern Europe by 300B.C. and USA in 16$^{th}$ and 17$^{th}$ Centuries from southern Europe, particularly West Africa along with the slave trade (Vishnu swaroop, 2012). In India cowpea is known since from the Vedic period. Cow pea is majorly grown in Africa, Latin America, United States. It occupies more areas in South Asiacountries mainly India, Bangladesh, Pakistan, Myanmar and Sri Lankabesides Southeast Asian countries like Indonesia, Malaysia,

Thailand and Philippines.In India, it is majorly grown in states of Karnataka, Kerala, Tamil Nadu, Andhra Pradesh, Telangana, Rajasthan, Maharashtra, Gujarat, Madhya Pradesh, Odisha, UP, West Bengal and North eastern states.

## Botany and Taxonomy

Cowpea (*Vigna unguiculata*L.)is a diploid with 2n=2x=22, 24, belongs to the family Fabaceae (Leguminosae). It is an annual herb that grows vigorously having strong tap root system. Cowpea has erect, semi-erect, spreading and prostrate (trailing) or climbing like growth habits with profuse branching. Cylindrical and twisting type of stem and inflorescence is an un-branched axillary raceme bearing several flowers at the terminal end of peduncles. peduncle length varies from 5-60 cm. Anthesis takes place early in the morning between 6.30 and 9.00 a.m. The process of opening corolla takes 45-60 minutes. Dehiscence of anthers is much earlier and it varies from 10.0 p.m. to 00.45 a.m. For the purpose of hybridization, cowpea flower emasculation is done 20 hours before flower opening. Best time for emasculation is during morning hours preceeding day of anthesis. Stigma becomes receptive from 12 hours before blooming to 6 hours after anthesis (Gopalakrishnan, 2007).

*Vigna Unguiculata* and different variants, wild and cultivated are classified as subspecies or varieties (Sampson,1936; Sellschop,1962; Verdcourt,1970). According to Verdcourt (1970), there are five subspecies of Vigna *Unguiculata.* Two subspecies like *dekindtiana* and *mensensis* are wild and other three subspecies are cultivated and widespread in India (Steele, 1976). Ng and Marechal (1985) renamed subspecies Unguiculata, *Cylindrica* and Sesquipedalis as cultigroups Unguiculata, *Biflora*and *Sesquipedalis* respectively. A brief description of the three subspecies are given below.

- *Vigna Unguiculata* Subsp. *Vigna Unguiculata* (Common cowpea): plants have viny, prostrate, erect or semi erect growth habit with linear, crescent, round, coiled, pendant and short pendant pod shapes, 20-30cm long, cylindrical or square shaped small seeds.
- *Vigna Unguiculata* Subsp.*Cylindrica* (Catjung bean): plants are more erect than subsp *Unguiculata.* Pods are small 7-13 cm long, more or less erect and small seeds 5-6 mm long and kidney shaped. *Biflora* is commonly known as catjang bean and used as dry seeds and fodder.
- *Vigna Unguiculata* Subsp.*Sesquipedalis* (Yard-long bean, asparagus bean, vegetable cowpea, snake bean): Plants are trailing or climbing growth habit. Pods are pendant, 30-90 cm long, fleshy and inflated tending to shrink when dry. Seeds are big elongated kidney shaped, 8-12mm long.

All these subspecies are interfertile.

## Nutritive value

Cowpea is low in saturated fat and very low in cholesterol and sodium. 100g of green tender pods contain protein (4.3g), fibre (2.0g), carbohydrates (8.0g), phosphorus (74mg), iron (2.5mg), vitamin-C (13.0mg), minerals (0.9mg) etc., (Gopalakrishnan, 2007). Cowpea is called as vegetable meat due to high amount of protein in grain with better biological value on dry weight basis. On dry weight basis, cowpea grain contains 24.8 %protein, 1.9 % fat and 63.6 % carbohydrates, 6.3 % fiber (Davis et al., 2000) and it is rich source of calcium and iron.

## Uses

The crop is used in many ways. Cowpea is mainly grown for its green pods, tender seeds are consumed as cooked or fried vegetable and dry seeds used as a pulse. In general, cowpea can be used same like other beans and it will be cocked with other vegetables. The tender green leaves are also an important food source in Africa and they prepare a pot herb like spinach from the tender leaves (Parthasarathy and Chakroborti, 2003). Besides its cultivation as a vegetable crop, it is also grown as a green manure and fodder crops.

**Varieties:** A brief description of the improved varieties across the country is given below.

| Institute | Varieties | Special features |
|---|---|---|
| Indian Institute of Horticultural Research (IIHR), Bengaluru (Karnataka) | Arka Garima | Plants tall, photo-insensitive. Pods light green, long, thick, round, fleshy and stringless. Suitable for vegetable purpose. Tolerant to heat and low moisture stress. Pod Yield: 18 t/ha in 70-75 days. |
| | Arka Suman | Plants erect, bushy and photo-insensitive. Pods medium long, tender, fleshy, crisp, without parchment with good cooking qualities. Pod Yield: 18 t/ha in 70-75 days. |
| | Arka Samrudhi | Plants erect, bushy and photo-insensitive. Pods green, medium thick, medium long round, tender, fleshy without parchment with good cooking qualities. Pod Yield: 19 t/ha in 70-75 days. |
| | Arka Mangala (Yard long bean) | Plants tall (3-4 m), pods are very long (80 cm), light green, stringless, round, tender with crisp texture and matures in 60 days. Suitable for kharif and rabi. Pod yield: 25 t/ha in 100 days. |
| IARI, New Delhi | Pusa Phalguni | It is selected from Canadian cultivar (Pal *et al.*, 1956).It is a bushy, dwarf cultivar suitable for sowing in February- March in northern India. Pods are dark green, erect 10-12 cm long and appear in two flushes. Matures in 60days and Pod yield:70-75q/ha |
| | Pusa Barsati | It is selected from collection of Philippines (Pal *et al.*, 1956).early variety and suitable growing in kharif season. It flowers in 35 days after sowing and matures at 45 days. Pods are light green colour, pendant, 26-28 cm long. Yield: 70-75q/ha |

| Institute | Varieties | Special features |
|---|---|---|
| | Pusa Dofasli | It is developed by crossing PusaPhalguni with a cultivar from the Philippines (Sing *et al.*, 1968). It is a bushy cultivar suitable for sowing in both spring- summer and rainy season. It bears flowers in 40days after sowing. Pods are light green with 18cm long. Yield: 75-80q/ha. |
| | Pusa Komal | It is widely adoptable, Photo-insensitive, indeterminate and bushy growth habit. It bears flowers in 40-45days after sowing. Pods are 20-23 cm long with synchronous bearing habit. Suitable for both spring summer and rainy season cultivation. Plants are resistant to bacterial blight. Average yield 100-115 q/ha. |
| | PusaSukomal | Resistant to golden yellow mosaic virus and leaf spot diseases. Average yield 60 and 65 q/ha during summer and kharif seasons, respectively. Maturity in 55-60 days. |
| | Pusa Rituraj | The variety can be grown in summer as well as kharif due to it's highly photo thermo insensitive nature, bushy type. Pods are 22-24cm long, thin and palatable. Dual purpose variety. Seeds brown. The harvest starts from 45-50 days. Pod yield is 8-9t/ha. |
| Central Horticultural Experiment Station (CHES)Vejalpur, Godhra, Gujarat, Regional Station of ICAR-CIAH, Bikaner | Thar Jyothi | It is a photo insensitive type and high yielding variety cultivated round the year. It is having short stature (bushy growth habit) grows up to of 50-56 cm height. It is an early flowering and early maturing variety. It takes 40-42 days for first flowering and 48-50 days for first picking. Average pod length of 25 to 26.50cm with pod girth of 2.5cm and pod weight of 9.65g. The total number of pods per plant varies between 120-150 and an average yield of 1.5 to 2.0 kg/plant of fresh pods and yield potential of 20-25 t/ha. |
| | Thar Prateeksha | The plants have climbing (pole type) growth habit and grow up to 3.5-4.20 m and they require stacking or support for higher yield and better quality pod production. It takes 38-40days for first flowering and 48-50 days after sowing for first harvesting of fresh tender light green colour pods. The pods have 66.0cm length, 3.4cm girth and 32.10g pod weight. The total number of pods per plant varies 80-120 pods/plant with an average yield of 2.0 to 2.5kg/plant of fresh pods was obtained with yield potential of 28-30 t/ha. |

| Institute | Varieties | Special features |
|---|---|---|
| | Thar Deeksha | The plants have climbing (pole type) growth habit and grow up to 4.5m and they require stacking or support for higher yield and better quality pod production. It takes 29-30days for first flowering and 36-38 days after sowing for first harvesting of fresh tender light green colour pods. The pods have 52.0-54.0cm length, 3.30cm girth and 28.0-30.0gpod weight. The total number of pods per plant varies 150-180 pods/plant with an average yield of 3.0kg/plant of fresh pods was obtained with yield potential of 34-38 t/ha |
| | Thar Surya | The plants have climbing (pole type) growth habit and grow up to 4.0-4.5m and they require stacking or support for higher yield and better quality pod production. It takes 35-38days for first flowering and 44-46 days after sowing for first harvesting of fresh tender pods. The pods have 52.0-54.0cm length, 2.6cm girth and 22.0-23.0g pod weight. The total number of pods per plant varies 180-200 pods/plant with an average yield of 2.0 to 3.0kg/plant of fresh pods was obtained with yield potential of 30-32 t/ha. It is rich in Anthocyanins (180-200mg/100g) |
| Kerala Agricultural University, Thrissur | Vyjayanthi | Trailing growth habit, long wine red coloured pods, brown seeds; Av. Length of pod 50.62 cm: Av. Fruit weight 16.17 g; No. of seeds / pod 20-22; Productivity 12.6 t/ha. |
| | Lola | Trailing growth habit, smooth and extra long pods; glossy light green-coloured long pods with purple tip, black seeds; Av. Length of pod 53.38 cm; Av. Pod weight 22 g; No. of seeds / pod 20-22; Productivity 20.0 t/ha. |
| | KMV1 | Trailing growth habit, long light green pods with brown tip, Immature seeds light green, Av. Length of pod 38 cm; No. of seeds / pod 19; Productivity 14.2 t/ha. |
| | Bhagyalakshmi | Early flowering, bushy growth habit, light green medium sized pods, mottled seeds; Av. Length of pod 27.0 cm; Av. Pod weight 7.13 g; Days to first harvest 48 days; Productivity 6.48 t/ha. |
| | Kanakamani | Semi trailing growth habit, pod-dark green, bold seeded and medium long, reddish brown seeds; Av. Pod length 17.8 cm; Av. Pod weight 2.0 g; Seeds / pod 17; Productivity 7.0 t/ha. |

| Institute | Varieties | Special features |
|---|---|---|
| | Kairali | Resistant to mosaic disease, semi trailing growth habit, pink coloured medium long pods, reddish brown seeds; Av. Length of pod 22.78 cm; Av. Pod weight 7.08 g; Days to first harvest 50, Productivity 7.13 t/ha. |
| | Anaswara | Semi trailing, pods-light green, bold seeded and medium long, purple flowers; Av. Pod length 28.13 cm; Av. Pod weight 12.5 g. Seeds / pod 19; Productivity 12.5 t/ha. |
| | Varun | Tolerant to mosaic, semi trailing growth habit, pink coloured medium long pods, reddish. Tolerant to mosaic, semi trailing growth habit, pink coloured medium long pods, reddish brown seeds; Av. Length of pod 27 cm; Av. Pod weight 11 g; Days to first harvest 50; Productivity 8.4 t/ha. |
| | KAU Deepika | Local collection from Kakkamoola, Thiruvananthapuram. High yielding variety, suitable for homesteads of Kerala, long, green pods, crop duration of 100-120 days, low incidence of pests and diseases and good cooking quality |
| | Vellayani Jyothika | Selection from Sreekaryam local. It is a high yielding(19.33t/ha) with long light green poded variety. |
| | Githika | High yielding, mosaic resistant yard long bean variety having long thick fleshy pods with light green colour having pod length 53.4 cm and reddish brown seeds, average yield : 27.6 t /ha. |
| | Manjari | Trailing habit, shade tolerant, ideal for intercropping and tolerant to mosaic. Light green pods (43.2 cm) with red seeds. Days to first harvest 48-50 days. Yield 5.6 t/ha. |
| | Mithra | Trailing growth habit, 90- 130 days duration, attractive long (78.6cm) light green pods and deep brown seeds with white speck at one end. Field tolerance to fusarium wilt and pythium rot. Suitable for riverine alluvium of Central Travancore , Yield 20.7 t/ha. |
| IIVR, Varanasi | Kashi Nidhi | Plants are dwarf, erect and bushy, with 20-25 peduncle per plant. Fruits are green, 25-30 cm long. Seed colour is reddish brown. Golden mosaic virus and Pseudocercospora cruenta tolerant with an average pod yield of 140-150 q/ ha. Better yield and keeping quality suitable for distant marketing. Recommended for release and cultivation in the states of Uttar Pradesh, Bihar, Haryana, Punjab and Jharkhand. |

| Institute | Varieties | Special features |
|---|---|---|
| | Kashi Sudha | Golden mosaic virus and Pseudocercospora cruenta tolerant, Identified for UP, Bihar, Jharkhand, Bihar, Andhra Pradesh, Orissa, Chattisgarh, Madhya Pradesh and Maharastra by AICRP-VC |
| | Kashi Kanchan | This is dwarf and bush type (height 50-60 cm), photo-insensitive, early flowering (40-45 days after sowing) and early picking (50-55 days after sowing) variety suitable for growing in both spring-summer and rainy seasons. Pods are about 30-35 cm long, dark green, soft, fleshy and free from parchment. The cultivar gives green pod yield of about 150-175 q/ ha and is resistant to golden mosaic virus and Pseudocercospora cruenta. It is recommended for cultivation in U.P., Punjab, Bihar, Chhattisagarh, Odisha, A.P., M.P. and states. |
| | Kashi Unnati | This is a photo-insensitive variety. Plants of this variety are dwarf and bushy, height 40-50 cm, branches 4-5 per plant, early flowering (30-35 days after sowing), first harvesting at 40-45 days after sowing, produces 40-45 pods per plant. Pods are 30-35 cm long, light green, soft, fleshy and free from parchment. The cultivar is resistant to golden mosaic virus and Pseudocercospora cruenta, and gives green pod yield of about 125-150 q/ ha. It is recommended for the cultivation in Punjab, U.P., and Jharkhand. |
| Tamil Nadu Agricultural University, Coimbatore | CO.2 | A semi spreading variety with 26 cm long pods having less fibre. Yield 9.4 t/ha. |
| | Vamban | Variety with long green pods. Yield 10.6 t/ha. |
| Punjab Agricultural University, Ludhiana | Sel-263* | Early maturing dwarf variety with green thick and fleshy pods of 20 cm length. Resistant to mosaic and golden mosaic virus. |
| BCKVV, Kalyani | Bidhan Barati-1 | Bushy variety resistant to cowpea mosaic and golden mosaic virus. Flowers white. Pods green, medium long (25.2 cm) borne on long stout peduncle. Yield 13.4 t/ha. |
| | Bidhan Barati-2 | Semi determinate variety with purple flowers. Pods light green, thick medium long (25.8 cm) borne on short peduncle. Yield 15.9 t/ha. |

Source: Gopalakrishnan (2007), Singh and Nath (2011), Parthasarathy and Chakroborti (2003) https://www.iihr.res.in , https://iivr.icar.gov.in and http://iari.res.in

## Cultivation

### Soil and climate

Cowpea is a warm season crop and comes up well under rainfed conditions where temperature ranging from 21 to 35 C (Gopalakrishnan, 2007). It is well adopted many areas of the humid tropics and subtropical zones, tolerate hardy conditions including high temperature, drought. Cowpea can be grown all types of soilor poor soil with 6 - 7 pH range. The soil should be ploughed to break the clods and remove the weeds completely. The shallow cultivation is practiced to bring the soil into fine tilth.

### Sowing

In India the crop is grown almost all the season mainly kharif with onset of monsoon ranging from early June to end of July, Rabi- October to November (Southern India), In summer it can be sown in the month of February- March (grain) in north Indian plains, December to January in the southern plains as a spring- summer crop and April-May in Hills. Sowing is done by broadcasting or line sowing method. The seeds of busy varieties are sown at a spacing of 35x15cm, for semi trailing varieties 45x30cm and trailing (yard long bean) varieties 60cmx1m. The seed rate of cowpea is 20-25 kg/ha for bush and semi trailing varieties and 4-5 kg for trailing varieties (Gopalakrishnan, 2007).

### Nutrient requirement

Although cowpea is a legume crop, it responds well to the application of fertilizers. About 25 t/ha of FYM is applied atleast fifteen days before sowing. A fertilizer dose of 25:75:60: kg NPK/ha is recommended (Prabhakara *et al.*, 2018). Half of the N, full P and K are applied as basal dose and remaining half N is applied 25 -30 days after sowing.

### Weed control

Weeds may be controlled mechanically or by using weedicides. The first weeding may be done at 20-30days after sowing. One hoeing should be done at one month's stage of crop is necessary to control the weeds and facilitates the earthing up. Fluchloralin (2liters/ha) will check the weed growth for 20-25 days (Parthasarathy and Chakroborti., 2003). Once crop is covered weeds will naturally controlled.

### Staking or support

Yardlong bean varieties need support, since the plants have twinning growth habit. After 10-15 days after sowing the supporting structure should be made with the help of GI-iron wires supported by bamboo poles or cement poles. All the plant vines should be twined with gunny threads or any other plastic threads and allowed them to twine and spread on the bamboo stakes for better growth and fruit set. Over-crowding of plant vines due to non-staking affects formation of pods and yield adversely. Staking improves plant spread and photosynthetic activity. As a result, there is a higher yield due to higher number of pods per plant.

### Water management

Cowpea is a hardy crop comes up well in rainfed conditions. It does not require water in the rainy season. However, one or two irrigations may be needed when there is delay in rains. Flowering and pod development periods are the critical stages. Depending upon the weather conditions irrigation may be given once in 10-12 days and avoid the water logging conditions (Parthasarathy and Chakroborti, 2003).The summer crop requires frequent irrigation at once in a week interval. Cowpea is a hardy crop comes up well in rainfed conditions.

### Harvesting

Tender pods are harvested after attaining full size, before they become fibrous. Harvesting starts 45- 60days after sowing. In bush type varieties 2-5 harvests are possible. Trailing varieties usually harvested at alternate days. While harvesting of cowpea care must be taken to avoid the injury to green tender pods and avoid the fibrous pods. The green pod yield of 4-5t/h in bush varieties, 7-8t/ha in semi trailing varieties and 15-18t/ha in trailing varieties (yard long bean) are obtained. In case of dry seed yield of 750-1250kg/ha in bush varieties, 1000-1500kg/ha in semi trailing varieties and 450-600kg/ha in trailing varieties are obtained (Gopalakrishnan, 2007).

## Plant Protection

### Powdery mildew (*Leveillula taurica var. macrospora*)

The infection on leaves appears initially as small darkened areas, later becoming white powdery growth which spreads to stems and other parts of the plant. In severe infections the leaves become chlorotic and fall off. Pods are reduced in size and distorted in shape. Plant growth is reduced due to defoliation. The pathogen has wide host range and can survive on other leguminous crops.

Spray with 0.5% wettables ulphur or with benlate or Bavistin 0.15%. If needed, repeat the spray at 14-day intervals.

### Leaf spot (*Cercospora dolichii*)

The affected leaves have circular to angular spots with grey centre and reddish border. The spots gradually cover the entire leaf.

Destruction of diseased plant materials and proper crop rotation has to be done. Spraying with benomyl @ 0.2% or Dithane M-45or Blitox @ 0.2% or Thiram @0.2%

### Ashy stem blight (*Macrophomina phaseoli*)

Black sunken cankers on cotyledons and stem of young seedlings which results in damping off and death of seedlings. Yellowing and drooping of foliage in older plants. The disease is seed borne. Warm and humid weather are favorable for the disease infection.

Destruction of diseased plant debris left from previous crop and follow phytosanitary measures. Use disease free seeds. Seed treatment with Thiram (2g/ kg). Avoid excessive soil moisture. Use organic manure preferred.

### Rust (*Uromyces phaseolivignae*)

The disease mostly attacks leaves and rarely stem and petioles. Rust pustules appear in the form of minute, slightly raised spots which enlarge to form reddish brown sori. These sori in advanced stage turn dark brown to black with formation of teliospores. Severely affected leaves turn yellow and may fall off.

To manage the disease, destroy infected plant debris and follow crop rotation and wider plant spacing. Spray the crop with Indofil M-45 @ 2.5 g per litre of water.

### Anthracnose (*Colletotrichum lindemuthianum*)

This is a seed borne fungal disease and attacks leaves, stem and pods. Infected spots show dark-brown sunken spots with raised reddish or yellowish margins. Tissues underneath the necrotic spots are thin and papery and produce abundant pinkish slimy mass in humid conditions.

Use disease free seed and follow crop rotations. Treat the seed with Captan/ Captaf orThiram @ 2-3 g per kilogram of seed. Spray the crop with Indofil M-45 @ 2g/l litre of water or copper oxychloride @ 0.25 percent.

### Bacterial blight (*Xanthomonas vignicola*)

Bacterial blight is a seed borne disease. The affected leaves show light yellow,irregular to circular spots with necrotic brown centre which later on changes to straw colour.Dark green water soaked lesions of variable shapes and sizes appear on pods,which later on turn, yellow and dry.

Uprooting of the diseased plant and also, avoid the irrigation through diseased plant to protect the healthy plants from further infection. Use disease free seeds and grow the resistant varieties. The disease spreads rapidly when moisture is present hence, avoid overhead watering. Drenching with Copper Oxychloride (COC) @ 4-5g/l of water along with Streptomycin (150ppm) is effectively control the disease.

### Cowpea mosaic virus disease

This virus disease is seed borne and also transmitted by aphids *(Aphis craccivora)*. A characteristic symptom of the mosaic virus is an intermixing of light and dark brown areas. Infected leaves are generally smaller than the healthy ones. The leaves show puckering and curling. The plants remain stunted and produce deformed pods.

Use the yellow sticky traps (10 numbers/ha). Spraying of neem oil @ 3g/l of water followed by the systemic insecticides such as., Dimethoate 30% EC (Rogor) or Imidacloprid(17.8% SL)@1ml/litre of water. If required, repeat the spray at fortnightly intervals.

### Cowpea yellow flecks

This virus disease is transmitted by whitefly (*Bemisia tabaci*). The Leaves of the diseased plants are reduced in size and develop bright yellow irregular patches interspersed with green areas. In severe cases, the entire leaf looks golden yellow in colour.

Remove and destroy the infected plants. Control the insect vector by spraying Diafenthiuron 50WP @1.5g/litre, Imidacloprid (17.8% SL) @ 1ml/litre or Acetamiprid 2ml/litre of water. Repeat the spray at 10-15days intervals based on the insect vector population.

## Major Insect and their Management

### Bean Aphid (*Aphis craccivora*)

It Suck the plant sap which results in stunted plant growth, curling of leaves, twisting of twigs and developing fruits and sometimes shedding of flowers. Aphids excrete honeydew, a sugary substance that causes sticky, shiny leaves and black sooty mold growth and it spreads plant diseases (a large number of viruses are vectored by aphids).

Use the yellow sticky traps (10 numbers/ha). Spraying of neem oil @ 3g/l of water followed by the systemic insecticides such as., Dimethoate 30% EC (Rogor)2ml/litre or Imidacloprid(17.8%SL) @1ml/litreof water. If required, repeat the spray at fortnightly intervals.

### Whitefly (*Bemisia tabaci*)

Whiteflies feed on the sap of cowpea plants using their piercing-sucking mouthparts. Whiteflies excrete a sticky substance known as honeydew, which can lead to the accumulation of honeydew promotes the growth of sooty mold, which can cover leaves and reduce photosynthesis. It is a vector for several plant viruses. The infested leaves may turn yellow and curl downward, indicating stress from feeding or virus infection. Infested plants may produce fewer pods, and the quality of the pods can be compromised, affecting harvest yield.

Remove and destroy the infected plants. Control the insect vector by spraying of Diafenthiuron 50WP @1.5g/litre, Imidacloprid (17.8% SL) @ 1ml/litre or Acetamiprid 2ml/litre of water. Repeat the spray at 10-15days intervals based on the insect vector population.

### Integrated Pest Management

(Arora *et al.*, 2017 and Rai *et al.*, 2014)

**Cultural control:**

- The irrigation must be given at critical stages of the crop growth
- Growing plants like sunflower and coriander are helpful for the attraction of natural enemies

**Physical control:**

- Use yellow sticky traps for aphids, whiteflies, and jassids @ 4-5 traps/acre.
- Installation of light trap @ 1/acre and operate between 6 pm and 10 pm to trap the nymphs and adult stages of the pests

**Mechanical control:**

- Shaking the infested plants over the vessels of oil and water or oily cloth gives the most effective control
- Collection and destruction of eggs, and larvae
- Collect and destroy insect-infested plant parts

**Biological control:**

- Conserve predators such as ladybird beetles (*Menochilussexmaculata, Hippodamia variegata*, Brumus *suturalis Cocciniella septumpunctata*, and *Cheilomones vicina*). Adult beetle feed @10 to15 adults/day.
- Release of natural enemies of bean aphid: Parasitoids: *Diaeretiella rapae*, Aphelinus abdominalis, etc., Parasitoids: *Encarsia formosa, Chrysocharis pentheus, Eretmocerus* spp. etc.
- Spray of NSKE 5% or neem oil 5%.

**Chemical control**

- Systemic insecticides such as., Dimethoate 30% EC (Rogor) 2ml/litre or Imidacloprid (17.8%SL) @ 1ml/litre of water or Diafenthiuron 50WP @1.5g/ litre.If required, repeat the spray at fortnightly intervals.

## Pod-borer (*Helicoverpa armigera* and *Adisuraatkinsoni*)

It is a polyphagous pest and is an important pest on dolichos bean, cowpea and other pulses. Caterpillar first feeds on leaves, later bores into pods and feeds on seeds thereby, reducing the marketable yields drastically. Larva is seen feeding with the head alone thrust inside the parts and the rest of the body hanging out. Boreholes on pods, absence of seeds on pods and defoliation in early stages are the symptoms of attack.

Both the species of pod borer are effectively controlled by spraying of NSKE 5g/litre and use of chemical insecticides like spinosad 45 EC @ 0.4 ml/litre or Emamectin Benzoate @ 0.4 g/litre reduces the incidence of pod borer effectively.

## Spotted pod borer (*Maruca testulalis*)

Adults prefer to lay eggs on flower buds, flowers, terminal shoots and tender pods. The larvae feed from inside a webbed mass of leaves, flowers, flower buds and pods. This concealed feeding habit protects the larvae from natural enemies and insecticides and causing severe damage throughout the reproductive cycle of the crop.

Use pheromone traps, neem oil 3g/l, NSKE 5g/l, use any of these chemical insecticides like chlorpyriphos @ 0.05%, Novaluron 10% E.C @ 1.0 ml/l, Profenophos 50% EC @ 2.0 ml/litre, Flubendiamide @ 0.2 ml/litre and Emamectin Benzoate @ 0.4 g/litre of water were effective against spotted pod borer.

## Integrated Pest Management (IPM)

(Arora *et al.*, 2017 and Rai *et al.*, 2014)

### Cultural control:

- Summer deep ploughing by exposing the larval and pupal stages of the insect pests to the UV radiation or predated by the birds
- Early sowing, short-duration varieties
- Avoid closer plant spacing.

### Physical control:

- Setting light traps (1 light trap/5 acre) to kill the moth population.

### Mechanical control:

- Collection and destruction of grown-up larva
- Installation of Pheromone traps for *Helicoverpa armigera* 12/ha

### Biological control:

- Release of natural enemies viz., Parasitoids: *Trichogramma* spp, *Bracon pectoralis* and *Phanerotoma planifrons*, etc. Predators: Lacewing, ladybird beetle, spider, fire ant, dragonfly, robber fly, reduviid bug, preying mantid, black drongo, wasp, common mynah, etc.
- Use of botanical insecticide: Azadirachtin 0.03%WSP 2500-5000 g/ha
- Usage of biopesticides:
  a. Ha NPV 3 x1012 POB/ha in 0.1% teepol
  b. *Bacillus thuringiensis* serovar kurstaki (3a,3b,3c) 5%WP 1000-1250 g/ha
  c. Application of NPV 250 LE /ha with teepol 0.1% and Jaggery 0.5% thrice at 10 – 15 days intervals commencing from the flowering stage.

### Chemical control:

- Once the pests cross or tend to cross the ETL i.e., 10% pod damage. Application of insecticides is recommended viz.,
- Emamectin benzoate 5%SG 220 g/ha
- Indoxacarb 15.8%SC 333 ml/ha
- Spinosad 45%SC 125-162 ml/ha

## References


Arora, R., & Dhawan, A. K. (2017). Integrated pest management: concept and approaches. Theory and Practice of Integrated Pest Management, 1.

Davis, D.W., Oelke, E.A., Oplinger, E.S., Doll, J.D., Hanson, C.V. and Putnam, D.H.2000. Alternative field crops manual. http://www.hort.purdue.edu/newcrop/afem/cowpea.html.

Dhaliwal, M.S.2017.Legume vegetable. pp:228-278. Hand book of Vegetable Crops. Kalyani Publishers, New Delhi.

Gaangadhara, K., Yadav, L.P., Apparao, V.V., Singh, A.K., Verma, A.K., Selvakumar, R and Gograj, S.J.2023. Genetic variability and principal component analysis in Indian bean (*Lablab purpureus* var. *Typicus* L.) genotypes under rainfed conditions of western India. Genet. Resour. Crop. Evol. https://doi.org/10.1007/s10722-023-01702-9

Gangadhara K,Lalu Prasad Yadav, V.V. Apparao A. K. Singh and B.D.Sharma. 2023a. Thar Jyothi: A new Photo insensitive and early variety of Vegetable cowpea for dryland. Indian Horticulture. **68**(4): 3-4.

Gangadhara.K,L.P.Yadav, V.V.Appa Rao, A.K.Singh, A.K.Verma, Vikas Yadav and Anita Meena. 2023b. Genetic diversity through D2 and cluster analysis for structural and economic traits in yardlong bean (Vigna unguiculata var sesquipedalis L.) under rainfed semi-arid conditions of Gujarat. Agril. Mecha. in Asia, Africa, L. America. **54** (08):15299-15308.

Gopalakrishnan, T.R.2007.Vegetable crops. In: "Horticulture Science Series-4", New India publishing house, New Delhi. pp:1-354.

Ng, N.Q. and Marechal, R. 1985. Cowpea taxonomy, origin and germplasm. In: Cowpea Research, Production and Utilization. Wiley, New York, pp 11–210.

Parthasarathy, V.A. and Chakroborti, A.K. 2003. Cowpea. In: "Vegetable Crops Vol. 2", Naya Udyog, Kolkata. pp:229-250.

Prabhakara, B.S., Naik, L.B., Mohanm, N. 2018. Cowpea (*Vigna Unguiculata* L).In: "Text book of Vegetables, Tuber Crops and Spices). Directorate of Knowledge Management in Agriculture, Indian Council of Agricultural Research, New Delhi, pp: 206-211.

Rai, A. B., Halder, J., & Kodandaram, M. H. (2014). Emerging insect pest problems in vegetable crops and their management in India: An appraisal. Pest Management in Horticultural Ecosystems **20**(2): 113-22.

Raj Kumar, Kanak Lata, J.K.Yadav, A.K.Roy, S.Khajuria and Gangadhara K. 2023. Poshan va aay ke liye lobiya ki khethi. *Phal Phool*. September-October: pp. 49-50.

Sellschop.1962.*Wealth of India*, vol.X, Council of Scientific and Industrial Research, New Delhi.

Singh, D.N. and Nath,Vishal. 2011.*Varieties and Hybrids of Vegetables*. Satish Serial Publishing House, New Delhi, pp.1-426.

Steele, W.M. 1976. *Evolution of Crop Plants*, Simmonds, N.W. (Ed.). Longman, London, New York.

Venkatachalam, M., Kshirsagar, H.H., Tiwari, R. and Sathe, S.K. 2002. Annual meeting and food Expo- Anaheim, California, Session 30C, Food Chemistry: Protiens.

Verdcourt, B.1970.Kew Bull. **24**:507-569.

Vishnu Swarup.2012.Vegetable Science and Technology in India. Kalyani Publishers, New Delhi. pp:1-656.

Web sites: https://www.iihr.res.in , https://iivr.icar.gov.in, http://iari.res.in

CHAPTER 26

# HARNESSING ENTOMOPATHOGENIC NEMATODES (EPNS) AND SYMBIOTIC BACTERIA FOR SUSTAINABLE PEST CONTROL

**Anil[1], A. K. Singh[1], Gangadhara K[1], V. V. Appa Rao[1], Anand Sahil, Ramesh K. B.[2], and Manjunath, P. P.[3]**

*[1] ICAR-Central Horticultural Experiment Station (CIAH), Vejalpur, Godhra, Gujarat*
*[2] ICAR-Indian Institute of Vegetable Research (Varanasi), RS-Sargatia, UP*
*[3] ICAR-Indian Institute of Groundnut Research, Junagadh, Gujarat*
*Corresponding author: aksbicar@gmail.com*

---

Arthropods represent one of the most diverse and widespread animal groups, with an estimated 4–6 million species globally (Novotny et al., 2002). Although only a small fraction of these species is classified as pests, their impact on agriculture is substantial, accounting for approximately 18% of global annual crop losses (Oerke and Dehne, 2004), nearly 20% of post-harvest grain losses (Bergvinson and Garcia-Lara, 2004). Economic damages exceed US$100 billion annually (Carlini and Grossi-de-Sá, 2002). Phytophagous insects and mite pests significantly threaten global food production, necessitating extensive pest management strategies. In the absence of chemical insecticides, substantial reductions in crop yields would be inevitable. In 2001 alone, approximately US$7.56 billion was spent on protecting crops from damage caused by invertebrate phytophagous pests (Beckmann and Haack, 2003). Among these, lepidopteran larvae are particularly destructive, with an estimated 40% of all chemical insecticides targeting heliothine species (Brooks and Hines, 1999). According to the Entomological Society of America, over 25% of the 300 primary insect pests in the United States belong to the order

Lepidoptera. Additionally, numerous species from the orders Acarina, Coleoptera, Diptera, Hemiptera, Orthoptera, and Thysanoptera also contribute significantly to agricultural losses. The growing global population will impose increasing pressure on agricultural production to mitigate rising malnutrition. With the current world population at 6.5 billion, projections by the U.S. Census Bureau estimate an increase to over 9.2 billion by 2050, representing a more than 40% rise (U.S. Census Bureau, 2006). However, nearly all fertile land is already under cultivation, leaving little room for significant expansion of arable land. Consequently, the global challenge lies in enhancing crop productivity while ensuring environmental sustainability. Advancing pest management strategies is a critical approach to improving both the quality and quantity of agricultural production.

The indiscriminate use of pesticides has led to significant environmental concerns, including groundwater contamination, pesticide residues on food, resistance development in pests, and harm to wildlife. These issues have prompted strict regulatory measures (Zimmerman and Cranshaw, 1990) and increased interest in safer alternatives. Biological control offers a sustainable approach by utilizing insects, fungi, bacteria, viruses, and nematodes as natural insecticides.

## Entomopathogenic Nematodes (EPNs)

Nematodes represent the most abundant group of multicellular organisms on earth, exhibiting remarkable diversity in their ecological roles. Numerous parasitic nematode species are of significant concern due to their role in causing diseases in plants, animals, and humans (Gaugler, 2002). A single handful of soil can harbor thousands of these microscopic worms, many of which function as parasites of insects, plants, or animals, highlighting their ecological and agricultural importance. The majority of nematodes are free-living organisms inhabiting soil or aquatic environments. However, approximately one-quarter of all nematode species are parasitic, infecting either plants or animals. Among animal-parasitic nematodes, several species exhibit diverse associations with insects, ranging from phoresy and commensalism to facultative or obligate parasitism. These interactions encompass both insect-parasitic nematodes, such as the natural parasite *Hexamermis* sp., and entomopathogenic nematodes (EPNs), which play a crucial role in biological pest control.

Entomopathogenic nematodes (EPNs) are obligatory parasites dwelling in soil and preying on various insects, found abundantly in cultivated and uncultivated soils worldwide. Their distribution spans across all continents except Antarctica, as documented by Bhat et al. (2020). The life cycle of EPNs consists of an egg stage, four juvenile stages, and an adult stage. The third juvenile stage of EPNs is commonly known as "infective juveniles" (IJs) (Poinar 1990). Upon penetrating the host, IJs release bacteria that live symbiotically within the EPNs' gut. Once inside the host, these bacteria proliferate rapidly under optimal conditions, leading to host mortality within 24 to 48 hours (Shapiro-Ilan and Gaugler 2019). EPNs are found worldwide and comprise over 16 species of Heterorhabditis and at least 60 species of Steinernema (Nguyen and Hunt 2007).

In recent times, sequencing various regions of the EPN genome has emerged as the optimal method, not only for establishing phylogenetic relationships but also for species delimitation, given the limitations of morphological identification (Stock 2009). DNA sequencing of various genomic regions including the nuclear ribosomal DNA (rDNA), internal transcribed spacer regions (ITS1 and ITS2), mitochondrial DNA (mtDNA) gene, the 18S rRNA gene, and the 5.8S and 28S rRNA genes, featuring both variable and conserved regions has demonstrated its utility in investigating speciation, phylogenetic relationships, and molecular evolution within the Heterorhabditidae (Adams et al. 1998).

## Classification of Entomopathogenic Nematodes

Entomopathogenic nematodes (EPNs) belonging to the families *Steinernematidae* and *Heterorhabditidae* have emerged as a model system in biological pest control. With growing recognition of their pivotal role in managing soil-dwelling insect pests across natural and agricultural ecosystems, research on their biology, ecology, and application has advanced at an accelerating pace. The Steinernematidae and Heterorhabditidae nematode families have received considerable attention among all nematodes studied for insect biological control. This focus stems from their possession of various attributes that render them effective biological control agents, as documented by Kaya and Gaugler (1993), Grewal et al. (2005a), and Koppenhöfer (2007). These nematodes have been widely employed as classical, conservational, and augmentative biological control agents. EPNs offer an alternative approach to insect control compared to insecticides, as they provide an environmentally safer option that poses no harm to humans, animals, plants, or earthworms (Le Vieux and Malan 2013).

## Why EPNs can be considered as effective biocontrol agent

An effective natural enemy is characterized by its ability to reduce pest populations within a specific ecosystem without causing harm to beneficial organisms. The general life cycle of entomopathogenic nematodes (EPNs) provides insight into their invasion of hosts by infective juveniles and the benefits derived from infecting insects (Georgis et al., 2006). (Adams *et al.*, 2009) observed that initially, infective juveniles (IJs) enter the insect host and release enterobacteria into the hemocoel, resulting in the death of the insect. Subsequently, a new generation of EPNs feeds on the insect cadaver and proliferates alongside the symbiotic bacteria. Once the resources within the cadaver are depleted, the infective juveniles are released into the soil, where they actively migrate in search of new live hosts. This cyclical process ensures the perpetuation of EPN populations and their effective utilization as biological control agents.

## Symbiotic bacteria associated with EPNs

Two evolutionarily distinct groups of entomopathogenic nematodes, *Heterorhabditidae* and *Steinernematidae*, maintain an obligate mutualistic association with enteric bacteria of the genera *Photorhabdus* and *Xenorhabdus*, respectively.

These nematodes rely on their bacterial partners to establish infections in insect hosts, while the bacteria utilize the nematodes as vectors for transmission into the insect hemocoel. This symbiotic relationship is essential for the survival and proliferation of both partners, as neither can complete its life cycle independently within the host (Poinar and Thomas, 1966; Poinar *et al.*, 1977). The bacteria aid the nematodes by suppressing the host immune response, preventing decomposition by opportunistic microbes, and providing a nutrient-rich environment for nematode reproduction. In the absence of their symbiotic bacteria, nematodes typically fail to cause insect mortality or, if death occurs, cannot successfully reproduce (Han and Ehlers, 1998). However, the bacterial symbionts can be cultured separately, and axenic nematodes can be re-associated with their bacterial partners under laboratory conditions, facilitating in-depth investigations of their interactions in both insect hosts and artificial media.

The bacterial genera *Photorhabdus* and *Xenorhabdus* belong to a monophyletic clade within the *Enterobacteriaceae* and share a closer evolutionary relationship than with non-EPN-associated bacteria (Table 1) (Blaxter *et al.*, 1998; Rainey et al., 1995). Both groups of entomopathogenic nematodes (EPNs) exhibit similar life cycles, in which the infective juvenile (IJ)—a modified third-stage juvenile analogous to dauer larvae—serves as the vector for transmitting symbiotic bacteria into the insect host's hemocoel. Despite these similarities, significant differences exist between the two nematode genera and their respective symbiotic associations.

**Table 1. Comparative studies of entomopathogenic nematode-bacteria symbioses**

| Distinct character | Complexes | |
|---|---|---|
| | Heterorhabditis<br>Photorhabdus | Steinernema<br>Xenorhabdus |
| Nematode family | Rhabditidae | Strongyloidoitidae |
| Infective Juvenile | Dauer (IJ3) | Dauer (IJ3) |
| Reproduction | Heterogonic | Gonochoristic |
| Symbiont essential for insect pathogenicity | Always | Oftenly |
| symbiont location | Anterior or mid-intestine | Intestinal vesicle |
| symbiont supports insect growth and reproduction | Always | Oftenly |
| Symbiont family | Enterobacteriaceae | Enterobacteriaceae |
| Symbiont genera | Photorhabdus | Xenorhabdus |
| Symbiont characteristics: | | |
| Bioluminescence | Yes | No |
| Pigments | Yes | Usually |
| Anti-deterrent factor(s) | Yes | Yes |
| Antibiotics | Yes | Yes |

| Distinct character | Complexes | |
|---|---|---|
| Tc toxins | Yes | Yes |
| Insect Pathogenicity | Yes | Usually |
| CipA, CipB | Yes | No |

Source: (Blaxter *et al.*, 1998)

A key difference among entomopathogenic nematodes lies in their reproductive strategies. *Heterorhabditis* spp. exhibit heterogonic reproduction, employing both hermaphroditic (automictic) and amphimictic (male and female) modes, which are influenced by environmental conditions (Strauch *et al.*, 1994; Kehel-Raifer and Glazer, 2000). Notably, *Heterorhabditis* infective juvenile (IJ) nematodes are hermaphroditic, enabling reproduction from a single IJ infection, while transitioning to sexual reproduction in subsequent generations (Han and Ehlers, 2000). In contrast, *Steinernema* spp. are primarily gonochoristic, relying exclusively on amphimictic reproduction (Poinar, 1990), although exceptions have been documented (Griffin *et al.*, 2001; Stock *et al.*, 2004).

Another distinguishing factor is the location of symbiotic bacteria within the infective juvenile (IJ) stage of the nematode. *Heterorhabditis bacteriophora* primarily harbours approximately 130 symbiotic bacteria in the anterior region of the intestine. In contrast, *Steinernema carpocapsae* carries between 30 and 200 symbiotic bacteria within a bi-lobed vesicle situated in the anterior intestine (Martens *et al.*, 2003b). In both species, these bacterial symbionts can remain viable within the free-living IJ stage for extended periods before being released into the insect hemocoel to establish infection. The mechanisms of bacterial release, however, differ between the two genera: *H. bacteriophora* releases its symbionts through the mouth, resembling a regurgitation process (Ciche and Ensign, 2003), whereas *S. carpocapsae* expels them via defecation (Martens *et al.*, 2004; Poinar, 1966; Nickle, 1984).

Microscopic analysis has shown that *H. bacteriophora* regurgitates in response to a specific arthropod factor in the hemolymph, which is a low molecular weight, heat- and protease-resistant component. This regurgitation process can last for up to 300 minutes, during which individual or clusters of *Photorhabdus* are released in a pulsatile and staggered manner, averaging one symbiont every two minutes (Ciche and Ensign, 2003). It remains unclear whether the defecation process in *S. carpocapsae* is similarly triggered by this hemolymph component. After the nematode-bacterium complex has killed the insect host, both the nematodes and bacteria proliferate within the cadaver. For *H. bacteriophora*, the presence of symbiotic bacteria is essential for nematode growth, as progeny are rarely produced in the absence of *Photorhabdus* or related species. This dependency is consistent across various nutrient-rich agar media (Han *et al.*, 1990, 1991). In contrast, *Steinernema* species can grow and reproduce on non-symbiotic bacteria, though with reduced yields, both in insect hosts and on different agar and liquid media (Poinar and Thomas, 1966). In typical insect infections, both nematode species go through 2–3 generations, during which most nematodes bypass the infective third-stage juvenile but eventually develop into infective juveniles that become colonized by symbiotic bacteria.

## Pathogenicity and Host Specificity of Entomopathogenic Nematodes

The successful establishment of nematodes and their symbiotic bacteria within the host results from a complex interaction with the host's immune defense mechanisms, which largely determine the specificity of the host-pathogen relationship. The infection processes and pathogenicity are key factors in defining the host specificity of entomopathogenic nematodes from the families *Steinernematidae* and *Heterorhabditidae*.

### Invasion

*Steinernematidae* and *Heterorhabditidae* are insect pathogens that target the host's hemocoel. The infection of the hemocoel is a crucial step in the pathogenesis process, which occurs in two stages: first, the nematode invades the hemocoelic cavity, and then it must evade the host's immune responses. This enables the nematode to establish itself within the host and continue its development.

The exact sites through which the infective juveniles (IJs) of *Steinernematidae* and *Heterorhabditidae* nematodes enter the insect body are not yet fully understood. Triggiani and Poinar (1976) noted that *S. carpocapsae* IJs typically enter through the spiracles, mouth, and anus. Additionally, *S. carpocapsae* IJs have been shown to penetrate the pneumostatic lobes of adult tsetse flies (Poinar *et al.*, 1977) and the pupae of the cerambycid *Monochamus scutellatus* (Schmiege, 1963). Mra ívek et al. (1988) observed that *S. carpocapsae* enters *Galleria mellonella* (Lepidoptera: Pyralidae) through the mouth and anus, a similar mode of entry also reported for *S. glaseri* infecting *P. japonica* grubs (Cui *et al.*, 1993). In some cases, nematodes can directly penetrate the insect cuticle, as seen with *S. feltiae* infecting *Tipula paludosa* (Peter & Ehlers, 1994). However, *S. feltiae* typically invades other insect hosts through natural openings, suggesting a host-specific invasion strategy for this nematode species.

### Evasion

After entering the hemocoel, the nematode must successfully establish itself, a critical step for its growth, maturation, and reproduction. It must evade or suppress the host's immune defenses to accomplish this. Upon reaching the hemocoel, the nematode must establish itself, a critical step for its survival. This process appears to involve a dynamic interaction between the nematode and the host, where the insect attempts to recognize and encapsulate the nematode. In contrast, the nematode employs strategies to overcome the insect's defenses. In *Steinernematidae* and *Heterorhabditidae*, two primary strategies have been identified to evade host defenses: mimicry and the destruction of antibacterial factors.

The ability of *S. carpocapsae* to evade recognition by the host immune system is evident in certain insect species. In *Galleria mellonella* larvae, no significant changes occur in the total haemocyte count or the ratio of granulocytes to plasmatocytes within 4–6 hours after the injection of *S. carpocapsae* infective juveniles (IJs). Moreover, granulocytes retain their ability to adhere to *Bacillus subtilis* and maintain their lysozyme and phenoloxidase activity, indicating that the nematode does not trigger a detectable immune response in the host (Dunphy & Webster, 1985).

## Factors Influencing Entomopathogenic Nematode-insect Interactions

Entomopathogenic nematodes (EPNs) interact with diverse micro-environments, encompassing the insect hemocoel, soil pore spaces, or river bottoms, as delineated by Webster (1973). In their natural habitats, EPNs engage with a myriad of hosts owing to their broad host range, rendering them highly effective as biological control agents. Climate and soil type predominantly influence the interaction of EPNs with insects, with moisture and soil texture playing pivotal roles. Ecological barriers may impede the infection process of insect hosts by EPNs, as observed by Susurluk (2003). During periods devoid of hosts or characterized by adverse seasonal conditions such as droughts, infective juveniles existing freely in the environment must endure until conditions become conducive again. Consequently, these infective juveniles must possess the capability to withstand desiccation as soil moisture levels diminish, as noted by Stuart et al. (2006). Considering these factors enables a comprehensive understanding of how nematodes interact with insects within their natural environment. This understanding is crucial for developing effective EPN products and identifying the most suitable seasons and weather conditions for applying EPNs.

### Pathogenicity Test for EPNs

Both filter paper in petri dish assays and soil microcosms have been extensively employed to investigate the pathogenicity of entomopathogenic nematodes (EPNs) against insect pests (Shapiro and McCoy, 2000; Koppenhofer and Fuzy, 2006). The latter, resembling field soil conditions more closely, offers a valuable simulation for studying EPN behavior against the pest *A. suspensa*. Those EPN species consistently excelling in both types of assays, especially in soil conditions, were identified as the most virulent against Carib fly. Additionally, discerning the more virulent EPNs from the initial procedures involved comparing their LD values or infectious juvenile densities, a method established in prior studies to pinpoint the most potent species (Ansari and Moens, 2006); Giron-Pablo et al., 2012). This study aimed to select promising nematode species for field trials, with the ultimate goal of devising effective management strategies for *A. suspensa*.

### Host Searching Behavior of EPNs

Some nematode species exhibit different strategies in searching for hosts. For instance, species such as *S. carpocapsae* and *S. scapterisci* may seek hosts near the soil surface, while others like *H. bacteriophora* and *S. glaseri* are adapted to search deeper within the soil profile. The former group, termed "ambushers," typically remains relatively sedentary, awaiting the movement of surface-dwelling hosts (Campbell

& Gaugler, 1993). Conversely, the latter group, termed "cruisers," is highly mobile and actively responds to long-range chemical cues emitted by hosts, making them adept at locating sedentary hosts (Grewal et al., 1994).

Entomopathogenic nematodes (EPNs) are distributed across diverse habitats and exhibit the ability to thrive either externally on the exoskeleton of insects or internally within their reproductive, respiratory, digestive, or excretory systems. Many nematodes inhabit arid regions such as deserts, showcasing a remarkable capacity to endure various environmental stresses, including anoxybiosis, thermobiosis, and desiccation (Grewal, 2000).The third-stage juvenile of these nematodes possesses a closed mouth and anus, rendering it incapable of feeding until it encounters an insect. Typically located in soil, it becomes activated by insect movement and subsequently tracks a CO2 gradient to locate insect larvae (Gaugler et al., 1997), ultimately entering the insect's bloodstream cavity to induce its demise.

The free-living stage of entomopathogenic nematodes (EPNs) is notably the most infective and effective phase during their life cycle, wherein the free-living infective juveniles actively search and migrate through the soil in pursuit of insects to infect (Spence et al., 2008). Upon infecting insects, EPNs utilize the insect cadaver as a nutrient source and also consume the microflora and microfauna associated with the cadaver (Flint et al., 1998). The application of molecular biology techniques and genomics has significantly enhanced the efficiency and precision in identifying EPNs suitable for biological control purposes, as well as understanding their ecology, behavior, and responses to climate change through the screening of relevant genes (Segal *et al.*, 2000).

## The storage and formulation of EPNs.

When creating a formulated product, it's crucial to recognize that EPNs are greatly affected by factors like ultraviolet radiation, high humidity, and desiccation. To enhance the efficacy of these nematodes, one must carefully consider the timing of application and the choice of formulations (Shapiro-Ilan, 2000). Past formulations utilized encompass clay, polyether polyurethane sponge, anhydrobiotic nematodes, bait, and activated charcoal (Grewal, 2000). For nematodes to serve as effective pest control agents, they must reach the end user in optimal condition. There's no value in commercializing them if they perish before application. Developing proper storage and formulation techniques is vital to ensure nematode survival and heightened infectivity maintenance. Hence, comprehending nematode ecology and analyzing the environmental factors influencing their activity and infectivity are critical (Lewis et al., 2006).

Moreover, an effective approach to formulation development would entail determining the medium in which infective juveniles can endure desiccation. Conducting experiments to evaluate the impact of different formulation materials on infectivity levels is crucial. Hence, desiccation studies are invaluable as they offer significant insights into this research area. The composition and content of the formulation are also pivotal for ensuring the long-term survival of nematodes. Additionally, the incorporation of preservatives can influence nematode viability, as can aerated water, which plays a vital role in their reproduction, growth, and survival (Shapiro-Ilan, 2000).

## Application technology and Environmental considerations

A single strain of a particular nematode species can serve as a versatile tool for controlling various insect pests across agricultural industries, forestry, and other relevant fields. The effectiveness of pest management relies heavily on the application methods of EPNs, which encompass the use of spray equipment, irrigation systems, and pumping systems (Shapiro-Ilan et al., 2006). The selection of equipment and application techniques significantly influences the likelihood of optimal contact between the nematode and the targeted insect (Jagdale et al., 2009). When biological control methods are applied correctly within suitable systems, economic and environmental success can be expected.

Critical analysis of the application environment is essential, considering factors that could significantly affect the infectivity of EPNs. Environmental variables such as soil moisture, water content, temperature, and ultraviolet radiation can profoundly impact the efficacy of EPNs.

## Categorization of Recently Identified EPN Association

According to the standards outlined above, *Caenorhabditis briggsae* may not fit the classification of an EPN. The infective juveniles obtained from deceased insects exhibit the capability to infect new hosts, albeit with reduced virulence in *Galleria mellonella* as a complex compared to injection of the bacteria alone. This suggests potential inefficiency in the pathogen's release or some form of antagonism by the nematode vector. This ambiguity could imply that *C. briggsae* falls somewhere between necromenic and entomopathogenic, indicating it may be an emerging entomopathogen not yet fully effective, or that *G. mellonella* is not an ideal host. However, the heritability of the symbiont has not been proven, leaving the nature of *C. briggsae's* bacterial association unresolved (Ye *et al.*, 2010; Abebe *et al.*, 2010; Abebe *et al.*, 2011). Therefore, until the heritability of the pathogenic bacteria is confirmed and more is understood about bacterial release and the speed of host mortality, *C. briggsae* should not be classified as an entomopathogenic nematode, whether facultative or otherwise. In contrast, *Oscheius chongmingensis* and *O. carolinensis* (Zhang *et al.*, 2008;Torres-Barragan *et al.*, 2011) have fulfilled the suggested criteria, thus warranting their classification as EPNs. Nonetheless, further investigation is necessary to ascertain the nature and heritability of their bacterial associations and to determine whether they are obligate or facultative EPNs.

## Future Prospects

The study of entomopathogenic nematodes (EPNs) has significantly advanced our understanding of their biology, symbiotic relationships, and potential applications in sustainable pest management. However, several key areas remain unexplored or underdeveloped, presenting exciting future research opportunities.

One of the most promising research directions lies in enhancing the biocontrol efficacy of EPNs through genetic and molecular engineering. Advances in genomics and transcriptomics have begun to unveil the genetic basis of virulence, host specificity, and environmental adaptability in EPNs and their symbiotic bacteria. Future studies focusing on genetic modifications of both nematodes and their bacterial partners could lead to the development of strains with improved virulence, enhanced stress tolerance, and greater persistence in diverse ecological conditions. Moreover, the integration of CRISPR-based genome editing could provide insights into host-pathogen interactions and lead to targeted improvements in EPN performance.

Another critical area for future research is the refinement of EPN formulations and delivery methods. Current formulations, including clay, sponge, and liquid suspensions, have limitations in terms of nematode survival, desiccation resistance, and field efficacy. Investigating novel encapsulation technologies, such as microencapsulation and hydrogel-based carriers, could significantly enhance EPN shelf-life and effectiveness. Additionally, optimizing application techniques by incorporating EPNs into precision agriculture technologies, such as automated irrigation systems and drone-based dispersal, could revolutionize their practical deployment in pest management.Ecological research on EPNs should also be prioritized to better understand their interactions within soil ecosystems. While EPNs are known to be highly effective against various insect pests, their non-target effects on beneficial soil fauna, microbial communities, and ecosystem services remain largely unquantified. Long-term field studies and metagenomic analyses could provide crucial insights into how EPNs influence soil biodiversity and ecosystem stability. Additionally, research into climate change resilience will be critical in determining how shifting temperature and moisture patterns may impact EPN effectiveness and survival in different agroecosystems.

In the realm of host-pathogen interactions, future studies should explore the co-evolutionary dynamics between EPNs, their symbiotic bacteria, and insect immune responses. While substantial progress has been made in understanding immune evasion mechanisms, gaps remain in elucidating the biochemical and molecular interactions governing host suppression. Identifying novel virulence factors and secondary metabolites from EPN-associated bacteria, such as Photorhabdus and Xenorhabdus species, could unlock new avenues for developing bioactive compounds with insecticidal and antimicrobial properties.Furthermore, expanding the taxonomic and ecological exploration of EPNs in underexplored regions, such as arid and semi-arid environments, could lead to the discovery of novel species with unique adaptations and enhanced biocontrol potential. Advances in environmental DNA (eDNA) sequencing and culture-independent techniques could greatly accelerate the identification of previously unknown EPN strains and their associated symbionts.

## Conclusion

Entomopathogenic nematodes represent a cornerstone of biological control, offering environmentally friendly alternatives to chemical pesticides. Their complex life cycles, symbiotic relationships, and diverse ecological interactions make them invaluable tools for integrated pest management (IPM). Despite significant advancements, challenges related to formulation, application technology, and ecological resilience must be addressed to maximize their effectiveness and commercial viability. Moving forward, interdisciplinary research integrating molecular biology, microbial ecology, and agricultural engineering will be instrumental in overcoming current limitations and unlocking the full potential of EPNs. Collaboration between academia, industry, and policymakers will be essential to translating fundamental discoveries into practical solutions for sustainable agriculture. As research continues to refine and expand the application of EPNs, their role in global pest management strategies is poised to become increasingly prominent, reinforcing their status as a vital component of next-generation biocontrol solutions.

## References


Abebe, E., Akele, F.A., Morrison, J., Cooper, V., and Thomas, K.T. (2011). An insect pathogenic symbiosis between Caenorhabditis and Serratia. *Virulence.***2**: 158–61

Abebe, E., Jumba, M., Bonner, K., Gray, V., Morris, K., and Thomas, W.K. (2010). An entomopathogenic *Caenorhabditis briggsae*. *Journal of Experimental Biology*.**213,** 3223– 29.

Adams BJ, Burnell AM, Powers TO (1998). A phylogenetic analysis of the genus Heterorhabditis (Nemata: Rhabditidae) based on internal transcribed spacer1 DNA sequence data. *Journal of Nematology* **30:** 22–39.

Ansari, M. A., & Moens, M. (2006). Diversity in pathogenicity of Steinernema carpocapsae and its symbiotic bacterium for Spodoptera spp. *Communications in Agricultural and Applied Biological Sciences* **71**: 689-700.

Beckmann, M., & Haack, K. J. (2003). Insektizide für die Landwirtschaft: Chemische Schädlingsbekämpfung. *Chemie in unserer Zeit* **37**(2): 88-97.

Bergvinson, D., &García-Lara, S. (2004). Genetic approaches to reducing losses of stored grain to insects and diseases. *Current Opinion in Plant Biology* **7**(4): 480-485.

Bhat, A. H., Chaubey, A. K., & Askary, T. H. (2020). Global distribution of entomopathogenic nematodes, Steinernema and Heterorhabditis. *Egyptian Journal of Biological Pest Control* **30**(1): 1-15.

Blaxter, M. L., De Ley, P., Garey, J. R., Liu, L. X., Scheldeman, P., Vierstraete, A., ... & Thomas, W. K. (1998). A molecular evolutionary framework for the phylum Nematoda. *Nature* 392(6671): 71-75.

Brooke, E., & Hines, E. (1999). Viral biopesticides for Heliothine control-fact of fiction. *Today's Life Sci.* 11: 38-45.

Campbell, J. F., and Gaugler, R. (1993). Nictation behaviour and its ecological implications in the host search strategies of entomopathogenic nematodes (Heterorhabditidae and Steinernematidae). *Behaviour*, pp. 155-169.

Carlini, C. R., & Grossi-de-Sá, M. F. (2002). Plant toxic proteins with insecticidal properties. A review on their potentialities as bioinsecticides. *Toxicon* **40**(11): 1515-39.

Ciche, T. A., & Ensign, J. C. (2003). For the insect pathogen Photorhabdus luminescens, which end of a nematode is out? *Applied and environmental microbiology* **69**(4): 1890-1897.

Cui, L., Gaugler, R., & Wang, Y. (1993). Penetration of steinernematid nematodes (Nematoda: Steinernematidae) into Japanese beetle larvae, Popillia japonica (Coleoptera: Scarabaeidae). *Journal of Invertebrate Pathology* **62**(1): 73-78.

Dunphy, G. B., & Webster, J. M. (1985). Influence of Steinernema feltiae (Filipjev) Wouts, Mracek, Gerdin and Bedding DD136 strain on the humoral and haemocytic responses of Galleria mellonella (L.) larvae to selected bacteria. *Parasitology***91**(2): 369-80.

Flint M.L and Dreistadt S.H, (1998). Natural Enemies' Handbook, the illustrated guide to biological pest control. 2-35.

Gaugler, R. (Ed.). (2002). *Entomopathogenic nematology*. CABI publishing.

Georgis R, Koppenho A.M, Lacey L.A, Belair G, Duncan L.W, Grewal P.S, Samish M, Tan L, Torr P and, Tol R.W.H.M, (2006). Successes and failures in the use of parasitic nematodes for pest control. Biological Control **38**: 103–23.

Girón-Pablo, S., Ruiz-Vega, J., Pérez-Pacheco, R., Sánchez-García, J. A., & Aquino-Bolaños, T. (2012). Isolation of entomopathogenic nematodes and control of Phyllophaga vetula Horn in Oaxaca, Mexico. *African Journal of Biotechnology* **11**(99): 16525-16531.

Grewal, P. S., Gaugler, R., & Wang, Y. I. (1996). Enhanced cold tolerance of the entomopathogenic nematode Steinernema feltiae through genetic selection. *Annals of applied biology* **129**(2): 335-341.

Grewal, P. S., NARDO, E. A. D., & Aguillera, M. M. (2001). Entomopathogenic nematodes: potential for exploration and use in South America. *Neotropical Entomology* **30**: 191-205.

Griffin, C. T., O'CALLAGHAN, K. M., & Dix, I. (2001). A self-fertile species of Steinernema from Indonesia: further evidence of convergent evolution amongst entomopathogenic nematodes? *Parasitology* **122**(2): 181-86.

Han, R. C., & Ehlers, R. U. (1998). Cultivation of axenic Heterorhabditis spp. dauer juveniles and their response to non-specific Photorhabdus luminescens food signals.

Han, R. C., Wouts, W. M., & Li, L. I. Y. I. N. G. (1990). Development of Heterorhabditis spp. strains as characteristics of possible Xenorhabdus luminescens subspecies. *Rev. Nematol* **13**(4): 411-415.

Han, R., & Ehlers, R. U. (2000). Pathogenicity, development, and reproduction of Heterorhabditis bacteriophora and Steinernema carpocapsae under axenic in vivo conditions. *Journal of invertebrate pathology* **75**(1): 55-58.

Han, R., Wouts, W. M., & Li, L. (1991). Development and virulence of Heterorhabditis spp. strains associated with different Xenorhabdus luminescens isolates. *Journal of Invertebrate Pathology* **58**(1): 27-32.

Jagdale G.B, Kamoun S and, Grewal P.S, (2009). Entomopathogenic nematodes induce components of systemic resistance in plants: Biochemical and molecular evidence. *Biological Control* **51**: 102–109.

Kahel-Raifer, H., & Glazer, I. (2000). Environmental factors affecting sexual differentiation in the entomopathogenic nematode Heterorhabditis bacteriophora. *Journal of Experimental Zoology* **287**(2): 158-166.

Kaya, H. K., and Gaugler, R. 1993. Entomopathogenic nematodes. *Annual Review of Entomology* **38**:181–206.

Koppenhöfer, A. M., & Fuzy, E. M. (2006). Effect of soil type on infectivity and persistence of the entomopathogenic nematodes *Steinernema scarabaei,Steinernema glaseri, Heterorhabditis zealandica*, and *Heterorhabditis bacteriophora*. *Journal of Invertebrate Pathology*. **92**(1): 11-22.

Le Vieux PD, Malan AP (2013) The potential use of entomopathogenic nematodes to control *Planococcus ficus* (Signoret) (Hemiptera: Pseudococcidae). *South African Journal of Ecology and Viticulture*.**34**(2):296–306.

Martens, E. C., Heungens, K., & Goodrich-Blair, H. (2003). Early colonization events in the mutualistic association between *Steinernema carpocapsae* nematodes and *Xenorhabdus nematophilus* bacteria. *Journal of Bacteriology*. **185**(10): 3147-3154.

Martens, E. C., Vivas, E. I., Heungens, K., Cowles, C. E., & Goodrich-Blair, H. (2004, January). Investigating mutualism between entomopathogenic bacteria and nematodes. In *Proceedings of the Fourth International Congress of Nematology, 8-13 June 2002, Tenerife, Spain*. 447-462. Brill.

Mrá ek, Z., Hanzal, R., & Kodrík, D. (1988). Sites of penetration of juvenile steinernematids and heterorhabditids (Nematoda) into the larvae of *Galleria mellonella* (Lepidoptera).

Nguyen K. B, and Hunt D. (2007). Entomopathogenic nematodes: systematics, phylogeny, and bacterial symbionts. 816 pp. Leiden, E.J. Brill.

Nickle, W. R. (Ed.). (1984). *Plant and insect nematodes* (pp. xiv+-925pp).

Novotny, V., Basset, Y., Miller, S. E., Weiblen, G. D., Bremer, B., Cizek, L., & Drozd, P. (2002). Low host specificity of herbivorous insects in a tropical forest. *Nature*. **416**(6883), 841-844.

Oerke, E. C., & Dehne, H. W. (2004). Safeguarding production—losses in major crops and the role of crop protection. *Crop protection* **23**(4): 275-285.

Peters, A., & Ehlers, R. U. (1994). Susceptibility of leatherjackets (Tipula paludosa and Tipula oleracea; Tipulidae; Nematocera) to the entomopathogenic nematode Steinernema feltiae. *Journal of Invertebrate Pathology* **63**(2): 163-171.

Poinar G O (1990) Biology and taxonomy of Steinernematidae and Heterorhabditidae. In: *Entomopathogenic Nematodes in Biological Control*, Gaugler, R. and Kaya, H.K. and Boca Raton, FL (Eds): CRC Press, pp. 23–61.

Poinar Jr, G. O., & Thomas, G. M. (1966). Significance of Achromobacter nematophilus Poinar and Thomas (Achromobacteraceae: Eubacteriales) in the development of the nematode, DD-136 (Neoaplectana sp. Steinernematidae). *Parasitology* **56**(2): 385-390.

Poinar Jr, G. O., Van der Geest, L., Helle, W., & Wassink, H. (1977). Experiments with organisms from hosts other than Glossina. *The Future for Biological Methods in Integrated Control. IDRC-077E, Canada*, 88-92.

Poinar, G. J., Thomas, G. M., & Hess, R. (1977). Characteristics of the specific bacterium associated with Heterorhabditis bacteriophora (Heterorhabditidae: Rhabditida).

Poinar, J., & GO, G. (1966). The presence of Achromobacter nematophilus in the infective stage of a *Neoaplectana sp.*(Steinernematidae: Nematoda).

Rainey, F. A., Ehlers, R. U., & Stackebrandt, E. (1995). Inability of the polyphasic approach to systematics to determine the relatedness of the genera Xenorhabdus and Photorhabdus. *International Journal of Systematic and Evolutionary Microbiology* **45**(2): 379-381.

Schmiege, D. C. (1963). The feasibility of using a neoaplectanid nematode for control of some forest insect pests. *Journal of Economic Entomology* **56**(4): 427-431.

Segal D and Glazer I, (2000). Genetics for Improving Biological Control agents: the case of entomopathogenic nematodes.*Crop Protection* **19**: 685-689.

Shapiro, D. I., & McCoy, C. W. (2000). Virulence of entomopathogenic nematodes to Diaprepesabbreviatus (Coleoptera: Curculionidae) in the laboratory. *Journal of Economic Entomology* **93**(4): 1090-1095.

Spence K.O, Lewis E.E, and Perry R.N, (2008). Host-Finding and Invasion by Entomopathogenic and Plant-Parasitic Nematodes: Evaluating the Ability of Laboratory Bioassays to Predict. *Journal of Nematology* **40**(2): 93-98.

Stock S P (2009) Molecular approaches and the taxonomy of insect-parasitic and pathogenic nematodes. pp. 71–100 in Stock, S.P., Vandenburg, J., Glazer, I. & Boemare, N. (Eds) Insect pathogens: *Molecular Approaches and Techniques*. Wallingford, Oxon, UK, CAB International Press.

Stock, S. P., Griffin, C. T., & Chaerani, R. (2004). Morphological and molecular characterization of Steinernema hermaphroditism n. sp.(Nematoda: Steinernematidae), an entomopathogenic nematode from Indonesia, and its phylogenetic relationships with other members of the genus. *Nematology*, **6**(3): 401-412.

Strauch, O., Stoessel, S., & Ehlers, R. U. (1994). Culture conditions define automictic or amphimictic reproduction in entomopathogenic rhabditid nematodes of the genus Heterorhabditis. *Fundamental and Applied Nematology*. **17**(6), 575-82.

Stuart R.J, Barbercheck M.E, Grewal P.S, Taylor R.J and Hoy W.C, (2006). Morphological and functional dimorphism in Xenorhabdus spp., bacteria symbiotically associated with the insect pathogenic nematodes Neoaplectana and Heterorhabditis. *Microbiology*.**121**: 303- 09.

Torres-Barragan, A., Suazo, A., Buhler, W.G., and Cardoza, Y.J. (2011). Studies on the entomopathogenicity and bacterial associates of the nematode *Oscheius carolinensis*. *Biological Control*.**59**, 123–29.

Triggiani, O., & Poinar, G. J. (1976). Infection of adult Lepidoptera by *Neoaplectana carpocapsae* (Nematoda).

U.S. Census Bureau 2006. World Population Information. U.S. Census Bureau, Population Division,International Programs Center. Last Revised: June 08, 2006.

Ye, W.M., Torres-Barragan, A., and Cardoza, Y.J. (2010). Oscheius carolinensis n. sp (Nematoda: Rhabditidae), a potential entomopathogenic nematode from vermicompost. *Nematology*.**12**, 121–35.

Zhang, C., Liu, J., Sun, J., Yang, S., An, X., Gao, G., Lin, M., Lai, R., He, Z., Wu, Y., et al. (2008). *Heterorhabditidoides chongmingensis* gen. nov., sp. nov. (Rhabditida: Rhabditidae), a novel member of the entomopathogenic nematodes. *Journal of Invertebrate Pathology* **98**: 153–68.

Zimmerman, R. J., & Cranshaw, W. S. (1990). Compatibility of three entomogenous nematodes (Rhabditida) in aqueous solutions of pesticides used in turfgrass maintenance. *Journal of Economic Entomology* **83**(1): 97-100.

# CHAPTER 28

# HORTICULTURE: AN INTEGRAL COMPONENT OF ARID ECOTOURISM

**Prof. (Dr) P. L. Saroj**

*ICAR-Central Institute for Subtropical Horticulture,*
*Rehmankhera, P.O. Kakori-226001, Lucknow (UP), INDIA*
*Email plsaroj@yahoo.co.in*

The term 'ecotourism' comprised eco (ecological) and tourism (traveling). As per Oxford Dictionary, the term ecotour was first recorded in 1973. Sometimes, 'tourism' 'ecotourism' 'responsible tourism' and 'sustainable tourism' are considered as synonyms; which are interrelated but exactly neither the same nor oxymoron. According to The International Ecotourism Society (TIES), ecotourism can be defined as "responsible travel to natural areas that conserve the environment, sustains the well-being of the local people, and involves interpretation and education" where horticulture can play vital role. The Global Ecotourism Network (GEN) defines ecotourism as "responsible travel to natural areas that conserve the environment, sustains the well being of the local people and creates knowledge and understanding through interpretation and education of involved". A conservationist by the name of Hector Ceballos-Lascurain is often credited with the first definition of ecotourism in 1987, i.e. "tourism that consists in travelling to relatively undisturbed or uncontaminated natural areas with the specific object of studying, admiring and enjoying the scenery and its wild plants and animals, as well as any existing cultural manifestations (both past and present) found in these areas." Since 1980s, ecotourism has been considered as an important endeavor by environmentalists, who have said they want future generations to experience destinations relatively untouched by human interventions (Martha, 2008). Ecotourism typically involves travel to destinations where flora, fauna and cultural heritage are the primary attractions. Thus, responsible ecotourism should minimize the negative aspects of conventional tourism on the environment and enhance the cultural integrity of local people. Therefore, in addition to evaluating environmental and cultural factors, an integral part of ecotourism is the promotion of recycling, energy efficiency, water conservation and the creation of economic opportunities for local communities (Randall, 1987).

Ecotourism is often misinterpreted and assumes that simply creating destinations in natural areas is ecotourism. However, true ecotourism must, above all, sensitize people to the beauty and fragility of nature. Also, there are various risks associated with ecotourism like, long-distance travel by air may add ill impact to climate, presence of travelers in a particular region or location may negatively impact the ecological integrity of protected areas, presence of tourists may potentially increase rent, rates and property values, thereby marginalizing local communities, ecotourism activities may also place travelers at risk of health problems or injuries as well as sometimes travelers may bring pathogens to ecologically sensitive areas putting wild life as well as local communities at risk etc. Moreover, negative impact can be mitigated through regulations and codes of conduct that effectively and persuasively impart messages about appropriate visitor behavior (Weaver, 2008).

The global ecotourism market produced $92.2 billion in 2019 and is forecasted to generate $103.8 billion by 2027. Ecotourism has largely benefited destinations like Madagascar, Ecuador, Kenya, and Costa Rica, and has helped economic growth in some of the world's most impoverished communities. In this context, India has a significant geographical advantage owing to its rich natural & Eco-tourism resources viz., 70 per cent of the Himalayas, 7,000 km of coastline, one of the three countries in the world with both hot and cold deserts, ranks $10^{th}$ in total area under forest cover, ranks $6^{th}$ in terms of the number of recognized UNESCO Natural Heritage sites etc. The natural attractions in India are almost unrivalled in the world, ranging from pristine forests, the snow-clad Himalayas, montane grasslands, golden and silver deserts, rivers, lakes, wetlands, mangroves, beaches, volcanoes and corals, and housing tremendous biodiversity. Besides, we have the mega-fauna consisting of iconic species such as tigers, lions, elephants, rhinos, leopards, wild buffaloes, Indian bison (gaur), etc. and more than 1200 species of birds. India has tremendously rich in species and ecosystem diversity, as it is a mega diverse country having only 2.4% of the world's land area but accommodates 7-8% species of flora and fauna. Over 1,03,258 species of fauna and 55,048 species of flora have been documented in the 10 biogeographic zones of the country.

## 1. Sustainable Ecotourism

- Need based involvement of stakeholders in the planning, development, implementation and monitoring phases.
- Recognizes that nature is a central element to the tourist experience.
- Low impact upon protected sites/ landscape of natural resources.
- Controlled visits of recreational sites during peak season.
- Support to conservation of natural areas.

- Orient customers about visiting site and regulations of ecotourism.
- Hire local people for different activities and manage to supply locally available products.
- Link with government programme 'one district one product'.
- Trained guides in interpretation of scientific or natural history.
- Ensure that wildlife should not be harassed.
- Respect the privacy and culture of local people.

The ecotourism is entirely different than mass tourism. Thereby, ecotourism must be promoted as nature friendly programme after proper acquaintance between mass tourism and ecotourism as given in Table 1.

**Table 1. Distinct characteristics between mass tourism and ecotourism (Dorobantu & Nıstoreanu, 2012)**

| Mass tourism | Ecotourism |
|---|---|
| Large groups of visitors | Small groups of visitors |
| Often urban | Rural destinations |
| Touristic general marketing activities | Eco-marketing activities |
| Average prices for purposes of market penetration | High price with purpose of filtering the market |
| Impact on natural environment | Little impact on the natural environment |
| Advanced control options | Limited possibilities of control |
| Management based on macroeconomic principles | Management based on local economic principles |
| Anonymous relationship between visitors and local community | Personalized relationships between visitors and local community |
| General development goals | Local development objectives |
| Behavior-oriented leisure activities/ entertainment, opponents to education and training actions | Loyalty in the process of training and education for appropriate conduct for the natural environment |
| Intensive development of tourism facilities | Reduced development of tourism facilities |

The general trend in ecotourism is to increase experiences by encouraging activities such as long-distance walking, camping, boating, hunting, sight-seeing, swimming, cultural activities, bicycling, observing wildlife and nature, skiing, visiting historical places, horse riding among wildlife observation, participation in festivals, cultural activities and nature landscapes, hiking, outdoor sports, picnic, paragliding are different areas of interest influence the preferences of many visitors. Moreover, ecotourism activities have been divided into following categories;

- Marine Ecotourism: Marine cruising including sailing, yacht and power cruising, sea kayaking tours
- Land based Ecotourism: Bicycle touring/ Mountain biking; Horseback trail riding; Backpacking/ Trekking
- Freshwater River Rafting, Canoeing and Kayaking
- Winter Tourism (Tour skiing, Dog sledding, Snow shoeing)
- Walking, camping, boating, hunting, sight-seeing, swimming, cultural activities, observing wildlife and nature, skiing, visiting historical places, and horse riding etc.

## 2. Potential Benefits of Ecotourism

Besides two ways learning, positive impact on environmental issues and economic opportunities to local communities are the basic philosophy of ecotourism. Some of the potential consequences of ecotourism are listed below:

### Ecological

- Protection of natural environments.
- Incentives to rehabilitate/ modify landscape.
- Habitat maintenance and enhancement.
- Sometimes, ecotourists serve as guardians who personally intervene in situations where the environment is perceived to be threatened.
- Communities experience change in environmental attitude and behavior.

### Socio-cultural

- Foster community stability and well being through economic benefits.
- Aesthetic and spiritual benefits and enjoyment for locals and tourists.
- The local people may also learn new skills from the ecotourists.
- Involvement of broad spectrum of the population.

### Economic

- Generation of revenue and creation of employment related to this sector.
- Economic opportunities for peripheral regions.

## 3. Potential Risk of Tourism

There are a lot of risk involved in tourism activities as given in Table 2.

**Table 2. Potential risks from tourism**

| Element | Risks from tourism activities |
|---|---|
| Ecosystems | The construction of accommodation, visitor centres, infrastructure, and other services has a direct impact on the environment, from vegetation removal, animal disturbance, elimination of habitats, impacts on drainage etc.<br>Wildlife habitat may be significantly changed (travel routes, hunting areas, breeding areas, etc.) by all kinds of tourist development and use. |
| Soils | Soil compaction can occur in certain well-used areas.<br>Soil removal and erosion also occurs, and may continue after the disturbance is gone. |
| Vegetation | Concentrated use around facilities has a negative effect on vegetation.<br>Transportation may have direct negative impacts on the environment (e.g. vegetation removal, weed transmission, animal disturbance).<br>Fire frequency may change due to tourists and park tourism management. |
| Water | Increased demands for fresh water.<br>Disposal of sewage or litter in rivers, lakes or oceans.<br>Release of oil and fuel from ships and smaller craft.<br>Propeller-driven watercraft may affect certain aquatic plants and species. |
| Air | Motorized transportation may cause pollution from emissions (from plane, train, ship or automobile). |
| Wildlife | Hunting and fishing may change population dynamics.<br>Hunters and fishers may demand the introduction of foreign species,<br>and increased populations of target animals.<br>Impacts occur on insects and small invertebrates, from effect of transportation, introduced species, etc<br>Disturbance by visitors can occur for all species, including that are not attracting visitors.<br>Disturbance can be of several kinds: noise, visual or harassing behaviour.<br>The impact can last beyond the time of initial contact (e.g. before heartrate returns to normal, or before birds alight, or mammals resume breeding or eating).<br>Marine mammals may be hurt or killed by boat impacts or propeller cuts.<br>Habituation to humans can cause changed wildlife behavior, such as approaching people for food. |

*Kiper (2013)*

## Ecotourism Certification Programme

Ecotourism means management of tourism and conservation of nature in a way to maintain a fine balance between the requirements of tourism and ecology on the one hand and needs of the local communities for jobs, new skills, income-generating employment, and a better status of women on the other. Therefore, ecotourism certification programme came into existence. There were various moves for national and international certification programme. As a result, national certification programme has been put in place in countries like Costa Rica, Australia, Kenya, Estonia and Sweden. In fact, ecotourism certification program certifies tourism products (tours, accommodations, attractions) with a primary focus on nature. It assures travelers that certified products are backed by a strong, well managed commitment to sustainable practices and provides high-quality nature-based tourism experiences.

There are various certification agencies for ecotourism certification. 'Earth Check' certificate is the world's leading science-based environmental certification and benchmarking program for the travel and tourism industry. Also, some organizations like Ohio University started course on ecotourism certificate programme as an interdisciplinary program focused on promoting sustainable tourism development through a balance of environmental, economic, and socio-cultural aspects of nature-based tourism. The certificate is intended to complement other areas of study at the university and is especially relevant to students majoring in outdoor recreation and education; restaurant, hotel, and tourism; recreation management; global studies; environmental studies; and, entrepreneurship and other business-related disciplines. The course include; Principles of Tourism, Introduction to Cultural Anthropology, Natural Resource Economics, Introduction to Entrepreneurship, Marketing for Hospitality and Tourism, Environmental Ethics, Parks and Protected Areas Management; besides practical in recreation and leisure, hospitality etc.

***Ecotourism customers***: Ecotourism certification schemes can help ecotourism customers make informed and responsible choices, enjoy quality and authentic experiences, and contribute to conservation and development. By choosing certified ecotourism products and services, customers can trust that they are supporting ethical and sustainable practices, learning from local cultures and nature, and minimizing their negative impacts. Certification can also increase customer satisfaction and awareness.

***Ecotourism destinations***: Ecotourism certification schemes can help ecotourism destinations to protect and enhance their natural and cultural assets, diversify and strengthen their economy, and empower their communities. By promoting and rewarding certified ecotourism products and services, destinations can increase their visibility and attractiveness, generate more income and employment, and foster collaboration and participation among stakeholders. Certification can also support policy and planning for sustainable tourism development.

## International Tourism

World tourism sector has emerged as major industry with total international tourism receipt during 2022 was 1031 US$ billion. Out of total international tourism receipt, 48.98 per cent share is by top ten countries (Table 3). The United States of America rank first position with 13.11 per cent share followed by Spain (7.07%),

United Kingdom 6.61%) and France (5.79%). The share of India in world tourism sector by number is below 2 per cent and by receipt is 2.08 per cent. To decide destinations for most of the international tourist are natural beauty (scenic sites), historical monuments, travel connectivity, cost, tourist places, logistics, law and order situations etc.

UN Tourism (UNWTO until 2023) is a specialized agency of the United Nations which promotes responsible, sustainable and universally accessible tourism. UN Tourism headquarters are based in Madrid (Spain). United Nations World Tourism Organization emphasized to promote sustainable tourism as part of the Sustainable Development Goals (SDGs), through programmes like the International Year for Sustainable Tourism Development in 2017. There is a direct link between sustainable tourism and Sustainable Development Goals. Tourism for SDGs focuses on how SDG 8 (decent work and economic growth), SDG 12 (responsible consumption and production) and SDG 14 (life below water) implicate tourism in creating a sustainable economy (Anonyms, 2021).

**Table 3. Top ten countries of the world in terms of international tourism receipt during 2022.**

| Source Country | International Tourism Receipt (in US$ billion) | Percentage share |
|---|---|---|
| United States of America | 135.20 | 13.11 |
| Spain | 72.90 | 7.07 |
| United Kingdom | 68.20 | `6.61 |
| France | 59.70 | 5.79 |
| Italy | 44.30 | 4.30 |
| Turkey | 41.20 | 4.00 |
| United Arab Emirates* | - | - |
| Germany | 31.50 | 3.06 |
| Mexico | 28.00 | 2.72 |
| Canada | 24.00 | 2.33 |
| Total of top 10 countries | 505.00 | 48.98 |
| India | 21.04 | 2.08 |
| Others | 504.60 | 48.98 |
| Grand total | 1031.00 | 100.00 |

*Data not available

Source: UNWTO Barometer May, 2023

## Indian Tourism Industry

India is one of the most culturally beautiful places, famous for its diversity, landscapes, spirituality and thousand years of heritage that it has preserved. With over 1.3 billion people, India is the top most populated country in the world. As far as the physical features of India are concerned, it ranges from plains to mountains to deserts and many other reliefs. To the north of the country lies the Himalayas, which contain some of the highest peaks in the world; to the south is the Indian ocean with some of the most beautiful beaches in the world; to the west, we have the Thar Desert, which is the seventh largest desert in the world and to the east, the country has seven diverse and beautiful states that have preserved their culture and tradition since the ancient times. The major attractions for tourists are Taj Mahal, the holy city of Varanasi, Golden Temple of Amritsar, Beaches of Goa, Ajanta and Ellora Caves, Akshardham Temple, Jaisalmer, Khajuraho etc. Some of the tourist's sites of India is given in Table 4.

**Table 4. Places of tourist interest in India**

| Type of tourist sites | Names and Location |
|---|---|
| One of the 7th wonders | Taj Mahal, Agra (UP) |
| UNESCO world heritage site | Hampi, Vijayanagara (Karnataka); Chittorgarh Fort , Rajasthan |
| Holly places | Ayodhya-Ayodhya; Varanasi-Varanasi; Mathura Vrindavan-Mathura; Rishikesh-Haridwar (UP) |
| Palaces | Mysore Palace (Karnataka); Jaipur Palace, Udaipur Palace, Jaisalmer Palace in Rajasthan |
| Forts | Red Fort, Delhi; Agra Fort, Agra; Amer Fort and Hawa Mahal, Jaipur; Junagarh Fort, Bikaner; Mehrangarh Fort, Jodhpur; Jaisalmer Fort, Jaisalmer; Gwalior Fort, Gwalior; Golkonda Fort, Hyderabad etc. |
| Temples | Golden Temple, Amritsar; Badrinath Temple; Kedarnath Temple; The Konark Sun Temple; Somnath Temple; Sanchi Stupa; Rameshwaram Temple; Vaishno Devi Mandir; Sri Anantha Padmanabhaswamy Temple, Thiruvananthapuram; ISKCON Temple, Bangalore; Shri Swaminarayan Mandir; Tirupati Balaji Temple, Andhra Pradesh: Shri Jagannath Temple, Puri; Meenakshi Temple, Madurai; Shree Mahakaleshwar Temple, Ujjain; Khajuraho Temples, Chhatarpur etc. |
| Caves | Ajanta and Ellora Caves; Elephanta Caves; Badami Caves; Tabo Caves; Dungeshwari Cave Temples; Khandagiri and Udayagiri Caves; Undavalli Caves etc. |
| Beaches | Goa Beaches; Puri Beach, Odisha; Pati Sonepur Sea Beach in Ganjam; Rushikonda Beach in Visakhapatnam; Kovalam beach, Chennai; Eden Beach in Chinna Veerampattinam, Puducherry; Radhanagar beach, Andaman and Nicobar; Minicoy Thundi beach, Lakshadweep; Kappad beach, Kozhikode & Kasargod Kerala; Padubidri Beach, Udupi, Karnataka; Shivrajpur beach, Gujarat etc. |
| Hill Stations | Sri Nagar and Pahalgam (Jammu & Kashmir); Shimla and Manali (Himachal Pradesh); Mussoorie and Nainital (Uttarakhand); Coorg/ Madikeri (Karnataka); Munnar, Kerala; Kodaikanal and Ooty (Tamil Nadu); Shillong, Meghalaya; Mount Abu, Rajasthan etc. |

| | |
|---|---|
| Valleys | Kashmir Valley, Jammu & Kashmir; Nubra Valley, Leh; Kangra Valley, Himachal Pradesh; Bhakra Nangal Valley, Bilaspur, Himachal Pradesh; Kaveri Valley, Karnataka and Tamil Nadu; Ketti Valley, Ooty, Tamil Nadu; Chambal Valley, Madhya Pradesh; Damodar Valley, West Bengal. |
| Monuments | India Gate and Qutub Minar, New Delhi; Gateway of India, Maharashtra; Statue of Unity, Vadodara; Char Minar, Hyderabad; Victoria Memorial, West Bengal etc. |
| Museums | Indian Museum, Kolkata; Victoria Memorial, Kolkata; Chhatrapati Shivaji Maharaj Vastu Sangrahalaya, Maharashtra; The National Museum, New Delhi; The Calico Museum of Textiles, Ahmedabad; The National Rail Museum New Delhi; The Government Museum, Chennai; The Salar Jung Museum Hyderabad, Telangana; The Napier Museum, Trivandrum etc. |

Tourism sector of India is experiencing a significant rebound with foreign tourist arrivals in 2023 surging 64 per cent compared with 2022. This translates to 92,36,108 visitors during January to December 2023, compared with 64, 37,467 in the same period of previous year. However, despite this impressive growth, the sector remains 15.5 per cent below pre-pandemic levels recorded in 2019 (Ministry of Tourism, Government of India). The major source of tourists was United States, Bangladesh, United Kingdom, Australia and Canada. The international tourist arrivals in India from 2010 to 2023 are depicted in fig. 1. About 66 per cent of total tourist arrivals in India are contributed by 15 countries i.e. Bangladesh, United State of America, United Kingdom, Canada, Russia, Australia, France, Germany, Malaysia, Sri Lanka, Thailand, Japan, Afghanistan, Nepal and China. The preferred mode of travel is by Air (79%) followed by Land rout mainly Bangladesh and Pakistan (19.5%) and least by sea rout (1.5%). Among top five entry points, New Delhi held the leading position with 30.29 per cent tourist arrivals followed by Mumbai (15.08%), Chennai (8.34%), and Bengaluru and Hyderabad (6.20 per cent each). The financial earning from expenditure during January to December 2023 stood at 2, 31, 927 crores, reflecting a substantial 65.74 per cent growth from previous year.

**Tourist arrivals (in millions)**

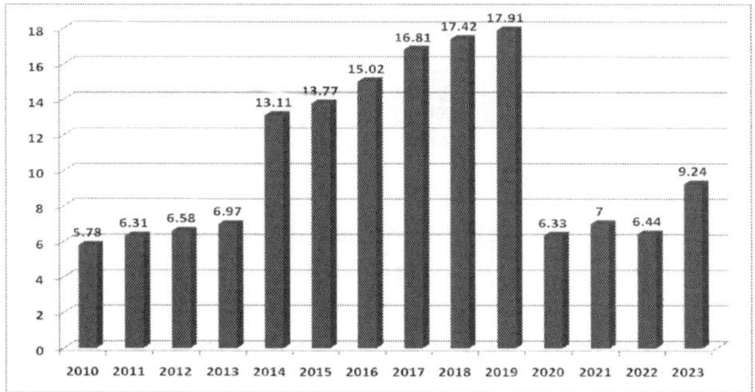


**Fig 1. International tourist arrivals in India from 2010 to 2023**

In India, the preference of international visitors in different states/ UTs is given in table 5 indicate that Gujarat followed by Maharashtra is the preferred states for foreign visitors. The share of Gujarat is maximum (20.70%) followed by Maharashtra (17.60%), West Bengal (12.08%) respectively, though it varies year to year. Similarly, Taj Mahal is most visited place followed by Mysore Palace.

**Table 5: Top ten states/ Union Territories of India in number of foreign tourist visits in 2022.**

| State/ UTs | Foreign tourist visitors (in million) | Percentage share |
|---|---|---|
| Gujarat | 1.78 | 20.70 |
| Maharashtra | 1.51 | 17.60 |
| West Bengal | 1.04 | 12.08 |
| New Delhi | 0.82 | 9.50 |
| Uttar Pradesh | 0.65 | 7.56 |
| Tamil Nadu | 0.41 | 4.74 |
| Rajasthan | 0.40 | 4.62 |
| Kerala | 0.35 | 4.02 |
| Punjab | 0.33 | 3.84 |
| Madhya Pradesh | 0.20 | 2.38 |
| Total of top ten | 7.47 | 87.03 |
| Others | 1.11 | 12.97 |
| Total | 8.59 | 100.00 |

## Indian Horticulture and Ecotourism

Indian horticulture has made commendable progress in past few decades and now contributes 30.4% of the agriculture Gross Domestic Product (GDP) using only 13.1% of gross cropped area. As per the 2$^{nd}$ advance estimates for 2022-23, the total horticulture production in India is estimated to be 351.92 million tons, surpassing the total food grain production of 329.69 million tons during the same year. Horticulture sector is making significant contribution to the Indian economy. Apart from ensuring nutritional security of the nation, it provides alternate rural employment opportunities, diversification in farm activities, earning of foreign of exchange, enhanced income of farmers and ecological security. India has emerged as world leader in the production of a variety of fruits like mango, banana, guava, papaya, sapota, pomegranate, lime and aonla and vegetables like okra and ginger. Besides, India has maintained its dominance in the production of spices, coconut and cashew nut. Among the new crops, kiwi, gherkins, kinnow, date palm, oil palm and dragon fruit have been successfully introduced for commercial cultivation in the country. This could happen only with proper policy planning, systematic research by scientists, devotion of farmers and other stakeholders. Moreover, the demand for horticultural produce is increasing day by day due to greater health awareness, rising income, export demands and increasing population poses the challenge for further increasing the production and productivity of horticultural crops.

The greater awareness towards health issues has further increased the demand of horticultural commodities, which was very well realized during pandemic period of covid-19. Therefore, not only sufficient food is required but balance diet is essential for proper growth and development. As per recent reports of Indian Council of Medicinal Research (ICMR), the per capita food requirement is 1200g which provides about 2000 calories. For maintaining optimum health, per capita requirement is 100g fruits, 400g green vegetable, 300ml milk or curd, 85g pulses or egg and 35g dry fruits. It was further stressed that taking fat >70g in a day is risky for health. For non-vegetarians, 70g meat or chicken in a day is good enough. The ideal diet can reduce about 80% of type 2 diabetic problem, hyper tension, heart problems and noninfectious diseases. Reports also indicated that consumption of salt is more and potassium is less in Indian diets. In this context, horticulture is playing a pivotal role. Thereby, policy framework must pay due attention not only on production of horticultural commodities more and more but also require huge infrastructure development for minimizing post harvest losses of horticultural commodities.

Horticulture is an integral component of ecotourism. Without integrating horticulture, ecotourism cannot be thought of. Even sometimes, in forest or desert ecology, ecotourism can be done but there too some kind of landscape development is required by fencing of area, making paths, natural or manmade water bodies, developing topiary, making hut site office etc; where some kind of beautification is required by growing fruits, vegetables, spices, medicinal and aromatic plants. Development of parks, grass lawn, avenue plantations, rockery, protected structures, nursery management, processing and value addition, packaging, labeling etc. are the areas in horticulture can give impetus to ecotourism. Now, there is a demand for ornamental and landscaping personnel in big way in metro cities and tourist places. This has great potential in generation of employment opportunities for rural and urban youth.

## Ecotourism in Arid Region

In India, about 12.02 per cent of the total geographical area is categorized as arid region which includes 31.71 million hectares under hot arid region and 7.84 million hectares under cold arid region. The hot arid region is mainly spread over north-western part of the country covering the state of Rajasthan (19.60 million hectares) followed by Gujarat (6.20 million hectares) and Punjab (1.50 million hectares) and only 3.10 million hectare is in the peninsular region of the country. The cold region is mainly in the States of Jammu and Kashmir, Himachal Pradesh and Uttarakhand. The major farming constraints of arid region are poor and erratic rainfall, extremes of temperature, sandy soil with poor fertility, high wind velocity, deep water table and brackish underground water resource, problem of wild animals etc. The major challenges in arid region are complex diverse and risk-prone farming situations owing to poor productivity and profitability, vulnerable to climate change, genetic erosion of important biological diversity, poor input use efficiency, limited and traditional post harvest management, loose linkage and coordination among public and private organizations, limited state of art facilities etc.

In spite of various biophysical farming constraints in arid region, there is ample opportunity of horticulture in the region which must be harnessed scientifically.

Moreover, the benefit of 'horticulture revolution' in India was not realized well in resource poor conditions like arid region. In general, attention on horticulture was given little late in arid region; but because of its own relevance in human nutrition, economic viability and ecological security; it started spreading silently in all eco-regions of the country and lastly reached in arid region too. After establishment of National Research Centre for Arid Zone Fruits, Bikaner in 1993 which was upgraded to ICAR-Central Institute for Arid Horticulture in w.e.f. 27th September, 2000 gave impetus to address the problems of arid horticulture.

The arid ecotourism has special significance in our country because of the vast land resource in form of deserts, large number of historical monuments, forts, museums, parks, lakes, hotels, variety of food and cloths etc. The camel safari is another point of attraction. The government of Rajasthan is promoting tourism in the state because; it supports the state economy, secures public welfare by creating more employment opportunities and increases the prestige of the state as an important cultural and historical site. The department of Rajasthan has developed tourism circuits in the state such as Sunda Mata temple in Jalore district, Bassi in Chittorgarh district, Kumbhalgarh in Rajsamand district, Hamirgarh and Menal, both in Bhilwara district to facilitate tourists. Padharo Mhare Desh is Rajasthan tourism's traditional logo and connects tourists with the state's tradition and culture. The previous BJP government had replaced the grand old campaign slogan 'Padharo Mhare Desh' (Welcome to my land) and introduced 'Jaane Kya Dikh Jaye' as the new slogan for Rajasthan tourism in 2016. The state government of Rajasthan launched 'Rajasthan Ecotourism Policy 2021' with an aim to generate economic benefits after conserving natural areas and attracting tourism. The forest department will be the nodal department for implementing the ecotourism policy. Therefore, horticulture centric role has to be played in proper planning, implementation and execution of tourism involving private sector to ensure sustainable ecotourism development.

Moreover, desert is not available to be experienced everywhere and in all the countries. If you are visiting from Delhi to desert destinations of western part of Rajasthan like Jaisalmer and Bikaner by air, you can see below very well miles and miles of sand with tiny green patches in between due to scarcity of water. A large segment of the population is dependent on rainfed agriculture and allied activities for its livelihood. Even then, Rajasthan attract tourists because of its varied climate and rich heritage. Also, visiting India is cheaper than going to most of the developed countries. Deserts have its own beauty. Sunsets, sunrise, local handicrafts, indigenous animals like camels, deers, donkeys, lizards, desert fox, wildcat, chameleons and birds; food specialties and languages, dance and multi-colour traditions. Western Rajasthan can boast of several forts, palaces, temples and havelis with especial history associated. Numerous temples such as Jain temple at Lodrava and Osiyan, Karni Mata temple of Bikaner, Temples at Kiradu, Sandeshwar temple and more are worth visiting. Other attractions are sand dunes, Khuri village Safari, Patwon li Haveli, Mandore Gardens, Marwar- the old capital of Jodhpur; colourful handicrafts, bedspreads and clothes with embroidery or traditional prints like Bandhani, Lahariya, etc, jutis and bags of camel skin and many such things are hunt worthy for the domestic as well as the foreign tourist. Vegetable of *Sangri* (pods of *Prosopis cineraria*) and *Dal-Bati-Churma* are mouth-watering traditional culinarians of the desert. However, threats to these areas are rising due to increasing accessibility, overuse of nature to various developmental activities and increasing population pressure. Therefore, such fragile environments must be protected properly.

A gradual shift to varied ecotourism viz; Wildlife Ecotourism, Weekend Ecotourism or Conference Ecotourism can be encouraged (Chandel, Kanga , 2020).

Value-added Desert Ecotourism which can include adventure, desert camps and safaris, dressing up in costumes of the locals, desert sports, star gazing in the night in the desert, desert landscape painting competitions, etc. can be the next step on this trajectory. There are local fairs organized in various parts of the desert (Table 6) which can be enjoyed by the domestic and foreign tourists in the season. However, careful planning that primarily involves the local community and a tourist friendly infrastructure is the basic requirement. The state government of Rajasthan needs to gear up its facilities and prepare a good basic infrastructure that can bear the added load of visitors. This will include well lighted roads with international standard road signs, hotels that can cater to tourists with varying paying capacity and good food, besides toilet facilities by the roadside, development of proper sewerage system. It must be borne in mind that tourists come to places which are high on security criteria. Such provisions need to be developed and regularly maintained. At the same time domestic tourism must be promoted, and government shops selling local products at standard prices should be set up so that the tourists don't feel cheated. It is also pertinent to mention that the positive behavior of all stakeholders makes a lot of difference. So, people must try to go out of their way and help or guide visitors, whenever desired. Harassing or misguiding the tourists should be made a punishable offence. After all, the centuries old tradition of welcoming the 'Athithi' is our tradition and also good sense of responsibility is expected from travelers (Meeta Garg, 2021).

**Table 6. Major festivals in Rajasthan**

| Festivals | Place | Period |
|---|---|---|
| Camel festival | Bikaner | January |
| Desert festival | Jaisalmer | January-February |
| Baneshwar fair | Baneshwar | January -February |
| Gangaur festival | Jaipur, Bikaner | March-April |
| Mewar festival | Udaipur | March-April |
| Elephant festival | Jaipur | March-April |
| Urs Ajmer Sharif | Ajmer | July-August |
| Summer festival | Mount Abu | June |
| Teej | Jaipur | July-August |
| Marwar festival | Jodhpur | October |
| Ramdeoji fair | Ramdevra | July-August. |

## Challenges in Ecotourism

Some of the challenges of ecotourism include the need to address CO2 emissions, water consumption, landscape degradation, and biodiversity loss; besides inadequate infrastructure such as transport, lack of adequate sanitation facilities, disposal of campsite sewage in rivers, contaminating the wildlife, livestock and people who draw drinking water from it, deforestation, disruption of ecological life systems and various forms of pollution. There is also a challenge in understandissng the dynamics of the sustainability concept and its connection to human well-being.

***Ecotourism providers*:** Ecotourism certification schemes can also pose some challenges for ecotourism providers, such as high costs, complex procedures, and low recognition. Obtaining and maintaining certification can require significant financial, human, and technical resources, especially for small and medium enterprises. The process of certification can be lengthy, demanding, and inconsistent, depending on the scheme and the context. The benefits of certification can be undermined by the lack of awareness and demand from customers and intermediaries.

***Ecotourism customers*:** Ecotourism certification schemes can also pose some challenges for ecotourism customers, such as limited availability, confusing information, and green washing. Finding and booking certified ecotourism products and services can be difficult, especially in remote or emerging destinations. The information and communication about certification can be unclear, inconsistent, or misleading, depending on the source and the medium. The credibility and value of certification can be diluted by the proliferation and variation of schemes and labels.

***Ecotourism destinations*:** Ecotourism certification schemes can also pose some challenges for ecotourism destinations, such as low participation, weak enforcement, and unintended consequences. Encouraging and supporting ecotourism providers to participate in certification can be challenging, especially in the absence of clear incentives and regulations. Monitoring and enforcing the compliance and performance of certified ecotourism products and services can be costly and difficult, especially in large or diverse destinations. The success of certification can also create pressures and trade-offs for the destination, such as overcrowding, displacement, or elitism.

## Policy Framework on Ecotourism

In order to regulate and encourage planned development of areas in and around protected areas, the Ministry of Forest, Environment and Climate Change has recently notified "Ecotourism Guidelines in and Around Protected Areas 2021". These guidelines are based on the key recommendations of the Tiger Task Force (2005), provisions contained in the Wildlife (Protection) Act 1972, Scheduled Tribes and Other Forest Dwellers (Recognition of Forest Rights) Act 2006. Accordingly, a document on "National Strategy for Ecotourism 2022" has been formulated by Ministry of Ecotourism, Government of India. The vision of National Strategy for Ecotourism was to position India as a preferred destination for Ecotourism globally with the mission to develop an enabling ecosystem for the growth of Ecotourism in the Country and create synergies amongst the Central and State Governments, Local communities, conservation NGOs and the private sector for the development of Ecotourism in the country.

***Pillars of Ecotourism*:** The following strategic pillars have been identified for the development of Ecotourism: (i) State Assessment and Ranking (ii) State

Strategy for Ecotourism (iii) Information, Education and Communication (IEC), Capacity Building and Certification (iv) Marketing and Promotion (v) Destination and Product Development (vi) Public-Private and Community Partnerships and (vii) Governance and Institutional Framework.

***Guiding Principles for Ecotourism*** : The Ministry of Environment, Forest and Climate Change has laid down the following principles for the development of Ecotourism: (i) Adopt low-impact wildlife tourism that protects the ecological integrity of forest and wildlife areas, secures the wildlife values of the destination and its surrounding (ii) Highlight the biodiversity richness, their values and their ecological services to people (iii) Highlight the heritage value of India's wilderness and protected areas (iv) Build environmental and cultural awareness and respect (v) Facilitate the sustainability of Ecotourism enterprises and activities (vi) Provide livelihood opportunities to local communities (vii) Use indigenous, locally produced and ecologically Sustainable materials for tourism activities.

In addition to the above regulatory guidelines, the Strategy for Ecotourism will also include marketing and promotion, ease of travel, tourism infrastructure, ICT infrastructure, training and capacity building etc.

***Information, Education and Communication***: There is a need for a well-planned Information, Education and Communication (IEC) campaign to create awareness and understanding amongst all stakeholders. The campaign will have to be spearheaded by the Central Government and State Governments with the full support of the industry and other stakeholders including local government functionaries, forest and wildlife officials, local communities and non-government organizations working in the field.

***Digitalization and Branding of Certification Schemes***: The entire process of adventure and Eco-tourism certification shall be an end-to-end digital. The Certification will have attractive branding and it will be aggressively promoted by the Ministry of Tourism through all its channels including the incredible India website and mobile application.

## Conclusion

Globally, ecotourism has emerged as a sunrise sector and clutch the shape of an industry. However, the benefits of ecotourism extend far beyond the individual traveler and must reaches the communities that thrive on tourism, the ecosystems that need protection, and the cultures that deserve appreciation. Furthermore, ecotourism can be crucial in building a more sustainable future by creating sustainable livelihoods, preserving biodiversity and fostering environmental stewardship. There must be a balance between limits and usage so that continuous changing, monitoring and planning ensure that tourism can be managed. Horticulture is playing crucial role in development of ecotourism industry. The arid ecotourism has special significance in our country because of the vast land resource in form of deserts, large number of historical monuments, forts, museums, parks, lakes, hotels, variety of food and cloths etc. On the other hand, threats to these areas are rising due to increasing accessibility, overuse of nature to various developmental activities and increasing population pressure. Therefore, such fragile environments must be protected properly. Certification of ecotourism programme and use of digital techniques can give further impetus to this industry. Educating young minds about ecotourism and including in modern education course curricula are need of the hour.

## References

Anonyms 2021. Tourism and Sustainable Development Goals-Tourism for SDGs (http:// tourism4sdgs.org/tourism-for-sdgs/tourism-and-sdgs). Retrieved 10-1-2021.

Chandel Rajeev Singh, Kanga Shruti (2020). Sustainable Management Of Ecotourism In Western Rajasthan, India: A Geospatial Approach, GeoJournal of Tourism and Geosites Year XIII, vol. 29( 2), p.521-533 ISSN 2065-1198, E-ISSN 2065-0817.

Dorobantu, M.R. and Nıstoreanu, P. (2012). Rural Tourism and Ecotourism – The main priorities in sustainable development orientations of rural local communities in Romania. *Economy Transdisciplinarity Cognition*, **15**(1): 259-266.

Martha, H. (2008). Ecotourism and sustainable development: Who owns paradise? (Second edition). Washington, DC: Island Press ISBN 978-1-59726-125-8.

Meeta Garg (2021). Ecotourism for western Rajasthan, *International Journal of Education, Modern Management, Applied Science & Social Science*, **02**(II):193-196

National Strategy for Ecotourism 2022, pp. 1-39.

Randall, A. (1987). Resource economics (Second edition). New York, USA, Pub. John Wiley and Sons.

Kiper, T. (2013) Role of Ecotourism in Sustainable Development, http://dx.doi.org/ 10.5772/ 55749.

Weaver, D. (2008). Ecotourism (Second edition). Pub. Wiley, p.8. ISBN 978-0470813041.

# CHAPTER 28

# UNDEREXPLOITED VEGETABLES FOR NUTRITIONAL SECURITY AND PROSPERITY IN BUNDELKHAND REGION

**Ajeet Singh*, Himanshu Singh, M.K. Singh, Ashutosh Rai, A.C. Mishra, Jayshree Singh and S.V. Dwivedi**

*Department of Vegetable Science, College of Horticulture, Banda University of Agriculture and Technology, Banda-210001*

***Corresponding author**: ajeetsinghbuat@gmail.com

---

Underutilized vegetables are appraise as treasure of nutrients and antioxidants. Despite being highly nutritive with medicinal attributes and their ability to grow under adverse soil and climatic conditions, they have not received the desired attention. The ability of these environmentally fragile lands to generate high-quality horticultural produce has just come to light, and this has not only created opportunities for the local population to enhance vegetable production in new places but also for them to provide an economic means of subsistence. Bundelkhand region in Uttar Pradesh is blessed with a plenty of underutilized fruit and vegetable crops. Underutilized crops have an important role to play in satisfying the demand for nutritious, delicately flavored and attractive natural foods of high therapeutic value. The underutilized vegetables like spine gourd, ivy gourd, pointed gourd, moringa, etc. are the main sources of livelihood for the marginal and small farmers and play an important role in overcoming the problem of malnutrition. There is a need to create demand for underutilized vegetable crops in the domestic and international markets. This, to some extent, can be achieved through developing suitable processing and marketing strategies for these underutilized vegetables.

India's diverse climate ensures the availability of all varieties of fresh fruits & vegetables. It ranks second in fruits and vegetable production in the world, after China. As per National Horticulture Database (Second Advance Estimates) published by National Horticulture Board, during 2020-21, India produced 200.45 million metric tonnes of vegetables. The area under cultivation of vegetables was 10.86 million hectares. Although, diverse agro climatic conditions of India permit to grow more than 60 cultivated and about 30 lesser-known vegetable crops, not much attention has been given on underutilized vegetables known (Kumar *et al.*, 2018)

Underutilised crops are those species which are grown in local production systems, where they are highly adapted to a range of ecological niches (Broin, 2006). Bundelkhand region of Uttar Pradesh is blessed with plenty of underutilized vegetable crops such as ivy gourd, amaranthus, spine gourd, kulfa, wild green brinjal etc. There is need for alternative approaches to ensure food and nutrition security. Such solutions should be sustainable, resilient and of practical solutions to facilitate marginal farmers of Bundelkhand region in respect of water scarcity due to climate change and variability. As such, Underutilized crops may have evolved to become adapted to adverse environmental conditions such as drought stress. It is within this context that underutilized crops have a role to play as possible future crops. If indeed underutilized crops have evolved to become drought tolerant, they may have a role to play in guaranteeing future food security either directly as alternative crops in areas that are predicted to become drought-prone or indirectly as germplasm resources for crop improvement.

## Prospects of underexploited vegetable

Bundelkhand region of Uttar Pradesh is low in productivity with limited choices of crops because of its too hot and semi humid climate. In spite of the scarcity of water the region possess a rich diversity of many species of vegetables crops e.g. ivy gourd, kachri, spine gourd, purslane, pointed gourd etc. which are used as vegetable by the people of rural communities. Some of the minor vegetables are basically collected from the wild and eaten by the local inhabitants while few are underutilized which exhibit rich diversity, but are not commercially utilized. Cultivation of most of these crops is restricted to small farms, backyards or marginal land. Being wild, most of these vegetables possess high nutritional quality. An additional feature of these crops is that these can grow well in harsh climates and on the marginal lands without any special care. These crops have the potential to contribute not merely to agriculture biodiversity but most importantly to the livelihood of poor and generating income. Bundelkhand with its wide variability of climate and soil, is highly favourable for growing a large number of horticultural crops. It is the fastest growing sector within agriculture. It contributes in poverty alleviation, nutritional security and have ample scope for farmers to increase their income and helpful in sustaining large number of agro-based industries which generate huge employment opportunities. Underutilized vegetable crops can easily be grown under adverse conditions and are also known for their therapeutic and nutritive value and can satisfy the demands of the health-conscious consumers.

There is a need to create demand for underutilized vegetable crops in the domestic and international markets. This, to some extent, can be achieved through developing suitable processing and marketing strategies for these underutilized vegetables. Post-harvest losses of vegetables in developing countries are more serious than those in well developed countries. The total losses from harvest to the consumer point are as high as 35-40%. Underutilized crops have an important role to play in satisfying the demand for nutritious, delicately flavored and attractive natural foods of high therapeutic value. The underutilized vegetables like spine gourd, ivy gourd, pointed gourd, moringa, etc. are the main sources of livelihood for the marginal and small farmers and play an important role in overcoming the problem of malnutrition. They are in general accepted as being rich in vitamins, minerals, and dietary fibre and therefore, are an essential ingredient of a healthy diet.

## Improvement of underutilized vegetables

A study is being conducted at Banda University of Agriculture and Technology, Banda, to collect, conserve, evaluate and ethnobotanical studies with nutritional qualities, for the improvement of underutilized vegetables. Some valuable nutrients such as vitamin C, ß-carotene, dietary fibre, pectin, minerals, antioxidants, and anti-nutritional factors is being analyzed in all the mentioned crops in order to get their comprehensive nutritional profile. The information regarding medicinal as well as household uses of underutilized fruit and vegetable crops is being obtained from folklore through interview schedule method and is being properly documented. Eventually the screened crops based on their nutritional profile is being recommended for their post-harvest processing treatments and development of various valued added products. Thus, the development of such products may help the consumers to harness the nutritional and medicinal properties of such underutilized/unexploited crops and may play pivotal role in providing Nutritional and Economic Security to the local people of the Bundelkhand region of U.P. by giving employment and by fetching good returns from their sale in raw form as well as value-added products.

## underexploited vegetable crops

### *Momordica dioca* (spine gourd)

The spine guard is a summer loving crop and required an optimum temperature of 27 to 33 °C for ideal growth and development. It is also known as kakrol or spine gourd belonging to the family Cucurbitaceae. It has high nutritional and medicinal with economic values. It is cultivated for its fruits, which is used as a vegetable has a high demand in the market because of nutritional, therapeutic value, long shelf-life, suitable for long-distance transportation. Young leaves, flowers and seeds are also edible. Fruits are also used in ulcers, piles, sores, and obstruction of liver and spleen. It also possesses several medicinal properties and is good for those suffering from cough and indigestion (Rai et al., 2005).With the development of different varieties of spine gourd for cultivation, the major bottlenecks in popularization of spine gourd cultivation is the lack of true-to-type planting materials, seed multiplication technique and their availability in the market.

**Pointed gourd (*Trichosanthes dioica*)**

This perennial plant grows best in warm to hot humid climates and does not tolerate frost & extreme cold conditions. Dry tropical weather after monsoon is preferred. The pointed gourd plants require temperature is 30 C to 35 C for optimum growth and yield. The water stress can be adaptable by the plant. Pointed gourd is known as 'King of gourd'. It is also known as Parwal, Pravar in Hindi.

In India, it is often called Green potato. Its leaves are very nutritive and used leafy as vegetable. It is rich source of carbohydrates, Vitamin A, Vitamin C and minerals. Its juice is used as tonic, alopecia and subacute cases of enlargement of liver. It is dioecious climber with perennial root stock. There is scope and need for promoting pointed gourd cultivation as an alternative source of income for small farmers in the Bundelkhand region.

**Kachri (*Cucumis callosus*)**

Kachri is a drought hardy and high temperature tolerant crop (up to 46 °C), found in the arid zones during rainy season. Unripe fruits are bitter in tase but at ripening the become edible. The mature fruits are usually cooked with various vegetable preparation, dried chips, chutney and pickles and are also used for garnishing vegetables or as salad. Kachri can be cultivated successfully both during rainy and spring summer seasons. It requires warm and dry weather with plenty sunshine for growth and production. The optimum temperature for seed germination 20-22 °C whereas, 32-38 °C is optimum for vegetative growth and fruit setting. Its growth is severely retarded below 12 °C and plants are killed instantly by frost or at temperature below 4 °C during severe winter.

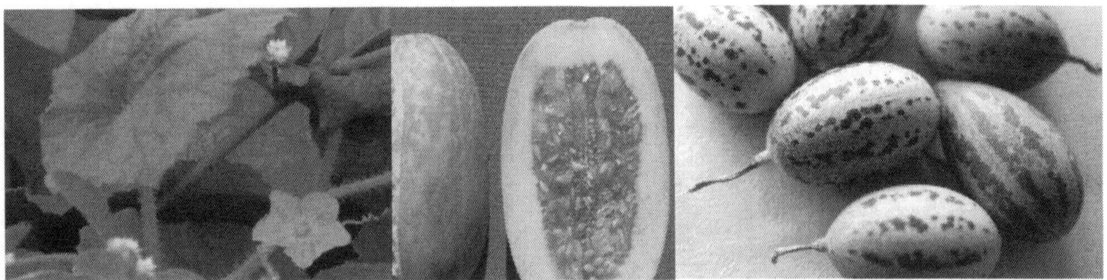

**Snapmelon (*Cucumis melo var. momordica*)**

It is important vegetable of hot humid region is popularly known as 'phoot'. A great extent of genetic variability exists in India with respect to vegetative growth, quality attribute and resistance to biotic and abiotic stresses. Its immature fruit are used as salad, vegetables and for other culinary preparation.

**Purslane (*Portulaca oleracea* L.)**

The plant *Portulaca oleracea* L. is a succulent summer annual plant, cosmopolitan in distribution, it occurs in riverine, riverbed, gardens, and dumpsite and prefers sunny and warm areas of Bundelkhand. It can be easily grown in warm moist places during the summer and spring season.

This species has heat, temperature, salt, and variable pH tolerance capacity so this can be well adapted to varied land conditions. This plant does not require a lot of agronomic expertise and can be easily grown compared to other nutritive plantspecies, therefore, this will be an ideal substitute for resource-poor agriculturalists in developing areas like Bundelkhand. Purslane is rich in beta-carotene, folic acid, vitamin C and essential fatty acid. 300-400 mg of omega-3 fatty acids, 12.2 mg of alpha- tocopherol, 26.6 mg of ascorbic acid, 1.9 mg of beta-carotene, and 14.8 mg of glutathione ((Simopoulos *et al.*,1992). This plant is recorded as one of the most valuable restorative plants and named "Global Panacea" by World Health Organization. $C_4$ type of photosynthetic metabolism occurs in Purslane that helps this plant to survive in the heat and intense sunlight stress conditions. It was reported by Lara *et al.* (2003) that Crassulacean acid-like metabolism induced after 21 days of drought stress in *P. oleracea*. Since ancient times in Egypt and England, this plant is used as a vegetable, spice, and medicinal plant.

**Amaranthus (*Amaranthus viridis*)**

Amaranth is distributed in the warmer parts of the world.In addition, the whole plant possesses analgesic and anti-pyretic properties and is used for the treatment of pain and fever respectively in traditional systems of medicine (Eluwa, 1977). Leaves are directly used to eczema, psoriasis and rashes etc. Other traditional uses range from an anti-inflammatory agent of the urinary tract, diuretic, venereal diseases vermifuge, analgesic, anti-rheumatic, antiulcer, antiemetic, improvement of appetite, laxative, anti lep-rotic, treatment of respiratory, eye problems and treatment of asthma.

**Snake gourd (*Trichosanthes cucumerina*)**

The plant does well in the humid low land tropics. Characteristically, snake tomatoes are annual tendril-climbers, which bear fruits 2-4 months after planting. It bears long, narrow and cylindrical fruits giving the shape of a snake. The pulp is very reddish in colour and can be used to improve the appearance of food as it can be blended and used to produce a paste for stew which tastes like, and serves the role of tomatoes (Enwere, 1998).The higher ascorbic acid content in snake gourd than that of tomato varieties suggest the possibility of utilizing snake gourd in the industrial production of tomato paste and puree.

**Ivy gourd (*Coccinea grandis*)**

It thrives best in humid and warm climatic conditions. The ideal temperature range of 20 C to 32 C is best for its growth, quality and good yield. This vegetable can be cultivated throughout the year. The tender green fruits are nutritious and are good source of protein, calcium, fiber and beta carotene, Vitamin-A. Fruits of *Coccinia grandis* contains, lupeol, taraxerol, β carotene, lycopene, cryptoxanthin, xyloglucen, carotenoids and β-sitosterol. Every part of this plant is important in medicine of as it is used in treatment of various skin diseases, bronchial catarrh, bronchitis and Unani systems of medicine for ring worm, psoriasis, small pox, and scabies and other itchy skin eruptions and ulcers (Deshpande, 2011).

**Water Spinach (*Ipomoea aquatic*)**

Water spinach (*Ipomoea aquatica*) is one of the important leafy vegetable grown in several parts of the country. The leaves of water spinatch is good source of minerals and vitamins especially carotene. Leaves are heart-shaped and 20-30 cm long. ICAR-Indian Institute of Vegetable Research, Varanasi is maintaining several diverse forms of water spinach. Water spinach is commonly grown in waterlogged areas, which makes the plant protection practices and harvesting cumbersome. ICAR-IIVR, Varanasi has made efforts for cultivation of water spinach in upland field condition and promising results were obtained for the same. Leaves are usually sold in the bunches of 500 g in the markets at the rate of ₹ 35- 40/Kg. In view of the importance, popularization of water spinach on large scale under upland field condition could make significant contribution towards economic upliftment of the society along with ensuring nutritional security.

## Challenges

- Lack of awareness among the farming community about the nutritional and medicinal value of underutilized Vegetable crops.
- Lack of awareness about nutritional and medicinal aspects among stake holders
- Lack of desirable seeds and planting material.

- Limited application of advance on-farm agro techniques.
- Lack of application of innovative and novel technologies such as biotechnology, plasticulture for enhancement of productivity.
- Lack of about post-harvest management practices.
- Limited and inadequate marketing supports & infrastructure facilities for transportation, storage and processing.
- Poor recognition of these crops in horticulture promotion programmes.
- Improper institutional arrangements and limited role played by financial Institutions in setting up of agro industrial and horticulture based industrial Units.

## Strategies for development -

Exploration of genotypes of underutilized vegetables is being done from Bundelkhand region. Simultaneously the collection of these vegetables will also be attempted from gene banks, institutes and farmers fields from all over the country to enrich the germplasm. Their ethno-botanical information will also be collected through survey and questionnaires. Evaluation and characterization is being done based on fruit characters, yield, nutritional qualities and processability.

Screening of germplasm is being done based on plant parameters, yield, fruits parameters, nutritional parameters, phytochemical profile and process ability. Phytochemical and nutritional profiling of the selectedgermplasm of above-mentioned crops is being done.

Value added products from the given crops is being developed and evaluated for physical, nutritional, functional,and sensory parameters. The storage life of the fresh as well as developed products will also be evaluated.

## Conclusion

Bundelkhand contains a wide variety of underused vegetables that are rich in nutrients and have medicinal benefits, as well as the ability to withstand harsh climatic conditions. The underutilised vegetables are still neglected due to its incomplete knowledge of their package and practices, a lack of planting materials, technical know how about production systems and marketability etc. To ensure future food and nutritional security, it is vital to boot up a programme on genetic resource investigation, management, nutritional profiling, usage, and enhancement of these underutilised vegetable crops. Additionally, the cultivation of underutilised vegetables will address the lack of per capita consumption availability, addressing the nutritional gap, while also creating jobs, raising rural residents' incomes, and possibly even boosting the country's economy.

## References

Deshpande S V. 2011. In vitro antioxidant study of petroleum ether chloroform and ethyl acetate fractions of *Coccinia grandis* stems extract. *International journal of chemical sciences* **9**: 80-86.

Eluwa M C. 1977. Studies on *G. rhomboidalis* (Boheman) (coleoptera: Curculionidea), a pest of *African spinach. Journal of Natural History* **11**: 417 – 424.

Enwere N J. 1998. Foods of Plant Origin. *Afroorbis Publications*, Nsukka: 153-168.

Kumar R, Rajashree V, Sagar L, Tripura U, Karthick K, Karthick R, and Prasath G. 2018. Underutilized vegetables as rich source of medicinal value: A boon for the Pharmaceutical industries and farmer income. *International Journal of Chemical Studies* **6**: 3320-23.

Lara M V, Disante K, Podestá F E, Andreo C S, and Drincovich M F. 2003. Induction of a Crassulacean acid like metabolism in the C4 succulent plant, *Portulaca oleracea* L.: physiological and morphological changes are accompanied by specific modifications in phosphoenolpyruvate carboxylase. *Photochem Res* **77**: 241–54.

Rai N, Asati B S, Patel R K, Patel K K, & Yadav D S. 2005. Underutilized Horticultural Crops in North Eastern Region. *ENVIS Bulletin: Himalayan Ecology* **13**: 10-16.

Simopoulos A P, Norman H A, Gillaspy J E, and Duke J A. 1992. "Common purslane: a source of omega-3 fatty acids and antioxidants," *Journal of the American College of Nutrition* **11**: 374-82.

# CHAPTER 29

# POTTED ORNAMENTALS: AN APPROACH FOR AIR QUALITY, HEALTH AND HAPPINESS

**Alka Singh and Shailesh Chaudhary**

*Department of Floriculture and landscape Architecture, ASPEE College of Horticulture, Navsari Agricultural University, Navsari, Gujarat*

## INTRODUCTION

Flowers and foliage have integrated relationship with mankind since ancient times as they are entwined with human culture. Flowers depict nature's true charisma, representing live beauty, focusing on aesthetic and positive side of life and have been associated with progress and prosperity from ancient time. Vital role of ornamental plants directly on environment, psychological and emotional status and indirectly on human feelings associated with work performance and work productivity has been recognized. Rapidly growing urbanization and population explosion with hasty life style have created space paucity for living, wherein pot plants offer some relief with great scope for urban green scaping. Consequently, the potted plants are becoming increasingly popular throughout the world. The segment of ornamental plants has been increasing in leaps and bounds in the world flower trade which comprised $ 24.1 billion including fresh cut and dry flowers, ornamental plants, flower bulbs and cut foliage (OEC World statistics, 2022). The global floriculture market size was valued at USD 53.6 Billion in 2022. The market size is growing at a CAGR of 6.6 % and is expected to reach USD 101.9 billion by 2032 as per the statistics of Spherical Insights.

People living in urban areas spend up to 85-90 % of their time indoors, without being aware of continuous exposure to indoor air pollution, as per the report of the United States Environmental Protection Agency (USEPA), 2023. According to the WHO, indoor air pollution has been ranked among the top five risks to public health. Around 3.11 million people died prematurely from illness attributable to household air pollution, in 2021 as per the report of the Institute for Health Metrics and Evaluation (IHME). Indoor air pollutants primarily originate from building products, human activities inside the building and infiltration of outdoor air. Indoor air pollutants include volatile organic compounds (VOCs), particulate matter, ozone, and biological contaminants.

Plants have the capacity to purify, revive and freshen up air in our homes and offices, shielding us from the negative effects of such common pollutants like ammonia, formaldehyde and benzene. Plants remove VOCs (Volatile Organic Compounds) from indoor air through stomatal uptake, absorption, and adsorption to plant surfaces.

Several indoor species have been screened for their ability to remove benzene, in addition to other VOCs (e.g., toluene, TCE, m-xylene, hexane). The efficiency of VOC removal varies substantially among species and with the molecular characteristics of each compound.

Besides, ornamental plants add freshness and cheerfulness, reduce allergies and act as stress buster. Viewing green plants and their proximity at home or office provide ample involuntary attention, increase positive feelings, block or reduce worrisome thoughts and promote restoration from stress. Further, some ornamental flowering plants are edible and consumption in form of petals teas, garnishing or in desserts adds vibrancy to the cuisine and serve to boost immunity and health. Living and working in an environment having potted ornamentals in the surroundings is refreshing and energies human force to work with more optimistic approach.

## Ambient Air Quality in the Present Era

Urban life is constantly deteriorating with exploding problems of atmospheric pollution and massive population explosion, resulting in a disastrous effect on our life style, health and psyche. Rapid urbanization accompanied with degraded urban environment and population explosion estimated at 1.42 billion inhabitants in India, account to approximately 17.31 % of the global population. The rapid urbanization that has accompanied this demographic shift poses significant challenges in the realms of environmental sustainability. Thus, rapid urbanization and growth of large cities with deforestation has been accompanied by expansion of highly vulnerable urban communities living in informal settlements, creating high environmental pressures with poor air quality indoors as well as outdoors.

A high proportion of the population and economic activities are at risk from climate change are in urban areas and a high proportion of global greenhouse gas emissions are generated by urban-based activities and residents. Suspended Particulate Matter (SPM) is the main problem of urban air pollution in India, caused mainly due to fast growing population of automobiles and poor traffic control, congested roads as well as factories in the peri urban areas. The most watched pollutants include particulate matter (PM), Nitrogen dioxide ($NO_2$), Sulphur dioxide

($SO_2$) and Carbon dioxide ($CO_2$), which are continuously degrading the ambient air quality in major cities in India. The annual average concentration is very high in Indian cities. Many cities like Delhi, Agra, Faridabad *etc.* have exceeded the officially designated critical levels of ambient air quality standards set by the World Health Organization (WHO) owing to high suspended particulate matter (SPM). The use of pesticides, though the most important factor in improving agricultural production to sustain the greater supply of food, but has been become hazardous owing to remarkably higher use with less than 0.1 per cent of pesticides reaching the targeted pest and the remainder adversely affecting human life and environment.

Indoor pollution refers to the physical, chemical and biological characteristics of air in the indoor environment within a home, building or an institution or commercial facility. Indoor air pollution is of greater concern in the highly developed urban vicinity, where energy efficiency improvements sometimes make indoor environment in houses as well as offices relatively airtight, reducing ventilation and raising pollutant levels. Indoor air pollution is one of the world's largest environmental problems. Indoor air pollution is caused by the inefficient use of solid fuels for cooking and cleaning. Around 47 % of the world's population i.e. about 3.6 billion people are exposed to household air pollution from the burning of solid fuels for cooking (coal, crop waste, charcoal, wood, among others) in open fires and leaky stoves (IHME) and as per a WHO report, smoke inhaled by women from unclean fuel is equivalent to burning 400 cigarettes in an hour.

In urban areas, exposure to indoor air pollution has increased due to a variety of reasons, including the construction of more tightly sealed buildings, reduced ventilation, the use of synthetic materials for building and furnishing and the use of chemical products, pesticides, and household care products. Indoor air problems can be subtle and do not always produce easily recognized impacts on health. Although hundreds of separate chemical agents have been identified in the smoke from biofuels, the four most serious pollutants are particulates, carbon monoxide, polycyclic organic matter, and formaldehyde (carpets, particle boards and insulation foams). Besides, these Nitrogen dioxide, Carbon monoxide, and lead (Pb) are a number of other pollutants that affect the air quality in an enclosed space. In buildings having lack of ventilation, asbestos and radon are also a cause of pollution. Many offices and new homes have tighter construction, sealing the building from the outside air that further raises the indoor air pollutants to dangerous levels.

## Pollutants – Ambient Air

Major pollutants of ambient air are mentioned below:

*Combustion Products:* In general, combustion or burning is a chemical reaction in which a fuel combines with oxygen. When there is insufficient oxygen, the fuel burns incompletely and produces by products of combustion. Examples of such by products are carbon monoxide, nitrogen dioxide, carbon dioxide and respirable particulates which are generated from automobile exhausts, furnaces, water heaters, space heaters, scented candles, oil lamps, wood stoves, fireplaces and gas stoves.

*Volatile Organic Compounds (VOCs):* VOCs consist of a large number of organic substances with diverse physical, chemical and biological properties, which will volatilize at normal room temperatures. Concentrations of VOCs in indoor air has been observed to be 2 to 5 times greater than in outdoor air and during and for several hours immediately after certain activities, such as paint stripping, levels may be 1,000 times background outdoor levels as per the report of The United States Environmental Protection Agency (EPA). House hold products like paints, paint strippers and other solvents, wood preservatives, aerosol sprays, cleansers and disinfectants, moth repellents and air fresheners, stored fuels and automotive products, hobby supplies, dry-cleaned clothing, pesticide and other products like building materials, furnishings and furniture coatings. Further, office equipment such as copiers and printers, correction fluids and carbonless copy paper, graphics and craft materials including glues, adhesives, permanent markers and photographic solutions contribute to the VOCs built up in indoors. There are different types of VOCs released through different sources, although each compound is likely to be present in very low concentrations but amalgation can result into more dangerous levels. Application of pesticides in indoors is highly dangerous as they last for many years in absence of sunlight specially in carpets, sofa seats, curtains. Human health is adversely affected with such VOCs exposure such as conjunctive irritation, nose and throat discomfort, headache and sleeplessness, allergic skin reaction, nausea, fatigue and dizziness. There are dangerous effects of long term exposure like loss of coordination, leukaemia, anaemia, cancer and damage to liver, kidney and central nervous system.

*Second Hand Smoke (SMS):* Passive smoking is the inhalation of smoke called Second Hand Smoke (SHS) or Environmental Tobacco Smoke (ETS) by persons other than the intended "active" smoker. Second Hand Smoke (SHS) is classified as a "known human carcinogen" by the International Agency for Research on Cancer, a branch of the World Health Organization. Tobacco smoke is a mixture of gases and particles. It contains more than 7,000 chemical compounds. More than 250 of these chemicals are known to be harmful and at least 69 are known to cause cancer. Long term exposure to SHS may increase the risk of lung cancer, cardiovascular diseases and respiratory diseases.

*Formaldehyde:* It is a colourless gas with a pungent odour that comes mainly from particle board, pressed wood, foam insulation, paper bags, waxed papers, facial tissues, stiffeners and wrinkle resisters, water repellents, fire retardants, binders in floor coverings, carpet backing, permanent press clothes, natural gas, kerosene and cigarette. It causes irritation of mucous membranes of the eyes, nose and throat, those allergic contact dermatitis, respiratory problems, eye irritation, headaches, asthma and carcinogenicity to the throat.

*Bioaerosols:* A bioaerosol is an airborne product of a biological contaminant. Biological contaminants in the home may include mold, bacteria, viruses, mites, and pollen. COVID -19 was most dangerous example of bio-aerosol. They can also cause allergic reactions, including hypersensitivity pneumonitis, allergic rhinitis, and some types of asthma. Symptoms caused by biological pollutants include sneezing, eye irritation, coughing, dizziness and respiratory infections. Lack of ventilation along with wet or moist building materials (carpeting, ceilings and walls) and poorly maintained humidifiers, dehumidifiers and air conditioners contribute to the growth and spread of biological contaminants.

*Asbestos:* It is often used in shingles, fireproofing, heating systems, floor tiles, and ceiling tiles. When asbestos containing material is damaged or disintegrates, microscopic fibres are dispersed into the air. Inhalation of these microscopic asbestos fibres may lead to lung cancer or mesothelioma.

*Radon:* It is an invisible, radioactive atomic gas that results from the radioactive decay of radium, which may be found in rock formations beneath buildings or in certain building materials themselves. Due to modern houses having poor ventilation, it is confined inside the house. Radon is a heavy gas and tends to accumulate at the floor level. The half life for radon is 3.8 days, indicating that once the source is removed, the hazard will be greatly reduced. After cigarette smoking, radon is the second leading cause of lung cancer.

*Particulate Matter (PM):* It is a complex mixture of solid and liquid particles of organic and inorganic substances suspended in the air. The major components of PM are sulphate, nitrates, ammonia, sodium chloride, black carbon, mineral dust and water. The most health damaging particles are those with a diameter of 10 microns or less, which can penetrate and lodge deep inside the lungs. Scientific studies have linked exposure to high concentrations of some types of PM with a variety of problems, including irregular heartbeat, aggravated asthma, decreased lung function, increased respiratory symptoms, such as irritation of the airways, coughing or difficulty in breathing, non fatal heart attacks and premature death in people with heart or lung disease.

## Potted Ornamentals for Indoor Air Quality

Indoor air pollution is a significant health risk, causing millions of premature deaths globally. However, plants have the capacity to purify, revive and freshen up the air in our homes and offices, shielding us from the negative effects of common pollutants like ammonia, formaldehyde, and benzene. A variety of ornamental plants having vibrant colours foliage can be grown as house plants. Colourful and lush green potted ornamentals can be well placed in window-box gardening, roof gardening, balconies and inter spaces with the home. They not only add beauty and biological comfort but also improve the aesthetic outlook.

Houseplants can purify and revitalise air in our homes and offices, protecting us from the negative effects of such common toxins as ammonia, formaldehyde and benzene. Plants remove VOCs (Volatile Organic Compounds) from indoor air through stomatal uptake, absorption, and adsorption to plant surfaces. Several indoor species have been screened for their ability to remove benzene, in addition to other VOCs (e.g., toluene, TCE, m-xylene, hexane). The efficiency of VOC removal varies substantially among species and with the molecular characteristics of each compound. Research has also revealed significant decrease in the levels of $CO_2$ and CO in indoor environment under AC as well as non AC conditions with the placement of potted indoor plants. Plants like *Aglaonema brevispathum*, *Pachira aquatic* and *Ficus benjamina* reduced the levels of VOCs like methane, toluene, ethyl benzene and formaldehyde. Many indoor plants species *viz.*, *Asplenium nidus*, *Ficus elastica*, *Hedera helix*, *Nephrolepis obliterata*, *Sinningia speciosa*, *Ocimum basilicum*, *Yucca massengena*, *Spathiphyllum wallisi*, *Dieffenbachia* spp., *Rhapis excelsa* can effectively reduce the VOCs (*viz.*, HCHO, Toluene, Xylene, Acetaldehyde etc.) and $CO_2$ from indoor air as well as increase aesthetic value.

Further plants like succulents *i.e. Z. zamifolia*, Haworthia, Sansevieria, *etc.* having CAM metabolism are highly beneficial for improving indoor air quality with the added benefit of requiring less water. Thus, plants in the home or office can filter out toxins, stale air, pollutants, harmful viruses and mould spores. Plants act as filters because they absorb the toxins through their leaves, especially those with the largest leaves.

## For Health

People in a natural environment having plants and flowers in the surroundings not only show faster physical recovery from stress, but also improve psychological, emotional and cognitive health. Viewing nature/ plants is linked to positive health outcomes of individuals, such as in pain reduction, less need for analgesics and a quicker recovery from surgery as revealed from scientific studies.

Research carried out at Kansas State University reveals that addition of plants in hospital rooms hastened recovery rates of surgical patients. Compared to patients in rooms without plants, patients in rooms with plants request less pain medication, had lower heart rates and blood pressure, experience less fatigue and anxiety and were discharged from the hospital sooner.

Ornamental plants directly or indirectly have health promoting benefits.

*Stress buster*: Proximity to live greenery makes us feel more at ease with our surroundings. We experience less stress when there are plants around us. House plants can contribute to a feeling of wellbeing, induce calmness and optimism.

*Relief from headache problems:* Pollution and excess carbon dioxide contribute to headache problems. Indoor plants generate oxygen, increasing its level in air and lower the carbon dioxide level. This reduces the possibilities of headache.

*Reduces allergies:* Pot plants in indoors can reduce allergies to some extent by increasing humidity levels and decreasing dust as plants are a natural humidifier and contribute in maintaining moisture content in the indoor atmosphere that relieves from dry and itchy skin.

*Concentration and Focusing:* Students have shown an increase in concentration and focusing with higher attendance in classrooms with indoor plants.

*Sick Building Syndrome:* Plants improve the indoor environment and are useful in fighting the modern phenomenon of Sick Building Syndrome (SBS) that results from poor ventilation, excessive disturbing noise, high VOCs with other pollutants, limited visibility of outdoor view *etc.*

Some of the potted ornamental plants *i.e.* Areca Palm, *Aloe vera*, Aglaonema, Dieffenbachia, *Chlorophytum, Hedera helix*, Ferns, Golden Pothos, Song of India, *Dracaena*, Peace Lily, *Sansevieria* and *Zamioculcas zamiifolia* improve ambient air quality and thus contribute to better health. They are suitable for hospital environments to improve air quality, making it cleaner and more breathable. Lush green foliage plants like Z. *zamiifolia*, Alpinia, Monstera, Aspidistra, Dieffenbachia, Rubber plant, Pothos, *Chlorophytum*, and Palms break the monotony of sick building syndrome in an enclosed room and freshness and make the environment live and healthy. White fragrant flowering plants, including Peace Lily, Jasmine, Clematis, and Gardenia, are known for emitting a soothing fragrance that promotes a sense of peace and calmness. Bright flowering plants like chrysanthemum, aster, rose, hibiscus, adenium, golden trumpet, portulaca *etc.* placed near windows for visual effect can induce positive vibes for faster health recovery.

## Edible and Medicinal Use

Flowers with their bright sparkling colours, also possess the ability of adding flavour, colour, nutrition and even medicine to our food. Flowers, the beautiful gift of nature have now attracted the scientists' attention for the presence of a large amount of phytochemicals like natural pigments viz., anthocyanins, carotenoids, flavonoids, etc. with broad bioactive edible potential and health benefit. Flowers can thus, serve as food for the body and soul and are recently gaining significance even in the culinary arts with medicinal use too. The most common application of flowers in human nutrition is the preparation of hot beverages (tisane or infusion), giving several types of flower teas, which are drunk with the aim of providing wellness due to the medicinal properties of each kind of flower. From a dietary point of view, a great advantage of teas made with edible flowers is the fact that they do not contain caffeine, whereas many types of tea contain stimulant drugs such as the alkaloid xanthene caffeine and theobromine. Studies reveal, flowers possess phytochemicals that can serve as anti-anxiety, anti-cancer, anti-diabetic, anti-inflammatory, anti-oxidant, diuretic, anthelmintic, immune-modulatory and anti-microbial properties. Pigments present in flowers play a major role to boost immunity, scavenging free radicals and toxins from the body, and helping us combating with major diseases. Flowers of rose, marigold, hibiscus, *Antigonon leptopus*, calendula, *etc.* possess high antioxidant compounds. Marigold and Calendula have been known for beneficial effects on vision and skin. The flowers of *Hibiscus rosa-chinensis, Geranium*, petunia, calendula, rose, lavender, sun flower, *etc.* are found to be high in DPPH and FRAP value, which could be applied as potential rich resource of natural antioxidants and colorants for functional foods. The bright pigments present in flowers forms the basic health boosting factor. Flowers are rich in nectar and pollen, and some are high in vitamins and minerals. Rose hips are very high in vitamin C, marigolds and nasturtiums contain vitamin C and dandelion blossoms contain vitamins A and C. Chrysanthemum flowers contain vitamin A, B and amino acids. Nasturtiums also supply small amounts of vitamin D. In edible flowers, an increased content in the level of flavonoids, phenols and pigments contribute to their high antioxidant activity. These flowers can be well grown in pots to serve health with beautiful and bright ambiance.

## For Happiness

Happiness is well associated with flowers and plants as people involved in gardening or having hobbies of flower arranging, bonsai making, home gardening and kitchen gardening are healthier and happier. The presence of flowers triggers happy emotions, heightens feelings of life satisfaction and affects social behaviour in a positive manner. Researchers who have assessed the impact of nature and plants on human health have suggested that people-plant/flower interactions provide physiological stress reduction. Flowers have a long-term positive effect on the mind frame.

## Psychologicals for Mental Health

Nature has a restorative impact on psychological well-being, reducing stress and enhancing mental health. Greening of urban spaces offers psychological and emotional benefits for the mental well-being of urban residents. Various sensory elements of flowers like fragrance, colour, texture, softness, freshness, *etc.* all directly influence human psyche and awaken his senses, making him sensitive and aware of the existing wonderful nature. Flowers have been termed to act as "super stimuli," and directly affect human psyche through multi-channel sensory interactions, according to a study at Psychology department, Rutgers, the State University of New Jersey, US. Influence of colours on human mind and health has long been studied, wherein warm colours like red and orange are known to evoke emotions ranging from feelings of warmth and comfort to feelings of anger and hostility while colours on the blue side of the spectrum are known as cool colours often described as calm. Further, different colours are associated with different feelings like red stimulates and excites brain-wave activity, pink has a soothing - tranquillizing effect, orange reduces fatigue, yellow is a memory stimulator while green marks spring and new beginnings and white represents peace. Similarly, scent has also been shown to have deep psychological influence. Smells can evoke strong emotional reactions as our olfactory receptors are directly connected to the limbic system, the most ancient and primitive part of the brain, which is thought to be the seat of emotion. Pleasant fragrances have been found to have positive effects on moods in all age group. Thus, vibrant colours of flowers, delicate texture and the floral scent tends to influence human psychology.

Red flowers of poinsettia, rose, adenium, etc adds excitement and energy, orange flowers of football lily, crossandra, nastrutium reduces fatique, pink flowers of rose,petunia, phlox induces soothing effect, while yellow flowers of sunflower, chrysanthemum, allamanda adds charm and freshness. Plants like the Sansevieria and crassula is known to have a swathed aura of good luck and positivity around it. Rubber plant is known to bring in good luck and adds a relaxed aesthetic charm by creating an effect of lightness and dilution in spaces. Peace lily brings purity through colour of its leaves and flower represents calmness, healing hope and prosperity. Lucky bamboo symbolizes prosperity and good luck. *Ocimum sanctum* (*tulsi*) is also known to have a positive vibration in the house. Lavender is known for its calming and relaxing aroma; lavender can help reduce anxiety and stress. Jasmine with its sweet and pleasant scent improve mood and alleviate stress. Rosemary a fragrant herb enhances concentration and cognitive function. Uplifting scent of geranium can help in reducing fatigue and promoting emotional wellness. Sunflowers with their bright and cheerful appearance can bring joy and positivity, uplifting mental well-being.

Interacting with the leaves and flowers of a plant can be a delightful sensory experience that uplifts your mood and contributes to your overall well-being. There are several plants with unique leaves and flower textures that provide a wonderful tactile experience. The African violet, with its heart-shaped leaves, offers a soothing and soft touch. The geranium, with its soft petals and fuzzy leaves, also provides a delightful tactile sensation. Plants such as acalypha, with its rough-textured leaves, offer a diverse tactile experience. The strawflower has a glossy, stiff, and papery texture, adding to its appeal. Succulents, with their thick and soft leaves, provide a pleasingly soft touch. Including plants with a variety of textures in leaves and flowers in your surroundings or in sensory activities can potentially enhance your mental well-being.

## Positive Emotions

The sight and fragrance of flowers combine to influence brain chemistry and stimulate more positive feelings, as reported by Dr. Jeannette Haviland-Jones, professor of psychology and director of the Human Emotions Lab at Rutgers University, New Jersey, US. Researchers who have assessed the impact of nature/plants on human health have suggested that people-plant/flower interactions provide physiological stress reduction. Flowers have a long-term positive effect on the mind frame. Flowers with sweet fragrance, colour and soft texture stimulate and awaken all the five senses in humans and thus significantly influence psychology to greater extent that serve as mood enhancer and stress buster.

Flowers display a strong effect on immediate emotional behavioural expression and display positive emotion in human beings. Presence of flowers, plants and greenspaces trigger happy emotions with feelings of life satisfaction that ultimately improve work performance. Further, it has been found that workers idea generation, creative performance and problem solving skills improved substantially in workplace environments that included flowers and plants.

Thus, Flowers contribute in developing congenial environment for facilitating intimate connections and the presence of flowers lead to increased contact within family and friends. Further, studies have revealed positive effects of the flowers not only on receivers but givers also. Flower givers are perceived as happy, capable, and emotionally intelligent and make impression of being compassionate, nature lover, with high expressive and aesthetic sense. Flowers can thereby, be used to influence and change the thought process in a significant way that can be exploited for enhancing personal as well as even business relationships. Potted ornamentals like Peace lily, Rose, Lilies, Poinsettia, Chrysanthemum, Calendula, Bougainvillea with vibrant flower colours displayed just near the doors, balconies or window boxes trigger a positive psychology. Rose flowers are particularly popular in inducing positive effect on people, wherein the yellow flowers were found to be more efficient in improving feelings of relaxation compared to the red and white flowers.

Flower meanings and symbolization have been the key elements for flower choice since time immemorial as they express the most sensitive, delicate and ever loving feelings eloquently. Flowers of different colours carry special meanings and emotions. Red roses, carnations, and tulips symbolize the expression of love. While, pink flowers like carnations, roses, tulips, lilies convey feelings of gratitude and appreciation. Whereas, roses, chrysanthemums, carnations, and lilies having white coloured flowers denote purity, truth and innocence. Understanding the symbolism behind each colour adds depth and significance to the act of giving and receiving flowers.

Besides, flower also adds meaning to the actual life and reinforces faith in nature by filling the whole atmosphere with its freshness, fragrance and daintiness. The enchanting and pleasant fragrance from the flowers of jasmine, peace lily, fragrant rose, gardenia, *Murraya exotica* (*Kamini*), Clematis, night jasmine (*Cestrum nocturnum*), coral jasmine (*Nyctanthus* arbor-*tristis*) helps in reducing stress and anxiety levels inducing peacefulness and improving sleep. Visually appealing vibrant flowers like Bougainvillea, Mussaenda, crepe jasmine (*Tabernaemontana*), symbolize passion and liveliness in surroundings. Flowers like chrysanthemum, lisianthus, gypsophila ignite creativity and uplift mood state.

## Trends in Potted Ornamentals

### Container Gardening

Container gardening has emerged as a ground breaking method for cultivating plants within confined spaces. This contemporary approach has witnessed a surge in popularity, particularly among urban residents facing spatial constraints. The portability of containers enables gardeners to modify the layout effortlessly and relocate plants as needed.

*Selection of containers:* containers were selected as per the suitability of the plant's need, environment, cost and availability, durability. There are various types of containers available in the market, including those made up of plastic, fibreglass, terracotta, ceramic, concrete, metal, wood, jiffy pots, cow dung, coir baskets *etc*. Various shapes of attractive containers, along with stands, can create beautiful views. Different shapes and sizes of containers like Some deep and narrow, others shallow and wide. Some painted in solid colours, others painted more decoratively.

Surfaces can be shiny or not, rounded or flat, and can come in just about any shape you can dream up. With all the choices in containers and plants, the designs are limitless, it can add a visually appealing element to the space. This type of display can be particularly eye-catching in settings such as gardens, event venues, or retail spaces. The combination of diverse containers and stands can contribute to a visually dynamic and aesthetically pleasing environment.

*Selection of plants:* For container gardening, plants with evergreen canopy, hardy and thornless characteristics are selected. High-value attractive ornamental foliage and flowering plants with compact growth are chosen for such gardens. Plants such as *Sansevieria trifasciata, Aglaonema costatum, Scindapsus aureus, Areca lutescens, Monstera deliciosa, Philodendron scandens, Spathiphyllum, Zamioculcas zamiifolia,* ferns and chlorophytum are most suitable for container gardening in urban landscapes.

Terrarium, window garden, miniature garden, hanging baskets and bonsai are known for its pot design and arrangements. Regular maintenance tasks, including pruning, deadheading, and repotting, are indispensable for a flourishing container garden. container gardening stands as a trailblazing approach to indoor landscape, catering to the needs of urban and space-constrained gardeners.

## Gardening in Hanging Basket

A group of attractive indoor plants in a hanging basket will indeed provide a beautiful display. Baskets are the perfect solution for soil-less balconies or small urban gardens where space is a constraint. Baskets are hung at the entrance of the house, hall, drawing room and in the bathroom under a fluorescent light. Baskets with attractive appearance, light weight and easy to hang & preferably with a drainage hole. Wire baskets made up of coconut coir with three hooks are a good container for hanging gardens. Many garden flowers like Begonia, Geranium, Portulaca, Pansy, Petunia *etc.* and foliage like Asparagus, Chlorophytum, *Callisia repens,* Coleus, Ferns, Ivy plant, Philodendron, Pilea, Pothos, Setcreasea, Tradescantia and Zebrina can be easily grown in hanging baskets.

## Window Gardening

Window garden also known as window box gardening is the making of garden in the window sill or any other attachment to it. It is novel method of gardening and is ideally suited for congested cities and flat-dwellers.

*Plant selection*: for window garden plants were selected based on growth habits, weather, and sunlight requirements. Sun-loving plants *i.e.* Alyssum, Geranium, Petunia, Cosmos, Snapdragon, Dahlia, French Marigold, Nasturtium, Succulents *etc.* Shade-loving plants *i.e.* Ferns, Begonias, Impatiens, Ivy, Caladium, Violas, Mint, Philodendron, *Spathiphyllum etc.* Creepers *i.e. Tradescantia, Hedera helix, Juniperus prostate,* Morning Glory *etc.* Perennials: *Lantana camara,* Verbena, *Russelia juncea etc.*

## Terrarium

Terrariums are a great addition to any indoor space. They are perfect gifts and add an intriguing natural element to any room. It is an entertaining and attractive way to integrate plants into home furnishings. A terrarium is like an aquarium but for plants instead of fish. It is made in just about any glass container. It is planted to look like a miniature garden or forest air-tight enclosed in its own little world and the growing requirements of plants like the $O_2$, $CO_2$ and water are constantly recycled within the container by the plant. The terrarium is a closed feature, hence, the plants do not require frequent watering, as the moisture developed from the transpiration of leaves and evaporation from the soil, again becomes available to the plants for their functions. This way by keeping the plants in enclosed containers, drought is excluded and high humidity can be maintained. Terrarium feature can be developed in different forms and sizes. It may be squire, round, flat etc., depending on their choice and availability of space in the house. A terrarium container must be made from clear glass or plastic. Plants to be selected for a terrarium are very limited. The plants should be such types which do not require pruning frequently. Only small rooted cuttings or offshoots should be planted. The suitable plants for the terrarium include Begonia-rex, Calathea, *Chamadorea,* Cryptanthus, *Dracaena sandariana,* ferns, *Fittonia, Hypostes, Pepromia, Pilea,* Sanseviera, *Selaginella, Zebrina, etc.*

## Miniature Gardens

Miniature garden is a garden planned in a miniature scale or sometimes referred to as showcase size. It is based on the creative concept wherein different garden features such as path, rocks, hills, pool, flowers, shrubs, trees are to be designed in miniature form. This type of garden is best laid in troughs of concrete, stone or wood. Gardening in troughs and dishes is generally done with indoor plants only. Dishes or bowls with well-arranged plants are used for decorating the dining table, central table and tea table. Suitable plants for such places are cacti, succulents, aralia, dracaena, sanderiana, chlorophytum, begonia *etc.*

## Kokedama

Kokedama is a Japanese gardening technique that translates to "moss ball" in English. This unique way of presenting plants involves wrapping the roots of a plant in a ball of soil and moss, tied with a string, creating a living and decorative arrangement. Originating from the art of bonsai, Kokedama has gained popularity worldwide for its aesthetic appeal and the symbiotic relationship it fosters between plant and artisan. It's unique presentation blends nature with art, appealing to those seeking distinctive plant decor. The growing interest in biophilic design has propelled its popularity, as it brings nature into indoor spaces.

## Conclusion

Vital role of ornamental plants directly on environment, psychological and emotional status and indirectly on human feelings associated with work performance and work productivity has been recognized and the business of pot plants is the new upcoming industry in the country. Ornamental Pot plant Industry holds great promise with immense investments and employment opportunities along with several challenges. This industry and despite of several constraints at production and marketing ends, is striving to flourish and escalate with the support of Government. Research based on pot plant technology, identification of indigenous plant species for pot, soilless based media standardization, packaging and transportation, need to be focused and innovative approach and ideas among growers need to be encouraged. Further, efforts are needed to generate authentic trade figures, popularize pot plants in offices as well as homes, market survey and market research. The potential of pot plant industry as money spinner is needed to be realized, recognized and further explored.

## References

Berger, J., Essah, E., Blanusa, T. and Beaman, C.P. 2022. The appearance of indoor plants and their effect on people's perceptions of indoor air quality and subjective well-being. *Building and Environment* **219**: 109151.

Duffield, G. and Bunn, S. 2023. Indoor Air Quality. Parliamentary Office of Science and Technology - POST Brief 54, UK Parliament.

Haviland-Jones, J., Rosario, H. H., Wilson, P. and McGuire, T. R. 2005. An Environmental Approach to Positive Emotion: Flowers. *Evolutionary Psychology* **3**:104–132.

Janakiram, T., Ritu Jain and Kumar, P. P. 2013. Potted plants keep indoors healthy. *Indian Horticulture* **58**(4):18-21.

Nisitha, S., Balasubramani, G. and Paul P. J. 2023. Systemic review on indoor plants as an alternative technique for reducing indoor air pollutants. *Journal of Xidian University* **17**(11):1074-1083.

Patel, N. L., Ahlawat, T. R. and Singh, Alka. 2016. *Handbook of Urban and Periurban Horticulture*, Ambika Book Agency, Jaipur.

Singh, Alka and Ahlawat, T. R. 2019. Flowers and greenscapes are nature's message to rejoice. *Indian Horticulture* **60**(4):16-18.

Singh, Alka. 2023. Indoorscaping: A novel venture for health and happiness. *Floriculture Today* 34-36.

Xie, J., Liu, B. and Elsadek, M. 2021. How can flowers and their colors promote individuals' physiological and psychological states during the COVID-19 Lockdown? *International Journal of Environmental Research and Public Health* **18**:10258.

# CHAPTER 30

# ORNAMENTAL EDIBLE FLOWERS

**S. L. Chawla, Vaidehee Ninama, Sudha Patil and Mallika Sindha**

*Department of Floriculture and Landscape Architecture, ASPEE College of Horticulture, Navsari Agricultural University, Navsari (Gujarat)*
*Corresponding author: shivlalchawla@nau.in*

Flowers are beautiful creations of nature. The origin of edible flowers dates back to ancient civilizations, where they had symbolic and cultural importance in everyday life and traditions. Ancient cultures such as the Greeks, Romans and Chinese incorporated edible flowers and left a legacy of using flowers in art, natural medicine as well as in cooking to colour, fragrance and adorn various dishes. Edible flowers that can be eaten without endangering the health of consumers. This joy can also be felt in meals such as edible flowers in salads or dishes decorated with them. Although use of flowers in natural medicine and food is an ancient practice, it has been observed in several countries in recent decades that use of flowers in cooking has been appreciated, higher due to changes in people's habits or functional properties that improve health and prevent disease (Santos *et al.*, 2021). Edible flower petals and buds can be eaten raw in salads. A refreshing tea can also be prepared from the flower petals (Lim, 2012). Regarding the nutritional value of edible flowers; petals are a source of vitamins, minerals and antioxidants, which help to increase enjoyment when consumed. In addition to their decorative and culinary properties, nutritional and chemical protective properties of some edible flowers have been well documented. Edible flowers can also be classified as nutritional foods (Mlcek and Rop, 2011).

Edible flowers have traditionally been used as human food in many different cultures. They improve the appearance, taste and aesthetic value of foods, aspects that are highly appreciated by consumers. Eating flowers is a common practice in Europe and is extremely popular in French and Swiss cuisine as well as in Asia. According to ancient scriptures, edible flowers have been used for culinary and medicinal purposes in many parts of the world. In India, edible flowers were used for culinary purposes by ancient Indians as early as 3000 BC. Flowers have been used in India for various purposes since ancient times, but recently the field of floriculture has gained commercial stature in India due to increasing demand. Edible flowers are consumed by the tribal people of India, in the area where these edible flowers are grown.

Deb *et al.*, (2019) reported that utilization of various parts of plants including flowers of *Dendrocalamus giganteus, Musa sikkimensis, Bauhinia purpurea, Amomum dealbatum, Curcuma augustifolia, etc.* for food and medical purposes by the local population of *Tuensang, Kohima* and *Phek* districts of North-Eastern's states of India.

Currently, there is an increasing consumers' demand for functional and healthy foods of natural origin, while an important market niche refers to edible flowers (Benvenuti and Mazzoncini, 2021). This demand is also driven by the pursuit of adding new colours and flavors in gourmet dishes by chefs that may diversify food products and make dishes and recipes more attractive to consumers (Cunningham *et al.*, 2021). However, apart from the aesthetic aspects and the visual improvement that edible flowers may confer to food, they are also highly appreciated for their nutritional and bioactive properties.

In Brazil, supermarkets, stores, and shops specializing in culinary products have been selling edible flowers, which are used in salads, soups, pizzas, canapes and jellies, both in sweet and savoury dishes (Franzen *et al.*, 2016). Currently, there are few studies that prove the edibility of flowers when they are related to compounds of nutritional interest since there is no tradition of using such flowers in food, in addition to the lack of research regarding the toxicity of some species.

***Flower fritters:*** The flowers of lilac, elder, marrows and squashes, and fruit blossoms condition all be dipped in batter and deep fried. The marrow flowers can also be stuffed with tied onion, bread crumbs and parsley, before deep-flying in batter.

***Flower vinegar:*** White wine or cider vinegar can be flavoured with flowers. Primrose, rose, violet, elderflower, nasturtium, lavender, rosemary and thyme can be used.

## Importance of Edible Flowers

The heterogeneity of botanical structure of several species of the flowers is closely connected with their respective chemical diversity. This phytochemical complexity and their nutritional value are provided by the pollen (rich in amino acid and proteins), nectar (rich in sugar) and corolla tissue (rich in pigment, vitamin and microelements). Mixture of primary nutrition (sugar, proteins, *etc.*) helps prevent nutritional deficiencies in the human diet (Pires *et al.*, 2021). Different varieties of flowers may provide varying levels of nutrients, but many are known to be good sources of vitamin C, vitamin A, and certain B vitamins. Additionally, they can contribute essential minerals such as potassium, calcium, and magnesium.

***Rich source of antioxidant:*** Antioxidants play a crucial role in protecting our cells from damage caused by harmful molecules called free radicals. Flowers are known to be rich sources of antioxidants, which help reduce oxidative stress and promote overall well-being. The vibrant pigments found in many edible flowers, such as anthocyanins and carotenoids, are powerful antioxidants that contribute to their health benefits. For example, Butterfly pea flower is rich in antioxidant and has a range of health benefits including weight loss, controlling blood sugar level or maintain healthy hair and skin. The ingredient (anthocyanin, carotenoids) are well known for its brilliant blue colour often used in cocktail, cosmetic and herbal tea blend.

***Potential of medicinal property:*** Beyond their nutritional value, some flowers have been traditionally used for their potential medicinal properties. For example, certain flowers like chamomile and lavender are known for their calming and soothing effects, making them popular choices for herbal teas and relaxation remedies. Calendula flowers are often used for their anti-inflammatory and wound-healing properties. The nutritional need required by the human body in health and disease states has been the object of intense research in recent years, as well as the concern about the chemical characterization of foods with economic and nutritional potential, especially those of low caloric value, since obesity and chronic-degenerative diseases become prominent in public health.

***Enhance diversity of nutrients:*** The diet of active people should take care of proper all nutrients, including dietary fibre and protein. Low content of nutrients may reduce performance and prolonged recovery time after physical activity. Variety of vitamins, minerals, and bioactive compounds that may not be as abundant in other foods. This diversity of nutrients is important for overall health and can contribute to a well-rounded and balanced diet. Embracing flowers in culinary adventures not only adds beauty and flavour to your dishes but also offers a range of potential health benefits. From their nutritional value and antioxidant content to their possible medicinal properties, edible flowers provide a unique opportunity to enhance consumers well-being while indulging in aesthetically pleasing and delicious meals.

***Storage of Edible Flowers:*** In the world of healthiest food, it often feels like the phrase "put a flower on it" reigns supreme. While flowers undeniably enhance the visual appeals of dishes. They can introduce unexpected texture and flavours, elevating culinary expertise.

There are several delicious way to have flowers, whether raw or processed in our diets. The simplest entry point of edible flower is to use their petals as a garnish like fresh herbs, but they are highly perishable because of their better moisture content. It isn't always feasible to preserve them for longer time without processing. Second of all the cases where a fruit or other part is used as such and requires processing. Drying, freezing or step of oil may prolong the edible flower's shelf life for further use. Candied flowers are made by crystalizing egg whites and sugar. They may be used for drinking or desserts, jam and other food.

## Nutraceutical potential

The biodiversity of ornamental and wild species with edible flowers is one of the most promising resources for gastronomic innovation aimed at offering new tastes and new combinations of flowers with other foods (including vegetables and meat, fish and types of beer and wines). Their nutraceutical potential benefits human health also in terms of a psychological impact by combining the beauty of flowers with their tastiness. Future studies will increasingly underline the nutraceutical potential of the various species of flowers that can prevent certain pathologies and thus be part of "personalized diets" aimed at specific health problems. As to which flowers can be eaten, there are no clear boundaries between edible species, and it depends on the taste of the consumer. The only objective characteristic is their non-toxicity, which will be examined in the next section. To limit the species with edible flowers there are various books (Barash, 1993).

Most species used as edible flowers are ornamental species. This derives both from their wide availability, as widely cultivated species, and due to the fact that their ornamentality is closely linked to their large corollas and/or the high number of flowers or inflorescences per plant. Species with large-sized flowers are, for example, *Hibiscus rosa-sinensis, Lilium bulbiferum, Petunia x hybrida, Tagetes erecta, Tropaeolum majus* and *Viola x wittrockiana*. Other species have smaller but extremely numerous flowers such as *Oxalis corymbosa, Impatiens walleriana, Dianthus x barbatus, Limonium sinuatum, Viola cornuta, Ageratum houstonianum*, and *Agastache foeniculum*. Some species with edible flowers, although sometimes also used as ornamentals, are typical medicinal crops such *as Matricaria chamomilla, Calendula officinalis, Taraxacum officinalis, Rosa chinensis, Caesalpinia pulcherrima* and *Lavandula angustifolia*. Some flowers of trees are gathered in inflorescences and are small but very aromatic, and can be used in chromatically and aromatically innovative gastronomic creations. In other cases, the edible flowers belong to common horticultural species which have been traditionally (*Cucurbita pepo*), or recently (*Cichorium intlybus*), used as food (Benuveti and Mazzoncicni, 2021).

## Flowers should be avoided?

Some plant species have evolved toxicity and/or poisonousness in order to defend themselves from both parasitic and/or pathogenic organisms and, above all, from herbivorous fauna. Consequently, great care must be taken not to confuse the beauty of the flowers with their usability as food since some may be toxic or poisonous for both humans and animals (Tamilselvan *et al.*, 2014). Do not eat flowers obtained form florist. Sometimes only petals of flowers are edible and pollen may cause allergies.

Phytochemicals are responsible for toxicity all to the chemical categories of alkaloids, saponins, terpenes and glycosides (Mithofer and Boland, 2012). These species mainly belong to the botanical families of Liliaceae, Amaryllidaceae, Apiaceae, Fabaceae, Scrophulariaceae, Solanaceae, Euphorbiaceae, Apocynaceae and Ranunculaceae. While in some botanical families, almost all species have frequently high toxicity and/or poisonousness. The toxic species of other botanical families such as Liliaceae, Fabaceae, Scrophulariaceae and Solanaceae are rare, while many others, belonging to these botanical taxa, are widely used as food or as medicine, including both the flowers and other parts of the plant.

Edible flowers which contain alkaloids with medicinal properties were generally considered safe for consumption whereas flowers bearing harmful alkaloids with psychedelic and hallucinatory properties were pondered as toxic and unsafe for human consumption. (Nicolau and Gostin, 2016).

## Precautions

- Harvest flowers in the morning after the dew have evaporated. Choose flowers at their peak for best flavour. Put long-stemmed flowers in water and keep in a cool place. Use short-stemmed blossoms within a few hours of harvest or store between layer of dump paper towelling or plastic bag.
- Remove the stamens, pistils and the green, leaf-like Sepals at the base of the flowers
- Do not use flowers exposed to pesticides or those growing by the roadside.

## Challenges

Some individuals may develop an allergic reaction to some edible flowers, which can result in side effects such as skin rash, itching and even more severe symptoms. Before these blooms are included in your diet, it is vital to be informed about possible allergic reactions. It is not safe for some flowers to be eaten. Some can be toxic and cause digestive issues, nausea, or more severe health problems. In order to correctly identify edible flowers and prevent their consumption, it is important not to eat any of them that may be injurious. Pesticide residues can be found in edible flowers bought from non organic sources. The risk of pesticide exposure can be reduced if these flowers are carefully washed or chosen organic options. Individual sensitivity to certain foods may differ, just like any other food. After ingestion of a certain edible flowers, some people may suffer from gastrointestinal symptoms or mild digestive problems.

## Opportunity

Flowers can be used to flavour drinks and make drinks such as herbal teas and wine. They are found in seasonings such as vinegars, marinades, and sauces, as well as spreads such as butter and fruit jam. It can be added to salads or consumed as part of a main meal. Edible flowers can be used to decorate cookies, cakes *etc.* for beautification and nutritional purposes. Due to their medicinal properties, some edible flowers can be used in the pharmaceutical industry. The presence of phenolic compounds in flower extracts is reflected in their excellent antioxidant, antibacterial, and anti acetylcholinesterase activities, which have been proven to be beneficial for human health. Some edible flowers can be used as new foods and medicines due to their non-toxic and nutraceutical properties.

## Conclusion

In order to meet the increasing demand of consumers for natural functional foods, edible flowers may play a very important role. In fact, edible flowers can be used as raw materials for extraction of valuable compounds with nutraceutical potential and health benefits in food industry, beverages, food colouring, floral hydrolates, syrups and jams or in the biomedical sector. On the other hand, edible flowers are very much in demand at a small scale. In order to industrialize and increase their production, there is a need to deal with their low lifetimes, their availability in a specific time of year and the need for the application of appropriate drying methods. In the past several centuries, edible flowers have been proven to contain large amounts of phytochemicals including phenolic acids, flavonoids, carotenoids, tocopherols and other substances that may be incorporated into traditional foods in order to improve their functionality. Although there are many studies of the phytochemicals content and antioxidative activity of edible flowers, their large number in the world requires further study. To further assess its inclusion in food products, as well as their stability during processing and storage, select edible flowers with a higher phytochemical content and function because not all of them contain significant amounts.

**Table 1 Edible flowers with their flavours and use**

| Flower | Flavour | Use |
|---|---|---|
| Nasturtium | Peppery and slightly tangy | Both the leaves and flowers are edible. They add a vibrant colour to salads and are often used as a garnish. |
| Rose | Floral and sweet | Petals can be used to make gulkand, syrups, jams or desserts. They are also commonly used in salads. |
| Lavender | Fragrant and slightly sweet | Often used in baking, desserts, and teas. Lavender pairs well with chocolate and lemon-flavored dishes. |
| Calendula | Mildly tangy with a hint of spice | Petals can be added to salads, soups, or used to colour and flavor rice. Calendula is sometimes referred to as "poor man's saffron." |
| Violets | Sweet and floral | Violets are often crystallized for cake decoration or added to salads for a burst of colour. They can also be used to make jellies and syrups. |
| Chrysanthemum | Bitter and slightly floral | Popular in Asian cuisine, chrysanthemum petals are used to make tea. They can also be added to salads. |
| Dandelion | Slightly bitter | Both the flowers and young leaves are edible. They can be added to salads, used in fritters, or brewed into a tea. |

| Hibiscus | Tart and cranberry-like | Dried hibiscus flowers are commonly used to make herbal teas. They can also be added to salads or used as a garnish. |
|---|---|---|
| Marigold | Citrusy and slightly spicy. | Petals can be used in salads, rice dishes, or as a colorful addition to soups. Marigolds are also used to make a saffron-like dye. |
| Borage | Mild cucumber-like taste | Both the flowers and leaves are edible. Borage flowers are often used in salads or as a garnish for cold soups. |
| Gladiolus | Similar to lettuce | Petals use as salad |
| Garlic vine | - | Flowers used as raw or in omlettes |
| Sunflower (*Helianthus annus*) | Leafy, slightly bitter | Unopened flower buds can be steamed like artichokes.<br>Lightly steamed petals to lessen bitterness |
| Geranium | Lemon to mint | Flower petals used in salad<br>Citronella variety is not edible |
| Tuberous begonia | Crisp, sour and lemony | Only hybrids are edible |
| Lilac | Lemony, floral and pungent | Garnishing |

Source: https://www.financialexpress.com accessed on 24 December 2023

**Table 2 Common Garden flower to avoid (Deepika *et al.*, 2014)**

| Name of flowers | Reason |
|---|---|
| Apple flower | Contain cyanide precursor |
| Borage and Daylily | Flowers are diuretic |
| Chrysanthemum | *Chrysanthemum coronarium* should be eaten; it was not advisable to eat other types of chrysanthemum |
| False Jasmine (*Gelsemium iempervirens*) | Only *Jasmine officinale* is edible' The false Jasmine (*Gelsemium iempervirens*) is completely different genus and is considered too poisonous for human consumption |
| Begonia | The flower and stems contain oxalic acid and should not be consumed by individual suffering from gout, kidney stones or rheumatism. |

**Table 3 Ornamental edible flowers with their form of use and properties**

| Flower | Part used | Form of use | Properties |
|---|---|---|---|
| Rose | Petals | Fresh and dried petal for tea | Antioxidant |
| | | Gulkand | Laxative and tonic |
| | Miniature rose varieties | Decorate pudding, ice cream or cakes | Culinary |
| | Large rose petal | Salad or used to perfume butter | |
| | Hips<br>*(Rosa brunonii)*<br>*(Rosa webbiana)*<br>*(Rosa rugosa)*<br>*(Rosa gallica)* | Jam, Jelly, Sherbat | Rich source of Vit-C and Vit-A |
| Nasturtium<br>(*Tropaeolum majus)* | Leaves and flower petals | Salad | Vit -C<br>Antibacterial<br>Antifungal |
| Lotus<br>(*Nelumbo nucifera*) | Petals | Garnishing | Dietary fibre, Vitamin B6, Copper, Manganese and low in fat. All can be eaten raw but there is risk of parasite transmission |
| | Smaller leaves and Rhizomes | As vegetable | |
| | Rhizome | Soup, deep fry and pickle | |
| | Seed | Raw, dried or as popped like Popcorn | |
| Water lilies<br>(*Nymphaea spp.*) | Young leaves and flower buds | As vegetables | Anti inflammatory<br>Regulating insulin level and promoting liver health |
| | Seeds | Seeds of *N. alba* is a coffee substitute | |
| *Hibiscus sabdariffa* | Flower, Calyx and Seed | Beverage, Jellies, tea and Jam | Decrease blood pressure<br>Reduce body fat |
| | In India (Fruit) | Beverage, Jellies and Jam | In Nigeria:- "*Zobo*" refreshing drink "*Pinnies*" are prepared from seeds |
| *Rhododendron arboreum* | Flower | Pickle, Sherbat, Jelly | Eaten raw as famine food in India |

| Bauhinia variegata | Flower bud | Making "*Raita*" | Anti diabetic<br>Antioxidant |
|---|---|---|---|
| | Flower | After boiling vegetable | |
| Garlic vine | Flower | Used as raw or in omlete | |
| | Leaves | In Decoction | Remedy for flu and Pneumonia |
| *Clitoria ternatea* | Flower | Tea | Antioxidant |
| *Tropaeolum majus* | Flower and leaves | Salad | Antioxidant |
| Sunflower<br>(*Helianthus annus*) | Flowers | Unopened flower buds can be steamed like artichokes | use as a remedy for pulmonary affections |

## REFERENCES


Benvenuti, S. and Mazzoncini, M (2021). The Biodiversity of Edible Flowers: Discovering New Tastes and New Health Benefits. *Front. Plant Sci.* **11**: 1-11. doi: 10.3389/fpls.2020.569499

Cunningham, S. A., Shaikh, N. I., Datar, A., Chernishkin, A. E. and Patil, S. S. (2021). Food subsidies, nutrition transition, and dietary patterns in a remote Indian district. *Global Food Security* **29**: 1-8.

Deb, C. R., Khruomo, N. and Paul, A. (2019). Underutilized edible plants of Nagaland: a survey and documentation from Kohima, Phek and Tuensang District of Nagaland, India. *American J. Plant Sci.* **10**(1): 162-66.

Deepika S.; Sowjanya L.; Laxmi S. K. and Sulakshana (2014). Edible flowers-A review article. *Int. J. Adv. Res, Sci. Techno.* **3**: 51-57.

Franzen D. L., Santos R. D., N. S. P., De Oliveira, M. S. R., Backes, F. A. A. L., Menegaes, J. F. and Zago, A. P. (2016). Characterization and nutritional quality of ornamental flower petals. *Acta Iguazu* **5**(3): 8-70.

Lim, T.K. (2012). Edible medicinal and non-medicinal plants Dordrecht, The Netherlands *Springer* **1**: 656-87. doi:10.1007/978-94-007-1764-0

Mithofer, A. and Boland, W. (2012). Plant defence against herbivores: chemical aspects. *Annual Review of Plant Biology* **63**: 431-50.

Mlcek J. and Rop O. (2011). Fresh edible flowers of ornamental plants – A new source of nutraceutical foods. *Trends Food Sci. Techno.* **22**(10): 561-69.

*Nicolau, A. I. and Gostin, A. I. (2016). Safety of edible flower. Regulating safety of traditional and ethnic foods. *Elsevier Inc.* 1016/B978-0-12-800605-4.00021-9.

Pires Jr, E. D. O.; Di Gioia, F.; Rouphael, Y.; Ferreira, I. C.; Caleja, C.; Barros, L. and Petropoulos, S. A. (2021). The compositional aspects of edible flowers as an emerging horticultural product. *Molecules* **26**(22): 1-31 doi: 10.3390/molecules26226940

Santos, Izabel C. D. and Simone Novaes Reis. (2021). Edible flowers: traditional and current use. *Ornamental Horticulture* **27**: 438-45.

Tamilselvan, N.; Thirumalai, T.; Shyamala, P. and David, E. (2014). A review on some poisonous plants and their medicinal values. *Journal of Acute Disease* **3**(2): 5-89.

## CHAPTER 31

# FARMER PRODUCERS' ORGANIZATION FOR BETTER LIVELIHOOD OF COCONUT FARMERS

**Prabhat Kumar, Hanumanthe Gowda. B and Mridula. K**

*Coconut Development Board, Kochi-682011*

India is often described as an agricultural economy with majority of the population depending on agriculture for their livelihood. But the agriculture sector often faces setbacks owing to diverse factors. The growth registered in the agriculture sector indicates a diminishing trend. (Ranjith kumar et al., 2022). The measures declared during various instances by the policy makers to aid the farmers like write off of loans, declaration of minimum support price etc often play the role of a pain killer, while the real disease, i.e. lack of profitability in farming sector due to lower prices and higher production expenses remains as a hidden systemic disease. Even though policy-makers are offering huge subsidy and higher procurement prices, the benefits are not reaching to real farmers which will keeps the farmers crippled ever. The only way out to address such situations is to empower the farmer to produce quality product and undertake joint efforts as a team in production, value-addition and marketing of the produce adhering to world standards.

Farmers Producers Organizations (FPOs) are considered to one of the most imminent tools of intervention for upliftment of farmers' condition in India. When more than 85 per cent of farmers are smallholders, it becomes quite challenging for them to access the modern production technologies, and use market information for their advantage, transact the commodities in input or output market on their own terms, and ultimately keep their farming profitable. The collectivization of farmers through FPOs help in bringing economy of scale in different on-farm as well as off-farm activities at all three stages, pre-production, production and post-production levels. The Government of India is promoting the formation of FPOs through different programmes and schemes in different sectors including coconut since last one decade.

## Coconut Scenario in India

The Coconut is eulogized as the ***'Kalpavriksha'***, the 'Tree of life', due to its multifarious utilization as food, fuel, medicine, timber and other utility purposes of different parts from root to leaves, from tender nut water to outer husk, etc. thus offering scope for sustaining the livelihood of growers, farm communities and industries in major coconut-growing countries of the world. The crop is intricately woven into the socio-economic and cultural backdrop of the Indian subcontinent. As per 2021 statistics of ICC, India is the largest coconut producing country in the world, with 31.10 % share of global production, followed by Philippines and Indonesia. India ranks second in terms of productivity (9,430 nuts per ha) next to Jamaica (10,348 nuts per ha) (Anonymous,2021).

Out of the total geographical area of 328.73 million hectares of India, the total agricultural land is 180.88 million hectare. About 2.18 million ha is covered under coconut which is about 1.20 % of the total agricultural area. As per the All-India 3$^{rd}$ estimate for the year 2022-23the area and production of coconut in the country is 2.29 million hectares and 20015.89 million nuts(2552' 000 MT of copra equivalent),respectively, spread over 19 States and 4 Union Territories located in varied agro-climatic zones. However, about 90 % of the area and production fall in the south peninsular region covering 4 States of Tamil Nadu, Kerala, Karnataka and Andhra Pradesh. Even after such a high rate of production, India's annual export earnings from coconut and coconut based products form just **0.1%** of total export earnings(Bharathi and Kumari,2024).

Coconut is predominantly a smallholders' crop in India. More than 30 million people including 12 million farmers in the country dependent on coconut for their livelihood and very meagre portion of them are big farmers. The crop contributes around Rs 280,210 million (US$ 3359.43 M) to the country's GDP and earns export revenue of around Rs. 75,464 million (US$ 940.49M).

## Coconut Development Board

Coconut Development Board (CDB) is a statutory body established in 1981, under the Ministry of Agriculture, Government of India, by an act of Parliament, for integrated development of coconut cultivation and industry in the country with focus on productivity increase and product diversification.

## CDB's Role in Farmer Producer Organizations

Even though Cooperative movement is deep rooted in many of the coconut-growing states, most of the coconut farmers are unorganized even now. As such the efforts taken for increasing the production and productivity of coconut from small and marginal holdings are showing poor results. It is felt imperative that a movement in the nature of society of small and marginal farmers will help to improve both the production and productivity. With this objective in view the Coconut Development Board has come forward to encourage the farmers to form Coconut Producers' Societies (CPS). The Board anticipates that these Societies will collectively enter into procuring, processing and marketing activities. The Board had planned to unite the small and marginal farm holders and to register them under the Charitable Societies Act.CDB had taken up the task of facilitating formation and hand holding of FPOs during its initial stages (Nair , 2011).

## Rationale of FPOs

Farmer Producer Organization is formed with the main objective of socio economic development of farmers through productivity improvement, cost reduction, efficient aggregation, processing for value addition, product diversification better by-product utilization and efficient marketing of the produce. It aims at providing a fair, steady and reasonable income to farmers by organizing the unorganized coconut sector through farmer collectives.

True empowerment happens only when farmers are involved in all stages of value addition supply chain, such as, production, aggregation, processing, marketing, distribution and sales. Only when farmers starts an enterprise of their own, they will learn to plan, collaborate, work in a team, take decisions and risks, interact with people from other spheres and sectors, thus learning and growing with the enterprise. This helps in inducing self-confidence within the farmers thus enabling their overall development.

The name 'Producer Company' is provided in order to indicate that the members have to necessarily be a primary producer. A new Part IXA, sections 581A to 581ZT of Indian Companies Act 1956 provide the features of a Producer Company. It is a hybrid between a Private limited Company and a Cooperative society as it minimizes interference from other external bodies thus ensuring that the real power rests with members of the company itself. Such a company functions like any other private limited company with only difference being that the profit incurred flows back to farmers who own the company(Thamban *et al* ,2020).

## Approach

The Coconut Farmers Producer Organizations thus formed have a three tier structure consisting of Coconut Producers Society (CPS), Coconut Producers Federation (CPF) and Coconut Producers Company (CPC).

**Coconut Producers Society (CPS):-** CPS is formed by associating **40-100 coconut growers** in a contiguous area with range of **4000- 6000 yielding palms**. Farmers with a minimum of **10 palms** are only eligible to be a part of this society. Once the society is formed, it is registered under charitable societies act and also with Coconut Development Board. All the societies have a common bye - law.

**Coconut Producers Federation (CPF):-**CPF is formed by combining **8-10 CPS**. A CPF would have around **1, 00,000 palms** under it. CPF is also registered as a charitable society and further registered with CDB.

**Coconut Producers Company (CPC):-8-10 CPFs** would join together to form a CPC. A CPC would consist of around **10, 00,000 yielding palms**. This company would be registered under section 581B of Indian Companies Act of 1956. The Producer Company is wholly and fully owned by the farmers.

## FPO Concept

### Genesis of FPOs

| Year | FPO formation |
|---|---|
| 2009-10 | CDB initiated formation of CPS, 112 CPS formed |
| 2011-12 | Scaling up of CPS formation to all districts of Kerala<br>1158 CPS established as on 31.3.2012 |
| 2012-13 | Scaling up of CPS formation in Tamil Nādu, Karnataka and Andhra Pradesh<br>Integration of CPS to form CPF in Kerala<br>CPS established as on 31.3.2013 - 2253 in Kerala, 62 in Karnataka, 1 in Andhra Pradesh and 28 in Tamil Nādu<br>CPF established as on 31.3.2013 - 103 in Kerala and 18 in Karnataka |
| 2013-14 | Initiated formation of Coconut Producer Companies.<br>17 Coconut Producer Companies established – 12 in Kerala, 4 in Karnataka and 1 in Andhra Pradesh |
| 2014-15 | Scaling up of CPS formation in non-traditional states of Assam, Odisha, Maharashtra, and West Bengal.<br>CPS-6487 in Kerala,Tamilnadu-317, AP-475, Karnataka-368, Odisha-01 and WB-32<br>CPF-365 in Kerala, TN-317, AP-35, Karnataka-111<br>25 Coconut Producer Companies established – 19 in Kerala, 4 in Karnataka and 2 in Andhra Pradesh |
| 2015-16 | Coconut Producer Companies established – 29 in Kerala, 16 in Tamil Nadu, 9 in Karnataka, 7 in AP |
| 2016-17 | CPS-9354 (7156 in Kerala, 553 in Tamil Nādu, 398 in Karnataka, 1013 in Andhra Pradesh, 234 in other states)<br>CPF-728 (460 in Kerala, 60 in Tamil Nādu, 124 in Karnataka, 84 in Andhra Pradesh)<br>Coconut Producer Companies established – 66 no's(29 in Kerala, 17 in TN, 12 in Karnataka, 8 in AP) |
| 2017-2023 | The process of formation of CPS, CPF and CPC continued and the current status of FPOs is shown in separate table. |

## Current Status of FPO

| Sl. No. | State | CPS | CPF | CPC |
|---|---|---|---|---|
| 1. | Kerala | 7233 | 467 | 29 |
| 2. | Tamil Nādu | 697 | 73 | 20 |
| 3. | Karnataka | 401 | 125 | 16 |
| 4. | Andhra Pradesh | 1157 | 82 | 9 |
| 5. | Odisha | 40 | 0 | 0 |
| 6. | West Bengal | 218 | 0 | 0 |
| 7. | Gujarat | 14 | 0 | 0 |
| 8. | Assam | 29 | 0 | 0 |
| 9. | Maharashtra | 1 | 0 | 0 |
|  | Total | 9790 | 747 | 74 |

## Benefits and Advantages of FPOs

- Providing awareness on better technical know-how to improve cultivation practices and thereby increasing income level and reducing the labour cost.
- Enhancing bargaining power in getting reasonable price for their produce by implementing group marketing concept.
- Reduced dependence on local moneylenders by implementing participatory credit and insurance packages.
- Inculcating a feeling of oneness among group members.
- Common platform for discussing and solving problems by themselves .
- Common purchases of inputs which reduces the total cost of cultivation.
- Availing participatory credit and insurance facilities and it is easier to approach banks and other institutions for loans and other financial support as a collective rather than as an individual businessman.
- On the other hand, a company so formed derives advantage from economies of scale, scope and integration. The **10 lakh palms** under a Producer Company provide ample raw material to conduct business on large scale. This in turn, leads to better utilization of fixed assets reducing the per unit cost for the product. Coconut is a zero wastage product. Each and every part of coconut can be utilized to produce a value added product. The same principle is used in Producer Company to gain maximum benefit by complete utilization of all parts of coconut. Finally in economies of integration the company integrates backward with farmers to procure raw material and then integrate forward to take value added products from coconut to markets. Apart from the above mentioned advantages, the company also stands benefited with better support from government and other institutions.

- Better pest &disease management by adopting community based plant protection measures.

## Benefits realised by FPOs

Through Friends of Coconut Tree (FoCT) programme by CDB, many people, both male and female, are trained in coconut tree climbing. They are also provided insurance and similar support schemes of CDB for one year. Further many farmers who wish to engage in some kind of business, but are unable to do so due to financial and other constraints see this as an opportunity to engage in some sort of business activity. While starting a business in a collective manner, the risk involved is distributed among its members and hence minimized. These collectives also benefitted from CDB schemes such as Replanting & Rejuvenation wherein farmers are provided with fertilizers. As a part of FPO, nurseries have also been established to provide farmers with high yielding varieties of coconut seedlings. Fertilizers and the nurseries are the major factors which pulls the farmers towards the collective(Vijayan,N, 2011).

Besides the FPOs have been supported by CDB in the following ways:

- ***For Technology*** : Technology is always a handholding support for agriculture production and marketing. The technical support in the form of machinery, digitalization and the process helped structure the FPOs for better agriculture marketing and increasing revenue. There is a need for identifying the right kind of support agencies with skilled technical experts for developing a sustainable FPO. In this context, CDB Institute of Technology (CIT), CDB's in-house research centre is set up for providing technical support, consultancy and efficient technologies for integrated coconut processing. Quality testing, training programmes, product development and technology transfer support from CIT help FPOs in adopting modern and advanced technologies as per their requirement.
- ***Technology Mission on Coconut (TMOC):*** TMOC supports FPOs for establishment of coconut processing units at **25% of project cost** subject to a maximum of **Rs.50 lakhs**. Apart from this, TMOC also does hand holding for market promotion, brand building, product processing and diversification.
- ***Marketing support:*** Under thisCDB facilitated market identification, development and product promotion for the companies. It also works on generating demand through consumer awareness by conducting coconut festivals, exhibitions, fairs etc. Use of potentials of advances in Information technology, exclusive coconut corners, generic product promotion etc. are some other ways in which market building is carried out by CDB.
- ***Capacity building and trainings:*** Friends of Coconut Tree (FoCT) training for coconut climbers, Neera technician training for Neera tapers etc. are some of the trainings initiated by CDB to build skilled laborers in coconut sector. This not only addresses the issue of labour scarcity and lack of efficiency, but also provides employment opportunity with good remuneration and

dignity for aspiring people. Trainings and certification courses in coconut processing and value addition are also started to attract younger generation to this sector. Capacity building is also done by conducting leadership and other trainings for directors of the companies.

- ***Technical and managerial professionals in coconut sector***: Absence of sufficient technical and managerial professionals is one of the most important reasons for the inadequate growth of this sector. In order to bridge this gap, CDB has been collaborating with reputed management, technical and research institutes to enable professional participation in this sector. CIT, the in-house research centre of CDB is an initiative in this direction. Through CIT, CDB intends to build up its technical and research side in coconut processing, value addition and related topics. Various trainings and courses by CIT aims at attracting younger generation and developing skilled professionals in this area. Similarly, in the management side CDB had collaborated with institutes such as Institute of Rural Management Anand (IRMA), National Institute of Rural Development (NIRD), Xavier Institute of Management and Entrepreneurship (XIME), School of Communication and Management studies (SCMS) etc. by providing the students internship as well as job opportunities with the organization. Such a technical and managerial professional base can provide the much required boost by bringing in new culture and face to the sector.
- **Quality planting material** :The first and foremost requirement for producing international standard coconut products is presence of high quality raw material base. Currently India faces lack of quality planting material required to build up this raw material base. CDB tried to address this issue by facilitating a de- centralized and farmer participatory nursery system. With emergence of FPOs, federations can take up the responsibility of delivering quality planting materials to all CPS under them by maintaining own nursery. This would not only give way to better productivity and quality, but also invoke among the farmers a sense of ownership.

Apart from the above mentioned points, CDB also acts as an interface to help FPOs collaborate with various local, state and central government institutions in matters of financial support and policy advocacy during this initial hand holding phase. CDB has proposed formation of Consortium of Producer Companies to take over the above mentioned role played by CDB and to establish itself as an independent entity which addresses the common needs of Producer Companies and act as a powerful force to build a more favourable business climate.

## Challenges faced by FPOs

- Lack of proper handholding support to FPOs. Handholding support, Easier Grants (Single Window System) of various Govt sponsored Schemes for Infrastructure, Finance, Export, Licensing, etc. is faced by almost all the FPOs.

- Financial support is regarded as one of the most important factors in improving the performance of FPOs. The FPOs established have also been unable to attract institutional finance. Lack of sufficient seed capital for the initial phase is one of the most difficult impeding factors, followed by lack of loan from banks and lack of knowledge about IPR issue is one of the main challenges faced by FPOs.
- Dynamic leadership is also one of the most important factor in facilitating the growth of FPC, followed by hard work of membersand skill training of members.To enable excellent performance of FPCs, these variables must be recognized and handled the factors like marinating financial records, filling GST returns on time, preparing budget and sharing it with stakeholders were important factors impacting the FPO performance.
- Lack of proper selection of CEOs is critical to the success of an FPO, and therefore, it needs to be ensured that a person with a good educational background and experience is selected as CEO.
- In addition to finance and governance, the role of marketing is also a big challenge before the FPOs. Membership in farmer groups and collective marketing significantly positively affected farm net revenues.
- Measures like Management Information System, Business Plan Preparation, Share Capital, Turnover, Infrastructure, Market Linkages, Technical Support, and Age of FPO play an important role in the performance of the FPOs and one of the biggest challenge.

## Way forward

Going forward, CDB planned to form 100 Producer Companies, 1000 CPFs and 20,000 CPS' in the country. The purpose behind the same is that at least 10% of total production happening in coconut sector should be through Producer Companies. Only in such a situation would the actual bargaining power reside in hands of farmers. Besides, this would also encourage new entrepreneurs to establish business in this sector as the issue of aggregation and organized availability of raw materials would be addressed through FPOs. Coconut cultivation is a sector with much potential and growth opportunities. As we have seen, there are many health benefits attached to coconut. Coconut being a zero wastage product, each and every part of coconut can be put to some use. There also exists a latent demand for many coconut products. Globally, we can see that there are many companies involved in this sector making huge profits and performing very well. But the sector also comes with lots of challenges. There is a huge lack of awareness among public on benefits of coconut and its products. People also carry some misconceptions regarding coconut which says that it is harmful for their health. The frequently fluctuating prices, low productivity, labour issues etc. have forced farmers to move from coconut to other plantation crops which are more profitable. Thus the need of the hour is to organize this sector in such a way that the bargaining power rests with farmers.

Farmer Producer Organizations intend to do the same. With the proper support from government, FPOs have complete potential to benefit all stakeholders namely the government, farmers and the consumers. The export of coconut produce can earn revenue for the government; it can ensure bargaining power and constant returns to farmers; and above all, deliver quality products to consumers. Hence the effort of CDB to form FPOs started since 2009-10 can be seen as the stepping stone for revival of coconut sector.

Besides some of the recommendations are suggested to the sustainability of the FPOs in the coconut sector:

- ***Proper support system*** which provides support in various forms to the FPO; mainly through the government and financial institutions'. Strengthen the role of CDB as an commodity organization to support and handhold the FPOs with the assistance of other government at state and central levels.
- ***The FPOs need four types of support which will drive growth and sustenance***: Managerial competence, financial support, technology and capacity building. The financial support to the FPOs may be channelized through state level federation from various institutions. These government institutions along with research institutes, management institutes and through nearby KVKs would provide capacity building to the FPOs. The FPO federation may play active role in marketing promotion of the FPOs' products at national and international level.
- In the current digital era, **technology helps** in improving operational efficiency and proper monitoring of the business activities. Data collection through technology can help the policy maker in taking appropriate decision. Hence, technological support to the FPOs in the form of ERP (Enterprise Resource Planning) can help achieve the above-mentioned goals.
- For managerial competence, the **FPOs should recruit agribusiness professional**. The CEO having graduation in agriculture or allied sciences with management experiences will bring professionalism in the FPO. They may be incentivized for the positive growth of the FPO by having performance-linked compensation. The financial requirement of the FPO, both for capital investment and working capital need may be met by financial institutions with the credit guarantee of the federation/ supporting agencies.
- The FPO would provide both **backward and forward linkage** to the farmers. Backward linkage would be in the form of input sale, provision of farm machinery, capacity building and advisory services to the farmers. FPO would acquire licence for input dealership and can procure from the manufacturers directly or their distributors. The CEO's expertise will bring professionalism in entire operation right from negotiation with the supplier, procurement quantity & frequency, inventory management and so on. FPO would provide a platform for the Agri-insurance companies to connect

with the farmers for crop & livestock insurance. In forward linkage, the FPO would explore value addition possibilities, while Government may recognize the FPO as collection centre as well as procurement centre for the designated commodities under different schemes.

- ***Common post-harvest infrastructure facilities for FPOs:*** Common infrastructure facilities like warehousing, cold-storage, sorting & grading, packaging, etc. may be created at district-level, which can be used by the registered FPOs of that district at initial stage on pay-and-use basis. This will also create rural micro-entrepreneurship opportunities and every FPOs need not to bother about setting up such facilities at initial stage of their lifecycle.
- ***Preference to the FPOs:*** As FPOs are facing difficulty in marketing of their produces, the FPOs should be given preference for procurement of agricultural commodities in all the government schemes, like MSP-based procurement and other such schemes of the government. It will not only ease the procurement burden by the government, at the same time FPOs will get an alternative for produce dispAAosal. FPOs can be encouraged to supply coconut and coconut value added products under *Mid-day Meal Scheme*of the state government. Similarly, private manufacturers of fertilizers and agro-chemicals may be nudged to encourage direct engagement with the FPOs.

## References

ICC. 2021, International Coconut Community: Statistics Year Book Published by International Coconut Community. Pp; 59.

Anonymous, 2023, *Country Paper of India,* presented during 59th ICC Session and Ministerial meeting, held during 5-7 November 2023 at Lampung, Indonesia, pp; 23

Bharati. N and Kumari. S, 2024, FPOs as a Sustainable Business: Factors Impacting the Performance of Agriculture Collectives in India Management **28**(1): 2-6.

Nair D, 2011, Unite to Triumph than Perspire to Succumb -, Coconut Development Board, *Indian Coconut Journal* **4**(5): 28-30.

Thamban, C, Jayasekha, Chandran, K.P., and Rajesh, M.K.,2020, Policy Paper-Sustainability of Farmer Producer Organisations - The case of producer organisations involved in the production and marketing of 'neera' in the coconut sector in Kerala, India, *Journal of Plantation Crops*, **48**(2):150-59.

Ranjit Kumar, Sanjiv Kumar, R.S. Pundir, V. Surjit, and Ch. Srinivasa Rao. 2022, FPOs in India: Creating Enabling Ecosystem for their Sustainability. ICAR-National Academy of Agricultural Research Management, Hyderabad, India, 22 .

Vijayan, N., 2011, Farmer Groups for Better Production and Marketing - Vegetable and Fruit Promotion Council of Kerala (VFPCK), Kakkanad, Kochi, Kerala, *Indian Coconut Journal* **4**(6): 34.

# CHAPTER 32

# MEDICINAL AND AROMATIC PLANTS IN SEED SECTOR: ISSUES AND FUTURE PROSPECTS

**Prawal Pratap Singh Verma**

*Division of Crop Production and Protection, CSIR-Central Institute of Medicinal and Aromatic Plants, Research Centre, Purara, PO Gagrigole, Bageshwar (Uttarakhand) India*
**Corresponding author: prawal.psv@cimap.res.in*

---

The challenges are faced by farmers and industries in cultivation of medicinal and aromatic plants (MAPs), with a particular emphasis on complexities of seed production. It addresses the difficulties industries encounter due to farmers' limited adoption of improved varieties developed through extensive research and development efforts. Additionally, the chapter presents practical solutions to enhance seed production for various MAPs, including *Withania somnifera, Hyoscyamus niger, Chamomilla recutita, Salvia sclarea, Plantago ovata, Andrographis paniculata, Cymbopogon martini, Chrysanthemum cinerariifolium, Asparagus racemosus, Cassia senna, Ocimum basilicum,* and *Catharanthus roseus*, among others. The objective is to efficiently meet the growing demand for MAP-based raw materials used in various industries. In the current landscape, industries striving to produce high-quality products for health-conscious consumers face significant challenges in sourcing premium raw materials. Addressing this issue requires strategic solutions, with a primary focus on selecting superior-quality seeds. This chapter examines the diverse roles, challenges, and opportunities for research institutions, farmers, and seed-producing organizations involved in the MAPs sector. Furthermore, it provides an in-depth discussion on future roadmaps for advancing seed production technologies to support both institutions and industries in ensuring sustainable growth.

Medicinal and aromatic plants have been a fundamental source of remedies for human ailments throughout history (Chhetri *et al.*, 2021). They serve as invaluable reservoirs for creating new drugs, playing a pivotal role in two distinct domains of healthcare globally. Firstly, they are integral to modern medicine. Secondly, they are at the heart of traditional and indigenous healthcare systems or folk medicine. In Europe, approximately 1,300 medicinal plants are harnessed for medicinal purposes, with 90% of these sourced from the natural environment (Frey, 2020). In developing countries, more than 80% of the population relies on herbal remedies as their primary healthcare solution, and over 25% of prescribed medicines in developed nations are derived from wild plant species. India, with its rich cultural heritage, has a long history of utilizing plants and plant-based products for their healing properties. References (Sucholas et al., 2021) to the medicinal properties of plants can be found in ancient texts such as the Rigveda (400-1500 B.C.), Atharvaveda (1500 B.C.), Upnishada (1000 B.C.), Mahabharta, and Puranas (700-400 B.C.). Notably, Charakasamhita and Susrutasamhita, authored between 1000 B.C. and 600 B.C., are monumental works that have contributed to the understanding of plant-based medicines. Influences from Greek and Muslim rulers have also played a significant role in shaping the use of plant-based medicines (Riaz et al., 2021). Additionally, the rise of Buddhism has spurred the study of herbal medicines in ancient times. This rich heritage has led to the development of Ayurveda and Unani, two important systems of medicine that continue to be widely practiced in the present Indian healthcare system (Novais et al., 2021). Traditional medicine in India operates through two distinct avenues. The first involves local, tribal, or folk practitioners, while the second encompasses codified and organized medical systems like Ayurveda, Unani, Siddha, and Amachi (Rahmawaty et al., 2019). MAPs hold tremendous significance in various aspects, encompassing health, culture, economics, and the environment.

These plants have played a pivotal role in traditional and alternative medicine, offering remedies for various ailments. Additionally, they contribute to the cultural and spiritual heritage of communities worldwide (Kalauni & Joshi, 2018; Marotti et al., 2023). MAPs have a substantial economic impact, providing livelihoods for countless individuals, especially in rural areas. The cultivation, collection, and processing of these plants create income opportunities for farmers, collectors, and local industries. The trade of MAPs and their derivatives, including essential oils, herbal products, and spices, supports economies globally. From an ecological perspective, MAPs are integral to biodiversity and ecosystems (Chaachouay et al., 2019). They represent a valuable component of biological heritage and contribute to conservation efforts. Maintaining genetic diversity within MAPs is crucial for the sustainable utilization of these plants and their preservation for future generations. MAPs also serve as sources of various products, such as natural dyes, essential oils, bio-pesticides, proteins, vitamins, and spices, further emphasizing their versatile nature. Moreover, they play a fundamental role in the pharmaceutical and cosmetic industries, providing essential ingredients for therapeutic and fragrance products.

The importance of seed production in the context of MAPs cannot be overstated. These seeds are the cornerstone of MAPs cultivation and, by extension, the entire industry, supporting a diverse array of crops, including *Withania somnifera, Hyoscyamus niger, Chamomilla recutita, Salvia scalarea, Plantago ovata, Andrographis paniculata, Cymbopogon martini, Chrysanthemum cinerarifolium, Asparagus racemosus, Cassia senna, Ocimum spp.*, and *Catharanthus roseus*. The quality of seeds directly influences the success and sustainability of MAPs farming, making it vital for consistent yields and maintaining product quality. By ensuring genetic diversity and the preservation of desirable traits in these various medicinal and aromatic species, high-quality seeds are integral to the survival of MAPs, especially in the face of changing environmental conditions and evolving pest pressures (Craker, 2007). This reliable source of plant material not only supports rural economies but also provides a foundation for further research and development across all these different crops. Seed production facilitates breeding programs that aim to improve characteristics like disease resistance, yield, and chemical composition in each of these diverse species, thus benefiting the entire industry (Muscalu et al., 2022). Additionally, the conservation of rare and endangered species among these various crops is made feasible through seed banks and controlled propagation, making it a crucial component of preservation efforts across the board. In sum, the role of seed production in MAPs goes beyond mere propagation; it is the linchpin that sustains the industry, supports livelihoods, fosters research, and aids in conservation across a wide spectrum of medicinal and aromatic plants crops.

In summary, this chapter delves into the complex landscape of seed production in the context of MAPs. It seeks to address a wide array of challenges and opportunities within the MAPs seed industry, with the goal of establishing sustainable seed production systems for various valuable MAPs. Emphasizing innovation, research, and collaborative efforts, this industry has the potential to play a vital role in meeting the ever-growing global demand for MAPs-derived products. Subsequent sections will provide a more in-depth exploration of these challenges and opportunities, offering valuable insights and strategies to enhance seed production in MAPs.

## Challenges in MAPs cultivation and opportunities

Cultivating MAPs presents a multifaceted set of challenges that extend across various dimensions of agriculture, from environmental considerations to economic viability (Rana et al., 2020). Successfully addressing the intricacies of MAPs cultivation necessitates a comprehensive grasp of the obstacles faced by both growers and the industry (Table 2). One of the foremost challenges emanates from environmental factors, most notably climate change, which is actively reshaping growing seasons and mandating adjustments in traditional agricultural practices (Spinozzi et al., 2021). Moreover, the vitality and productivity of MAPs hinge on soil quality and land utilization. Sustainable cultivation practices demand immediate attention to counteract soil degradation and implement effective land management methods (Solomou et al., 2019). Pests and diseases pose yet another formidable challenge. MAPs are susceptible to a gamut of potential threats that can adversely affect crop yields and undermine the overall quality of the produce. It is imperative to institute judicious strategies for pest and disease management. Sustainability serves as a cornerstone for the long-term viability of MAPs cultivation (Rai, 2012). Embracing environmentally responsible practices, such as organic and agroecological methods, not only mitigates the industry's ecological footprint but also elevates the quality of the end products. Beyond the sphere of environmental concerns, the economic sustainability of MAPs cultivation is intrinsically tied to market dynamics, fluctuations in prices, and shifts in consumer demand. Skilfully navigating these economic challenges is paramount for ensuring the well-being of MAPs farmers and the industry (Oztruk & Hakeem, 2018). These multifaceted challenges are profoundly interconnected and represent linchpins for the successful cultivation of MAPs. As we delve deeper into these complexities, our objective is to provide a comprehensive understanding of the multifarious obstacles faced by growers and the broader industry. A pivotal challenge within the realm of MAPs cultivation is the constrained acceptance of improved varieties and cultivars (Schippmann et al., 2005). These advanced plant varieties, meticulously developed through extensive research and development endeavors, offer the promise of elevated yields, superior quality, and heightened resilience against environmental adversities. Yet, their underutilization not only curtails these advantages but also hinders the progress of the entire MAPs industry. It is imperative to magnify our focus on this issue to fully comprehend its implications and explore potential remedies for the betterment of MAPs cultivation.

The cultivation of MAPs presents a promising landscape with several opportunities that can be harnessed for sustainable growth in the industry. These opportunities hold the potential to address the existing challenges and pave the way for a more vibrant and robust MAPs seed production sector. Firstly, the growing global demand for MAPs presents a significant opportunity. The world's increasing recognition of the therapeutic and aromatic properties of these plants has led to a rising global demand. This presents a significant opportunity for seed producers to cater to the expanding market, both domestically and internationally. By focusing on producing quality seeds, the industry can capitalize on this heightened demand and contribute to the export market. Secondly, the adaptability of MAPs to local climates offers a unique advantage (Kalamartzis et al., 2023). These plants often thrive in diverse climatic conditions. Seed producers have the opportunity to adapt their offerings to suit local climates and specific growing regions. By developing climate-resilient seed varieties, they can enhance crop productivity and support local farmers in optimizing their yields. Investment in R&D activities forms a critical avenue for seed production. Advancements in breeding, genetics, and cultivation techniques can lead to the development of superior MAPs seed varieties. Collaborative efforts between research institutions, government agencies, and the private sector can further innovation and enhance seed quality. Furthermore, promoting sustainable farming practices not only benefits the environment but also offers a competitive advantage. Seed producers can support and advocate for sustainable cultivation methods among farmers. By incorporating organic and agroecological approaches, the industry can ensure the long-term viability of seed production and meet the increasing demand for eco-friendly products. Diversification of MAPs also stands as an opportunity. The wide variety of MAPs available provides an opportunity for diversification in seed production. By expanding the range of MAPs cultivated and offered as seeds, producers can tap into niche markets and cater to specific industries, including pharmaceuticals, cosmetics, and herbal remedies. Government initiatives and support provide another avenue for growth. Indian government recognize the economic potential of MAPs and offer support in the form of subsidies, incentives, and favorable policies. Seed producers can leverage these government initiatives to expand their operations, enhance seed quality, and explore new markets. Increasing awareness about the value of MAPs in various industries can boost demand. Seed producers can play a crucial role in educating the market about the benefits of MAPs and their applications, thereby creating a more receptive environment for their products. Lastly, access to high-value markets is a noteworthy opportunity. Certain MAPs, such as saffron, truffles, and vanilla, command high prices in international markets. By focusing on these high-value crops and ensuring quality seed production, the industry can tap into lucrative markets and enhance profitability.

## Specific problem and solutions

The cultivation of medicinal and aromatic plants (MAPs) poses unique challenges as these plants are not as commonly grown among farmers, high-quality seeds are often scarce, and specific seed preparation methods are required (Fig 2). Additionally, the lack of knowledge regarding nursery management after seed germination exacerbates the issues faced by farmers in the production of MAPs. Following are some comprehensive solutions to address these challenges.

### Indian ginseng *(Withania somnifera)*

The *Withania somnifera*, commonly known as Ashwagandha or Indian Ginseng, has long been cherished for its therapeutic properties rooted in the Ayurvedic tradition. However, its propagation through seed-based methods faces distinctive challenges. One challenge is the inconsistent germination of *Withania somnifera* seeds. These seeds exhibit irregular germination patterns, and the seeds of different accessions display varying germination behaviours. Another challenge is the suboptimal field conditions. Under natural field conditions, the germination of *Withania somnifera* seeds is often poor, and seedling mortality is high. In the quest for addressing the challenges associated with *Withania somnifera* seed germination, several solutions have been explored (Kalamartzis et al., 2023). Notably, the application of growth regulators, specifically gibberellic acid ($GA_3$), has proven to be highly effective. An optimal concentration of 150 µg/ml of $GA_3$, combined with a temperature of 25 C, significantly promotes germination. Temperature indeed plays a pivotal role in the germination of *Withania somnifera*. Research has established that an ideal temperature for germination is 25 C, which provides a substantial boost to germination rates. Additionally, the role of light conditions cannot be underestimated in this process. Continuous light has been found to foster germination, underscoring the significance of photoperiod as a determinant in the germination process.

## Black henbane *(Hyoscyamus niger)*

*Hyoscyamus niger* L., commonly known as Khurasani-ajvayan or black henbane, holds a significant place among medicinal plants due to its diverse applications. The leaves of this plant serve as a valuable source of medicinal drugs, particularly for their sedative, narcotic, and therapeutic properties in conditions like asthma and whooping cough. *Hyoscyamus niger* L. seeds can be quite challenging to germinate. They exhibit variability in germination rates, and sometimes, germination may not occur efficiently. These seeds have a hard coat, making them a bit stubborn in their germination behavior. To overcome the problem of low germination rates in seeds, the researchers conducted experiments involving various treatments and environmental conditions (Verma et al., 2014). These treatments included soaking the seeds in water for different durations (24 and 48 hours), using gibberellic acid ($GA_3$) at various concentrations (200 and 500 ppm), applying potassium nitrate ($KNO_3$) at 0.2%, and briefly exposing the seeds to concentrated sulfuric acid ($H_2SO_4$) for 15 seconds. The study revealed that the most effective treatment for enhancing germination was soaking the seeds in $GA_3$ (200 ppm). This treatment resulted in the highest percentage of normal seedlings (73.0%) under specific conditions, such as substrate and temperature, after 12 days.

## Chamomile *(Chamomilla recutita)*

In a study on chamomile (*Matricaria chamomile* L.) seed quality conducted at Ethiopia (Garkebo & Melese, 2023), several significant data values were obtained. The study found that the germination percentage of chamomile seeds declined as storage periods increased. Initially, the germination percentage was 78.2%, but after 12 months of storage, it dropped to 42%. This represents a substantial reduction in seed quality over time. The research suggested that the choice of storage method significantly impacted seed quality. For instance, seeds stored in polypropylene bags recorded the lowest germination percentages after 12 months, while those stored in bottles maintained higher germination percentages. The data showed that bottle storage resulted in a 60% germination rate after 12 months, which was better compared to polypropylene bags. Furthermore, the moisture content of seeds stored in polypropylene bags was significantly higher after 12 months, reaching 55%. In contrast, the bottles maintained lower moisture content (47%) over the same period.

## Clary sage *(Salvia scalarea)*

This study investigates the effect of temperature on seed germination and emergence of *Salvia sclarea* L. in sub-tropical climatic conditions. *Salvia sclarea* L. has been introduced as a winter season crop in sub-tropical regions, but the problem identified is the variation in seed germination and emergence under different temperature conditions in these areas. The study highlights (Javaid et al., 2018) the crucial role of temperature in regulating seed germination and emergence of *Salvia sclarea* L., emphasizing that understanding the optimal temperature conditions is essential for successful cultivation in these regions. The research revealed that temperature significantly influenced the germination of *Salvia sclarea* L. seeds. The maximum seed germination (78.50%) occurred at 25 C, followed by 20 C (74.44%) and 15 C (65.80%), while the lowest germination was observed at 5 C (2.33%). Similarly, the first seed emergence occurred at four days at 25 C. In conclusion, the study recommends a temperature range of 15 C to 25 C as suitable for seed germination of *Salvia sclarea* L. in sub-tropical conditions. To ensure successful seed germination and emergence in sub-tropical climates, it is advisable to sow the seeds during the favorable temperature conditions, typically in the months of October and November, when the temperature falls within the optimal range.

## Isabgol *(Plantago ovata)*

Seed shattering poses a significant challenge in the cultivation of *Plantago ovata*. This natural phenomenon refers to the plant's tendency to release mature seeds from the seed heads, making it a complex issue for farmers. To combat this problem, precise timing is critical. Harvesting Isabgol seeds must be done when they are just mature but before they fully shatter. This requires close monitoring and a deep understanding of the plant's growth stages. Gentle harvesting techniques should be employed to minimize seed loss. Whether through hand harvesting or specialized equipment designed to reduce seed shattering, the objective is to handle the plants with care and minimize mechanical stress, which can trigger seed release. Additionally, selecting non-shattering varieties can be a worthwhile consideration (H Patel, A U Amin, 1970). Breeding and choosing Isabgol cultivars with reduced seed shattering tendencies can offer a long-term solution to this issue. Some growers may explore the use of growth regulators or chemicals that inhibit seed abscission to minimize seed shattering, although the safety and regulations surrounding their use should be thoroughly evaluated. Furthermore, after the harvest, it is crucial to ensure that the seeds are properly dried and stored to prevent further shattering. Excess moisture can lead to premature seed release, affecting the overall yield and quality. Continuous research and development efforts are also essential in identifying and developing solutions to mitigate seed shattering in Isabgol cultivation. This persistent problem can have a significant impact on crop yield and quality, making it vital for agricultural researchers and practitioners to work together to find effective strategies for addressing this challenge. By implementing these strategies and remaining dedicated to ongoing improvement, farmers can reduce the impact of seed shattering in Isabgol cultivation, leading to higher yields and improved seed quality.

## Kalmegh *(Andrographis paniculata)*

*Andrographis paniculata* (Burm. F.) Wall ex., Nees, a valuable medicinal herb, faces a challenge of low seed germination rates due to its hard seed coat. To overcome this issue, researchers tested various treatments (Zuffo et al., 2017).

They found that soaking the seeds in 50 C hot water for 5 minutes significantly improved germination, with a maximum germination percentage of 93%. This treatment helped soften the hard seed coat, facilitating germination. In contrast, hormonal treatments with Gibberellic acid and other growth regulators did not yield significant improvements. The study provides a practical solution for enhancing seed germination in *Andrographis paniculata* (Burm. F.) Wall ex., Nees, which is essential for its propagation and medicinal use.

## Palmarosa *(Cymbopogon martini)*

*Cymbopogon martini* is a fragrant grass species of economic and medicinal significance due to its essential oil. However, seed production in *Cymbopogon martini* faces several notable challenges. One primary concern is the low seed set, where the plants produce a limited number of seeds, hampering the availability of seeds for cultivation and essential oil extraction. Low seed set can be attributed to factors like inadequate pollination and genetic variations. Another challenge is the species' dependence on cross-pollination, making it essential to maintain a diverse plant population to facilitate effective pollination and mitigate reduced seed yields. Furthermore, variability in seed viability can lead to inconsistent germination rates, affecting crop establishment. The susceptibility of Palmarosa seeds to viability loss during storage complicates the preservation of seed quality over time. Additionally, seed dormancy can hinder germination under suitable conditions, necessitating methods to break dormancy for successful crop establishment. To tackle these issues in *Cymbopogon martini* seed production, various strategies can be employed. These include improved pollination management, such as maintaining diverse plant populations and ensuring the presence of pollinators. Regular seed viability testing helps select high-quality seeds for sowing and crop establishment (Pinzón-Torres et al., 2014). Effective seed storage protocols should be developed to prolong seed viability and maintain a consistent seed supply. Additionally, exploring methods to break seed dormancy can improve germination rates. Genetic improvement through breeding programs can address specific challenges, such as increasing seed set and reducing seed dormancy. Addressing these challenges is essential to ensure a stable and reliable seed supply for Palmarosa cultivation and essential oil extraction, thereby supporting the economic and medicinal value of this plant species.

## Pyrethrum *(Chrysanthemum cinerarifolium)*

*Chrysanthemum cinerarifolium* seed production encounters various challenges that hinder the efficient and effective generation of quality seeds. Key issues include low germination rates, which often lead to poor stand establishment and reduced crop yields, impacting overall farm productivity. Additionally, maintaining genetic purity is crucial, but cross-pollination with other Tanacetum species can introduce genetic impurity, resulting in inconsistent *C. cinerarifolium* plants. Susceptibility to diseases, such as powdery mildew and foliar pathogens, further affects seed quality and germination rates. Growing pyrethrum in regions susceptible to environmental stressors, including drought, salinity, and temperature fluctuations, compounds these problems. To address these challenges and improve pyrethrum seed production, several solutions can be considered (Jawhari et al., 2022; Nikolova, 2016). Selective breeding programs can be implemented to develop pyrethrum varieties with higher germination rates and genetic purity, necessitating stringent monitoring and crop isolation to prevent cross-pollination. Effective disease management strategies, such as using disease-resistant varieties, proper crop rotation, and suitable fungicides, can prevent and control diseases that affect seed quality. The development and promotion of stress-tolerant pyrethrum varieties can ensure more consistent seed production, even in challenging environmental conditions (Li et al., 2011). Utilizing seed priming techniques like hydropriming can enhance seed quality and germination rates, improving seedling establishment. Implementing best practices in farm management, including soil preparation, irrigation, and pest control, creates optimal conditions for pyrethrum seed production. Rigorous quality control measures at every stage of seed production, from planting to harvesting and processing, help ensure that only high-quality seeds are available to farmers.

### Shatavari *(Asparagus racemosus)*

The production of seeds for *Asparagus racemosus* presents various challenges and opportunities that are crucial for the cultivation and conservation of this medicinal plant. One of the primary challenges is the low and erratic germination rate of *A. racemosus* seeds. Achieving synchronized and early germination is vital for successful cultivation. To address this challenge, researchers have explored pre-sowing seed treatments, such as soaking in hot water and dilute sulfuric acid. These treatments have shown promise in overcoming seed dormancy and improving germination rates (Parveen et al., 2020). Another significant problem associated with *A. racemosus* is the high demand in the herbal medicine market, leading to overharvesting and habitat destruction. To meet market demand while conserving the plant, mass cultivation is proposed as a solution. Improving seed germination is a key step in enabling large-scale cultivation efforts. Physiological dormancy in *A. racemosus* seeds has been identified as a barrier to achieving high germination rates. Study suggests the use of hot water and dilute sulfuric acid treatments to break this dormancy effectively. These treatments are not only practical but also cost-effective and can be implemented under field conditions, making them valuable tools for seed production.

### Senna *(Cassia auriculata)*

Breaking seed dormancy in *Cassia auriculata*, presents a critical challenge for successful cultivation. To address this issue, a comprehensive study explored various treatments, including sulfuric acid scarification, hot water soaking, and gibberellic acid ($GA_3$) application (Khan, 2013). Among these treatments, soaking the seeds in hot water (100 C) for 15 minutes, followed by a 3-hour immersion in 100 ppm $GA_3$, proved to be the most effective method, resulting in a remarkable increase in germination rates to 98%, significant growth in root and shoot lengths, and improved dry matter production and vigour index. This approach was particularly successful in weakening the seed coat, a primary dormancy factor, and the application of $GA_3$ further promoted germination by stimulating crucial metabolic processes. The findings of this study offer valuable insights into addressing seed dormancy in *Cassia auriculata*, contributing to enhanced crop establishment and productivity in the cultivation of this medicinal plant.

### Basil (*Ocimum basilicum*)

To overcome problems in *Ocimum basilicum* seed production, it's essential to isolate varieties to prevent cross-pollination, monitor for and address pests and diseases, *O. basilicum* densely for self-pollination, attract pollinators through companion planting or hand-pollination, control temperature and start seeds indoors in unfavourable climates, wait for plant maturity before seed collection, ensure seeds are fully dried before storage, store them in a cool, dry place in airtight containers, and select healthy, robust parent plants for the best seed quality, thus promoting successful basil seed production. To tackle problems in *O. basilicum* seed production, consider these solutions: First, address cross-pollination issues by isolating basil varieties or resorting to hand-pollination techniques. Second, actively monitor plants for pests and diseases, promptly implementing organic or chemical remedies when necessary, and explore companion planting to deter common pests. Third, ensure sufficient plant density by properly spacing basil plants for enhanced self-pollination. By implementing these solutions, you can improve your chances of successful basil seed production (Thongtip et al., 2022).

### Periwinkle (*Catharanthus roseus*)

The seed production of *Catharanthus roseus*, may encounter several challenges that can be effectively addressed with appropriate solutions. One common issue is the low germination rate of the seeds, which can be improved by using fresh seeds and subjecting them to a period of cold stratification. Cross-pollination is another concern, and to ensure true-to-type seeds, isolation of different varieties and hand-pollination can be employed. Additionally, managing seed quality and viability by harvesting from healthy plants and storing seeds in optimal conditions is crucial. Proper seed collection, sowing, and growth conditions are essential for maximizing germination, while regular disease control and crop rotation help maintain seed health (Nabaei & Amooaghaie, 2019). These measures collectively contribute to the successful seed production of *C. roseus*.

## Role of quality seeds in industry

High-quality MAPs seeds are of paramount importance in fulfilling the demands of multiple industries, including pharmaceuticals, herbal products, cosmetics, and traditional medicine. These seeds serve as the cornerstone for a consistent and dependable supply of medicinal and aromatic plants with specific properties. First and foremost, high-quality MAPs seeds are the backbone of the pharmaceutical and herbal medicine sectors. They are essential for cultivating plants with distinct medicinal attributes, such as analgesic, anti-inflammatory, or antimicrobial properties. These cultivated plants serve as the primary source of raw materials to produce drugs and herbal remedies that address the healthcare needs of the population. Crucially, these seeds contribute to consistency in active compounds found in medicinal plants (Nagaraju et al., 2018; Medar, 2018). These active compounds have therapeutic effects, and high-quality seeds ensure that the plants consistently contain the desired levels of these compounds. This consistency is vital for the pharmaceutical industry to manufacture standardized medicines with predictable efficacy and safety. Moreover, high-quality MAPs seeds play a pivotal role in ensuring adherence to quality control standards and regulatory compliance. Regulatory authorities mandate that pharmaceutical and herbal product manufacturers meet stringent quality requirements. High-quality seeds are fundamental in achieving the standards necessary to comply with regulations related to product efficacy and safety. The cosmetic and fragrance industry also relies heavily on aromatic plants grown from high-quality seeds. These plants provide the essential oils and fragrances used in cosmetics, perfumes, and aromatherapy products. The fragrance industry depends on the consistency of fragrance profiles and essential oil content, which can only be achieved using high-quality seeds. Traditional and alternative medicine systems, such as Ayurveda and Traditional Chinese Medicine, depend on specific medicinal plants. High-quality seeds are instrumental in the cultivation of plants with the right properties for traditional remedies, ensuring the continued effectiveness of these treatments. Furthermore, the herbal products and dietary supplements industry requires a reliable supply of medicinal plants (Sinjushin, 2015). High-quality seeds are vital for meeting the demand for herbal teas, dietary supplements, and other wellness products that incorporate plant-based ingredients. Sustainability is a growing concern in these industries. The pharmaceutical and herbal sectors are increasingly focused on sustainable sourcing practices. High-quality seeds enable the cultivation of medicinal plants in a sustainable manner, reducing the impact on natural plant populations and ecosystems.

Research and development in botany, pharmacology, and traditional medicine rely on high-quality MAPs seeds. These seeds are essential for conducting studies on plant properties, active compounds, and their effects on health and well-being. They underpin innovation and product development, driving progress in these fields. Moreover, high-quality MAPs seed also play a role in the conservation of threatened or endangered medicinal plant species. By cultivating these species, industries can contribute to their preservation while sustainably meeting demand. Finally, the MAPs seed industry contributes to economic growth by providing a sustainable source of raw materials for various sectors. It supports job creation, value addition, and export opportunities, particularly in regions with rich biodiversity. High-quality MAPs seeds are indispensable for fulfilling the diverse demands of industries relying on medicinal and aromatic plants. These seeds ensure the production of consistent, effective, and safe products in the pharmaceutical, herbal, cosmetic, and traditional medicine sectors. They also contribute to sustainability, research, and economic development while promoting the responsible use and conservation of medicinal plant resources.

## Collaborative efforts for MAPs seed production

### Research institutes

Research institutes play a pivotal role in advancing the quality and quantity of MAPs seed production. Their involvement is multifaceted, encompassing various critical aspects. Firstly, research institutes are instrumental in conducting extensive studies on MAPs, including their genetic characteristics, growth patterns, and optimal cultivation techniques. Through research, they identify the most suitable species and varieties for cultivation and develop innovative strategies for enhancing seed quality and yield (Inoue et al., 2019). These institutes are also at the forefront of germplasm conservation, safeguarding the genetic diversity of MAPs. By maintaining seed banks and genetic repositories, they ensure that valuable genetic resources are preserved for future use, preventing the loss of potentially important plant varieties. Additionally, research institutes engage in breeding programs aimed at developing high-yielding and disease-resistant MAPs varieties. These efforts contribute to increased seed production while reducing the vulnerability of crops to pests and diseases. Through selective breeding, they can also enhance the content of specific bioactive compounds in MAPs, making them more valuable for various industries (Amin et al., 2022; Inoue & Craker, 2014a). Research institutes facilitate knowledge dissemination by conducting training programs and providing technical assistance to farmers and seed producers. They share their findings on best practices, seed treatment methods, and sustainable cultivation techniques (Table 1 & Fig 1), empowering stakeholders in the MAPs seed production chain.

### Farmers in seed production

Farmers are the backbone of MAPs seed production. Their active involvement in cultivation is vital for ensuring a consistent and high-quality seed supply (Fig 3-4). They are responsible for the hands-on work of sowing, nurturing, and harvesting MAPs plants to obtain seeds (Shukla et al., 2022). To fulfill this role effectively, farmers must receive adequate training and support. Collaborative efforts with farmers involve providing them with access to quality seeds, which are the starting point of the entire process. Equipped with high-quality seeds, farmers can establish healthy MAPs crops that serve as the source of seeds for subsequent generations. They should also be educated on sustainable farming practices that consider soil health, water conservation, and pest management to ensure the longevity of seed production (Inoue & Craker, 2014b). To promote the cultivation of specific MAPs varieties that align with industry demands, collaboration with farmers can include incentive programs and contracts that guarantee the purchase of seeds at fair prices. Such initiatives encourage farmers to focus on the production of sought-after MAPs varieties, fostering a more responsive seed supply chain.

### Seed-producing organizations

Seed-producing organizations are central to the collaborative efforts aimed at MAPs seed production. These organizations serve as intermediaries connecting research institutes, farmers, and industry stakeholders (Fig 3-4). They play a coordinating role in facilitating the exchange of knowledge, expertise, and resources. Seed-producing organizations often function as seed banks and seed certification authorities. They are responsible for maintaining a repository of high-quality MAPs seeds and ensuring their authenticity and quality. This quality assurance is essential to meet the stringent standards set by industries like pharmaceuticals, cosmetics, and herbal products. Furthermore, these organizations may engage in research and development activities to improve seed quality and production techniques continuously. They work closely with research institutes to stay updated on the latest advancements in MAPs cultivation and incorporate these innovations into seed production processes. In summary, collaborative efforts for MAPs seed production require a tripartite approach involving research institutes, farmers, and seed-producing organizations. Research institutes contribute expertise, knowledge, and genetic resources, while farmers provide the essential cultivation and harvesting work. Seed-producing organizations act as intermediaries, ensuring quality control and facilitating the flow of information and resources. This synergy is essential to meet the growing demand for high-quality MAPs seeds across various industries.

## Enhancing seed quality

Enhancing seed quality is crucial for the cultivation of medicinal and aromatic crops in India, given their economic and cultural importance. These crops often face challenges like dormancy and germination inhibitors, making seed enhancement techniques essential (Nagaraju et al., 2018). These techniques involve treating seeds with specific substances under precise conditions to improve germination (Table 3). Seed quality is vital for successful cultivation, especially for crops like medicinal and aromatic plants (MAPs) with diverse applications. A significant challenge is their often-poor seed germination, which can be attributed to various factors. Accelerating the germination process through techniques like seed priming can help overcome these issues. Seed priming offers advantages such as improved germination rates and enzyme activity stimulation, enhancing seedling development. Various seed priming methods are available to meet MAPs' specific germination requirements.

***Hydro-priming:*** In hydro-priming, seeds are soaked in water for a specified duration. This method allows seeds to imbibe water and initiate metabolic processes, preparing them for germination. Hydro-priming is often used for seeds that require hydration to break dormancy.

***Halo-priming:*** Halo-priming involves the hydration of seeds using inorganic salt solutions. This technique can be particularly effective for MAPs with hard seed coats or those that are sensitive to salinity.

***Osmo-priming:*** Osmo-priming utilizes solutions with substances like polyethylene glycol (PEG) to facilitate seed hydration. It is a valuable method for improving germination in seeds that require osmotic adjustments to break dormancy or enhance germination rates.

***Thermo-priming:*** Thermo-priming is based on subjecting seeds to specific temperature regimes, either high or low, before sowing. This technique can stimulate germination in some MAPs by breaking dormancy or promoting quicker and more uniform germination.

***Matri-priming:*** In matri-priming, seeds are treated with solid matrices like hydrated sand, peat, or vermiculite. This method ensures uniform hydration of seeds and can be especially useful for small or irregularly shaped seeds.

***Bio-priming:*** Bio-priming involves coating seeds with beneficial microorganisms such as *Trichoderma spp.* or *Pseudomonas aureofaciens*. These microorganisms can assist in enhancing seed germination and protecting seeds from soilborne pathogens.

***Chemical priming:*** Some MAPs may benefit from chemical priming using substances like gibberellic acid ($GA^3$) or other plant growth regulators. These chemicals can help break dormancy and improve germination rates.

## Conclusion

The climate in our country is well-suited for the cultivation of medicinal and aromatic plants. Currently, the cultivation of these plants is limited due to low demand compared to traditional crops like grains and vegetables. However, looking ahead, it is essential to consider expanding the cultivation of medicinal and aromatic plants. This is because health-conscious individuals are increasingly turning to naturopathy, and meeting this demand requires an expansion of the cultivation areas for these plants. By increasing the cultivation of these plants, our country can become a leading provider of raw materials, while also generating revenue through value addition and product manufacturing. Realizing this potential necessitates collaboration among farmers, traders, product manufacturers, and governing bodies. It is only through coordinated efforts, addressing current challenges, and planning for the future that we can overcome the issues related to seed production and cultivation of medicinal and aromatic plants.

**Table 1. Detailed information on cultivation parameters for seed production of *Withania somnifera, Hyoscyamus niger, Chamomilla recutita, Salvia scalarea, Plantago ovata, Andrographis paniculata, Cymbopogon martini, Chrysanthemum cinerarifolium, Asparagus racemosus, Cassia senna, Ocimum basilicum.,* and *Catharanthus roseus.***

| | Name of crop | Plant spacing | Seed requirement/ ha | Fertilizers requirement/ ha | Irrigation during growing season | Isolation distance for seed production | Rouging | Seed yield/ha (kg) |
|---|---|---|---|---|---|---|---|---|
| | Withania somnifera | 50X50 cm | 5-7 kg | 60-80:40-60:40-60 kg/ha | 2-4 | 400-800 meters | Frequent inspection to remove weeds, off-types, and diseased plants throughout the growing season. | 100 - 300 |
| | Hyoscyamus niger L. | 30X30 cm | 2-3 kg | 60-80:40-60:40-60 kg/ha | 1-2 | 400-800 meters | Early rouging to eliminate unwanted plants and maintain purity. | 50 - 200 |
| | Chamomilla recutita | 30X40 cm | 1-2 kg | 30-40:20-30:20-30 kg/ha | 3-5 | 200-400 meters | Start early and maintain uniformity by removing unwanted plants. | 150-200 |
| | Salvia scalarea L. | 50X50 cm | 4-6 kg | 30-40:20-30:20-30 kg/ha | 2-4 | 400-800 meters | Begin rouging as plants emerge to remove weeds and off-types. | 150- 200 |
| | Plantago ovata | 40X40 cm | 3-5 kg | 40-50:20-30:20-30 kg/ha | 4-6 | 200-400 meters | Early rouging to ensure purity and eliminate undesired plants. | 400 - 500 |
| | Andrographis paniculata (Burm. F.) Wall ex., Nees | 30X30 cm | 1-2 kg | 40-60:20-30:20-30 kg/ha | 2-4 | 400-800 meters | Continuous rouging to eliminate weeds and maintain purity. | 200-250 |

| | Name of crop | Plant spacing | Seed requirement/ ha | Fertilizers requirement/ ha | Irrigation during growing season | Isolation distance for seed production | Rouging | Seed yield/ha (kg) |
|---|---|---|---|---|---|---|---|---|
| | Cymbopogon martini | 100X100 cm | 4-6 kg | 40-60:30-40:30-40 kg/ha | 3-5 | 200-400 meters | Early rouging to maintain a pure stand by removing unwanted plants. | 100- 150 |
| | Chrysanthemum cinerarifolium | 50X50 cm | 500-600 gm | 50-60:40-50:40-50 kg/ha | 3-4 | 400-800 meters | Early rouging to maintain the purity of the Pyrethrum variety. | 180- 200 |
| | Asparagus racemosus | 30X30 cm | 1-2 kg | 80-100:40-60:40-60 kg/ ha | 4-6 | 400-800 meters | Ongoing rouging throughout the season to remove weeds and off-types. | 200- 250 |
| | Cassia auriculata | 60X60 cm | 15-20 kg | 40-60:30-40:30-40 kg/ha | 2-4 | 400-800 meters | Early rouging to eliminate undesired plants and maintain purity. | 200 - 300 |
| | Ocimum basilicum | 40X40 cm | 1-2 kg | 60-80:30-40:30-40 kg/ha | 4-6 | 200-400 meters | Early and continuous rouging to remove weeds and off-types. | 200- 250 |
| | Catharanthus roseus | 50X50 cm | 200-400 gm | 50-70:30-40:30-40 kg/ha | 3-5 | 200-400 meters | Early rouging to maintain variety purity. | 250-300 |

**Table 2. Roadmaps for future MAPs seed production technologies.**

| Opportunity | Key actions and considerations |
| --- | --- |
| Crop diversity | Explore the demand for various crop varieties, both traditional and novel.<br>Assess regional adaptability of different crops. |
| Genetic research | Invest in genomic research to develop high-yield and resilient varieties.<br>Collaborate with research institutions for genetic mapping. |
| Sustainability | Promote sustainable agriculture practices, including organic and regenerative farming.<br>Implement eco-friendly seed production techniques. |
| Technological advancements | Embrace modern technologies such as precision farming, IoT, and blockchain for traceability.<br>Train farmers in tech adoption. |
| Market research | Conduct market surveys to identify emerging trends and consumer preferences.<br>Analyze competitors and assess market potential. |

**Table 3. Seed treatment methods and their impact on seed quality in various crops.**

| Treatment | Enhancing seed quality through treatments | Crop | Ref |
| --- | --- | --- | --- |
| Seed dormancy breaking treatments | Immerse seeds in a 1% sodium hypochlorite solution to overcome seed dormancy issues. | *Chrysanthemum cinerarifolium* | (Somrug et al., 2021) |
| $GA_3$ treatment | Soak seeds in $GA_3$ (Gibberellic Acid) to enhance germination rates. | *Saussurea lappa* | (Li et al., 2022) |
| Seed pelleting | Apply hydro priming for 12 hours, resulting in the highest seedling vigor, germination percentage, and seedling dry weight. | *Ocimum basilicum* | (Nawrot-Chorabik et al., 2021) |
| Seed soaking in cold water | Submerge seeds in cold water to achieve the highest germination percentage. | *Catharanthus roseus* | (Odoi et al., 2021) |
| Water soaking | Soak seeds in water for 18-24 hours to enhance the quality of aged seed lots. | *Tagetes erecta* | (Asropi et al., 2019) |
| Hydro priming | Implement hydro priming for 12 hours, enhancing germination and promoting dry matter production. | *Trigonella Foenum-Graecum* | (Kalita et al., 2021) |

| | | | |
|---|---|---|---|
| Acid scarification | Fresh seeds undergo scarification by immersing them in concentrated sulfuric acid ($H_2SO_4$). | *Rheum webbianum* | (*Šoch* et al., 2023) |
| Seed priming with growth regulators | Utilize seed priming with growth regulators by soaking seeds in 100 ppm $GA_3$ for 6 hours to enhance germination. | *Andrographis paniculata* | (Qamar et al., 2022) |
| | Soaking seeds in 50 ppm $GA_3$ for 3 hours maximizes seedling vigor. | *Asparagus sprengeri* | (Elhindi et al., 2016) |
| | Soak seeds in 200 ppm IAA for 6 hours to achieve higher germination rates. | *Cassia obtusifolia* | (Rout et al., 2017) |
| | Soak seeds in 200 ppm Ethrel for 6 hours to maximize seedling growth.<br>Soaking seeds in a 1% $KH_2P0_4$ solution for 1 hour is beneficial for *Withania somnifera.* | *Gymnema sylvistre (Gymnema)* | (Nagaraju et al., 2018) |
| Priming with inorganic salts | Apply priming with inorganic salts, such as soaking seeds in 50 to 100 mM $KNO_3$, to achieve maximum germination compared to untreated seeds. | *Aconitum balfourii (Indian Atees)* | (Dhillon et al., 2021) |
| | Soaking seeds in 130 mm thiourea leads to early germination. | *Aconitum heterophyllum* | (Pandey et al., 2000) |
| | Soaking seeds in a 1% solution of sodium nitrate enhances plant height, root length, and other yield parameters. | *Withania somnifera* | (Suryawanshi et al., 2013) |
| Seed priming with organic products | Apply seed priming with organic products, such as soaking seeds in 25% coconut water for 6 hours to enhance nursery emergence. | *Bixa orellana (Achiote)* | (Lakmali & Seran, 2022) |
| Hydration Treatments | Implement hydration treatments, including soaking in pre-chilling conditions at 4 C for 10 days. This approach stimulates germination and reduces the time required for germination.<br>Similarly, pre-chilling at 4 C for 10 days significantly improves germination in Descurainia Sophia (Flixweed). | *Descurainia sophia* | (Asgharpour et al., 2022) |

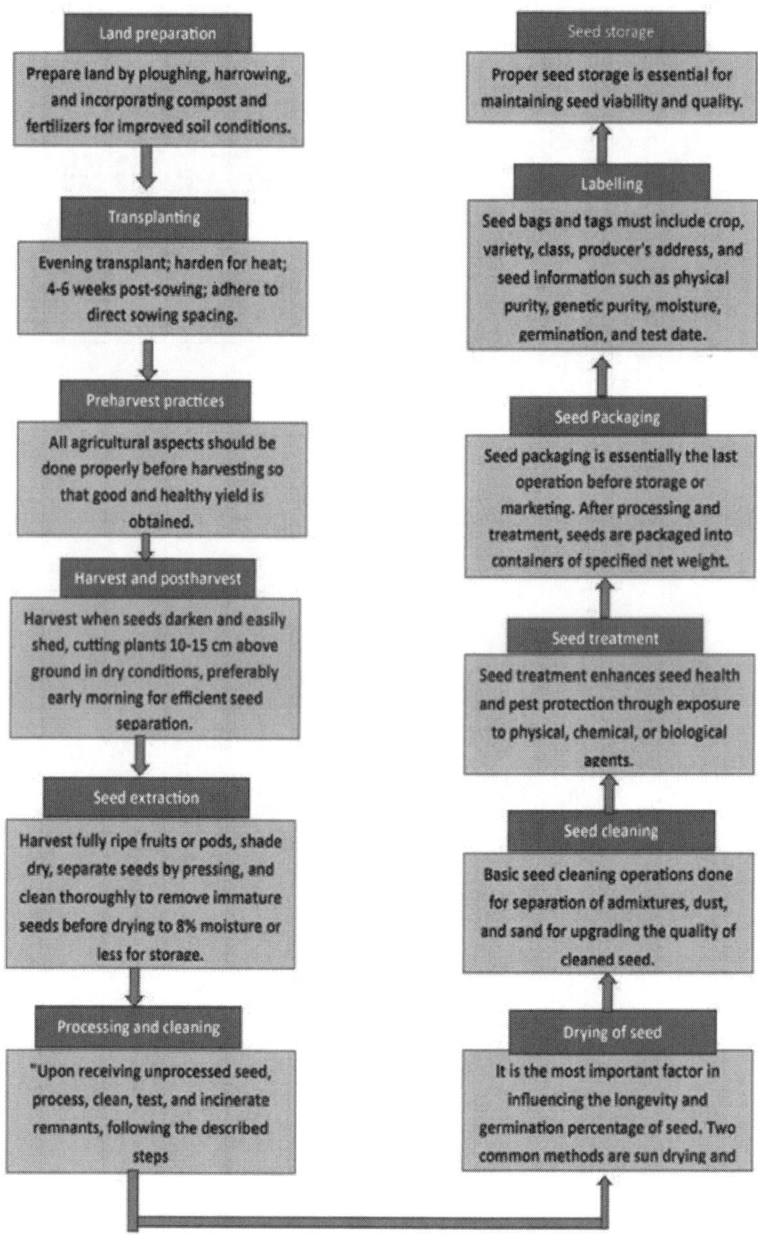


**Fig 1. Comprehensive process for achieving successful seed production of** *Withania somnifera, Hyoscyamus niger, Chamomilla recutita, Salvia scalarea, Plantago ovata, Andrographis paniculata, Cymbopogon martini, Chrysanthemum cinerarifolium, Asparagus racemosus, Cassia senna, Ocimum basilicum.,* **and** *Catharanthus roseus*

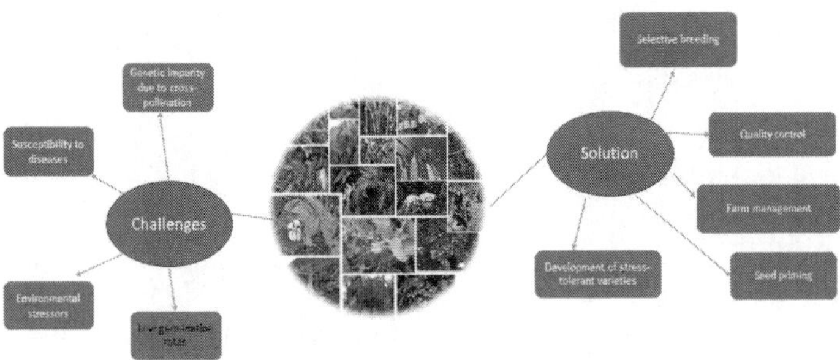


**Fig 2. Challenges and solutions in seed production of various medicinal and aromatic plants (MAPs).**

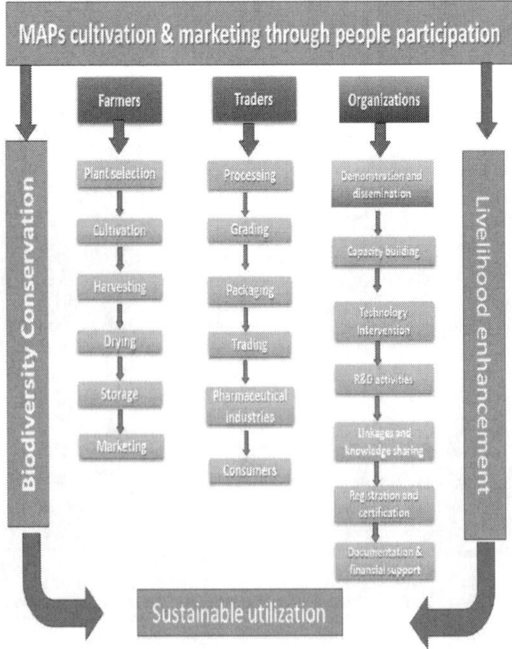


**Fig 3. Cultivation, conservation, and marketing strategies involving farmers, traders, and organizations for medicinal and aromatic plants (MAPs)**

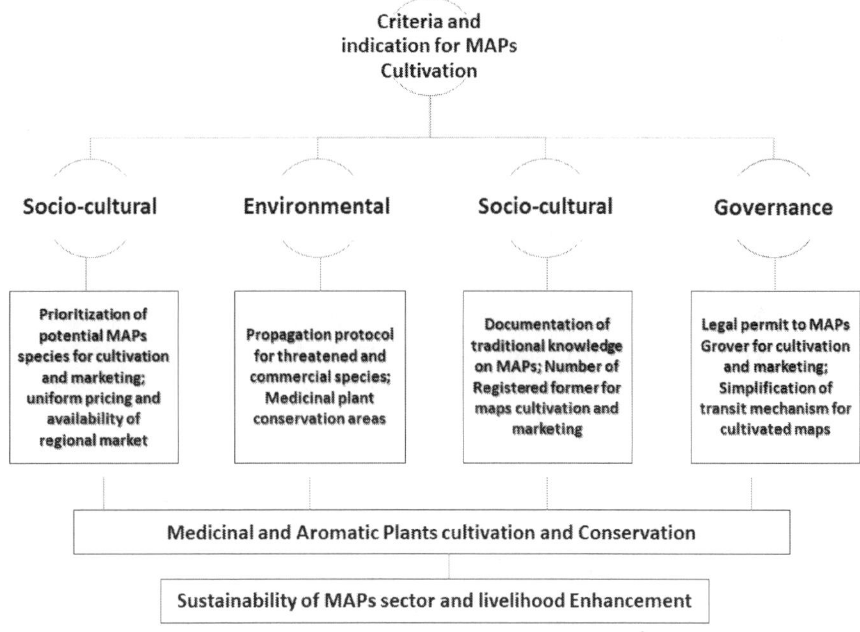


**Fig 4. Criteria and indicators for medicinal and aromatic plants (MAPs) cultivation.**

## References

Abbasi Khalaki, M., Moameri, M., Asgari Lajayer, B., & Astatkie, T. (2021). Influence of nano-priming on seed germination and plant growth of forage and medicinal plants. In *Plant Growth Regulation* (Vol. 93, Issue 1). https://doi.org/10.1007/s10725-020-00670-9

Amin, M., Mahmood, K., Bodlah, I., & Qasim, M. (2022). Assessing diversity of aphids (Hemiptera: Aphididae) along elevation gradient in Azad Jammu and Kashmir-Pakistan. *International Journal of Tropical Insect Science*, *42*(1). https://doi.org/10.1007/s42690-021-00573-w

Asgharpour, M., Enayati, N., Rezaei Sadrabadi, M., Mohamadi Afrakati, M., Khavandegar, A., Mardi, P., Alirezaei, A., Taherinia, A., & Bakhtiyari, M. (2022). Effects of Descurainia sophia on Oxidative Stress Markers and Thirst Alleviation in Hemodialysis Patients: A Randomized Double-Blinded Placebo-Controlled Cross-Over Clinical Trial. *Evidence-Based Complementary and Alternative Medicine, 2022*. https://doi.org/10.1155/2022/2001441

Asropi, A., Bintoro, N., Karyadi, J. N. W., Rahayoe, S., & Saputro, A. D. (2019). Kinetika Perubahan Sifat Fisik dan Kadar Tanin Biji Sorgum (Sorghum Bicolor L.) Selama Perendaman. *AgriTECH*, *39*(3). https://doi.org/10.22146/agritech.42610

Chaachouay, N., Benkhnigue, O., Fadli, M., El Ibaoui, H., & Zidane, L. (2019). Ethnobotanical and ethnopharmacological studies of medicinal and aromatic plants used in the treatment of metabolic diseases in the Moroccan Rif. *Heliyon*, *5*(10). https://doi.org/10.1016/j.heliyon.2019.e02191

Chhetri, V. T., Shrestha, S., Thapa, S., & Timilsina, S. (2021). Status and Role of Medicinal and Aromatic Plants (MAPs) in Nepalese Livelihood. *International Journal of Environment*, *10*(1). https://doi.org/10.3126/ije.v10i1.38405

Craker, L. E. (2007). Medicinal and Aromatic Plants — Future Opportunities. *Nutrition Business Journal, Flexner 1910.*

Dhillon, B. S., Kumar, V., Sagwal, P., Kaur, N., Singh Mangat, G., & Singh, S. (2021). Seed priming with potassium nitrate and gibberellic acid enhances the performance of dry direct seeded rice (Oryza sativa l.) in north-western india. *Agronomy*, *11*(5). https://doi.org/10.3390/agronomy11050849

Elhindi, K. M., Dewir, Y. H., Asrar, A. W., Abdel-Salam, E., El-Din, A. S., & Ali, M. (2016). Improvement of seed germination in three medicinal plant species by plant growth regulators. *HortScience*, *51*(7). https://doi.org/10.21273/hortsci.51.7.887

Frey, K. M. (2020). Structure activity relationship (SAR) maps: A student-friendly tool to teach medicinal chemistry in integrated pharmacotherapy courses. *Currents in Pharmacy Teaching and Learning*, *12*(3). https://doi.org/10.1016/j.cptl.2019.12.014

Gantait, S., Kundu, S., Ali, N., & Sahu, N. C. (2015). Synthetic seed production of medicinal plants: a review on influence of explants, encapsulation agent and matrix. In *Acta Physiologiae Plantarum* (Vol. 37, Issue 5). https://doi.org/10.1007/s11738-015-1847-2

Garkebo, H., & Melese, B. (2023). Effects of Storage Periods and Methods on Seed Quality of Chamomile (Matricaria chamomile.L) in Wondogenet, Ethiopia. *Global Academic Journal of Agriculture and Biosciences*, *5*(03). https://doi.org/10.36348/gajab.2023.v05i03.001

H Patel, A U Amin, D. G. P. H. B. P. P. T. P. (1970). Evaluation of stable and non-shattering isabgol cultivar - Gujarat isabgol 4. *Journal of Spices and Aromatic Crops*. https://doi.org/10.25081/josac.2018.v27.i1.1022

Inoue, M., & Craker, L. E. (2014a). Medicinal and aromatic plants-uses and functions. In *Horticulture: Plants for People and Places* (Vol. 2). https://doi.org/10.1007/978-94-017-8581-5_3

Inoue, M., & Craker, L. E. (2014b). Medicinal and aromatic plants–uses and functions. In *Horticulture: Plants for People and Places, Volume 2: Environmental Horticulture*. https://doi.org/10.1007/10.1007/978-94-017-8581-5_3

Inoue, M., Hayashi, S., & E. Craker, L. (2019). Role of Medicinal and Aromatic Plants: Past, Present, and Future. In *Pharmacognosy - Medicinal Plants*. https://doi.org/10.5772/intechopen.82497

Islam, M. S., & Bari, M. A. (2012). In vitro regeneration protocol for artificial seed production in an important medicinal plant Mentha arvensis L. *Journal of Bio-Science, 20*(0). https://doi.org/10.3329/jbs.v20i0.17722

Javaid, M. M., Florentine, S., Ali, H. H., & Weller, S. (2018). Effect of environmental factors on the germination and emergence of Salvia verbenaca L. cultivars (verbenaca and vernalis): An invasive species in semi-arid and arid rangeland regions. *PLoS ONE, 13*(3). https://doi.org/10.1371/journal.pone.0194319

Jawhari, F. Z., Imtara, H., El Moussaoui, A., Khalis, H., Es-Safi, I., Al Kamaly, O., Saleh, A., Parvez, M. K., Guemmouh, R., & Bari, A. (2022). Reproductive Biology of the Two Varieties of Anacyclus pyrethrum L.—Anacyclus pyrethrum var. pyrethrum (L.) Link and Anacyclus pyrethrum var. depressus (Ball.) Maire—An Endemic Endangered Species. *Plants, 11*(17). https://doi.org/10.3390/plants11172299

Kalamartzis, I., Menexes, G., & Dordas, C. (2023). Effect of Low Water Availability on Seed Yield and Seed Quality of Basil (Ocimum basilicum). *Plants, 12*(5). https://doi.org/10.3390/plants12051094

Kalauni, D., & Joshi, A. (2018). Status of Medicinal and Aromatic Plant (MAPs) and Socio-Economic Influence in Nepalese Livelihood-A Review Research. *Acta Scientific Agriculture, 2*(9).

Kalita, D., Jain, S., Srivastava, B., & Goud, V. V. (2021). Sono-hydro priming process (ultrasound modulated hydration): Modelling hydration kinetic during paddy germination. *Ultrasonics Sonochemistry, 70*. https://doi.org/10.1016/j.ultsonch.2020.105321

Kavusi, E., Shahi Khalaf Ansar, B., Dehghanian, Z., Asgari Lajayer, B., Nobaharan, K., Ma, Y., & Glick, B. R. (2023). Delivery of Beneficial Microbes via Seed Coating for Medicinal and Aromatic Plant Production: A Critical Review. In *Journal of Plant Growth Regulation* (Vol. 42, Issue 2). https://doi.org/10.1007/s00344-022-10597-2

Khan, M. R. (2013). The effect of pretreatments- on seed germination of the Cassia auriculata Linn. and Cassia tora Linn. *Advances in Plant Sciences, 26*(1).

Lakmali, K. N. D., & Seran, T. H. (2022). Impact of Seed Priming with King Coconut Water on Growth and Yield of Okra (<em>Abelmoschus esculents</em> L.). *Journal of Food and Agriculture, 15*(2). https://doi.org/10.4038/jfa.v15i2.5271

Li, J., Bai, Y., Xie, Y., & Gao, J. (2022). Ultrastructure change and transcriptome analysis of GA3 treatment on seed germination of moso bamboo(Phyllostachys edulis). *Plant Signaling and Behavior, 17*(1). https://doi.org/10.1080/15592324.2022.2091305

Li, J., Yin, L. Y., Jongsma, M. A., & Wang, C. Y. (2011). Effects of light, hydropriming and abiotic stress on seed germination, and shoot and root growth of pyrethrum (Tanacetum cinerariifolium). *Industrial Crops and Products, 34*(3). https://doi.org/10.1016/j.indcrop.2011.05.012

Marotti, I., Whittaker, A., Ba dat, R. B., Akin, P. A., Ergün, N., & Dinelli, G. (2023). Intercropping Perennial Fruit Trees and Annual Field Crops with Aromatic and Medicinal Plants (MAPs) in the Mediterranean Basin. In *Sustainability (Switzerland)* (Vol. 15, Issue 15). https://doi.org/10.3390/su151512054

Muscalu, A., Tudora, C., Soric , C., Mîrzan, O., Naie, M., & Barcanu-Tudor, E. (2022). MECHANISED HARVESTING OF INFLORESCENCES OF MEDICINAL AND AROMATIC PLANT SPECIES CULTIVATED ON A SMALL SCALE. *INMATEH - Agricultural Engineering*, *67*(2). https://doi.org/10.35633/inmateh-67-42

Nabaei, M., & Amooaghaie, R. (2019). Interactive Effect of Melatonin and Sodium Nitroprusside on Seed Germination and Seedling Growth of Catharanthus roseus under Cadmium Stress. *Russian Journal of Plant Physiology*, *66*(1). https://doi.org/10.1134/S1021443719010126

Nagaraju, K., Medar, V. S., & K, A. (2018). Seed quality enhancement techniques in medicinal and aromatic crops. *Journal of Pharmacognosy and Phytochemistry*, *3*.

Nawrot-Chorabik, K., Osmenda, M., Słowi ski, K., Latowski, D., Tabor, S., & Woodward, S. (2021). Stratification, scarification and application of phytohormones promote dormancy breaking and germination of pelleted scots pine (Pinus sylvestris l.) seeds. *Forests*, *12*(5). https://doi.org/10.3390/f12050621

Nikolova, I. M. (2016). Effects of organic products with insecticidal action on key insect pests in alfalfa seed production. *Emirates Journal of Food and Agriculture*, *28*(9). https://doi.org/10.9755/ejfa.2016-04-381

Novais, C., Pereira, C., Molina, A. K., Liberal, Â., Dias, M. I., Añibarro-Ortega, M., Alves, M. J., Calhelha, R. C., Ferreira, I. C. F. R., & Barros, L. (2021). Bioactive and nutritional potential of medicinal and aromatic plant (Map) seasoning mixtures. *Molecules*, *26*(6). https://doi.org/10.3390/molecules26061587

Odoi, J. B., Mugeni, D., Kiiza, R., Apolot, B., & Gwali, S. (2021). Effect of Soaking Treatment on Germination of Hard Coated Tropical Forest Tree Seeds. *Uganda Journal of Agricultural Sciences*, *19*(2). https://doi.org/10.4314/ujas.v19i2.1

Oztruk, M., & Hakeem, K. R. (2018). Plant and Human Health, Volume 1: Ethnobotany and Physiology. In *Springer International Publishing* (Vol. 1).

Pandey, H., Nandi, S. K., Nadeem, M., & Palni, L. M. S. (2000). Chemical stimulation of seed germination in Aconitum heterophyllum Wall. and A. balfourii Stapf.: Important Himalayan species of medicinal value. *Seed Science and Technology*, *28*(1).

Parveen, Kumar, A., Barthwal, S., & Kant, R. (2020). Identification of seed sources of Asparagus racemosus Willd. For higher productivity. *Acta Horticulturae*, *1301*. https://doi.org/10.17660/ActaHortic.2020.1301.12

Ph D Scholar, A. K., & Medar, V. S. (2018). National conference on "Conservation, Cultivation and Utilization of medicinal and Aromatic plants" Seed quality enhancement techniques in medicinal and aromatic crops. *~ 104 ~ Journal of Pharmacognosy and Phytochemistry*, *3*.

Pinzón-Torres, J. A., Porras, N. J. C., Garcia, D. C. D., Morales, J. R. M., & Stashenko, E. (2014). Green biomass production and quality of essential oils of palmarosa (Cymbopogon martini) with application of synthesis fertilizers and organic fertilizers. *Acta Agronomica, 63*(4). https://doi.org/10.15446/acag.v63n4.42840

Prasuna, V. S., Reddy, C. N., Rajagopal, M., Purna, G. S., & Srinivas, B. (2022). In vitro Flowering and Production of Seeds in Oxalis corniculata L. - an Important Medicinal Plant. *Plant Tissue Culture and Biotechnology, 32*(2). https://doi.org/10.3329/ptcb.v32i2.63553

Qamar, R., Khan, S., Safdar, M. E., ur-Rehman, A., Rehman, A., Javeed, H. M. R., Nadeem, M. A., Al-Yahyai, R., & Alkahtani, J. (2022). Seed priming with growth regulators modulates production, physiology and antioxidant defense of Indian squash (Praecitrullus fistulosus) under semi-arid conditions. In *PLoS ONE* (Vol. 17, Issue 4 April). https://doi.org/10.1371/journal.pone.0265694

Rahmawaty, Samosir, J. B., Batubara, R., & Rauf, A. (2019). Diversity and distribution of medicinal plants in the Universitas Sumatera Utara arboretum of deli serdang, North Sumatra, Indonesia. *Biodiversitas, 20*(5). https://doi.org/10.13057/biodiv/d200539

Rai, P. P. (2012). Status of Medicinal and Aromatic Plants in Papua New Guinea: a Review Prem. *Pacific Journal of Medical Sciences, 9*(2).

Rana, S. K., Rana, H. K., Ranjitkar, S., Ghimire, S. K., Gurmachhan, C. M., O'Neill, A. R., & Sun, H. (2020). Climate-change threats to distribution, habitats, sustainability and conservation of highly traded medicinal and aromatic plants in Nepal. *Ecological Indicators, 115*. https://doi.org/10.1016/j.ecolind.2020.106435

Riaz, U., Iqbal, S., Sohail, M. I., Samreen, T., Ashraf, M., Akmal, F., Siddiqui, A., Ahmad, I., Naveed, M., Khan, N. I., & Akhter, R. M. (2021). A Comprehensive Review on Emerging Importance and Economical Potential of Medicinal and Aromatic Plants (MAPs) in Current Scenario. In *Pakistan Journal of Agricultural Research* (Vol. 34, Issue 2). https://doi.org/10.17582/journal.pjar/2021/34.2.381.392

Rout, S., Khare, N., Patra, S. S., Beura, S., & Nayak, S. (2017). Effect of seed pre-treatment with different concentrations of gibberellic acid (GA3) on seed germination and seedling growth of Cassia fistula L. ~ *135 ~ Journal of Medicinal Plants Studies, 5*(6).

Schippmann, U., Cunningham, A. B., Leaman, D. J., & Walter, S. (2005). Impact of cultivation and collection on the conservation of medicinal plants: Global trends and issues. *Acta Horticulturae, 676*. https://doi.org/10.17660/actahortic.2005.676.3

Shukla, S., Singh, R., Bhutani, R., Tripathi, A., & Shukla, S. K. (2022). Restoration and Conservation of Plant Genetic Resources via Molecular Techniques: An Important Measure for Sustainable Agriculture. In *Agro-biodiversity and Agri-ecosystem Management*. https://doi.org/10.1007/978-981-19-0928-3_12

Sinjushin, A. (2015). Cytogenomics and mutagenomics in plant functional biology and breeding. In *PlantOmics: The Omics of Plant Science*. https://doi.org/10.1007/978-81-322-2172-2_5

Šoch, J., Šonka, J., & Ponert, J. (2023). Acid scarification as a potent treatment for an in vitro germination of mature endozoochorous Vanilla planifolia seeds. *Botanical Studies, 64*(1). https://doi.org/10.1186/s40529-023-00374-z

Solomou, A. D., Skoufogianni, E., Mylonas, C., Germani, R., & Danalatos, N. G. (2019). Cultivation and utilization of "Greek mountain tea" (Sideritis spp.): Current knowledge and future challenges. *Asian Journal of Agriculture and Biology, 7*(2).

Somrug, K., Hemwong, S., & Jorjong, S. (2021). Effect of seed dormancy breaking treatment and storability on seed quality of indigo (Indigofera tinctoria l.). *Asian Journal of Plant Sciences, 20*(3). https://doi.org/10.3923/ajps.2021.389.395

Spinozzi, E., Maggi, F., Bonacucina, G., Pavela, R., Boukouvala, M. C., Kavallieratos, N. G., Canale, A., Romano, D., Desneux, N., Wilke, A. B. B., Beier, J. C., & Benelli, G. (2021). Apiaceae essential oils and their constituents as insecticides against mosquitoes—A review. In *Industrial Crops and Products* (Vol. 171). https://doi.org/10.1016/j.indcrop.2021.113892

Sucholas, J., Ukhanova, M., Greinwald, A., & Luick, R. (2021). Wild collection of medicinal and aromatic plants (MAPs) for commercial purposes in Poland - a system's analysis. *Herba Polonica, 67*(3). https://doi.org/10.2478/hepo-2021-0014

Suryawanshi, J. B., Bhingarde, M. T., Waghmode, S. U., & Deshmukh, D. V. (2013). Effect of pre-soaking seed treatments on seed quality in Ashwagandha (Withania somnifera Dunal.). *Ecology, Environment and Conservation, 19*(3).

Thongtip, A., Mosaleeyanon, K., Korinsak, S., Toojinda, T., Darwell, C. T., Chutimanukul, P., & Chutimanukul, P. (2022). Promotion of seed germination and early plant growth by KNO3 and light spectra in Ocimum tenuiflorum using a plant factory. *Scientific Reports, 12*(1). https://doi.org/10.1038/s41598-022-11001-5

Verma, S. S., Verma, P. K., & Verma, U. (2014). Standardization of seed germination technique in Khurasani-ajavayan (Hyoscyamus niger) seeds. *Indian Journal of Agricultural Sciences, 84*(11). https://doi.org/10.56093/ijas.v84i11.44643

Zuffo, A. M., Steiner, F., Zuffo Júnior, J. M., Bush, A., da Silva, J. R. M., Limede, A. C., & Oliveira, C. E. da S. (2017). Non-chemical methods to break seed dormancy of canafistula [Peltophorum dubium (Sprengel) Taubert (Fabaceae)]. *Australian Journal of Crop Science, 11*(12). https://doi.org/10.21475/ajcs.17.11.12.pne730

# Index

F



G



H



I



J



L



M



N



A



P

## Q



## R



## S



## T



## V



## W



## X



## Z